KU-633-719

Victor E. Borisenko
Stefano Ossicini
What is What in the Nanoworld

Related Titles

Vedmedenko, E.

Competing Interactions and Patterns in Nanoworld

2007
ISBN: 978-3-527-40484-1

Wolf, E. L.

Nanophysics and Nanotechnology

An Introduction to Modern Concepts in Nanoscience

2006
ISBN: 978-3-527-40651-7

Poole, C. P., Owens, F. J.

Introduction to Nanotechnology

2003
ISBN: 978-0-471-07935-4

Balzani, V., Credi, A., Venturi, M.

Molecular Devices and Machines

A Journey into the Nanoworld

2003
ISBN: 978-3-527-30506-3

Victor E. Borisenko and Stefano Ossicini

What is What in the Nanoworld

A Handbook on Nanoscience and Nanotechnology

Second, Completely Revised and Enlarged Edition

WILEY-VCH

WILEY-VCH Verlag GmbH & Co. KGaA

The Authors

Victor E. Borisenko
University of Informatics
and Radioelectronics
Minsk, Belarus

Stefano Ossicini
University of Modena and Reggio Emilia
Faculty of Engineering
Reggio Emilia, Italy

Cover
Silver tip for scanning near-field optical
microscopy, sharpened by focused ion beam
milling. (SEM image).
Experiment: Gian Carlo Gazzadi, S3 Center
(INFM-CNR), Modena and Pietro Gucciardi,
CNR-IPCF, Messina.
Artwork: Lucia Covi. From "Blow-up. Images
from the nanoworld" (www.s3.infm.it/blowup);
Copyright S3, 2007.

All books published by Wiley-VCH are
carefully produced. Nevertheless, authors,
editors, and publisher do not warrant the
information contained in these books,
including this book, to be free of errors.
Readers are advised to keep in mind that
statements, data, illustrations, procedural
details or other items may inadvertently be
inaccurate.

Library of Congress Card No.: applied for
British Library Cataloguing-in-Publication Data
A catalogue record for this book is available
from the British Library.

**Bibliographic information published by
the Deutsche Nationalbibliothek**
The Deutsche Nationalbibliothek lists this
publication in the Deutsche National-
bibliografie; detailed bibliographic data are
available in the Internet at
<http://dnb.d-nb.de>.

© 2008 WILEY-VCH Verlag GmbH & Co.
KGaA, Weinheim

All rights reserved (including those of
translation into other languages). No part of
this book may be reproduced in any form – by
photoprinting, microfilm, or any other
means – nor transmitted or translated into a
machine language without written permission
from the publishers. Registered names,
trademarks, etc. used in this book, even when
not specifically marked as such, are not to be
considered unprotected by law.

Typesetting Laserwords Private Ltd,
Chennai, India
Printing Strauss GmbH, Mörlenbach
Binding Litges & Dopf GmbH, Heppenheim

Printed in the Federal Republic of Germany
Printed on acid-free paper

ISBN: 978-3-527-40783-5

D
620·1129
BOR

Contents

What is What in the Nanoworld. Victor E. Borisenko and Stefano Ossicini
Copyright © 2008 Wiley-VCH Verlag GmbH & Co. KGaA, Weinheim
ISBN: 978-3-527-40783-5

Contents

Preface to the Second Edition

This is the second, enlarged and updated edition of our book. From more than 1400 entries in the first edition we have now reached about 2000 entries. Moreover a large number of the old entries have been extended. The gallery of illustrations is enriched by new figures, and new tables are added throughout the book.

The presented terms, phenomena, regulations, experimental and theoretical tools are very easy to consult since they are arranged in alphabetical order, with a chapter for each letter. The great majority of the terms have additional information in the form of notes such as *"First described in: ..."*, *"Recognition: ..."*, *"More details in: ..."*, thus giving a historical retrospective of the subject with references to further sources of extended information, which can be articles, books, review articles, or web sites. In particular, in this second edition we have tried, for the overwhelming majority of the items, to find out who was the initiator and when and where the term was born, defined or first discussed. We think that all these additional notes are quite useful, since they give the possibility to all the readers to start independently their personal research.

Only four years separate this second edition from the first; nevertheless we have seen a true explosion of research in nanoscience and developments in nanotechnologies. One measure of the emergence of these fields is the growth of the literature dedicated to the new disciplines. Nanoscience and nanotechnology have, in the last years, witnessed not only an explosive growth in the number of relevant and important "classical" scientific journals, which have devoted, more and more, an increasing proportion of their published papers to "nano"-related research, but also in the number of new journals which contain the stem "nano" in their title. A list of 62 "nano" journals has been added to the Appendix of the book. Only a few of them appeared before 2000 and most of them started their activity in the last four years.

In reviewing the first edition of this book Professor Vincenzo Balzani correctly pointed out that *"the actual Nanoworld is very large and comprises at least four regions that can be labelled Physics, Chemistry, Biology and Engineering. ... The four component regions of the nanoscience and nanotechnology realm partly overlap but often ignore one another. Even worse, in the overlapping territories they do not speak the same language to such an extent that, in some cases, they seem even to obey different laws. Clearly,*

What is What in the Nanoworld. Victor E. Borisenko and Stefano Ossicini
Copyright © 2008 WILEY-VCH Verlag GmbH & Co. KGaA, Weinheim
ISBN: 978-3-527-40783-5

Preface to the Second Edition

cooperation among physicists, chemists, biologists and engineers, which is of course essential for the progress of nanoscience and nanotechnology, is often hampered by such language barriers". Looking at the titles of the new journals listed in the Appendix, we see that the actual nanoworld is even larger. The concepts like nanoethics, nanoeducation, nanomedicine, nanotoxicology, and so on are new actors in the nanoarena. Moreover, fine art has now entered the nanoworld. The image on the front page of this edition is part of a book "Blow-up. Images from the nanoworld" born from the collaboration between Lucia Covi, an Italian photographer, and the CNR-INFM Research Center S^3-*nanoStructures and bioSystems at Surfaces* in Modena, Italy, where one of the authors of this book is active. Commenting on this picture (a silver tip for near-field scanning optical microscopy obtained by focused ion milling) in the foreword to the "Blow-up" book, Professor Roald Hoffmann, the Nobel Laureate in chemistry 1981, has labeled it *"a digital Tower of Babel"*. The Tower of Babel has to do with the myth of the birth of all the different languages; we hope this book can help in breaking these language barriers.

Minsk   *Victor E. Borisenko*
Modena/Reggio Emilia   *Stefano Ossicini*
January 2008

Preface to the First Edition

There's Plenty of Room at the Bottom

Richard P. Feynman 1959

There's even more Room at the Top

Jean-Marie Lehn 1995

Nanotechnology and nanoscience are concerned with material science and its application at, or around, the nanometer scale (1 nm $= 10^{-9}$ m, 1 billionth of a meter). The nanoscale can be reached either from the top down, by machining to smaller and smaller dimension, or from the bottom up, by exploiting the ability of molecules and biological systems to self-assemble tiny structures. Individual inorganic and organic nanostructures involve clusters, nanoparticles, nanocrystals, quantum dots, nanowires, and nanotubes, while collections of nanostructures involve arrays, assemblies, and superlattices of individual nanostructures.

Rather than a new specific area of science, nanoscience is a new way of thinking. Its revolutionary potential lies in its intrinsic multidisciplinarity. Its development and successes strongly depend on efforts from, and fruitful interactions among, physics, chemistry, mathematics, life sciences, and engineering. This handbook intends to contribute to a broad comprehension of what are nanoscience and nanotechnology.

It is an introductory, reference handbook that first summarizes terms and definitions, most important phenomena, regulations, and experimental and theoretical tools discovered in physics, chemistry, technology and the application of nanostructures. We present a representative collection of fundamental terms and most important supporting definitions taken from general physics and quantum mechanics, material science and technology, mathematics and information theory, organic and inorganic chemistry, solid-state physics and biology. As a result, fast progressing nanoelectronics and optoelectronics, molecular electronics and spintronics, nano-fabrication and -manufacturing, bioengineering and quantum processing of information, an area of fundamental importance for the information society of the twenty-first century, are covered. More than 1300 entries, from a few sentences to a page in length, are given for readers ranging from beginners to professionals.

What is What in the Nanoworld. Victor E. Borisenko and Stefano Ossicini
Copyright © 2008 WILEY-VCH Verlag GmbH & Co. KGaA, Weinheim
ISBN: 978-3-527-40783-5

Preface to the First Edition

The book is organized as follows. Terms and definitions are arranged in alphabetic order. Those written in an article body with bold letters have extended details arranged alphabetically. Each entry in the book interprets the term or definition under consideration and briefly presents the main features of the phenomena behind it. The great majority of the terms are accompanied with additional information in the form of notes such as *"First described in: . . ."*, *"Recognition: . . ."*, *"More details in: . . ."*, thus giving a historical perspective of the subject with reference to further sources of extended information, which can be articles, books, review articles, or web sites. This makes it easier for the willing reader to reach a deeper insight. Bold characters in formulas symbolize vectors and matrices, while normal characters are scalar quantities. Symbols and constants of a general nature are handled consistently through the book (see **Fundamental Constants Used in Formulas**). They are used according to the IUPAP convention.

The book will help undergraduate and PhD students, teachers, researchers and scientific managers to understand properly the language used in modern nanoscience and nanotechnology. It will also appeal to readers from outside the nanoworld community, in particular to scientific journalists.

Comments and proposals related to the book will be appreciated and can be sent to borisenko@dsuir.unibel.by and/or to ossicini@unimore.it.

It is a pleasure for us to acknowledge our colleagues who have supported this work. Their contribution ranges from writing and correction of particular articles to critical comments and useful advice. In particular, we wish to thank (in alphabetical order) F. Arnaud d'Avitaya, L. J. Balk, C. M. Bertoni, V. P. Bondarenko, E. Degoli, J. Derrien, R. Di Felice, P. Facci, H. Fuchs, N. V. Gaponenko, S. V. Gaponenko, L. I. Ivanenko, G. F. Karpinchik, S. Y. Kilin, S. K. Lazarouk, E. Luppi, F. Manghi, R. Magri, M. Michailov, D. B. Migas, V. V. Nelaev, L. Pavesi, N. A. Poklonski, S. L. Prischepa, V. L. Shaposhnikov, G. Treglia, and A. Zaslavsky.

Minsk
Modena/Reggio Emilia
April 2004

Victor E. Borisenko
Stefano Ossicini

Sources of Information

Besides their personal knowledge and experience and the scientific journals and books cited in the text, the authors also used the following sources of information:

Encyclopedias and Dictionaries

1. *Encyclopedic Dictionary of Physics,* edited by J. Thewlis, R. G. Glass, D. J. Hughes, A. R. Meetham (Pergamon Press, Oxford 1961).

2. McGraw–Hill *Dictionary of Physics and Mathematics,* edited by D. N. Lapedes (McGraw–Hill Book Company, New York 1978).

3. Landolt-Bornstein. *Numerical Data and Functional Relationships in Science and Technology,* v. 17, edited by O. Madelung, M. Schultz, H. Weiss (Springer, Berlin 1982).

4. McGraw–Hill *Encyclopedia of Electronics and Computers,* edited by C. Hammer (McGraw–Hill Book Company, New York 1984).

5. *Encyclopedia of Semiconductor Technology,* edited by M. Grayson (John Wiley & Sons, New York 1984).

6. *Encyclopedia of Physics,* edited by R. G. Lerner, G. L. Trigg (VCH Publishers, New York 1991).

7. *Physics Encyclopedia,* edited by A. M. Prokhorov, vols. 1–5 (Bolshaya Rossijskaya Encyklopediya, Moscow 1998) – in Russian.

8. *Encyclopedia of Applied Physics,* Vols. 1–25, edited by G. L. Trigg (Wiley VCH, Weinheim 1992–2000).

9. *Encyclopedia of Physical Science and Technology,* Vols. 1–18, edited by R. A. Meyers (Academic Press, San Diego 2002).

10. *Handbook of Nanotechnology,* edited by B. Bhushan (Springer, Berlin 2004).

What is What in the Nanoworld. Victor E. Borisenko and Stefano Ossicini
Copyright © 2008 WILEY-VCH Verlag GmbH & Co. KGaA, Weinheim
ISBN: 978-3-527-40783-5

Sources of Information

Books

1. G. Alber, T. Beth, M. Horodecki, P. Horodecki, R. Horodecki, M. Rötteler, H. Weinfurter, R. Werner, A. Zeilinger, *Quantum Information* (Springer, Berlin, 2001).

2. G. B. Arfken, H. J. Weber, *Mathematical Methods for Physicists* (Academic Press, San Diego, 1995).

3. P. W. Atkins, J. De Paula, *Physical Chemistry* (Oxford University Press, Oxford, 2001).

4. V. Balzani, M. Venturi, A. Credi, *Molecular Devices and Machines: A Journey into the Nanoworld* (Wiley–VCH, Weinheim, 2003).

5. F. Bassani, G. Pastori Parravicini, *Electronic and Optical Properties of Solids* (Pergamon Press, London, 1975).

6. D. Bimberg, M. Grundman, N. N. Ledentsov, *Quantum Dot Heterostructures* (John Wiley & Sons, London, 1999).

7. W. Borchardt-Ott, *Crystallography*, Second edition (Springer, Berlin, 1995).

8. V. E. Borisenko, S. Ossicini, *What is What in the Nanoworld* (Wiley–VCH, Weinheim, 2004).

9. M. Born, E. Wolf, *Principles of Optics*, Seventh (expanded) edition (Cambridge University Press, Cambridge, 1999).

10. J. H. Davies, *The Physics of Low-Dimensional Semiconductors* (Cambridge University Press, Cambridge, 1995).

11. *DNA based Computers* edited by R. Lipton, E. Baum (Am. Math. Soc., Providence, 1995).

12. M. S. Dresselhaus, G. Dresselhaus, P. Eklund, *Science of Fullerenes and Carbon Nanotubes* (Academic Press, San Diego, 1996).

13. D. K. Ferry, S. M. Goodnick, *Transport in Nanostructures* (Cambridge University Press, Cambridge, 1997).

14. *Frontiers in Surface Nanophotonics*, edited by D. L. Andrews and Z. Gaburro (Springer, Berlin, 2007).

15. S. V. Gaponenko, *Optical Properties of Semiconductor Nanocrystals* (Cambridge University Press, Cambridge, 1998).

16. W. A. Harrison, *Electronic Structure and the Properties of Solids* (W. H. Freeman & Company, San Francisco, 1980).

17. H. Haug, S. W. Koch, *Quantum Theory of the Optical and Electronic Properties of Semiconductors* (World Scientific, Singapore, 1994).

18. S. Hüfner, *Photoelectron Spectroscopy* (Springer, Berlin, 1995).

19. Y. Imri, *Introduction to Mesoscopic Physics* (Oxford University Press, Oxford, 2002).

20. L. E. Ivchenko, G. Pikus, *Superlattices and Other Heterostructures: Symmetry and other Optical Phenomena* (Springer, Berlin, 1995).

21. C. Kittel, *Elementary Solid State Physics* (John Wiley & Sons, New York, 1962).

22. C. Kittel, *Quantum Theory of Solids* (John Wiley & Sons, New York, 1963).

23. C. Kittel, *Introduction to Solid State Physics*, seventh edition (John Wiley & Sons, New York, 1996).

24. L. Landau, E. Lifshitz, *Quantum Mechanics* (Addison–Wesley, London, 1958).

25. O. Madelung, *Semiconductors: Data Handbook* (Springer, Berlin, 2004).

26. G. Mahler, V. A. Weberrus, *Quantum Networks: Dynamics of Open Nanostructures* (Springer, New York, 1998).

27. L. Mandel, E. Wolf, *Optical Coherence and Quantum Optics* (Cambridge University Press, Cambridge, 1995).

28. *Molecular Electronics: Science and Technology* edited by A. Aviram, M. Ratner (Academy of Sciences, New York, 1998).

29. *Nanobiotechnology. Concepts, Applications and Perspectives*, edited by C. M. Niemeyer and C. A. Mirkin (Wiley–VCH, Weinheim, 2004).

30. *Nanoelectronics and Information Technology*, edited by R. Waser (Wiley–VCH, Weinheim, 2003).

31. *Nanostructured Materials and Nanotechnology*, edited by H. S. Nalwa (Academic Press, London, 2002).

32. R. C. O'Handley, *Modern Magnetic Materials: Principles and Applications* (Wiley & Sons, New York, 1999).

33. S. Ossicini, L. Pavesi, F. Priolo, *Light Emitting Silicon for Microphotonics*, Springer Tracts on Modern Physics **194** (Springer, Berlin, 2003).

34. J. Pankove, *Optical Processes in Semiconductors* (Dover, New York, 1971).

35. N. Peyghambarian, S. W. Koch, A. Mysyrowicz, *Introduction to Semiconductor Optics* (Prentice Hall, Englewood Cliffs, New Jersey, 1993).

36. C. P. Poole, F. J. Owens, *Introduction to Nanotechnology* (Wiley–VCH, Weinheim, 2003).

37. P. N. Prasad *Nanophotonics* (Wiley–VCH, Weinheim, 2004).

38. C. N. Rao, P. J. Thomas, G. U. Kulkarni, *Nanocrystals: Synthesis, Properties and Applications* (Springer, Berlin, 2007).

39. S. Reich, C. Thomsen, J. Maultzsch, *Carbon Nanotubes* (Wiley–VCH, Weinheim, 2004).

40. E. Rietman, *Molecular Engineering of Nanosystems* (Springer, New York, 2000).

Sources of Information

41. *Roadmap of Scanning Probe Microscopy*, edited by S. Morita (Springer, Berlin, 2007).

42. K. Sakoda, *Optical Properties of Photonic Crystals* (Springer, Berlin, 2001).

43. *Silicon Photonics*, edited by L. Pavesi and D. J. Lockwood (Springer, Berlin, 2004).

44. S. Sugano, H. Koizumi, *Microcluster Physics* (Springer, Berlin, 1998).

45. *The Chemistry of Nanomaterials. Synthesis, Properties and Applications*, edited by C. N. Rao, A. Müller, A. K. Cheetham (Wiley–VCH, Weinheim, 2004).

46. L. Theodore, R. G. Kunz, *Nanotechnology. Environmental Implications and Solutions* (Wiley–VCH, Weinheim, 2005).

47. J. D. Watson, M. Gilman, J. Witkowski, M. Zoller, *Recombinant DNA* (Scientific American Books, New York, 1992).

48. E. L. Wolf, *Nanophysics and Nanotechnology – Second Edition* (Wiley–VCH, Weinheim, 2006).

49. S. N. Yanushkevich, V. P. Shmerko, S. E. Lyshevski, *Logic Design of NanoICs* (CRC Press, Boca Raton, 2004).

50. P. Y. Yu, M. Cardona, *Fundamentals of Semiconductors* (Springer, Berlin, 1996).

Web sites

http://www.britannica.com	Encyclopedia Britannica
http://www.Google.com	Scientific Search Engine
http://www.wikipedia.com/	Encyclopedia
http://scienceworld.wolfram.com/	Science world. World of physics and mathematics. Eric Weisstein's World of Physics
http://www.photonics.com/dictionary/	PHOTONICS DIRECTORY
http://www.nobel.se/physics/laureates/index.html	The Nobel Prize Laureates
http://www-history.mcs.st-and.ac.uk/history/	MATHEMATICS ARCHIVE
http://www.chem.yorku.ca/NAMED/	Named Things in Chemistry and Physics
http://www.hyperdictionary.com/	HYPERDICTIONARY
http://www.wordreference.com/index.htm	WordReference.com. French, German, Italian and Spanish Dictionary with Collins Dictionaries
http://web.mit.edu/redingtn/www/netadv/	The Net Advance of Physics. Review Articles and Tutorials in an Encyclopedic Format

Sources of Information

Fundamental Constants Used in Formulas

$a_B = 5.29177 \times 10^{-11}$ m   Bohr radius

$c = 2.99792458 \times 10^{8}$ m/s   light speed in vacuum

$e = 1.602177 \times 10^{-19}$ C   charge of an electron

$h = 6.626076 \times 10^{-34}$ J·s   Planck constant

$\hbar = h/2\pi = 1.054573 \times 10^{-34}$ J·s   reduced Planck constant

$i = \sqrt{-1}$ imaginary unit

$k_B = 1.380658 \times 10^{23}$ J/K $(8.617385 \times 10^{5}$ eV/K)   Boltzmann constant

$m_0 = 9.10939 \times 10^{-31}$ kg   electron rest mass

$n_A = 6.0221367 \times 10^{23} \text{mol}^{-1}$ Avogadro constant

$R_0 = 8.314510$ J/(K·mol)   universal gas constant

$r_e = 2.817938$ m   radius of an electron

$\alpha = \mu_0 c e^2 / 2h = 7.297353 \times 10^{-3}$ fine structure constant

$\varepsilon_0 = 8.854187817 \times 10^{-12}$ F/m   permittivity of vacuum

$\mu_0 = 4\pi \times 10^{7}$ H/m   permeability of vacuum

$\mu_B = 9.27402 \times 10^{24}$ A·m^2 Bohr magneton

$\pi = 3.14159$

$\sigma = 5.6697 \times 10^{-5}$ erg/(cm^2·s·K)   Stefan–Boltzmann constant

From Abbe's principle to Azbel'–Kaner cyclotron resonance

Abbe's principle states that the smallest distance that can be resolved between two lines by optical instruments is proportional to the wavelength and inversely proportional to the angular distribution of the light observed ($d_{min} = \lambda/n \sin \alpha$). It establishes a prominent physical problem, known as the "diffraction limit". That is why it is also called **Abbe's resolution limit**. No matter how perfect an optical instrument is, its resolving capability will always have this diffraction limit. The limits of light microscopy are thus determined by the wavelength of visible light, which is 400–700 nm; the maximum resolving power of the light microscope is limited to about half the wavelength, typically about 300 nm. This value is close to the diameter of a small **bacterium**, and **virus**es, which cannot therefore be visualized. To attain sublight microscopic resolution, a new type of instrument would be needed; as we know today, accelerated electrons, which have a much smaller wavelength, are used in suitable instruments to scrutinize structures down to the 1 nm range.

The diffraction limit of light was first surpassed by the use of **scanning near-field optical microscopes**; by positioning a sharp optical probe only a few nanometers away from the object, the regime of far-field wave physics is circumvented, and the resolution is determined by the probe–sample distance and by the size of the probe which is scanned over the sample.

Also, fluorescence light microscopy based techniques have been developed in order to break the diffraction barrier, as in the case of **fluorescence nanoscopy**.

First described in: E. Abbe, *Beiträge zur Theorie des Mikroskops und der mikroskopischen Wahrnehmung*, Schultzes Archiv für mikroskopische Anatomie **9**, 413–668 (1873).

Abbe's resolution limit → Abbe's principle.

aberration – any image defect revealed as distortion or blurring in optics. This deviation from perfect image formation can be produced by optical lenses, mirrors and electron lens systems. Examples are astigmatism, chromatic or lateral aberration, coma, curvature of field, distortion, and spherical aberration.

In astronomy, it is an apparent angular displacement in the direction of motion of the observer of any celestial object due to the combination of the velocity of light and of the velocity of the observer.

What is What in the Nanoworld. Victor E. Borisenko and Stefano Ossicini
Copyright © 2008 WILEY-VCH Verlag GmbH & Co. KGaA, Weinheim
ISBN: 978-3-527-40783-5

ab initio (approach, theory, calculations)

***ab initio* (approach, theory, calculations)** – Latin meaning "from the beginning". It supposes that primary postulates, also called first principles, form the background of the referred theory, approach or calculations. The primary postulates are not so directly obvious from experiment, but owe their acceptance to the fact that conclusions drawn from them, often by long chains of reasoning, agree with experiment in all of the tests which have been made. For example, calculations based on the **Schrödinger** wave **equation**, as well as on the basis of **Newton equations** of motion or any other fundamental equations, are considered to be *ab initio* calculations.

Abney's law states that the shift in apparent hue of spectral color that is desaturated by addition of white light is toward the red end of the spectrum if the wavelength is below 570 nm and toward the blue if it is above.
  First described in: W. Abney, E. R. Festing, *Colour photometry*, Phil. Trans. Roy. Soc. London **177**, 423–456 (1886).
  More details in: W. Abney, *Researches in colour vision* (Longmans & Green, London, 1913).

Abrikosov vortex – a specific arrangement of lines of a magnetic field in a **type II superconductor**.
  First described in: A. A. Abrikosov, *An influence of the size on the critical field for type II superconductors*, Doklady Akademii Nauk SSSR **86**(3), 489–492 (1952) – in Russian.
  Recognition: in 2003 A. A. Abrikosov, V. L. Ginzburg, A. J. Leggett received the Nobel Prize in Physics for pioneering contributions to the theory of superconductors and superfluids.
  See also www.nobel.se/physics/laureates/2003/index.html.
  More details in: A. A. Abrikosov, Nobel Lecture: *Type-II superconductors and the vortex lattice*, Rev. Mod. Phys. **76**(3), 975–979 (2004).

absorption – a phenomenon arising when electromagnetic radiation or atomic particles enter matter. In general, two kinds of attenuation accompany the passage of radiation and particles through matter, which are absorption and scattering. Both obey the law $I = I_0 \exp(-\alpha x)$, where I_0 is the intensity (flux density) of radiation entering the matter, and I is the intensity depth x. In the absence of scatter, α is the **absorption coefficient**, and in the absence of absorption, α is the scattering coefficient. If both forms of attenuation are present, α is termed the total absorption coefficient → **dielectric function**.

acceptor (atom) – an impurity atom, typically in semiconductors, which accepts electron(s). Acceptor atoms usually form electron energy levels slightly higher than the uppermost field energy band, which is the valence band in semiconductors and dielectrics. An electron from this band is readily excited into the acceptor level.

The consequent deficiency in the previously filled band contributes to the hole conduction.

achiral → **chirality**.

acoustic phonon – a quantum of excitation related to an acoustic mode of atomic vibrations in solids → **phonon**.

actinic – pertaining to electromagnetic radiation capable of initiating photochemical reactions, as in photography or the fading of pigments.

actinodielectric – a dielectric exhibiting an increase in electrical conductivity when electromagnetic radiation is incident upon it.

activation energy – an energy in excess over a ground state, which must be added to a system to allow a particular process to take place.

adatom – an atom adsorbed on a solid surface.

adduct – a chemical compound that forms from the addition of two or more substances. The term comes from Latin meaning "drawn toward". An adduct is a product of the direct addition of two or more distinct molecules, resulting in a single reaction product containing all atoms of all components, with formation of two chemical bonds and a net reduction in bond multiplicity in at least one of the reactants. The resultant is considered a distinct molecular species. In general, the term is often used specifically for products of addition reactions.

adiabatic approximation is used to solve the **Schrödinger equation** for electrons in solids. It assumes that a change in the coordinates of a nucleus passes no energy to electrons, that is the electrons respond adiabatically, which then allows the decoupling of the motion of the nuclei and electrons → **Born-Oppenheimer approximation**.

adhesion – the property of a solid to cling to another solid controlled by intermolecular forces at their interface.

adiabatic principle – perturbations produced in a system by altering slowly external conditions resulting, in general, in a change in the energy distribution in it, but leaving the phase integrals unchanged.

adiabatic process – a thermodynamic procedure which take place in a system without an exchange of heat with surroundings.

adjacent charge rule states that it is possible to write formal electronic structures for some molecules where adjacent atoms have formal charges of the same sign. In the Pauling formulation (1939), it states that such structures will not be important owing to instability resulting from the charge distribution.

adjoint operator – an operator **B** such that the inner products $(\mathbf{A}x,y)$ and $(x,\mathbf{B}y)$ are equal for a given operator **A** and for all elements x and y of the **Hilbert space**. It is also known as **associate operator** and **Hermitian conjugate operator**.

adjoint wave functions – functions in the Dirac electron theory which are formed by applying the **Dirac matrix** to the **adjoint operators** of the original wave functions.

admittance – a measure of how readily alternating current will flow in an electric circuit. It is the reciprocal of **impedance**. The term was introduced by Heaviside (1878).

adsorption – a type of **absorption**, in which only the surface of a matter acts as the absorbing medium. **Physisorption** and **chemisorption** are distinguished as adsorption mechanisms.
    Term coined by: H. Kayser *Über die Verdichtung von Gasen an Oberflächen in ihrer Abhängigkeit von Druck und Temperatur*, Ann. Phys. **12**, 526–547 (1880).

AES – an acronym for **Auger electron spectroscopy**.

affinity → **electron affinity**.

AFM – an acronym for **atomic force microscopy**.

Aharonov–Bohm effect – the total amplitude of electron waves at a certain point oscillates periodically with respect to the magnetic flux enclosed by the two paths due to the interference effect. The design of the interferometer appropriate for experimental observation of this effect is shown in Figure A.1. Electron waves come from the waveguide to left terminal, split into two equal amplitudes going around the two halves of the ring, meet each other and interfere in the right part of the ring, and leave it through the right terminal. A small solenoid carrying magnetic flux Φ is positioned entirely inside the ring so that its magnetic field passes through the annulus of the ring. It is preferable to have the waveguide sufficiently small in order to restrict a number of possible coming electron modes to one or a few.
    The overall current through the structure from the left port to the right one depends on the relation between the length of the ring arms and the inelastic mean free path of electrons in the ring material. If this relation meets the requirements for quasi-ballistic transport, the current is determined by the phase interference of the electron waves at the exit (right) terminal. The vector potential **A** of the magnetic

field passing through the ring annulus is azimuthal. Hence electrons travelling in either arms of the ring move either parallel or antiparallel to the vector potential. As a result, there is a difference in the phases of the electron waves coming to the exit port from different arms. It is defined to be $\Delta\Phi = 2\pi(\Phi/\Phi_0)$, where $\Phi_0 = h/e$ is the quantum of flux. The interference of the electron waves appears to be periodic in the number of flux quanta passing through the ring. It is constructive when Φ is a multiple of Φ_0 and destructive halfway between. It produces a periodic modulation in the transverse conductance (resistance) of the ring by the magnetic field, which is known as the magnetic Aharonov–Bohm effect. It is worthwhile to note here that real devices hardly meet the requirements for observation of "pure" Aharonov–Bohm effect. The point is that the magnetic field penetrates the arms of the interferometer, not just the area enclosed by them. This leads to additional current variations at high magnetic fields, while the enclosed flux dominates at low magnetic fields.

First described in: Y. Aharonov, D. Bohm, *Significance of electromagnetic potentials in the quantum theory*, Phys. Rev. **115**(3), 485–491 (1959).

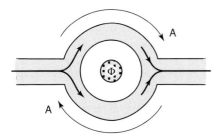

Figure A.1 Schematic layout of the interferometer for observation of the Aharonov–Bohm effect. Small solenoid inside the ring produces the magnetic field of the flux Φ enclosed between the two arms and characterized by the vector potential **A**.

Aharonov–Casher effect supposes that a beam of neutral particles with magnetic dipole moments passing around opposite sides of a line charge will undergo a relative quantum phase shift. The effect has a "duality" with the **Aharonov–Bohm effect**, where charged particles passing around a magnetic solenoid experience a phase shift despite, it is claimed, experiencing no classical force. It is pointed out that a magnetic dipole particle passing a line charge does indeed experience a classical electromagnetic force in the usual electric-current model for a magnetic dipole. This force will produce a relative lag between dipoles passing on opposite sides of the line charge, and the classical lag then leads to a quantum phase shift. Thus, the effect has a transparent explanation as a classical lag effect.

First described in: Y. Aharonov, A. Casher, *Topological quantum effects for neutral particles*, Phys. Rev. Lett. **53**(4), 319–321 (1984).

Airy equation

Airy equation – the second order differential equation $d^2y/dx^2 = xy$, also known as the **Stokes equation**. Here x represents the independent variable and y is the value of the function.

First described in: G. B. Airy, Trans. Camb. Phil. Soc. **6**, 379 (1838); G. B. Airy, *An Elementary Treatise on Partial Differential Equations* (1866).

Airy functions – solutions of the **Airy equation**. The equation has two linearly independent solutions, conventionally taken as the Airy integral functions $Ai(x)$ and $Bi(x)$. They are plotted in Figure A.2. There are no simple expressions for them in terms of elementary functions, while for large absolute values of x: $Ai(x) \sim \pi^{-1/2}x^{-1/4}\exp[-(2/3)x^{3/2}]$, $Ai(-x) \sim (1/2)\pi^{-1/2}x^{-1/4}\cos[-(2/3)x^{3/2} - \pi/4]$. Airy functions arise in solutions of the **Schrödinger equation** for some particular cases.

First described in: G. B. Airy, *An Elementary Treatise on Partial Differential Equations* (1866).

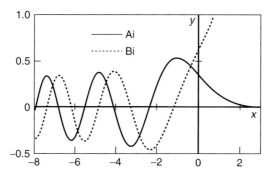

Figure A.2 Airy functions.

Airy spirals – spiral interference patterns formed by quartz cut perpendicularly to the axis in convergent circularly polarized light.

Recognition: in 1831 G. B. Airy received the Copley Medal of the Royal Society for their studies on optical subjects.

aldehydes – organic compounds that have at least one hydrogen atom bonded to the **carbonyl group** ($>C=O$). These may be RCHO or ArCHO compounds with R representing an **alkyl group** ($-C_nH_{2n+1}$) and Ar representing an **aromatic ring**.

algorithm – a set of well-defined rules for the solution of a problem in a finite number of steps.

aliphatic compound – an organic compound in which carbon atoms are joined together in straight or branched chains. The simplest aliphatic compound is methane

(CH$_4$). Most aliphatic compounds provide exothermic combustion reactions, thus allowing their use as a fuel.

alkanes $\rightarrow$ hydrocarbons.

alkenes $\rightarrow$ hydrocarbons.

alkyl groups $\rightarrow$ hydrocarbons.

allotropy – the property of a chemical element to exist in two or more different structural modifications in the solid state. The term **polymorphism** is used for compounds.

alternating current Josephson effect $\rightarrow$ Josephson effects.

Al'tshuler–Aronov–Spivak effect – occurs when the resistance of the conductor in the shape of a hollow cylinder oscillates as a function of the magnetic flux threading through the hollow with a period of $hc/2e$. This effect was predicted for the diffusive regime of the charge transport where the mean free path of the electrons is much smaller than the sample size. The conductance amplitude of the oscillations is of the order of e^2/h and depends on the phase coherence length over which an electron maintains its phase coherence. Coherent backscattering of an electron when there is interference in a pair of backscattered spatial waves with time-reversal symmetry causes the oscillations.
   First described in: B. L. Al'tshuler, A. G. Aronov, B. Z. Spivak, *Aharonov–Bohm effect in non-ordered conductors*, Pis'ma Zh. Eksp. Teor. Fiz. **33**(2), 101–103 (1981) – in Russian.

amides – organic compounds that are nitrogen derivates of **carboxylic acids**. The carbon atom of a carbonyl group ($>$C$=$O) is bonded directly to a nitrogen atom of an $-$NH$_2$, $-$NHR or $-$NR$_2$ group, where R represents an **alkyl group** ($-$C$_n$H$_{2n+1}$). The general formula of amides is RCONH$_2$.

amines – organic compounds that are ammonia molecules with hydrogen substituted by **alkyl groups** ($-$C$_n$H$_{2n+1}$) or **aromatic rings**. These can be RNH$_2$, R$_2$NH, or R$_3$N, where R is an alkyl or aromatic group.

amino acid – an organic compound containing an amino group (NH$_2$), a carboxylic acid group (COOH), and any of various side groups that are linked together by **peptide bonds**. The basic formula is NH$_2$CHRCOOH. Amino acids are building blocks of **proteins**.
   There are twenty standard amino acids used in protein biosynthesis. These are presented in Figure A.3.

Amontons' law

Figure A.3 Amino acids found in proteins. Their symbols are shown in parentheses.

Just as the letters of the alphabet can be combined to form an almost endless variety of words, amino acids can be linked in varying sequences to form a huge variety of proteins.

More details in: //en.wikipedia.org/wiki/Amino_acid.

Amontons' law currently supposes the statement that the friction force between two bodies is directly proportional to the applied load (normal), with a constant of proportionality that is the friction coefficient. This force is constant and independent of the contact area, the surface roughness and the sliding velocity.

In fact, this statement is a combination of a few laws: the law of Euler and Amontons stating that friction is proportional to the loading force, the law of Coulomb → **Coulomb law (mechanics)** stating that friction is independent of the velocity, and the law of Leonardo da Vinci stating that friction is independent of the area of contact. In particular, Leonardo da Vinci arrived (1500) at the result that on an inclined plane a slider would move if the ratio between the tangential and normal components of the gravitational force exceeded one-fourth.

First described in: G. Amontons, De la résistance causée dans les machines, Mem. Acad. Roy. Sci. A, 206–222 (1699).

*More information in: R. Schnurmann Amontons' Law, "traces" of frictional contact, and experiments on adhesion, J. Appl. Phys. **13**(4), 235 (1942).*

amorphous solid – a solid with no long-range atomic order.

Ampère currents – molecular-ring currents postulated to explain the phenomenon of magnetism as well as the apparent nonexistence of isolated magnetic poles.

Ampère's law, as amended by Maxwell, states that magnetomotive force round any closed curve equals the electric current flowing through any closed surface bounded by the curve. The force appears clockwise to an observer looking in the direction of the current. It means that $\int \mathbf{H} dl = I$, where $\mathbf{H}$ is the magnetic field strength, I is the current enclosed. The linear integral is taken round any closed path. If the current is flowing in a conducting medium, $I = \int \mathbf{J} ds$, where $\mathbf{J}$ is the current density. Finally, it may be shown that $\nabla x \mathbf{H} = \mathbf{J}$, which is a statement of Ampère's law at a point in a conducting medium.

First described in: A. M. Ampère *Mémoire sur les effets du courant électrique*, Annales de chimie et de physique **15**, 59–118 (1820).

More details in: André-Marie Ampère, *Exposé méthodique des phénomènes électrodynamiques et des lois de ces phénomènes* (Plasson, Paris, 1822).

Ampère's rule states that the direction of the magnetic field surrounding a conductor will be clockwise when viewed from the conductor if the direction of current flow is away from the observer.

First described in: A. M. Ampère, *Mémoire sur les effets du courant électrique*, Annales de chimie et de physique **15**, 59–118 (1820).

More details in: André-Marie Ampère, *Exposé méthodique des phénomènes électrodynamiques et des lois de ces phénomènes* (Plasson, Paris, 1822).

Ampère's theorem states that an electric current flowing in a circuit produces a magnetic field at external points equivalent to that due to a magnetic shell whose bounding edge is the the conductor and whose strength is equal to the strength of the current.

First described in: A. M. Ampère, *Mémoire sur les effets du courant électrique*, Annales de chimie et de physique **15**, 59–118 (1820).

More details in: André-Marie Ampère, *Exposé méthodique des phénomènes électrodynamiques et des lois de ces phénomènes* (Plasson, Paris, 1822).

amphichiral → **chirality**.

AND operator → **logic operator**.

Andersen-Nose algorithm – a method used in **molecular dynamics simulation** for numerical integration of ordinary differential equation systems based on a quadratic presentation of time-dependent atom displacement.

First described in: S. Nose, F. Yonezawa, *Isothermal–isobaric computer simulations of melting and crystallization of a Lennard–Jones system*, J. Chem. Phys. **84**(3), 1803–1812 (1986).

Anderson localization

Anderson localization means that electron wave function becomes spatially localized and the conductivity vanishes at zero temperature when the mean free path of electrons is short comparable to the Fermi wavelength ($\lambda_F = 2\pi/k_F$); multiple scattering becomes important. Metal-insulator transition takes place due to disorders. In the localized states, the wave function decays exponentially away from the localization center, that is $\psi(r) \sim \exp(-r/\xi)$, where ξ is called the localization length. Anderson localization depends strongly on dimensionality.

First described in: P. W. Anderson, *Absence of diffusion in certain random lattices*, Phys. Rev. **109**(5), 1492–1505 (1958).

Recognition: in 1977 P. W. Anderson, N. F. Mott and J. H. van Vleck received the Nobel Prize in Physics for their fundamental theoretical investigations of the electronic structure of magnetic and disordered systems.

See also www.nobel.se/physics/laureates/1977/index.html.

Anderson rule, which is also called the **electron affinity rule**, states that vacuum levels of two materials forming a **heterojunction** should be lined up. It is used for construction of energy band diagrams of **heterojunctions** and **quantum wells**.

The **electron affinity** χ of the materials is used for the lining up procedure. This material parameter is nearly independent of the position of the Fermi level, unlike the **work function**, which is measured from the Fermi level and therefore depends strongly on doping.

Figure A.4 shows the band alignment at the interface between small band gap material A with the electron affinity χ_A and large band gap material B with the electron affinity χ_B supposing $\chi_A > \chi_B$. According to the rule the offset of the conduction band $\Delta E_c = \Delta E_{cB} - \Delta E_{cA} = \chi_A - \chi_B$. Correspondingly, the offset of the valence band ΔE_v can be predicted from the above diagram accounting for both electron affinities and band gaps of the materials. At a temperature above absolute zero the misalignment of the Fermi levels, if there is any, is eliminated by redistribution of free charge carriers at the interface between the barrier and well regions.

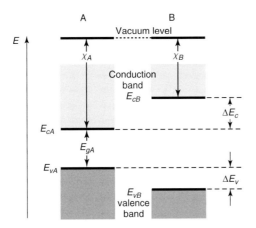

Figure A.4 Alignment of the bands at a heterojunction according to Anderson's rule.

The validity of the rule was discussed by H. Kroemer in his paper *Problems in the theory heterojunction discontinuities* CRC Crit. Rev. Solid State Sci. **5**(4), 555–564 (1975). The hidden assumption about the relation between the properties of the interface between two semiconductors and those of the much more drastic vacuum-to-semiconductor interface is a weak point of the rule.

First described in: R. L. Anderson, *Germanium-gallium arsenide heterojunction*, IBM J. Res. Dev. 4(3), 283–287 (1960).

Andreev process – reflection of a **quasiparticle** from the potential barrier formed by normal **conductor** and **superconductor** when the barrier height is less than the particle energy. It results in the temperature leap at the barrier if a heat flow takes place there. The conductor part of the structure can be made of a metal, **semimetal** or degenerate **semiconductor**.

The basic concept of the process is schematically illustrated in Figure A.5 for an electron crossing the interface between a conductor and superconductor.

There is a superconducting energy gap opened up for a single electron on the superconductor side. Thus, an electron approaching the barrier from the metal side with the energy above the **Fermi level**, but still within the gap, cannot be accommodated in the superconductor as a single particle. It can only form a **Cooper pair** there that needs an additional electron from the metal side with the energy below the Fermi level to come. This removed electron leaves behind a hole in the Fermi sea. If the incident electron had a momentum $\hbar k$, the generated hole has the momentum $-\hbar k$. It traces the same path as the electron, but in the opposite direction. Describing the phenomenon one says that the incident electron is reflected as a hole.

First described in: A. F. Andreev, *Thermal conductivity of the intermediate state of superconductors*, Zh. Exp. Teor. Fiz. **46**(5), 1823–1928 (1964) – in Russian.

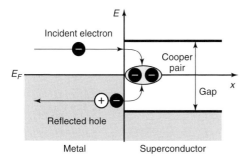

Figure A.5 Andreev reflection process.

Ångstrom – a metric unit of length measurements that corresponds to 10^{-10} m. The atomic diameters are in the range of 1–2 Å. It is named in honor of the nineteenth-century physicist Anders Jonas Ångstrom, one of the founders of modern spectroscopy.

angular momentum – the energy of a rotating particle. It is quantized for quantum particles as $L^2 = l(l + 1)\hbar^2$, where $l = 0, 1, 2, n - 1$, where n is the principal quantum number. In an atom electrons with $l = 0$ are termed s states, $l = 1$ (p states), $l = 2$ (d states), $l = 3$ (f states), $l = 4$ (g states). The letters s, p, d were first used to describe characteristic features of spectroscopic lines and stand for "sharp", "principal", and "diffuse". After d the letters run alphabetically.

anisodesmic structure – a structure of an ionic crystal in which bound groups of ions tend to be formed → **mesodesmic** and **isodesmic structures**.

anisotropy (of matter) – different physical properties of a medium in different directions. The alternative is **isotropy**.

anisotropic magnetic resistance – the difference in **magnetoresistance** when the resistance of a **conductor** is measured by the current passing either parallel or perpendicular to the material → **giant magnetoresistance effect**.
First described in: W. Thomson (Lord Kelvin), *On the electro-dynamic qualities of metals: effects of magnetization on the electric conductivity of nickel and of iron*, Proc. R. Soc. London **8**, 546–550 (1856).

anodizing = anodic oxidation, is the formation of an adherent oxide film on the surface of a metal or semiconductor when it is anodically polarized in a suitable electrolyte or plasma of an electric discharge in a gas.

anomalous Hall effect – an additional voltage proportional to the magnetization arising in **Hall effect** measurements in **ferromagnetic** materials. Unlike the ordinary **Hall effect**, this contribution is strongly temperature dependent.
The related transverse resistivity ρ_{xy} in ferromagnetics contains the contribution due to the magnetization M in addition to the usual **Hall effect**: $\rho_{xy} = R_0 B + 4\pi R_a M$, where B is the magnetic field induction, R_0 is the usual Hall coefficient, and R_a is the anomalous Hall coefficient. This expression can be used as an experimental tool to measure the magnetization as a function of temperature.
More details in: *The Hall Effect in Metals and Alloys*, edited by C. Hurd (Plenum, New York, 1972).
See also www.lakeshore.com/pdf_files/systems/Hall_Data_Sheets/Anomalous_Hall1.pdf.

anomalous Zeeman effect → **Zeeman effect**.

antibody – an inducible immunoglobulin **protein** produced by B lymphocytes of the immune system, in humans and other higher animals, which recognizes and binds to a specific **antigen** molecule of a foreign substance introduced into the organism. When antibodies bind to corresponding antigens they set in motion a process to eliminate the antigens.

antibonding orbital – the orbital which, if occupied, raises the energy of a molecule relative to the separated atoms. The corresponding wave function is orthogonal to that of the bonding state → **bonding orbital**.

antiferroelectric – a dielectric of high permittivity, which undergoes a change in crystal structure at a certain transition temperature, usually called the antiferroelectric **Curie temperature**. The antiferroelectric state in contrast to a **ferroelectric** state possesses no net spontaneous polarization below the Curie temperature. No hysteresis effects are therefore exhibited by this type of materials. Examples: $BaTiO_3$, $PbZrO_3$, $NaNbO_3$.

antiferromagnetic → **magnetism**.

antigen – any foreign substance, such as virus, bacterium, or **protein**, which, after introduction into an organism (humans and higher animals), elicits an immune response by stimulating the production of specific **antibodies**. It also can be any large molecule, which binds specifically to an antibody.

anti-Stokes line → **Raman effect**.

anti-dot – a **quantum dot** made of wider band gap semiconductor in/on a smaller band gap semiconductor, for example Si dot in/on Ge substrate. It repels charge carriers rather than attracting them.

anti-wires – the **quantum wires** made of wider band gap semiconductor in/on a smaller band gap semiconductor. They repel charge carriers rather than attracting them.

APFIM – an acronym for **atom probe field ion microscopy**.

a priori – Latin meaning "before the day". It usually indicates some postulates or facts known logically prior to the referred proposition. It pertains to deductive reasoning from assumed axioms or self-evident principles.

approximate self-consistent molecular orbital method – the **Hartree-Fock theory** as it stands is too time consuming for use in large systems. However, it can be used in a parametrized form, and this is the basis of many of the semi-empirical codes used like **Complete Neglect of Differential Overlap (CNDO)** and **Intermediate Neglect of Differential Overlap (INDO)**.

In the **CNDO** method all integrals involving different atomic orbitals are ignored. Thus, the overlap matrix becomes the unit matrix. Moreover, all the two-center electron integrals between a pair of atoms are set equal and the resonance integrals are set proportional to the overlap matrix. A minimum basis set of valence orbital is chosen using **Slater type orbitals**. These approximations strongly simplify the Fock equation.

In the **INDO** method the constraint present in CNDO that the monocentric two-electron integrals are set equal is removed. Since INDO and CNDO execute on a computer at about the same speed and INDO contains some important integrals neglected in CNDO, INDO performs much better than CNDO especially in prediction of molecular spectral properties.

It is interesting to note that the first papers dealing with the CNDO method appear in a supplementary issue of the Journal of Chemical Physics that contains the proceedings of the International Symposium on Atomic and Molecular Quantum theory dedicated to R. S. Mulliken → **Hund–Mulliken theory**, held in USA on 18–23 January 1965.

First described in: J. A. Pople, D. P. Santry, G. A. Segal, *Approximate self-consistent molecular orbital theory. I. Invariant procedures*, J. Chem. Phys. **43**(10), S129–S135 (1965); J. A. Pople, D. P. Santry, G. A. Segal, *Approximate self-consistent molecular orbital theory. II. Calculations with complete neglect of differential overlap*, J. Chem. Phys. **43**(10), S136–S151 (1965); J. A. Pople, D. P. Santry, G. A. Segal, *Approximate self consistent molecular orbital theory. III. CNDO results for AB_2 and AB_3 systems*, J. Chem. Phys. **44**(9), 3289–3296 (1965).

More details in: J. A. Pople, *Quantum chemical models*, Rev. Mod. Phys., **71** (5), 1267–1274 (1999).

Recognition: in 1998 J. A. Pople shared with W. Kohn the Nobel Prize in Chemistry for his development of computational methods in quantum chemistry.

See also www.nobel.se/chemistry/laureates/1998/index.html.

APW – an acronym for an **augmented plane wave**.

archaea – are single-celled organisms thriving in a variety of habitats. Most of the archaea prefer extreme environments. Archaea form together with bacteria and eucarya the three domains in life.

argon laser – a type of **ion laser** with ionized argon as the active medium. It generates light in the blue and green visible light spectrum, with two energy peaks: at 488 and 514 nm.

armchair structure → **carbon nanotube**.

aromatic compounds → **hydrocarbons**.

aromatic ring → **hydrocarbons**.

Arrhenius equation – the equation in the form $V = V_0 \exp(-E_a/k_B T)$, which is often used to describe temperature dependence of a process or reaction rate V, where V_0 is the temperature independent pre-exponential factor, E_a is the activation energy of the process or reaction, and T is the absolute temperature. The plot representing $\log(V/V_0)$ as a function of $1/k_B T$ or $1/T$ is called the **Arrhenius plot**. It is used to extract the activation energy E_a as the slope of a linear part of the curve.

First described by J. H. van't Hoff in 1884; in 1889, S. Arrhenius provided a justification and interpretation for it. See S. A. Arrhenius, *Über die Reaktiongeschwindigkeit der Inversion vor Rohrzucker durch Säuren*, Z. Phys. Chem. 4, 226 (1889).

Recognition: in 1901 J. H. van't Hoff received the Nobel Prize in Chemistry in recognition of the extraordinary services he has rendered by the discovery of the laws of chemical dynamics and osmotic pressure in solutions. In 1903 S. Arrhenius received the Nobel Prize in Chemistry in recognition of the extraordinary services he had rendered to the advancement of chemistry by his electrolytic theory of dissociation.

See also www.nobel.se/chemistry/laureates/1901/index.html.

See also www.nobel.se/chemistry/laureates/1903/index.html.

artificial atom(s) → **quantum confinement**.

associate operator → **adjoint operator**.

atomic engineering – a set of techniques used to built atomic-size structures. Atoms and molecules may be manipulated in a variety of ways by using the interaction present in the tunnel junction of **scanning tunneling microscope (STM)**. In a sense, there is a possibility to use the proximal probe in order to extend our touch to a realm where our hands are simply too big.

Two formal classes of atomic manipulation processes are distinguished: parallel processes and perpendicular processes. In the class of parallel processes an adsorbed atom or molecule is forced to move along the substrate surface. In the class of perpendicular processes the atom or molecular is transferred from the surface to the STM tip or vice versa. In both processes the goal is the purposeful rearrangement of matter on the atomic scale. One may view the act of rearrangement as a series of steps that results in the selective modification or breaking of chemical bonds between atoms and subsequent creation of new ones. It is equivalent to a procedure that causes a configuration of atoms to evolve along some time-dependent potential energy hyper-surface from an initial to a final configuration. Both points of view are useful in understanding physical mechanisms by which atoms may be manipulated with a proximal probe.

In the class of parallel processes, the bond between the manipulated atom and the underlying surface is never broken. This means that the adsorbate always lies within the absorption potential well. The relevant energy scale for these processes is the energy of the barrier to diffusion across the surface. This energy is typically in the range of 1/10 to 1/3 of the adsorption energy and thus varies from about 0.01 eV for weakly bound physisorbed atoms on a close-packed metal surface to 1 eV for strongly bound chemisorbed atoms. There are two parallel processes tested for atomic manipulation: field-assisted diffusion and sliding process.

The field-assisted diffusion is initiated by interaction of the spatially inhomogeneous electric field of an STM tip with the dipole moment of an adsorbed atom. The inhomogeneous electric field leads to a potential energy gradient at the surface resulting in a field-assisted directional diffusion motion of the adatom. In terms of the potential energy the process can be presented as follows.

An atom in an electric field $E(r)$ is polarized with a dipole moment $p = \mu + \overleftrightarrow{\alpha}\, E(r) + \cdots$, where μ is the static dipole moment, $\overleftrightarrow{\alpha}\, E(r)$ the induced dipole moment, and $\overleftrightarrow{\alpha}$ the polarizability tensor. The related spatially dependent energy of the atom is given by $U(r) = -\mu E(r) - 1/2\,\overleftrightarrow{\alpha}\,(r)E(r)E(r) + \cdots$. This potential energy is added to the periodic potential at the substrate surface. Weak periodic corrugation of the energy occurs. The resulting potential reliefs are shown in Figure A.6. A broad or sharp potential well is formed under the STM tip depending on the particular interaction between the tip, adatom and substrate atoms. The interaction of the electric field with the adsorbate dipole moment gives rise to a broad potential well. The potential energy gradient causes the adatom to diffuse toward the potential minimum under the tip. When there is a strong attraction of the adsorbate to the tip by chemical binding, it leads to a rather steep potential well located directly below the tip apex. The adsorbate remains trapped in the well as the tip is moved laterally.

Realization of field-assisted diffusion needs the substrate to be positively biased. At a negative substrate polarity the static and induced dipole terms being opposite in sign compensate each other. In this case no potential well and related stimulating energy gradient for diffusion are produced.

The sliding process supposes pulling of an adsorbate across the surface by the tip of a proximal probe. The tip always exerts a force on an adsorbate bound to the surface. One component of this force is due to the interatomic potential, that is, the chemical binding force, between the adsorbate and the outermost tip atoms.

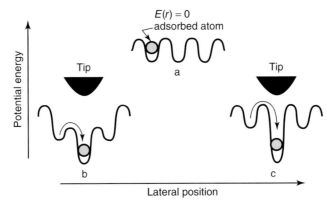

Figure A.6 Schematic of the potential energy of an adsorbed atom as a function of its lateral position on a surface above which there is located the STM tip.

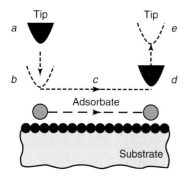

Figure A.7 Schematic of the sliding process: *a* and *e*–imaging, *b*–connecting, *c*–sliding, *d*–disconnecting.

By adjusting the position of the tip one may tune the magnitude and the direction of the force exerted on the adsorbate, thus forcing it to move across the surface.

The main steps of atomic manipulation via the sliding process are depicted in Figure A.7. The adsorbate to be moved is first located with the STM in its imaging mode and then the tip is placed near the adsorbate (position "*a*"). The tip–adsorbate interaction is subsequently increased by lowering the tip toward the adsorbate (position "*b*"). This is achieved by changing the required tunnel current to a higher value and letting the feedback loop move the tip to a height which yields the higher demanded current. The adsorbate–tip attractive force must be sufficient to keep the adsorbate located beneath the tip. The tip is then moved laterally across the surface under constant current conditions (path "*c*") to the desired destination (position "*d*"), pulling the adsorbate along with it. The process is terminated by reverting to the imaging mode (position "*e*"), which leaves the adsorbate bound to the surface at the desired location.

In order for the adsorbate to follow the lateral motion of the tip, the tip must exert enough force on the adsorbate to overcome the lateral forces between the adsorbate and the surface. Roughly speaking, the force necessary to move an adsorbate from site to site across the surface is given by the ratio of the corrugation energy to the sep-aration between atoms of the underlying surface. However, the presence of the tip may also cause the adsorbate to be displaced normal to the surface relative to its un-perturbed position. The displaced adsorbate would have an altered in-plane interac-tion with the underlying surface. If the tip pulls the adsorbate away from the surface causing a reduction of this in-plane interaction, then we would expect our estimate to be an upper bound for the force necessary to move the adsorbate across the surface.

The manipulation of an adsorbate with the sliding process may be characterized by a threshold tip height. Above this height the adsorbate-tip interaction is too weak to allow manipulation. At the threshold this interaction is just strong enough to allow the tip to pull the adatom along the surface. The absolute height of the STM tip above the surface is not directly measured. But resistance of the tunnel junction strongly correlated to the tip–surface separation is accurately controlled. An increasing resistance corresponds to greater tip–surface separation, and hence to their weaker interaction. The threshold resistance to slide an adsorbate depends on the particular arrangement of atoms at the apex of the tip. But for that reason it can vary by not more than a factor of 4. The resistance is more sensitive to the chemical nature of the adatom and surface atoms ranging from tens kΩ to a few MΩ. The ordering of the threshold resistances is consistent with the simple notion that the corrugation energy scales with the binding energy and thus greater force must be applied to move adatoms that are more strongly bound to the surface.

In perpendicular processes an atom, molecule or group of atoms is transferred from the tip to the surface or initially from the surface to the tip and then back to a new site on the surface. In order to illustrate the main regularities of these processes we discuss transferring an adsorbed atom from the surface to the tip. The relevant energy for such process is the height of the potential barrier that the adsorbate should come through to go from the tip to the surface. The height of this barrier depends on the separation of the tip from the surface. It approaches the adsorption energy in the limit of large tip–surface separation and goes to zero when the tip is located close enough to the adsorbate. By adjusting the height of the tip one may tune the magnitude of this barrier. Electrical biasing of the tip with respect to the substrate, as it is usually performed in STM, controls the transfer process. Three approaches distinguished by the physical mechanisms employed have been proposed for perpendicular manipulations of atoms. These are transfer on- or near-contact, field evaporation and electromigration.

The transfer on- or near-contact is conceptually the simplest among the atomic manipulation processes. It supposes the tip to be moved toward the adsorbate until the adsorption well on the tip and surface sides of the junction coalesce. That is, the energy barrier separating the two wells is gone and the adsorbate can be considered simultaneously bound to the tip and the surface. The tip is then withdrawn, carrying the adsorbate with it. For the process to be successful the adsorbate's bond to the surface must be broken when the tip is moved out. One might expect that the adsorbate would "choose" to remain bound to the side of the junction on which it has the greatest binding energy. However, the "moment of choice" comes when the adsorbate has strong interactions with both tip and surface, so the binding energy argument may be too simple. It does not account for the simultaneous interaction of the adsorbate with the tip and the surface.

At a slightly increased separation between the tip and sample surface, the adsorption well of the tip and surface atom are close enough to significantly reduce the intermediate barrier but have it still remain finite, such that thermal activation is sufficient for atom transfer. It is called transfer-near-contact. This process has a rate proportional to $v\exp(-E_a/k_B T)$, where v is the frequency factor, and E_a the reduced energy barrier between the tip and the sample. The transfer rate exhibits an anisotropy if the depth of the adsorption well is not the same on each side of the barrier. It is important to distinguish this transfer-near-contact mechanism from field evaporation, which requires an intermediate ionic state.

In its simplest form, the transfer on- or near-contact process occurs in the complete absence of any electric field, potential difference, or flow of current between the tip and the sample. Nevertheless, in some circumstances it should be possible to set the direction of transfer by biasing the junction during contact.

The field evaporation uses the ability of ions to drift in the electric field produced by an STM probe. It is a thermally activated process in which atoms at the tip or at the sample surface are ionized by the electric field and thermally evaporated. Drifting in this field they come more easily through the potential Schottky-type barrier separating the tip and the surface because this barrier appears to be decreased by the electric field applied. Such favorable conditions are simply realized for positively charged ions by the use of a pulse voltage applied to the tip separated from a sample surface by about 0.4 nm or smaller. Field evaporation of negative ions meets difficulties associated with the competing effect of field electron emission, which would melt the tip or surface at the fields necessary for negative ion formation.

The electromigration in the gap separating an STM tip and sample has much in common with the electromigration process in solids. There are two components of the force driving electromigration. The first is determined by the electrostatic interaction of the charged adsorbates with the electric field driving the electron current through the gap. The second, which is called the "wind" force, is induced by direct scattering of electrons at the atomic particles. These forces are most strongly felt by the atoms in the immediate vicinity of the tunnel junction formed by the tip of a proximal probe and sample surface. The highest electric field and current density are here. Within the electromigration mechanism the manipulated atoms always move in the same direction as the tunneling electrons. Moreover, "heating" of adsorbates by tunnel current stimulates electromigration as soon as a "hot" particle may more easily jump to a neighboring site. Atomic electromigration is a reversible process.

Summarizing the above-presented physical mechanisms used for manipulation of individual atoms with proximal probes one should remember that there is no universal approach among them. Applicability of each particular mechanism is mainly determined by the physical and chemical nature of the atoms supposed to be manipulated, by the substrate and to some extent by the probe material. An appropriate choice of the adsorbate/substrate systems still remains a state-of-art point.

More details in: *Handbook of Nanotechnology*, edited by B. Bhushan (Springer Verlag, Berlin Heidelberg, 2004).

atomic force microscopy (AFM) originated from **scanning tunneling microscopy (STM)**. Atomic and molecular forces, rather than a tunneling current, are monitored and used for the surface characterization at the atomic scale. The forces are detected by a probe tip mounted on a flexible cantilever, as shown in Figure A.8. Deflection of the cantilever, to a good approximation, is directly proportional to the acting force. It is optically or electronically monitored with high precision. The deflection signal is used to modulate the tip–sample separation in the way it is done in STM with the tunneling current. While scanning, one can obtain a profile of atomic and molecular forces over the sample surface. The sensitivity of AFM to the electronic structure of the sample surface, inherent in STM, is largely absent. Therefore it allows characterization of nonconducting materials.

atomic force microscopy (AFM)

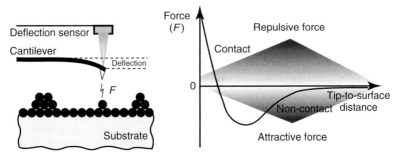

Figure A.8 Tip–sample geometry and registered effect in atomic force microscopy.

There are three principle types of imaging modes of the sample surface in AFM: contact, tapping, and non-contact modes. In the contact mode, the probing tip is always in contact with the sample surface, and surface structure is obtained from the deflection of the cantilever. The force on the tip is repulsive with a mean value of about 10^{-9} N. This force is set by pushing the cantilever against the sample surface with a piezoelectric positioning element. In the tapping mode, the probe tip is periodically in contact with the sample surface, and surface structure is obtained from the change in the vibration amplitude or phase of the oscillating cantilever. In the non-contact mode, the probe tip is not in contact with the sample surface, and surface structure is obtained from the change in the the vibration amplitude or resonant frequency of the oscillating cantilever.

In the contact mode, there is a high possibility that the strong repulsive force acting between the sample surface and the probe tip will destroy the sample surface and/or the tip apex. So, the tapping and non contact modes are widely used because these modes are more gentle than the contact mode.

In the tapping mode, the cantilever is driven at a fixed frequency near resonance with large vibration amplitude. When the probe tip is far from the surface, the vibration amplitude of the oscillating cantilever is held constant. When the probe tip is close to the surface, the probe tip is periodically in contact with the sample surface, and the vibration amplitude of the oscillating cantilever decreases because of cyclic repulsive contact between tip and surface with loss of the energy stored in the oscillating cantilever. The surface structure is obtained by maintaining the vibration amplitude at a constant level using the feedback circuit. The loading force acting between the probe tip and the sample surface can be greatly reduced in the tapping mode.

The atomic force microscopy technique has also been developed to detect electrostatic and magnetic forces as well as friction forces at atomic scale → **electrostatic force microscopy, magnetic force microscopy, friction force microscopy**.

First described in: G. Binning, C. F. Quate, Ch. Gerber, *Atomic force microscope*, Phys. Rev. Lett. **56**(9), 930–933 (1986).

More details in: Handbook of Nanotechnology, edited by B. Bhushan (Springer Verlag, Berlin Heidelberg, 2007); *Roadmap of Scanning Probe Microscopy*, edited by S. Morita (Springer Verlag, Berlin Heidelberg, 2007).

atomic number – the number of protons in the atomic nucleus, and hence the nuclear charge.

atomic orbital – a wave function of a hydrogenic (hydrogen-like) atom. This term expresses something less definite than "orbit" of classical mechanics. When an electron is described by one of the wave functions, one says that it occupies that orbital. It defines the spatial behavior of an electron of a given energy level in a particular atom. An overlap of orbitals in solids produces bands.

In an atom all orbitals of a given value of principal quantum number n form a single shell. It is common to refer to successive shells by the letters: K($n = 1$), L($n = 2$), M($n = 3$), N($n = 4$). The number of orbitals in a shell of principal number n is n^2. In a hydrogenic atom each shell is n^2-fold **degenerate**.

The orbitals with the same value of n but different **angular momentum**, which corresponds to different values of l, form the subshell of a given shell. The subshells are referred to by the letters: s($l = 0$), p($l = 1$), d($l = 2$), f($l = 3$). Thus, the subshell with $l = 1$ of the shell with $n = 3$ is called the 3p subshell. Electrons occupying these orbitals are called 3p electrons. The number of orbitals for different n and l is listed in the Table A.1.

Table A.1 Number of orbitals as a function of the quantum numbers n and l.

| $l \rightarrow$ | 0 | 1 | 2 | 3 | Total |
$n \downarrow$	s	p	d	f	number of orbitals
1	1				1
2	1	3			4
3	1	3	5		9
4	1	3	5	7	16

s orbitals are independent of angle (the angular momentum is zero), so they are spherically symmetrical. The first s orbitals are schematically shown in Figure A.9.

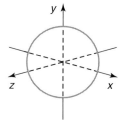

Figure A.9 The form of hydrogenic atomic s orbitals.

p orbitals are formed by electrons with the angular momentum $L^2 = 2\hbar^2$. These orbitals have zero amplitude at $r = 0$. This can be understood in terms of the

atomic orbital

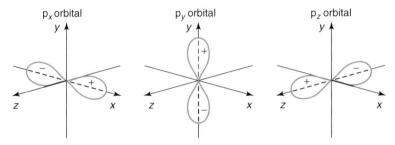

Figure A.10 Three hydrogenic atomic p orbitals, each directed along a different axis.

centrifugal effect of the angular momentum, which flings the electron away from the nucleus. The same effect appears in all orbitals with $l > 0$.

The three 2p orbitals are distinguished by the three different values that m_l can take when $l = 1$, where m_l represents the angular momentum around an axis. They are presented in Figure A.10. Different values of m_l denote orbitals in which the electron has different angular momenta around an arbitrary axis, for instance the z-axis, but the same magnitude of momentum because l is the same for all three. In this case the orbital with $m_l = 0$ has zero angular momentum around the z-axis. It has the form $f(r)\cos\theta$. The electron density, which is proportional to $\cos^2\theta$, has its maximum on either side of the nucleus along the z-axis (for $\theta = 0°$ and $\theta = 180°$). For this reason, the orbital is also called a p_z orbital. The orbital amplitude is zero when $\theta = 90°$, so the xy-plane is a nodal plane of the orbital. On this plane the probability of finding an electron occupying this orbital is zero.

The orbitals with $m_l = \pm 1$, which are the p_x and p_y orbitals, do have angular momentum about the z-axis. These two orbitals are different in the direction of the electron motion, which are opposite to each other. Nevertheless, they both have zero amplitude at $\theta = 0°$ and $\theta = 90°$ (along the z-axis) and maximum amplitude where $\theta = 90°$, which is in the xy-plane. The p_x and p_y orbitals have the same shape as the p_z orbital, but are directed along x- and y-axis, respectively. Their combinations are standing waves with no net angular momentum around the z-axis, since they are composed of equal but opposite values of m_l.

d orbitals appear when $n = 3$. There are five orbitals in this case with $m_l = 0, \pm 1, \pm 2$, which correspond to five different angular momenta around the z-axis but with the same magnitude of the momentum. The orbitals with opposite values of m_l (and hence opposite senses of motion around the z-axis) may combine in pairs to produce standing waves. An important feature of d orbitals is that they are concentrated much more closely at the nucleus than s and p orbitals are. An example of the d orbital is depicted in Figure A.11.

d orbitals are more strongly concentrated near the nucleus and isolated from neighboring atoms than others. They are important in studying properties of **rare-earth** metals.

For the definitions of σ and π orbitals → **molecular orbital**.

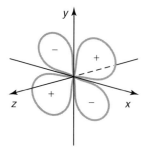

Figure A.11 The d orbital of the xy/r^2 form.

atom probe field ion microscopy (APFIM) – the technique originated from field ion microscopy. The analyzed sample is prepared in the form of a sharp tip. A voltage pulse is applied to the tip causing atoms on the surface of the tip to be ejected. The atoms travel down a drift tube where their time of arrival can be measured. The time taken for the atom to arrive at the detector is a measure of the mass of that atom. Thus, compositional analysis of the sample can be carried out on a layer by layer basis.

The technique enables one to determine the chemical identity along with the position of surface atoms with atomic structural resolution. It has no elemental mass limitations, making it unique among analytical instruments.

First described in: E. W. Müller, J. A. Panitz, S. B. McLane, *The atom probe microscope*, Rev. Sci. Instrum. **39**(1), 83–86 (1968).

More details in: T. T. Tsong, *Atom-Probe Field Ion Microscopy: Field Ion Emission, and Surfaces and Interfaces at Atomic Resolution* (Cambridge University Press, Cambridge, 1990).

atto- – a decimal prefix representing 10^{-18}, abbreviated a.

aufbau principle – states that in any atom the lowest energy orbitals fill first. In conjunction with the **Pauli exclusion principle** and the **Hund rules**, it gives the correct electron configuration for an atom or ground state ion.

First described in: N. Bohr, *Der Bau der Atome und die physikalischen und chemischen Eigenschaften der Elemente*, Z. Phys. **9**, 1–67 (1922); E. C. Stoner, *The distribution of electrons among atomic levels*, Phil. Mag. **48**, 719–736 (1924).

Auger effect – formation of non-radiative re-arrangement of atomic electrons after the atom has been ionized in one of its inner shells.

Classically, if an atom is ionized in its S shell, the radiative mode of de-excitation is that in which a transition occurs involving an electron falling from a less tightly bound shell T to S with emission of a quantum of radiation with the energy $E_S - E_T$, which is the difference between the binding energies of the S and T shells, respectively. In the Auger transition, an electron, called the **Auger electron**, is emitted with kinetic energy $E = E_S - E_T - E_U$, with U being the same shell as

Auger electron

T or another one less tightly bound than S. For the transition to be energetically possible, of course $E_S - E_T - E_U = 0$. The process can be interpreted as the T electron falling into the S shell, the energy released being used to eject the U electron. The binding energy of the U electron is dashed because when it is ejected the atom is already ionized.

Most observable Auger transitions originate either by primary ionization in the K or L shells of an atom, because transitions due to ionization in the higher shells cause the emission of electrons with energies too low to be detected. The emitted electrons are called Auger electrons. Their energy spectrum is a fingerprint of the chemical nature of the atom. It is widely used for analysis of chemical compositions of matter by **Auger electron spectroscopy**.

First described in: P. Auger, *Sur les rayons β secondaires produits dans un gaz par des rayons*, C. R. **177**, 169–171 (1923).

Auger electron – an electron that is expelled from an atom in the **Auger effect**.

Auger electron spectroscopy (AES) – a technique of nondestructive elemental analysis of matter by examining energy distributions of secondary electrons emitted by the **Auger effect**. If a material is bombarded by electrons with an energy sufficient to ionize inner orbits of the atoms, the energy released when the ionized atom rearranges itself to fill the ionized level is characteristic of the atom. This energy may appear as an X-ray photon or may instead go to an outer orbit electron, ejecting it by the radiationless Auger process. Thus, the energy distribution of secondary Auger electrons contains peaks localized at energies which serve to identify the atoms producing them. The depth at which this information can be obtained is 0.5–2 nm for metals, 1.5–4 nm for oxides, and 4–10 nm for polymers.

More details in: M. De Crescenzi, M. N. Piancastelli *Electron Scattering and Related Spectroscopies* (World Scientific Publishing, Singapore, 1996).

Auger recombination – a transition of an electron from the conduction band to the valence band by transfer of the energy to another free electron or hole. No electromagnetic radiation is emitted during such process.

Auger scattering – one of the interacting charge carriers gives up its potential energy to another and hence relaxes down its energy level.

augmented plane wave (APW) – piecewise defined function consisting of the solution to the **Schrödinger equation** for an isolated atom within a sphere of given radius and a **plane wave** outside this region.

augmented-plane-wave method supposes solving the **Schrödinger equation** for electrons in an atom in terms of a set of fabricated functions that combine the oscillations inside the core with plane waves elsewhere. The potential is assumed to be spherically symmetrical within spheres centered at each atomic nucleus and constant in the interstitial region. Wave functions in the form of augmented plane waves are constructed by matching solutions of the **Schrödinger equation** within each sphere with plane-wave solutions in the interstitial region. Linear combinations of these wave functions are then determined by the **variational method**.

First described in: J. C. Slater, *Wave functions in a periodic potential*, Phys. Rev. **51**(10), 846–851 (1937).

autocorrelation function – the measure of the dependence of time series values at one time on the values at another time. The term autocorrelation means self-correlation. However, instead of correlation between two different variables, the correlation between two values of the same variable at times x_n and x_{n+k} is analyzed. For a time series $x(n), n = 1, 2, \ldots N$, the autocorrelation function is defined as: $R(k) = \frac{1}{(N-k)} \sum_{n=1}^{N-k} x(n)x(n + k)$, where k is the correlation distance and N is the length of the series. The argument of the autocorrelation function is the correlation distance k, and the function value at k expresses the average correlation between numbers separated by the distance k in the series.

The autocorrelation function is used to detect non-randomness in data and to identify an appropriate time series model if the data are not random. It is a powerful tool to find weak periodic signals in noisy data.

First described in: G. E. P. Box and G. Jenkins, *Time Series Analysis: Forecasting and Control* (Holden-Day, 1976).

autoelectronic emission – an emission of electrons from the surface of a conductor by application of an external electric field in the temperature range where conventional thermo-stimulated electron emission is rather small.

Aviram–Ratner model – a model for unimolecular electrical rectification employing a molecule with electrically isolated electron donor and acceptor groups.

The molecule is postulated to be initially in an electronically neutral state and placed between two metal electrodes. The current flow is proposed to be a three-step process, upon application of a bias. First, an electron passes from the cathode to the acceptor. The Fermi level of the anode is sufficiently lowered under the bias to allow the second step of the process, electron tunneling from the donor to the anode, to occur. Finally, the electron inelastically tunnels from the **acceptor** to the **donor**.

A demonstration **molecular rectifier** consists of a donor π system (tetrathiafulvalene) and an acceptor π system (tetracyanoquinodimethane), separated by a σ bonded (methylene) tunneling bridge.

First described in: A. Aviram, M. A. Ratner, *Molecular rectifiers*, Chem. Phys. Lett. **29**(2), 277–283 (1974).

Azbel'–Kaner cyclotron resonance

Azbel'–Kaner cyclotron resonance – the method used to measure **cyclotron frequencies** in metals and thus useful in **Fermi surfaces** studies. In a large static magnetic field, a semiconductor permeated by a radiofrequency electromagnetic field produces sharp peaks in the radiofrequency energy absorption when the frequency coincides with the cyclotron resonance frequency ω_c. In metals this is impeded by the **skin-depth effect** that prevents the radiofrequency field penetration. In the Azbel'–Kaner method one accepts the limitation that the radiofrequency field can accelerate the electrons only within a very thin surface layer and arranges the geometry of the experiment such that the electrons return to this layer frequently. In order to obtain this the static magnetic field is oriented parallel to the surface of the sample. Thus, any electron at its general helical orbit around the magnetic field lines approaches the surface within the same distance on each cycle. If the radiofrequency coincides with the cyclotron frequency, the electron can resonantly absorb energy from the field. Here the condition for the resonance is $\omega = n\omega_c (n = 1, 2, 3, \ldots)$.

First described in: M. Ya. Azbel', E. A. Kaner, *Cyclotron resonance in metals*, J. Phys. Chem. Solids **6** (2–3), 113–135 (1958); M. Ya. Azbel', E. A. Kaner, Zh. Eksp. Teor. Fiz. **30**, 811 (1956) and **32**, 896 (1957) – in Russian.

More details in: L. M. Falicov, *Fermi surface studies*, in: *"Electrons in Crystalline Solids"* (IAEA, Vienna, 1973), pp. 207–280.

B

From B92 protocol to Burstein–Moss shift

B92 protocol – the protocol for quantum cryptographic key distribution developed by Charles **B**ennett in 19**92** (the acronym uses the bold characters). It works like the **BB84** protocol, but instead of using a system with four pairwise orthogonal states, only two non-orthogonal states are involved. This makes the protocol simpler.

First described in: C. H. Bennet, *Quantum cryptography using any two non-orthogonal states*, Phys. Rev. Lett. **68**(21), 3121–3124 (1992) – theory; A. Muller, J. Breguet, N. Gisin, *Experimental demonstration of quantum cryptography using polarized photons in optical fiber over more than 1 km*, Europhys. Lett. **23**(6), 383–388 (1993) – experiment.

Back–Goudsmit effect – breakdown of the coupling between the nuclear spin **angular momentum** and the total angular momentum of the electrons in an atom at relatively small magnetic field.

First described in: S. Goudsmit, E. Back, *Die Koppelung der Quantenvektoren bei Neon, Argon und einigen Spektren der Kohlenstoffgruppe*, Z. Phys. **40**(7), 530–538 (1927).

bacteriophage → **virus**.

bacterium – a single-celled organism. Bacteria form one of the three major branches of life. They lack the nucleus and **organelles** of the more complex cells which are **eukaryotes**. Nevertheless, they, like the cells of plants, mostly possess a carbohydrate-based cell wall.

Badger rule – a simple empirical relationship between the stretching force constant for a molecular bond and the bond length discovered by Badger in 1934, $k_e = A (R_e - B)^{-3}$, where k_e is the force constant and R_e is the equilibrium bond length.

First described in: R. M. Badger, *A relation between internuclear distances and bond force constants*, J. Chem. Phys. **2**(3), 128–131 (1934).

ballistic conductance – a characteristic of ideal **ballistic transport** of charge carriers in nanostructures. It is deduced from the fundamental constants, so it is material independent.

What is What in the Nanoworld. Victor E. Borisenko and Stefano Ossicini
Copyright © 2008 WILEY-VCH Verlag GmbH & Co. KGaA, Weinheim
ISBN: 978-3-527-40783-5

ballistic magnetoresistance

Figure B.1 Two reservoirs with electrons connected by a perfect conducting channel.

The simplest device appropriate for illustration this fact is a two-terminal conductor. It is schematically shown in Figure B.1, where constriction between two reservoirs of electrons acts as a conducting quantum wire. The size of the constriction must be close to the Fermi wavelength of electrons in order to observe the role of the wave-like nature of electrons in their transport behavior. No irregularities and related carrier scattering are expected along the conducting channel. Moreover, this perfect conducting tube is assumed to be tied to the reservoirs via tapered nonreflecting connectors. It means that carriers approaching a reservoir inevitably pass into it.

Take the zero temperature case and let the reservoirs be filled to with electrons up to the level characterized by electrochemical potentials μ_1 and μ_2, and $\mu_1 > \mu_2$. If electronic states in the range between μ_1 and μ_2 are fully occupied, there is a current between the reservoirs

$$I = (\mu_1 - \mu_2)ev(dn/d\mu),$$

where e is the electronic charge, v is the velocity component along the conducting channel at the Fermi surface, and $dn/d\mu$ is the density of states in the channel (allowing for spin degeneracy). In a quantum wire $dn/d\mu = 1/\pi hv$. Substituting $(\mu_1 - \mu_2) = e(V_1 - V_2)$, where V_1 and V_2 are the electrical potentials inducing the difference in the electrochemical potentials of the reservoirs, one can calculate the conductance of the quantum wire as

$$G = I/(V_1 - V_2) = e^2/(\pi h) = 2e^2/h.$$

This is the conductance of an ideal one-dimensional conductor operating in the ballistic transport regime. It is evident that it is determined only by fundamental constants. The quotient $e^2/h = 38.740\ \mu S$ is referred to as the quantum unit of conductance. The corresponding resistance is $h/e^2 = 25812.807\ \Omega$.

ballistic magnetoresistance – the electrical resistance of a structure with the ballistic transport of charge carriers measured in the presence of a magnetic field. The size of the ballistic magnetoresistance is expressed as a percentage representing the change of the resistance of the ballistic conducting channel between two electrodes when the relative magnetic alignment of the electrodes changes from being parallel to antiparallel.

First described in: N. García, M. Muñoz, Y.-W. Zhao, *Magnetoresistance in excess of 200% in ballistic Ni nanocontacts at room temperature and 100 Oe*, Phys. Rev. Lett. **82**(14), 2923–2926 (1999).

ballistic transport (of charge carriers) – charge carriers pass through a structure without scattering.

Balmer series → **Rydberg formula**.
First described by: Johann Jacob Balmer on June 25, 1884, during a lecture to the *Naturforschende Gesellschaft* in Basel, Switzerland. J. J. Balmer *Notiz über die Spektrallinien des Wasserstoffs*, Ann. Phys. **25**, 80 (1885).

Banach space – the extended version of the **Hilbert space**, which is a vector space with a norm, but not necessarily given by an inner product. The space should be complete. Every **Cauchy sequence** converges to a point of the space.
First described in: S. Banach, *Sur le Problème de la Mesure*, Fund. Math. **4** (1), 7–33 (1923).
More details in: S. Banach, *Theorie des Operations Linéaires* (Warszawa, 1932).

band structure → **electronic band structure**.

Bardeen–Cooper–Schrieffer (BCS) theory explains theoretically the superconductivity phenomenon in metals, metallic compounds and alloys → **superconductivity**.
Recognition: in 1972 J. Bardeen, L. N. Cooper and J. R. Schrieffer received the Nobel Prize in Physics for their jointly developed theory of superconductivity, usually called the BCS-theory.
See also www.nobel.se/physics/laureates/1972/index.html.

Barkhausen effect – when a ferromagnetic material is subjected to a changing external magnetic field, part of the resultant change in its magnetization occurs via series of discontinous steps, even though the rate of the magnetic field change may be extremely slow. The discontinuities are produced by irreversible changes in the domain structure of the material.
First described in: H. G. Barkhausen, *Zwei mit Hilfe der neuen Verstärker entdeckte Erscheinungen*, Phys. Z. **20**, 401–403 (1919).

Barlow rule states that the volume occupied by the atoms in a given molecule is proportional to the valence of the atoms, using the lowest valency values.

Barnett effect – the magnetization of an initially unmagnetized specimen acquired when it is rotated in the absence of any external magnetic field.
First described in: S. J. Barnett, *An investigation of the electric intensities and electric displacement produced in insulators by their motion in a magnetic field*, Phys. Rev. **27**(5), 425–472 (1908).

Barnett–Loudon rule

Barnett–Loudon rule states that allowed modifications of the spontaneous emission rate for an electric dipole transition, caused by the atomic environment, are constrained by a sum rule

$$\int_0^\infty d\omega_a \frac{\Gamma_m(\mathbf{r}, \omega_a) - \Gamma_0(\omega_a)}{\Gamma_0(\omega_a)} = 0,$$

where $\Gamma_0(\omega_a)$ is the **spontaneous emission** rate in free space and $\Gamma_m(\mathbf{r}, \omega_a)$ is the emission rate of an atom or molecule at position $\mathbf{r}$ as modified by the environment, whose effect is assumed to vary by a negligible amount across the extent of the emitting object. It follows that any reduction in spontaneous emission rate over some range of frequencies ω_a must necessarily be compensated by increases over some other range of transition frequencies. This rule is derived on the basis of **causality** requirements as expressed in the **Kramers–Kronig dispersion relation**.

It should be emphasized that the rule applies to the spontaneous emission by an atom or molecule in a completely arbitrary environment, which may include, for example, metallic mirrors, Bragg reflectors, photonic band gap materials and absorptive dielectrics or semiconductors. In all cases, the sum rule is used for model calculations of the modified spontaneous emission rates only if they use dielectric functions that conform to general causality and asymptotic requirements.

First described in: S. M. Barnett, R. Loudon, *Sum rule for modified spontaneous emission rates*, Phys. Rev. Lett. **77**(12), 2444–2446 (1996).

baryon – a particle belonging to the class of elementary particles that have a mass greater than or equal to that of the **proton**, participate in strong interactions, and have a spin of $\hbar/2$.

base pair – two complementary nitrogenous bases in a **DNA** molecule, such as the nucleotide coupling of adenine with thymine (A:T) and guanine with cytosine (G:C). It is also used as a unit of measurement for DNA sequences.

basis states – the states comprising the set in which the wave function is expanded.

BB84 protocol – the first protocol for quantum cryptographic key distribution developed by Charles **B**ennett and Gilles **B**rassard in 19**84** (the acronym uses the bold characters). This scheme, which is a four state scheme, uses the transmission of non-orthogonal states of single-photon qubits, with security derived from the impossibility of an eavesdropper distinguishing the two states without being detected (on average). It works as follows.

Alice (conventional name of a sender) and Bob (conventional name of a receiver) are connected by two channels, one quantum and another public and classic. If photons are the vehicle carrying the key, the quantum channel is usually an optical fiber. The public channel can also be so, but with one difference: in the quantum channel there is in principle only one photon per bit to be transported, while in the public channel, in which eavesdropping by any non authorized person does not matter, the intensity is hundreds of times bigger.

Step 1. Alice prepares photons with linear polarizations randomly chosen among the angles $0°$, $45°$, $90°$, $135°$, which she sends "in a row" through the quantum channel, while keeping a record of the sequence of the prepared states and of the associated sequence of logic 0's and 1's obtained representing by 0 the choices of 0 and 45 degrees, and by 1 otherwise. This sequence of bits is clearly random.

Step 2. Bob has two analyzers, one "rectangular" (+type), the other "diagonal" (xtype). Upon receiving each of Alice's photons, he decides at random what analyzer to use, and writes down the aleatory sequence of analyzers used as well as the result of each measurement. He also produces a bit sequence associating 0 to the cases in which the measurement produces a $0°$ or $45°$ photon, and 1 in cases of $90°$ or $135°$.

Step 3. Next they communicate with each other through the public channel the sequences of polarization basis and analyzers employed, as well as Bob's failures in detection, but never the specific states prepared by Alice in each basis nor the resulting states obtained by Bob upon measuring.

Step 4. They discard those cases in which Bob detects no photons, and also those cases in which the preparation basis used by Alice and the analyzer type used by Bob differ. After this distillation, both are left with the same random subsequence of bits 0, 1, which they will adopt as the shared secret key.

First described in: C. Bennett, G. Brassard, *Quantum cryptography: public key distribution and coin tossing*, Proc. IEEE Int. Conf. Comp. Syst. Signal Proc. **11**, 175–179 (1984) – theory; C. H. Bennett, G. Brassard, *The dawn of a new era for quantum cryptography: the experimental prototype is working!* SIGACT News **20**(4), 78–82 (1989) – experiment.

Becker–Kornetzki effect – a reduction in the internal friction of a ferromagnetic substance when it is subjected to a magnetic field that is large enough to produce magnetic saturation.

First described in R. Becker, M. Kornetzi, *Einige magneto-elastische Torsionsversuche*, Z. Phys. **88**(9–10), 634–646 (1934).

Beest–Kramer–Santen potential – the interatomic many force field Coulombic potential including a modified **Buckingham potential** for the repulsion and dispersion interactions, respectively, in the form

$$V(r_{ij}) = \sum_{i>j} \left[A_{ij} exp(-b_{ij} r_{ij}) - \frac{C_{ij}}{r_{ij}} \right] + \sum_{i>j} \frac{q_i q_j}{r_{ij}},$$

where r_{ij} is the distance between atoms i and j, the parameter A_{ij}, b_{ij}, and C_{ij} are fit to *ab initio* and empirical data, q_i and q_j are fixed partial charges of the atoms.

First described in: B. W. H. van Beest, G. J. Kramer, R. A. van Santen, *Force fields for silicas and aluminophosphates based on ab initio calculations*, Phys. Rev. Lett. **64**(16), 1955–1958 (1990).

Bell's inequality

Bell's inequality – the inequality for the outcomes of measurement of a bipartite system which must be satisfied if the system state were completely described by local **hidden variables**. For a pair of spins A and B, the inequality $P(1,2) + P(1,3) + P(2,3) \geq 1$ must hold, where $P(i,j)$ is the probability to get the same result when measuring the spin A along the direction i and the spin B along the direction j, the directions being defined in Cartesian coordinates as $\mathbf{n}_1 = (0,0,1), \mathbf{n}_2 = (\sqrt{3},0,-1)/2, \mathbf{n}_3 = (-\sqrt{3},0,-1)/2$. It shows that statistical predictions of quantum mechanics are incompatible with any local **hidden variables** theory apparently satisfying only the natural assumptions of locality. The inequality is violated if the joint state of two spins is a **Bell's state**.

First described in: J. S. Bell, *On the Einstein–Podolsky–Rosen paradox*, Physics **1**(3), 195–200 (1964); J. S. Bell, *On the problem of hidden variables in quantum mechanics*, Rev. Mod. Phys. **38**(3), 447–452 (1966).

More details in: J. S. Bell, *Speakable and Unspeakable in Quantum Mechanics* (Cambridge University Press, Cambridge, 1987).

Bell's state – the joint entangled state $|\psi\rangle = \frac{1}{\sqrt{2}}(|A-\uparrow\rangle|B-\downarrow\rangle - |A-\downarrow\rangle|B-\uparrow\rangle)$ of a pair of spins A and B. It is known as a spin singlet. This state violates the **Bell's inequality**.

Bell's theorem establishes that **hidden variables** theories can only reproduce the statistical predictions of quantum mechanics if the hidden variables in one region can be affected by measurements carried out in a space-like separated region. In other words it states that in the quantum theory there are states which violate **Bell's inequality**, showing the incompatibility of the quantum theory with the existence of local variables. Thus, some correlations predicted by quantum mechanics cannot be reproduced by any local theory.

First described in: J. S. Bell, *On the Einstein–Podolsky–Rosen paradox*, Physics **1**(3), 195–200 (1964); J. S. Bell, *On the problem of hidden variables in quantum mechanics*, Rev. Mod. Phys. **38**(3), 447–452 (1966).

More details in: J. S. Bell, *Speakable and Unspeakable in Quantum Mechanics* (Cambridge University Press, Cambridge, 1987).

Benedicks effect supposes that a voltage is induced between two points in a solid that are at the same temperature if the temperature profile between these points is asymmetric. It has been observed in semiconductors, while in metals it is expected to be negligibly small.

First described in: C. Benedicks, *Jetziger Stand der grundlegenden Kenntnisse der Thermoelektrizität*, Ergeb. Exakten Naturwiss. **8**, 26–67 (1929).

More details in: G. D. Mahan, *The Benedicks effect: nonlocal electron transport in metals*, Phys. Rev. B **43**(5), 3945–3951 (1991).

Berry phase arises in a system carried through a series of adiabatic changes, by varying a set of parameters, and then returned to its initial state when the parameters get their initial values. But the final state is multiplied by a phase factor that depends on the entire history of the parameters, and is thus nonlocal. This phase difference is called the Berry phase. Its occurrence typically indicates that the system's parameter dependence is undefined for some combination of parameters.

The Berry phase forms a crucial concept in many quantum mechanical effects, including for example the motion of vortices in superconductors, the electron transport in nanoelectronic devices, quantum computing.

First described in: M. V. Berry, *Quantal phase factors accompanying adiabatic changes*, Proc. R. Soc. A **392**, 45–57 (1984).

More details in: M. V. Berry, *Anticipations of the geometric phase*, Phys. Today **43**(12), 34–40 (1990).

Berthelot rule defines the relationship between energy attraction constants for a mixture of like $(\varepsilon_{ii}, \varepsilon_{jj})$ and unlike (ε_{ij}) species in the form $\varepsilon_{ij} = (\varepsilon_{ij} + \varepsilon_{ij})^{1/2}$.

Berthelot–Thomsen principle states that of all chemical reactions possible, the one developing the greatest amount of heat will take place, with certain obvious exceptions such as changes of state. This principle was formulated in two slightly different versions firstly by the Danish chemist Julius Thomsen in 1854 and then by the French chemist Marcellin Berthelot in 1873.

First described in: J. Thomsen, *Die Grundzüge eines thermochemischen Systems*, Ann. Phys. **47**, 34 (1854); M. Berthelot, *Sur la statique de dissolutions salines*, Bull. Soc. Chim. Paris **19**, 160 (1873).

Bessel function – the general part of the solution of the second order linear ordinary differential equation, known as **Bessel's equation,**

$$x^2 \frac{d^2 y}{dx^2} + x \frac{dy}{dx} + (x^2 - v^2)y = 0,$$

in which both x and the parameter v can be complex. A complete solution of the equation depends upon the nature of the parameter v. It can be found as a combination of the functions $B_v(x)$ with appropriate coefficients. The function

$$B_v(x) = \sum_{m=0}^{\infty} \frac{(-1)^m (0.5x)^{v+2m}}{m! \Gamma(v + m + 1)}$$

is a Bessel function of order v and argument x of the first kind.

First described by: Friedrich Bessel in 1817.

More details in: I. N. Sneddon, *Special Functions of Mathematical Physics and Chemistry* (Oliver and Boyd, Edinburgh, 1956).

Bessel's equation → **Bessel function.**

Bethe–Salpeter equation

Bethe–Salpeter equation – an integral equation for the relativistic bound state wave function of two interacting Fermi-Dirac particles. It is a direct application of the **S-matrix** formalism of Feynman. Starting from the Feynman two-body kernel one can prove that its usual power expansion can be re-expressed as an integral equation; this equation in the extreme non-relativistic approximation and to lowest order in the power expansion reduces to the appropriate **Schrödinger equation**. Consider two Fermi-Dirac particles of masses m_a and m_b, respectively, capable of interacting with each other through the virtual emission and absorption of quanta; let us denote the bound-state 4-momentum by P, while the Jacobi relative 4-momentum is denoted by

$$q = \frac{m_b p_a}{m_a + m_b} - \frac{m_a p_b}{m_a + m_b},$$

while the irreducible 2-body kernel is given by $G(q, q'; P)$. The Bethe–Salpeter equation takes the form:

$$\left(\frac{m_a \mathbf{P}}{m_a + m_b} + \mathbf{q} - m_a \right) \left(\frac{m_b \mathbf{P}}{m_a + m_b} - \mathbf{q} - m_b \right) \psi(q) = i \int d^4 q' \, G(q, q'; P) \psi(q').$$

The Bethe–Salpeter equation was proven within the quantum field theory by Gell-Mann and Low. In recent years the Bethe–Salpeter equation has been widely used to go beyond the simple one-particle picture in the description of the electronic excitation processes in nanostructures (electron-hole interaction); these electronic excitations lie at the origin of most of the commonly measured spectra. Typical Feynman graphs representing the Bethe-Salpeter equation for the polarizability χ are shown in Figure B.2.

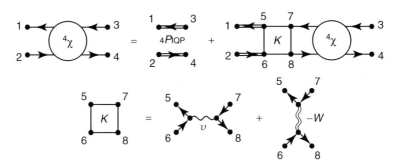

Figure B.2 Feynman diagrams representing the Bethe–Salpeter equation for χ. After G. Onida, L. Reining and A. Rubio, *Electronics excitations: density functional vs. many-body Green's-function approaches*, Rev. Mod. Phys. **74**(2), 601–659 (2002).

First described in: H. A. Bethe and E. E. Salpeter, *A Relativistic Equation for Bound State Problems*, Phys. Rev. **82**, 309–310 (1951), where the abstract of a paper presented at the 303-rd Meeting of the America Physical Society is reported. In the same year the equation was published in the paper with the same title: E. E. Salpeter and H. E. Bethe, *A Relativistic Equation for Bound State Problems*, Phys. Rev. **84**, 1232–1242 (1951); for the proof within quantum field theory see: M. Gell-Mann and F. Low, *Bound States in Quantum Field Theory*, Phys. Rev. **84**, 350–354 (1951).

More details in: G. Onida, L. Reining and A. Rubio, *Electronics excitations: density functional vs. many-body Green's-function approaches*, Rev. Mod. Phys. **74** (2), 601–659 (2002).

biexciton = excitonic molecule, which is a bound state of two electron-hole pairs (**excitons**) similar to the hydrogen molecule.

binary decision diagram (BDD) – a directed acyclic graph with exactly one root, whose sinks are labeled by the constants 1 and 0, and whose internal nodes are labelled by a Boolean variable → **Boolean algebra**. The *nodes* have only two outgoing edges, a 0-edge and 1-edge. Such diagrams are suitable for representation of logic functions and computing.

First described in: C. Y. Lee, *Representation of switching circuits by binary-decision programs*, Bell Systems Technical Journal **38**, 985–999 (1959).

More details in: R. Drechsler, B. Becker, *Binary Decision Diagrams–Theory and Implementation* (Kluwer Academic Publishers, Boston, 1998).

biophysics – the physics of living organisms and vital processes. It studies biological phenomena in terms of physical principles.

bifurcation – a qualitative change in the properties of a system.

bioengineered materials – materials employed in biomedical technology designed to have specific, desirable biological interactions with the surroundings. Materials scientists are increasingly deriving new lessons from naturally occurring materials about useful composition–structure property relationships that might be mimicked with synthetic materials → **biomimetics**.

More details in: M. Tyrrell, E. Kokkoli, M. Biesalski, *The role of surface science in bioengineered materials*, Surface Sci. **500**, 61–83 (2002).

biological surface science – the broad interdisciplinary area where properties and processes at interfaces between synthetic materials and biological environments are investigated and biofunctional surfaces that are capable of directing and controlling a desired biological response are fabricated.

More details in: B. Kasemo *Biological surface science*, Surface. Sci. **500**, 656–677 (2002).

biomimetics

biomimetics – the concept of taking ideas from nature (imitating, copying and learning) and implementing them into another technology.

bioremediation – a use of biological agents to reclaim soils and waters polluted by substances hazardous to human health and/or environment. It is an extension of biological treatment processes that have been used traditionally to treat wastes in which micro-organisms typically are used to biodegrade environmental pollutants. In the case of nanomaterials the driver would be to design systems capable of fixing heavy metals, cyanides and other environmentally damaging materials.

Biot law states that an optically active substance rotates plane polarized light through an angle inversely proportional in its wavelength.
First described by J. B. Biot in 1815.

bipyridine – the organic compound $C_{10}H_8N_2$ shown in Figure B.3.
It consists of two **pyridine** molecules linked by a single bond.

Figure B.3 Bipyridine molecule.

Bir–Aronov–Pikus mechanism – the electron spin relaxation mechanism in semiconductors caused by exchange and annihilation interaction between electrons and holes.

This mechanism is important for p-doped semiconductors, in which the electron-hole exchange interaction gives rise to fluctuating local magnetic fields flipping electron spins. For electron-spin relaxation due to exchange with nondegenerate holes the relaxation time is

$$\frac{1}{\tau_s} = \frac{2}{\tau_0} N_a a_b^3 \frac{v_k}{v_B} \left[\frac{p}{N_a} |\psi(0)|^4 + \frac{5}{3} \frac{N_a - p}{N_a} \right],$$

where t_0 is the exchange splitting parameter, N_a is the concentration of acceptors, p is the concentration of free holes, a_B is the exciton Bohr radius, the velocity $v_B = \frac{\hbar}{m_c a_B}$ (with m_c denoting the conduction electron mass), $|\psi(0)|^2$ is the **Sommerfeld's factor** enhancing the free hole contribution.

The exchange splitting parameter is defined from $\frac{1}{\tau_0} = \frac{3\pi}{64} \frac{\Delta_{ex}^2}{\hbar E_B}$ with the Bohr exciton energy $E_B = \frac{\hbar^2}{2m_c a_B^2}$ and Δ_{ex} to be the exchange splitting of the excitonic ground state.

If holes are degenerate and the velocity of electrons v_k is greater than the Fermi velocity of the holes, then

$$\frac{1}{\tau_s} = \frac{3}{\tau_0} pa_B^3 \frac{v_k}{v_B} \frac{k_B T}{E_{Fh}},$$

where k_B is the Boltzmann constant, T is the absolute temperature, and E_{Fh} is the hole Fermi energy.

For degenerate holes $|\psi(0)|^2$ is of order of 1. If electrons are thermalized, v_k has to be replaced by the thermal velocity $v_e = \left(\frac{3k_B T}{m_c}\right)^{\frac{1}{2}}$.

The temperature dependence of the relaxation time τ_s is dominated by the temperature dependence of the Sommerfeld's factor as well as by the hole concentration p. The influence of the acceptor concentration N_a is evident from the above equations, and is essentially $\frac{1}{\tau_s} \sim N_a$ for nondegenerate/bound holes and $\frac{1}{\tau_s} \sim N_a^{\frac{1}{3}}$ for degenerate holes.

First described in: G. L. Bir, A. G. Aronov, G. E. Pikus, *Spin relaxation of electrons scattered by holes*, Zh. Eksp. Teor. Fiz. **69**(4), 1382–1397 (1975) – in Russian.

birefringence – splitting of light passing through a plate made of an optically anisotropic material into two refracted waves with two different directions and polarizations. It is illustrated in Figure B.4.

Because of the different **refractive indices** in different directions, the material of the plate supports two different modes of distinctly different phase velocities. Therefore, each incident wave splits into two orthogonal components.

Anisotropic crystal plates, e.g. quartz or rutile, are used as polarizing light splitters creating two laterally separated rays with orthogonal polarizations.

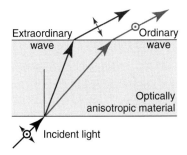

Figure B.4 Birefringence of light passing through optically anisotropic uniaxial plate.

Birge–Mecke rule states that the product of the equilibrium vibrational frequency and the square of the internuclear distance is a constant for various electronic states of a diatomic molecule.

First described in: R. Mecke, *Zur Systematik der Bandenspektra*, Z. Phys. **28**(1), 261–277 (1924); R. Mecke, Phys. Z. **26**, 217–225 (1925); R. T. Birge, *The structure of molecules*, Nature **117**, 300–302 (1926).

bit – an acronym for a **binary digit**. It is a unit of information content equal to one binary decision or the designation of one of two possible and equally likely values or states of anything used to store or convey information.

Bitter pattern

Bitter pattern – a pattern produced when a drop of colloidal suspension of **ferromagnetic** particles is placed on the surface of a ferromagnetic crystal. The particles collect along domain boundaries at the surface.

First described in: F. Bitter *On inhomogeneities in the magnetization of ferromagnetic materials*, Phys. Rev. **38**(10), 1903–1905 (1931).

black-and-white groups = **Shubnikov groups**.

bleaching – a loss of **fluorescence**. It often occurs as a result of photochemical reactions.

Bloch equation provides a simple method for detecting change in the orientation of nuclear moments through the normal methods of radio reception. The signals to be detected would be due to the electromagnetic induction caused by nuclear reorientation and should appear as a voltage difference between the terminals of an external electric circuit.

Consider a great number of nuclei contained in a macroscopic sample of matter and acted upon by two external magnetic fields: a strong constant field in the z direction with strength H_0 and a comparatively weak radio-frequency field in the x direction of amplitude $2H_1$ and frequency ω, so that the total external field vector **H** has components $H_x = 2H_1 \cos(\omega t)$; $H_y = 0$; and $H_z = H_0$. If the thermal agitation does not essentially affect the nuclei (i.e. the time of establishment of thermal equilibrium or relaxation time is long compared to the considered time intervals), the angular momentum vector **A** satisfies the classical equation $d\mathbf{A}/dt = \mathbf{T}$, where **T** is the total torque acting upon the nuclei and is $\mathbf{T} = [\mathbf{MH}]$, where **M** represents the nuclear polarization, i.e. the magnetic moment per unit volume. The parallelity between the magnetic moment μ and the angular momentum a for each nucleus implies that $\mu = \gamma a$, where γ is the gyromagnetic ratio. Thus, we have $\mathbf{M} = \gamma \mathbf{A}$. Combining the previous equation gives the Bloch equation for the temporal variation of the nuclear polarization vector **M** : $d\mathbf{M}/dt = \gamma[\mathbf{MH}]$.

Consequently a *radio-frequency* field at right angles to constant field causes a forced precession if the total polarization around the constant field, with a latitude that decreases as the frequency of the oscillating field, approaches the **Larmor** frequency. For frequencies near this magnetic resonance frequency one can expect an oscillating induced voltage in a pick-up coil with axis parallel to the y direction.

The relaxation of the condition about the influence of thermal agitation provides a way to the introduction of time constants or relaxation times t_1 and t_2. The system now performs a damped (relaxation acts like friction) precession in which the x and y components of M decay to zero with the time constant t_2 (called the transverse relaxation time) whereas M_z approaches its equilibrium value M_0 with the time constant t_1 (called the thermal or longitudinal relaxation time).

First described in: F. Bloch, *Nuclear induction*, Phys. Rev. **70**(7/8), 460–474 (1946), F. Bloch, W. W. Hansen, M. Packard, *The nuclear induction experiment*, Phys. Rev. **70**(7/8), 474–483 (1946).

Recognition: in 1952 F. Bloch and E. M. Purcell received the Nobel Prize in Physics for their development of new methods for nuclear magnetic precision measurements and discoveries in connection therewith.

See also www.nobel.se/physics/laureates/1952/index.html.

Bloch function – a solution of the **Schrödinger equation** for a periodic potential in the form $\psi_k = u_k exp(\mathbf{ikr})$, where u_k has the period of the crystal lattice **T**, so that $u_k = u_k(\mathbf{r} + \mathbf{T})$. A one-electron wave function of the above form is called a **Bloch function** or a **Bloch wave**. In fact, it is a plane wave with amplitude modulation.

Bloch oscillation – a phenomenon of external electric field induced periodic motion of an electron within a band rather than uniform acceleration.

Bloch sphere represents the three-dimensional simulation and visualization through a vector (Bloch vector) of the dynamics of a two level system under near-resonant excitation. It is well known in nuclear magnetic resonance and quantum optics. A famous application is due to Feynman, Vernon and Hellwarth, who demonstrated that the behavior of any quantum mechanical two-level system can be modeled by classical torque equations, that is there is a one-to-one correspondence between the dynamics of a two level system and the dynamics of a spinning top. The Bloch sphere is also widely used in quantum computing. In the computational basis, the general one **qubit** state is represented as a linear combination of the basis states $|a\rangle$ and $|b\rangle$ and is denoted by $|\psi\rangle = a|0\rangle + b|1\rangle$, where $|a|^2 + |b|^2 = 1$. Thus, the qubit state can be written in the form $|\psi\rangle = exp(i\gamma)[\cos(\theta/2)|0\rangle + exp(i\phi)\sin(\theta/2)|1\rangle]$ and by ignoring the global phase factor, because it has no observable effects, one has $|\psi\rangle = \cos(\theta/2)|0\rangle + exp(i\phi)\sin(\theta/2)|1\rangle$.

The angles ϕ and θ define a point (Bloch vector) on the unit three dimensional sphere, which is the Bloch sphere, as illustrated in Figure B.5.

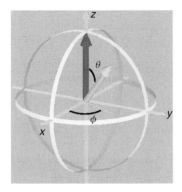

Figure B.5 Bloch sphere representation of a qubit state. The yellow arrow represents a pure $|\psi\rangle$ qubit state. The Euler angles, θ and ϕ are indicated. The red arrow along the z axis represents the vector for $|0\rangle$.

Bloch–Torrey–Kaplan equations

First described in: F. Bloch, A. Siegert, *Magnetic resonance for nonrotating field,* Phys. Rev. **57**(6), 522–527 (1940).

More details in: R. P. Feynman, F. L. Vernon Jr., R. W. Hellwarth, *Geometrical representation of the Schrödinger equation for solving maser problems,* J. Appl. Phys. **28**(1), 49–52 (1957).

Bloch–Torrey–Kaplan equations describe magnetization dynamics accounting for spin precession, spin relaxation and spin dephasing of a spin ensemble. Electronic magnetization **M** in an applied magnetic field $\mathbf{B}(t) = B_0 \mathbf{z} + \mathbf{B}_1(t)$, with a static longitudinal component B_0 (in the **z** direction) and a transverse oscillating part $\mathbf{B}_1$ (perpendicular to **z**) varies as follows:

$$\frac{\partial M_x}{\partial t} = \gamma (\mathbf{M} \times \mathbf{B})_x - \frac{M_x}{T_1} + D\nabla^2 M_x,$$

$$\frac{\partial M_y}{\partial t} = \gamma (\mathbf{M} \times \mathbf{B})_y - \frac{M_y}{T_1} + D\nabla^2 M_y,$$

$$\frac{\partial M_z}{\partial t} = \gamma (\mathbf{M} \times \mathbf{B})_z - \frac{M_z - M_z^0}{T_2} + D\nabla^2 M_x,$$

where $\gamma = \frac{\mu_B g}{h}$ is the electron gyromagnetic ratio (with g denoting the electronic g factor), T_1 is the spin dephasing time, T_2 is the spin relaxation time, D is the diffusion coefficient, $M_z^0 = \chi B_0$ (with χ denoting the system's static susceptibility) is the thermal equilibrium magnetization.

The time T_1 is the time it takes for an ensemble of transverse electron spins, initially precessing in phase about the longitudinal field, to lose their phase due to spatial and temporal fluctuations of the precessing frequencies. The time T_2 is the time it takes for the longitudinal magnetization to reach equilibrium. Equivalently, it is the time of thermal equilibration of the spin population with the lattice. In T_2 processes the energy has to be taken from the spin system, usually by phonons, to the lattice. Note that spin decoherence and relaxation processes in many spin systems are generally too complex to be fully described by only two parameters T_1 and T_2. Nevertheless these parameters are extremely robust and convenient measures for quantifying such processes are in many cases of interest.

First described in: F. Bloch, *Nuclear induction,* Phys. Rev. **70**(7/8), 460–474 (1946); H. C. Torrey, *Bloch equations with diffusion terms,* Phys. Rev. **104**(3), 563–565 (1956); J. I. Kaplan, *Application of the diffusion-modified Bloch equation to electron spin resonance in ordinary and ferromagnetic metals,* Phys. Rev. **115**(3), 575–577 (1959).

Bloch wall – a transition layer or sheet which separates adjacent domains magnetized in different directions in **ferromagnetic** or **ferrimagnetic** materials. It is also known as a **domain wall**.

First described in: F. Bloch, *Theory of the exchange problem and residual ferromagnetism,* Z. Phys. **74**(5/6), 295–335 (1932).

Bloch wave = Bloch function.

blue shift – a systematic shift of an optical spectrum toward the blue end of the optical range.

body centered cubic (bcc) structure – a cubic lattice with one lattice point in the center of the **unit cell** in addition to the eight corner points. It is schematically shown in Figure B.6.

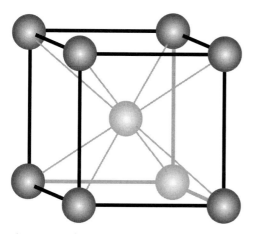

Figure B.6 Body-centered cubic structure.

Bohr correspondence principle states that quantum mechanics has a classical limit in which it is equivalent to classical mechanics. It should occur in the limit of vanishing **de Broglie wavelength**, or equivalently, at large quantum numbers.

First described in: N. Bohr, *Atomic structure*, Nature **108**, 208–209 (1921).

Recognition: in 1922 N. Bohr received the Nobel Prize in Physics for his services in the investigation of the structure of atoms and of the radiation emanating from them.

See also www.nobel.se/physics/laureates/1922/index.html.

Bohr frequency relation states that the frequency of radiation emitted or absorbed by a system when E_2 and E_1 are the energies of the states among which the transition takes place is $\nu = (E_2 + E_1)/h$.

First described in: N. Bohr *On the constitution of atoms and molecules*, Phil. Mag. **26**, 1–25 (1913).

Bohr magneton – the atomic unit of magnetic momentum, usually denoted by $\mu B = e\hbar/2m$, where m is the electron mass. It is the magnetic momentum of a single free electron spin.

Bohr magneton number – a magnetic momentum of an atom expressed in **Bohr magnetons**.

Bohr potential – the potential describing strong screened Coulomb collisions of two atoms in the form:

$$V(r) = \frac{Z_1 Z_2 e^2 a^{n-1}}{r^n} \left(\frac{n-1}{2.718} \right)^{n-1}.$$

For $n = 2$, it is

$$V(r) = \frac{Z_1 Z_2 e^2 a}{r^2} \left(\frac{1}{2.718} \right),$$

where r is the interatomic distance, and Z_1 and Z_2 are the atomic numbers of the interacting atoms,

$$a = \frac{a_B k}{(Z_1^{2/3} + Z_2^{2/3})^{1/2}}, \quad \text{where } k = 0.8 - 3.0$$

is the empirical coefficient used to fit experimental data.

First described in: N. Bohr, *The penetration of atomic particles through matter*, Kgl. Danske Videnskab. Selskab. Mat.-fys. Medd. **18**(8), 1–144 (1948).

Bohr radius – the radius of the lowest energy orbit in a hydrogen atom, denoted by $a_B = \hbar^2 / m e^2$, where m is the electron mass.

Boltzmann distribution (statistics) gives the number of noninteracting (or weakly interacting), distinguishable, classical particles with the energy E in equilibrium at a temperature T in the form $n(E) = A \exp[-E/(k_B T)]$, where A is the normalization constant.

First described by: L. Boltzmann in 1868–1871. See L. Boltzmann, *Einige allgemeine Sätze über Wärmegleichgewicht*, Sitzungsber. Akad. Wiss. Wien **63**, 679–711 (1871).

Boltzmann transport equation is used in the classical theory of transport processes: $\partial f / \partial t + \mathbf{v} \mathrm{grad}_v f + \alpha \mathrm{grad}_v f = (\partial f / \partial t)_{coll}$, where $f(r, v)$ is the classical distribution function of particles in six-dimensional space of the Cartesian coordinates r and velocity v. It is defined as $f(r, v) drdv = $ number of particles in $drdv$. $\alpha = dv/dt$ is the acceleration. In many problems the collision term is treated by the introduction of a relaxation time τ: $(\partial f / \partial t)_{coll} = -\frac{f - f_0}{\tau}$, where f_0 is the distribution function in thermal equilibrium. In the steady state $\partial f / \partial t = 0$ by definition.

bonding orbital – an atomic or molecular orbital which, if occupied by an electron, contributes to lowering of the energy of the molecule → **antibonding orbital**. Bonding is characterized by a bond order $b = 0.5(n - n^*)$, where n is the number of electrons in bonding orbitals and n^* the number in antibonding orbitals.

bond order → **bonding orbital**.

Boolean algebra (logic) – an algebraic system with two binary operations and one unitary operation important in representing a two-valued logic. Boolean calculus is useful for elucidating situations that can be *true* or *false*, making appropriate reasonings to draw conclusions correctly. They are therefore helpful in building practical computers and in programming.

First described in: G. Boole, *Mathematical Analysis of Logic* (1847) and *An Investigation of the Laws of Thought on which are founded the Mathematical Theories of Logic and Probabilities* (Walton and Maberly, London 1854).

Born approximation is based on an application of the perturbation theory to scattering problems in quantum mechanics. It consists of assuming that the scattering potential $V(r)$ can be regarded as small, so that the deviation of the wave function $\psi(r)$ from the incident plane wave is also small. So, one can write $V(r)\psi(r) \approx V(r)exp(\mathbf{ikr})$. This simplifies finding the solution of an appropriate **Schrödinger equation**. The approximation is most useful when the energy of colliding particles is large compared to the scattering potential.

First described in: M. Born, *Zur Quantenmechanik der Stossvorgänge*, Z. Phys. **37**(12), 863–867 (1926).

Recognition: in 1954 M. Born received the Nobel Prize in Physics for his fundamental research in quantum mechanics, especially for this statistical interpretation of the wavefunction.

See also www.nobel.se/physics/laureates/1954/index.html.

Born equation – an equation for determining the free energy of solvation of an ion. The equation identifies the variation in the **Gibbs free energy** for the process of solvation with the electrical work of transferring an ion from vacuum into the solvent as

$$\Delta G = -1/2 \left(\frac{Z^2 e^2 n_A}{4\pi\varepsilon_0 R} \right) [1 - (1/\varepsilon_r)],$$

where Z is the ion's electric charge number, R is the ionic radius, and ε_r is the relative dielectric constant of the solvent.

First described in: M. Born, *Volumen und Hydratationswärme der Ionen*, Z. Phys. **1**(1), 45–48 (1920).

Born–Haber cycle – a sequence of chemical and physical processes by means of which the cohesive energy of an **ionic crystal** can be deduced from experimental qualities. It leads from an initial state in which a crystal is at zero pressure and 0 K to a final state, which is an infinitely dilute gas of its constituent ions in the same pressure and temperature conditions.

First described by: M. Born, Ver. Deut. Phys. Ges. **21**, 13 (1919), F. Haber, Ver. Deut. Phys. Ges. **21**, 750 (1919).

Born–Mayer potential

Born–Mayer potential – the interatomic pair potential in the form:

$$V(r) = A \exp\left(-\frac{r}{a}\right),$$

where r is the interatomic distance, and A and a are the empirical parameters depending on the charge of the nucleus. It is used for the simulation of closed electron ion shells repulsion.

First described in: M. Born, J. E. Mayer, *Zur Gittertheorie der Ionenkristalle*, Z. Phys. **75**(1/2), 1–18 (1932).

More details in: I. G. Kaplan, *Intermolecular Interactions: Physical Picture, Computational Methods and Model Potentials* (John Wiley & Sons, Chichester, 2006).

Born–Mayer–Bohr potential – the interatomic pair potential in the form of the **Born–Mayer potential** taking into account the Coulomb repulsion:

$$V(r) = A \exp\left(-\frac{r}{a}\right) + \frac{Z_1 Z_2 e^2}{r} \exp\left(-\frac{r}{a}\right),$$

where r is the interatomic distance, A and a are the empirical parameters depending on charge of the nucleus, and Z_1 and Z_2 are the atomic numbers of the interacting atoms.

Born–Mayer–Huggins potential – the interatomic pair potential in the form:

$$V(r) = A \exp\left(-\frac{r}{a}\right) + \frac{q_1 q_2}{4\pi\varepsilon_0} \mathrm{erfc}\left(\frac{r}{b}\right),$$

where the first term corresponds to repulsion and the second to a Coulomb interaction. Here r is the interatomic distance, A, a, and b are the empirical parameters, and q_1 and q_2 are formal charges of the atoms.

First described in: M. L. Huggins, J. E. Mayer, *Interatomic distances in crystals of the alkali halides*, J. Chem. Phys. **1**(9), 643–646 (1933); M. L. Huggins, *Molecular constants and potential energy curves for diatomic molecules*, J. Chem. Phys. **3**(8), 473–479 (1935).

Born–Oppenheimer approximation separates electronic and nuclear motions. It supposes that the nuclei, being so much heavier than the electron, move relatively slowly and may be treated as stationary as electrons move around them → **adiabatic approximation.**

First described in: M. Born, R. Oppenheimer, *Zur Quantum Theorie der Molekeln*, Ann. Phys. **84**(4), 457–484 (1927).

Born–von Kármán boundary conditions – periodic boundary conditions used in quantum mechanical calculations. They suppose that the system under consideration can be constructed by repeating translations of a subsystem with the same wave function in each subsystem. It means that for a period of translation T the wave function $\psi(x)$ at $x = T$ must match smoothly to that at $x = 0$, which requires $\psi(0) = \psi(T)$ and $\frac{\partial \psi}{\partial x}|_{x=0} = \frac{\partial \psi}{\partial x}|_{x=L}$.

First described in: M. Born, T. von Karman, *Über die Verteilung der Eigenschwingungen von Punktgittern*, Phys. Z. **14**, 65–71 (1913).

Bose–Einstein condensate – an ensemble of particles, which are in the same quantum ground state, that is with zero momentum. To create such a state from atoms they must be cooled until their **de Broglie wavelength** becomes comparable with the interatomic spacing. This can be achieved at microkelvin temperatures. The process is similar to that when drops of liquid form from a gas; hence the term condensation is used to describe the process peculiarities. Bose–Einstein condensates are considered to be a new state of matter.

First described in: A. Einstein, *Quantentheorie des einatomigen idealen Gase. Zweite Abhandlung*, Sitzungber. Preuss. Akad. Wiss. Berlin **I**, 3–14 (1925) – theory; M. H. Anderson, J. R. Ensher, M. R. Matthews, C. E. Wieman, E. A. Cornell, *Observation of Bose–Einstein condensation in a dilute atomic vapor*, Science **269**, 198–201 (1995); C. C. Bradley, C. A. Sackett, J. J. Tollett, R. G. Hulet, *Evidence of Bose–Einstein condensation in an atomic gas with attractive interactions*, Phys. Rev. Lett. **75**(9), 1687–1690 (1995); K. B. Davis, M. O. Mewes, M. R. Andrews, N. J. van Druten, D. S. Durfee, D. M. Kurn, W. Ketterle, *Bose–Einstein Condensation in a gas of sodium atoms*, Phys. Rev. Lett. **75**(22), 3969–3973 (1995) – experiment.

Recognition: in 2001 E. A. Cornell, W. Ketterle and C. E. Wieman received the Nobel Prize in Physics for the achievement of Bose–Einstein condensation in dilute gases of alkali atoms, and for early fundamental studies of the properties of the condensates.

See also www.nobel.se/physics/laureates/2001/index.html.

More details in: W. Ketterle, *Dilute Bose–Einstein condensates – early predictions and recent experimental studies*,

www.physics.sunysb.edu/itp/symmetries-99/scans/talk06/talk06.html.

Bose–Einstein distribution (statistics) gives the number of photons with an energy E at a temperature T:

$$n(E) = \frac{1}{exp(E/k_B T) - 1}.$$

In general, it is relevant to particles which are indistinguishable and occupy definite quantum states, that is **bosons**.

First described in: S. N. Bose, *Plancks Gesetz und Lichtquantenhypothese*, Z. Phys. **26**(3), 178–181 (1924) and A. Einstein, *Quantentheorie des einatomigen idealen Gases. Erste Abhandlung*, Sitzungsber. Preuss. Akad. Wiss. Berlin **22**, 262–267 (1924).

bosons – quantum particles whose energy distribution obeys **Bose–Einstein statistics**. They have an integral or zero spin.

bottom-up approach – one of two ways to fabricate nanometer size elements of integrated electronic circuits. This is by building nanostructures from atoms and molecules by their precise positioning on a substrate. Hence a single device level is constructed upward.

Self-regulating processes, like **self-assembling** and **self-organization**, and **atomic engineering** are used for this. The alternative is a **top-down approach**. The bottom-up approach has the potential to go far beyond the limits of top-down technology by producing nanoscale features through synthesis and subsequent assembly.

Bouguer–Lambert law – a radiation power P of a parallel beam of monochromatic radiation entering an absorbing medium is described at a constant rate by each infinitesimally thin layer of thickness dx. It is expressed mathematically as $dP/P = -\alpha dx$, where α is the parameter depending on the nature of the absorber, energy of radiation, and for some materials (indeed for semiconductors) on the temperature. An integration gives $P = P_0 exp(-\alpha x)$ for a uniform medium, where P_0 is the power of the incident beam. Here α is the absorption coefficient of the medium.

First described by P. Bouger in 1729 – experiment; J. H. Lambert in 1760 – theory.

bound state (of an electron in an atom) – a state in which the energy of the electron is lower than it is when infinitely far away and at rest (which corresponds to the zero of energy). Bound states always have negative energies → **unbound state**.

Bowden–Tabor law takes into account the problem of **real contact area** between bodies to explain the independence of **Amonton's law** on the apparent contact area. When two rough surfaces are in contact, the apparent contact area, that is the projected area of one body on the other, is much bigger (several orders of magnitude) than the real contact area between the surface asperities. The force friction thus becomes $F_f = \tau A_r$, where τ is the **shear modulus**, the friction force per unit area, and A_r is the real contact area. This equation is considered to be the fundamental equation for friction.

First described in: F. P. Bowden, D. Tabor, *The area of contact between stationary and moving surfaces*, Proc. R. Soc. Lond. A **169**, 391–413 (1939).

bra → **Dirac notation**.

Bragg equation defines conditions for an interference maximum, when X-rays with a single wavelength λ are diffracted by crystals as in Figure B.7.

It has the form $n\lambda = 2d \sin \theta$, where n is an integer representing the order of the interference, d is the spacing between the lattice planes reflecting the X-rays, θ is the angle between the falling X-rays and lattice planes. The reflection in conditions meeting the above requirement is called the **Bragg reflection**, and θ is the **Bragg reflection angle**. In fact, the Bragg reflection providing the interference of the reflected rays is not restricted to X-rays only, but is a general phenomenon for interaction of waves with periodic structures.

First described in: W. L. Bragg, *Diffraction of short electromagnetic waves by a crystal,* Proc. Camb. Phil. Soc. **17**(1), 43–57 (1913).

Recognition: in 1915 W. H. Bragg and W. L. Bragg received the Nobel Prize in Physics for their services in the analysis of crystal structures by means of X-rays.

See also www.nobel.se/physics/laureates/1915/index.html.

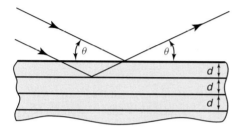

Figure B.7 Scheme of Bragg diffraction.

Bragg reflection → **Bragg equation.**

Bragg reflection angle → **Bragg equation.**

braking radiation → **Bremsstrahlung.**

Bravais lattice – a distinct type of lattice, which being translated completely fills the space. There are 5 Bravais lattices in two dimensions shown Figure B.8.

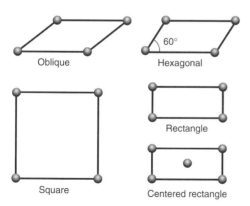

Figure B.8 Bravais lattices in two dimensions.

Breit–Wigner formula

There are 14 lattices in three dimensions shown in Figure B.9. Only these fundamentally different lattices are possible.

First described by: A. Bravais, *Études cristallographiques,* Paris 1866.

Figure B.9 Bravais lattice in three dimensions (angles are indicated where they are different from 90°).

Breit–Wigner formula → **Breit–Wigner resonance.**

Breit–Wigner resonance – a nuclear reaction induced by an incident particle with energy E corresponding to that required to form a discrete resonance level E_{res} of the component nucleus. The cross section is

$$\sigma(e) = \sigma_{res} \frac{\Gamma^2/4}{(E_{res}^2 - E^2) - \Gamma^2/4},$$

which is called the **Breit–Wigner formula**. Here Γ denotes the energy width over which the reaction occurs and σ_{res} is the resonant cross section.

First described in: G. Breit, E. Wigner, *Capture of slow neutrons*, Phys. Rev. **49**(7), 519–531 (1936).

Bremsstrahlung – German equivalent of the radiation also called **collision radiation** or **deceleration radiation** or **braking radiation**. This radiation arises when an electron is deflected by the Coulomb field of an atomic nucleus. In bremsstrahlung, a continuous spectrum with a characteristic profile and energy cutoff, that is wavelength minimum, is produced. In addition, lines can appear superimposed, corresponding to the ejection of K and L shell electrons knocked out of atoms in collisions with the high-energy electrons. This radiation is used for the analysis of solid, when it is known as the Bremsstrahlung isochromat spectroscopy technique.

First described theoretically in: H. Bethe, W. Heitler, *On the stopping of fast particles and on the creation of positive electrons*, Proc. Roy. Soc. A **146**, 83–112 (1934).

Bridgman effect – when an electric current passes through an anisotropic crystal, there is an absorption or liberation of heat that results from the nonuniform current distribution.

Bridgman relation – in a metal or semiconductor placed in a transverse magnetic field at a temperature T: $P = QT\sigma$, where P is the **Ettingshausen coefficient**, Q is the **Nernst coefficient**, and σ is the thermal conductivity of the material.

First described in: P. W. Bridgman, *Thermoelectric phenomena and electrical resistance in single metal crystals*, Proc. Am. Acad. Arts Sciences **63**, 351–99 (1929).

Brillouin function – a function of the form $f(x) = [(2n + 1)/2n] \coth[(2n + 1)x/2n] - (1/2n) \coth(x/2n)$, where n is the parameter. It appears in the quantum mechanical theories of paramagnetism and ferromagnetism.

Brillouin scattering – scattering of light in solids by acoustic waves.

First described in: L. Brillouin, *Diffusion of light and X-rays by a transparent homogeneous body. The influence of thermal agitation*, Ann. Phys. (France) **17**, 88–122 (1922) – in French, and independently L. I. Mandelstam, *On light scattering by an inhomogeneous medium*, Zh. Russkogo Fiz.-Khim. Obshch. **58**(2), 381–386 (1926) – in Russian.

Brillouin zone

Brillouin zone – a symmetrical unit cell constructed in **reciprocal lattice** space. The construction procedure is the same as for a **Wigner-Seitz primitive cell** except that the points are now reciprocal lattice points. The Brillouin zones and their points of high symmetry for some often met crystal lattices are shown in Figure B.10.

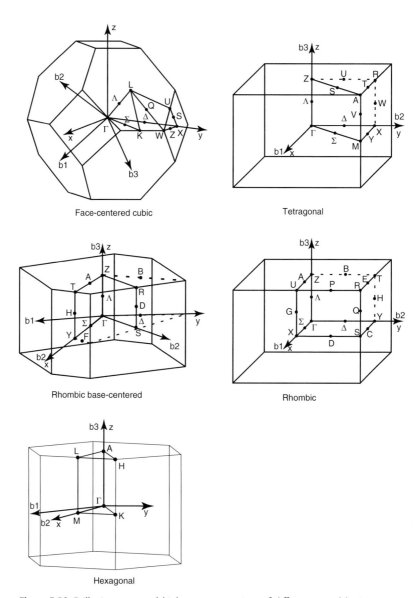

Figure B.10 Brillouin zones and high-symmetry points of different crystal lattices.

The first Brillouin zone is the smallest polyhedron centered at the origin and enclosed by perpendicular bisectors of reciprocal lattice vectors. In one dimension it is a region of the k-space defined by $[-\pi/R, \pi/R]$, where R is the translation period used to construct the lattice.

First described in: L. Brillouin, *Die Quantenstatistik und ihre Anwendung auf die Elektronentheorie der Metalle* (Springer, Berlin, 1931).

Brownian motion – a random walk whose probability distribution is governed by the **diffusion** equation. The trajectory is **fractal** with the derivative (velocity) not well defined.

First described by: R. Brown, *Mikroskopische Beobachtungen über die im Pollen der Pflanzen enthaltenen Partikeln, und das allgemeine Vorkommen activer Molecüle in organischen und anorganischen Körpern*, Ann. Phys. **14**, 294 (1828) (although he was not the first to see this phenomenon).

Buckingham potential – the interatomic pair potential in the form $V(r_{ij}) = A_{ij}exp$ $(-\frac{r_{ij}}{\sigma_{ij}}) - B_{ij}(\frac{\sigma_{ij}}{r_{ij}})^6$. Here r_{ij} is the interatomic distance and A_{ij}, B_{ij} and σ_{ij} are the parameters fitted to an experiment. The repulsive interaction is represented by the exponential term, and emanates from the repulsion of electron gas clouds around ions as they are brought together. The attractive part of the potential derives from the r_{ij}^{-6} term. It is a response to the dispersion forces sometimes called **van der Waals** attraction.

First described in: R. A. Buckingham, *The classical equation of state of gaseous helium, neon and argon*, Proc. R. Soc. Lond. A **168**, 264–283 (1938).

More details in: I. G. Kaplan, *Intermolecular Interactions: Physical Picture, Computational Methods and Model Potentials* (John Wiley & Sons, Chichester, 2006).

Buckingham's theorem states that if there are n physical quantities $x_1, x_2, \ldots, x_n$ which can be expressed in terms of m fundamental quantities and if there exists one and only one mathematical expression connecting them which remains formally true no matter how the units of the fundamental quantities are changed, namely $f(x_1, x_2, \ldots, x_n) = 0$, then the relation f can be expressed by a relation of the form $F(\pi_1, \pi_2, \pi_n - m) = 0$, where the πs are $n - m$ independent dimensionless products of $x_1, x_2, \ldots, x_n$. It is also known as the **pi or Π theorem**.

First described in: E. Buckingham *On physically similar systems; illustrations of the use of dimensional equations*, Phys. Rev. 4(4), 345–376 (1914).

Buckminsterfullerene = **fullerene**. The name is related to the architect R. Buckminster Fuller (1895–1983) who pioneered the use of geodesic domes in architecture.

Bunsen–Roscoe law states that the amount of chemical change produced in a light-irradiated reacting system is proportional to the amount of light absorbed. Actually, the change is also dependent on the intensity of light, which is named the

Burger's vector

reciprocity failure. Named after the German chemist R. W. E. Bunsen (1811–99) and the English chemist H. E. Roscoe (1833–1915).

Burger's vector shows the direction in which a region of a crystal is displaced relative to the neighboring region, thus forming a dislocation.

Burstein–Moss shift – a blue shift (the shift to the shorter wavelength range) of the absorption age in semiconductors due to band filling. It is observed in heavily doped materials with filled near band gap states.
   First described in: E. Burstein, *Anomalous optical absorption limit in InSb*, Phys. Rev. **93**(3), 632–633 (1954); T. S. Moss, *The interpretation of the properties of indium antimonide*, Proc. R. Phys. Soc. Lond. B **67**(10), 775–782 (1954).

From cage compound to cyclotron resonance

cage compound = clathrate compound.

Caldeira–Leggett model deals with the problem of dissipation in a quantum system. The model consists of an independent oscillator bath linearly coupled to the system of interest through a coordinate-coordinate coupling. Originally it was used for the study of the quantum **Brownian motion** showing that in the classical limits the formalism reduces to the classical **Fokker–Planck equation**. The model has been widely used for study of the influence of damping on quantum interference with special attention to quantum tunneling and quantum coherence. Effective decoherence as a result of the interaction of a quantum system with an environment can provide a natural mechanism for the transition from quantum to classical behavior.

First described in: A. O. Caldeira, A. J. Leggett, *Path integral approach to quantum Brownian motion*, Physica A: Statistical and Theoretical Physics, **121**(3), 587–616 (1983).

calixarene → **electron-beam lithography**.

Callier effect – selective scattering of light as it passes through a diffusing medium.
First described in: A. Callier, Photogr. J. **49**, 200 (1909).

canonical – relating to the simplest or most significant form of a general function, equation, statement, rule or expression.

canonical equations of motion → **Hamiltonian equations of motion**.

capacitance – a ratio of the charge Q on one of the plates of a capacitor (there being an equal and opposite charge on the other plate) to the potential difference V between the plates: $C = Q/V$.

carbohydrate – an organic compound containing carbon, hydrogen and oxygen atoms in a $1:2:1$ molecular ratio. Carbohydrates have the general formula $C_n(H_2O)_n$.

What is What in the Nanoworld. Victor E. Borisenko and Stefano Ossicini
Copyright © 2008 WILEY-VCH Verlag GmbH & Co. KGaA, Weinheim
ISBN: 978-3-527-40783-5

carbon dioxide laser

Carbohydrates are naturally produced by plants and animals. Glucose ($C_6H_{12}O_6$) is one of the most important among them, being used as an energy source in living bodies.

carbon dioxide laser – a type of **ion laser** with carbon dioxide gas as the active medium that produces infrared radiation at 10.6 μm.

carbon nanotube – naturally self-organized nanostructure in the form of a tube composed of carbon atoms with completed bonds. There are two general forms, which are single-wall and multiwall nanotubes. A single-wall carbon nanotube can be considered as a single sheet of graphite, which is called **graphene**, that has been rolled up into a tube. The resulting nanotube can have metallic or semiconducting properties depending on the direction in which the sheet was rolled up.

Graphene is a two-dimensional honeycomb net made up of sp^2 bonded carbon atoms as sketched in Figure C.1.

In a tube, some small contribution of sp^3 states are mixed in as a result of the tube curvature. For a seamless tube, the circumference or **chiral vector** $\mathbf{C}_h$ must be a linear combination of the unit vectors $\mathbf{a}_1$ and $\mathbf{a}_2$: $\mathbf{C}_h = n\mathbf{a}_1 + m\mathbf{a}_2$, where n and m are integers. These integers placed in brackets, for example (n, m), are used to identify a particular structure of a tube. The angle between $\mathbf{C}_h$ and $\mathbf{a}_1$ is called the **chiral angle**. The chiral vector also defines the propagation vector, which represents the periodicity of the tube structure in the direction parallel to its axis.

Typical structures of carbon nanotubes are shown in Figure C.2. If $n = m$, the chiral angle is $30°$ and the structure is called **armchair**. If either n or m is zero, the chiral angle is $0°$ and the structure is called **zig-zag**. All other nanotubes with nonzero $n \neq m$ have chiral angles between $0°$ and $30°$. They are said to have a **chiral structure**. There is a mirror image of their structure upon an exchange of n and m. The average diameter of single-wall nanotubes is 1.2–1.4 nm.

Figure C.1 Arrangement of carbon atoms in graphene.

armchair
$n=m$ - metal

zig-zag
n or $m = 0$
$n-m = 3i$ -metal
$n-m \neq 3i$ - semiconductor

chiral
$n-m = 3i$ -metal
$n-m \neq 3i$ - semiconductor

Figure C.2 Carbon nanotubes with different chiral vectors defined by n and m (by courtesy of Professor L. Kavan).

Electronic properties of single-wall nanotubes may be understood when those of graphene are analyzed. Graphene does not conduct electric current, except along certain directions where cones of electronic states at carbon atoms exist. This gives rise to the anisotropy of electronic properties of graphene in two principal directions, namely in the direction parallel to a C–C bond and in the direction perpendicular to a C–C bond. The electronic band structure of graphene is determined by electron scattering from the carbon atoms. Electrons moving along a C–C bond are backscattered by carbon atoms periodically appearing on their way. As a result, only electrons with certain energy can propagate in this direction. Thus, the material shows an energy gap like that in a semiconductor. Electrons traveling in the direction perpendicular to a C–C bond are scattered by different atoms. They interfere destructively, which suppresses backscattering and leads to metallic-like behavior. Thus, single-wall carbon nanotubes rolled up around an axis parallel to a C–C bond have semiconducting properties, while those rolled up around an axis perpendicular to the bonds are metallic in their electronic properties.

A particular electronic behavior of the nanotube is defined by its chiral vector $\mathbf{C}_h$ or equivalently by the relation of n and m. Armchair tubes always show metallic properties. This is also the case for zig-zag and chiral tubes, but when $n - m = 3i$, where i is an integer. In contrast, zig-zag and chiral tubes with $n - m \neq 3i$ demonstrate semiconducting behavior.

The fundamental band gap in semiconducting nanotubes ranges from 0.4 eV to 0.7 eV, being dependent on small variations in the diameter and bonding angle. In general, the band gap decreases with an increase of the tube diameter. There is quantum confinement in the radial direction of the tube provided by the monolayer thickness of its wall. In fact, single-wall nanotubes behave as one-dimensional structures. Electrons can travel along the tube for long distances without being backscattered. Moreover, metallic carbon nanotubes have been found to behave like quantum dots.

carbon nanotube

Multiwall nanotubes consist of several concentrically arranged single-wall nanotubes. They are typically 10–40 nm in diameter. Intertube coupling within a multiwall nanotube has a relatively small effect on its electronic band structure. As a consequence, semiconducting and metallic tubes retain their character if they are a part of a multiwall nanotube. By statistical probability, most of multiwall carbon nanotubes demonstrates overall metallic behavior because at least one metallic tube is sufficient to short circuit all semiconducting tubes. The phase coherence length in multiwall nanotubes (at 4.2 K) is about 250 nm, the elastic scattering length being about 60 nm.

Nanotubes have an impressive list of attributes. They can behave like metals or semiconductors, can conduct electric current better than copper, can transmit heat better than diamond, and rank among the strongest materials known-not bad for structures that are just a few nanometers across.

Carbon nanotubes are fabricated by vaporizing carbon electrodes using arc discharge, laser ablation, or chemical vapor deposition. The nanotubes grown by the arc discharge and laser ablation techniques produce tangled bunches. They are randomly arranged on electrodes or on a substrate, which makes them difficult to sort and manipulate for subsequent building of nanoscale electron devices. For studies and applications the tubes are extracted into suspension in dichloroethane by a short sonication and then dispersed and dried onto silicon or SiO_2 covered substrates. An **atomic force microscope (AFM)** operating in tapping mode is used to select appropriate nanotubes and locate them at a desired place on the substrate. In contrast, controlled growth of the tubes can be achieved by the use of the **chemical vapor deposition (CVD)** technique.

For the chemical vapor deposition of carbon materials, a hydrocarbon gas is passed over a heated catalyst. The action of the catalyst causes the hydrocarbon to decompose into hydrogen and carbon atoms, which provide the feedstock for the carbon tube growth. The key parameters controlling the growth are the type of hydrocarbons and catalysts used as well as the temperature at which the reaction is initiated. For fabrication of multiwall nanotubes one can use ethylene or acetylene gas as the carbon precursor, and iron, nickel or cobalt as the catalyst. The growth temperature is typically in the range of 500–700°C. At these temperatures carbon atoms dissolve in the metal, which eventually becomes saturated. The carbon then precipitates to form solid tubes, the diameters of which are determined by the size of the metal particles in the catalyst. The relatively low temperature of the process makes difficult the fabrication of tubes with a perfect crystalline structure. Thus, the use of methane and processing temperature as high as 900–1000°C is appropriate to produce virtually defect-free tubes, in particular single-wall ones. Methane is the most stable of all the hydrocarbon compounds with respect to the decomposition at high temperatures. It is a crucial property that prevents formation of amorphous carbon, which poisons the catalyst and coats the tubes.

Well-positioned and oriented tower-like bundles of nanotubes can be fabricated by CVD onto silicon substrates patterned with catalyst materials. Useful approaches are schematically shown in Figure C.3.

Figure C.3 Principles of controlled growth of carbon nanotubes: (a) the use of silicon pillars covered with a catalyst material, (b) the use of catalyst islands on SiO_2 film.

One of them includes fabrication of silicon pillars, tops of which are covered by the catalyst material (Figure C.3a). Individual nanotubes or bundles of them grow during methane decomposition on the tops of the pillars. They are suspended by the pillars and form networks that are oriented according to the pattern of the pillars. As the nanotubes grow, the flow of the gas keeps them floating and waving in the "wind". This prevents the nanotubes from touching the substrate and being caught by the bottom surface. The nearby pillars provide fixing points for the growing tubes. When the waving tubes touch the nearby pillars, the van der Waals interaction between the pillar and the tube are strong enough to catch the nanotubes and hold them aloft.

Controlled growth of nanotubes can be also performed at specific sites on an SiO_2-covered substrate patterned with catalyst islands (Figure C.3b). The mechanism of directional growth in this case has much in common with that previously discussed. A nanotube bridge forms when a tube growing from one catalyst island falls on and interacts with another island. When bundles of nanotubes are grown, an AFM tip is used to cut them mechanically or electrically until a single tube remains between the islands. The interactions tube/island and tube/substrate are strong enough to allow the nanotube to withstand mechanical forces in subsequent lithographic processing. Most methods produce nanotubes as long as 1–10 μm; however, lengths up to 200 μm can be reached.

Doping of the nanotube material, determining the type of charge carriers in semiconducting nanotubes, is achieved by substituting carbon atoms with boron (**acceptor** impurity) or nitrogen (**donor** impurity). Attaching alkali or halogen atoms to the outside of the tube can also be used for these purposes, but it is difficult to control.

Metal contacts connecting nanotubes with other electronic devices on the substrate can be fabricated in a number of different ways. One way is to make the electrodes and then drop the tubes onto them. Another is to deposit the tubes on to a substrate, locate them with a **scanning tunneling microscope (STM)** or AFM, and then attach leads using a lithography technique. More advanced approaches include the possibility of growing the tubes between electrodes made of a catalyst metal or of attaching the tubes to the surface in a controllable fashion using either electrostatic or chemical forces. Titanium has been found to give the lowest contact resistance compared with other metals, like gold and aluminum. The reason is that there is a strong chemical interaction between titanium and carbon atoms giving rise to carbide formation at their interface. This process leads to an intimate carbonyl

electrical coupling between the two materials. Gold and aluminum do not form stable carbides, and thus their contact resistance appears to be higher.

By combining different nanotubes, and supplementing them with gate electrodes, there is the potential to make a wide variety of nanoelectronic devices. The prototypes that have been fabricated include rectifying **diodes** using junctions of metallic and semiconducting tubes, single-electron transistors based on metallic nanotubes, and field-effect transistors employing semiconducting nanotubes.

As carbon atoms have very strong bonds in nanotubes, it is very difficult to displace them. Thus, carbon nanotubes appear to be much more resistant to electromigration than conventional copper and aluminum interconnections in integrated circuits. The electrical current that could be passed through a nanotube reaches 10^{13} A cm^{-2}. Small diameter carbon nanotubes and high current densities make them attractive in field electron emission applications. Mechanical deformations dramatically change the electronic properties of carbon nanotubes. It allows their use as building blocks of nanoscale electromechanical devices.

First described in: S. Iijima, *Helical microtubules of graphitic carbon*, Nature **354**, 56–58 (1991).

More details in: R. Saito, G. Dresselhaus, M. S. Dresselhaus, *Physical Properties of Carbon Nanotubes* (Imperial College Press, Singapore, 1998); P. J. F. Harris, *Carbon Nanotubes and Related Structures* (Cambridge University Press, 1999); R. E. Smalley, M. S. Dresselhaus, G. Dresselhaus, P. Avouris, *Carbon Nanotubes: Synthesis, Structure, Properties and Applications* (Springer, Berlin, 2001); S. Reich, C. Thomsen, J. Maultzsch, *Carbon Nanotubes* (Wiley-VCH, Weinheim, 2004).

See also www.pa.msu.edu/cmp/csc/nanotube.html; http://shachi.cochem2.tutkie.tut.ac.jp/Fuller/Fuller.html.

carbonyl – a chemical compound containing the **carbonyl group** ($>C=O$).

carbonyl group – a carbon-oxygen double bonded group $>C=O$.

carboxylate – a salt or an anion of carboxylic acid (COOH). In general it is RCOO$^-$, where R is hydrogen or an **alkyl group**.

carboxyl group – the functional group of the **carboxylic acids** represented as –COOH.

carboxylic acids – organic acids containing the functional group –COOH. They can be either aliphatic (RCOOH) or aromatic (ArCOOH), where R represents an **alkyl group** ($-C_nH_{2n+1}$) and Ar stands for an aromatic ring. The carboxylic acids with even numbers of carbon atoms, n, ranging from 4 to about 20 are called fatty acids (e.g. $n = 10, 12, 14, 16$ and 18 are called capric acid, lauric acid, myristic acid, palmitic acid, and stearic acid, respectively).

Car–Parrinello algorithm – an efficient method to perform **molecular dynamics simulation** for covalent crystals by first-principles quantum mechanical calculations where the forces on the ions are calculated directly from the total energy of the ionic and electronic systems without needing a parameterized interatomic few-body potential. For the electronic contribution to the total energy, one actually solves the **Schrödinger equation** within the **local density approximation** with a **plane-wave** basis set.

The electron density in terms of occupied single-particle orbitals $\Psi_i(r)$ is $n(r) = \Sigma_i |\Psi_i(r)|^2$. The point of the potential energy surface is given by the minimum of the energy functional

$$E[\{\Psi_i\}, \{R_I\}\{a_v\}] = \Sigma_i \int_\Omega d^3 r \Psi_i^*(r)[-(\hbar/2m)\nabla^2]\Psi_i r) + U[n(r), \{R_I\}\{a_v\}].$$

Here R_I indicate the nuclear coordinates and a_v are all possible external constraints imposed on the system like the volume Ω, the strain $\varepsilon_{\mu v}$, etc. The functional U contains the internuclear Coulomb repulsion and the effective electronic potential energy, including external nuclear, Hartree, exchange and correlation contributions.

Minimization of the energy functional $E[\Psi_i, R_I a_v]$ with respect to the orbitals $\Psi_i(r)$, subjected to the orthogonal constraint, leads to the self-consistent equations of the form:

$$\left[-\frac{\hbar^2}{2m}\nabla^2 + \frac{\delta U}{\delta n(r)} \right] \Psi_i(r) = \varepsilon_i \Psi_i(r).$$

Solution of the equations involves repeated matrix diagonalization with a computational effort rapidly growing with the size of the system.

First described in: R. Car, M. Parrinello, *Unified approach for molecular-dynamics and density-functional theory*, Phys. Rev. Lett. **55**(22), 2471–2474 (1985).

Cartesian axis – a set of mutually perpendicular lines which all pass through a single point. They are used to define a Cartesian coordinate system.

Casimir–du Pré theory describes spin lattice relaxation treating the lattice and spin system as distinct thermodynamic systems in thermal contact with one another.

First described in: H. B. G. Casimir, F. K. du Pré, *Note on the thermodynamic interpretation of paramagnetic relaxation phenomena*, Physica **5**(6), 507–511 (1938).

Casimir effect – attraction of two mirror plates facing each other in a vacuum due to fluctuations of surrounding electromagnetic fields exerting radiation pressure on the mirrors. The vacuum fluctuations were theoretically shown to result in the external pressure being, on average, greater than the internal pressure. as it is illustrated in Figure C.4.

Casimir force

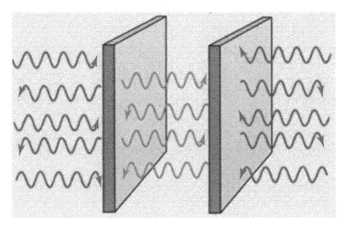

Figure C.4 Vacuum fluctuations in the Casimir effect.

The mirror plates are mutually attracted to each other by what is termed the **Casimir force**. This force has been calculated to be $F_C = \frac{\pi^2 \hbar c}{240} S/d^4$, where S is the cross sectional area of the two mirrors and d is the distance between the mirrors. At separation of 10 nm, the Casimir effect produces the equivalent of about 1 atmosphere of pressure. The Casimir force attracting a sphere of radius r to a plane mirror surface is $F_C = \frac{\pi^3 \hbar c}{360} r/d^3$, where d is the separation between the sphere and the mirror.

The Casimir effect can influence the operation of microelectromechanical **(MEMS)** and nanoelectromechanical systems **(NEMS)**.

First described: H. B. G. Casimir, *On the attraction between two perfectly conducting plates*, Proc. K. Ned. Akad. Wet. **51**(7), 793–796 (1948) – theoretical prediction; M. J. Sparnaay, *Measurements of attraction forces between flat plates*, Physica **24**(9), 751–764 (1958) – experimental confirmation.

More details in: K. A. Milton, *The Casimir Effect* (World Scientific, Singapore, 2001).

Casimir force → **Casimir effect.**

Casimir operator plays a crucial role in the classification of group representations. The discussion of the commutators of infinitesimal operators that generate infinitesimal transformations led to **Lie algebra**, while the generation of finite transformations formed the global groups, the **Lie groups**. Casimir and van der Waerden proposed the subgroups and subalgebras of the Lie algebras and Lie groups, with Casimir returning to Lie's idea of invariance, thus developing the Casimir operator, which is an operator $\Gamma = \Sigma_{i=1}^{m} e_i^R u^{iR}$ on a representation R of a Lie algebra. This operator commutes with the action of any representation; in other words commutes with all generators of a group.

First described in: H. Casimir, *Über die Konstruktion einer zu den irreduzibelen Darstellungen halbeinfacher gruppengehörigen Differentialgleichung*, Proc. Roy. Acad.

Sci. (Amsterdam) **34**, 844–846 (1931); H. Casimir, B. L. van der Waerden, *Algebraischer Beweis der vollständigen Reduzibilität der Darstellungen halbeinfacher Liescher Gruppen*, Mathematische Annalen **111**(1), 1–12 (1934).

catalyst – a substance that increases the rate of a chemical reaction but which is left unchanged by the reaction. It works by providing an extended convenient surface for the reaction and by lowering the activation energy for the reaction. The reacting particles gather on the catalyst surface and collide more frequently with each other thus increasing the reaction yield.

catenane – a mechanically interlocked molecular architecture consisting of two or more interlocked molecular macrocycles. The families of catenanes are schematically shown in Figure C.5.

The interlocked rings cannot be separated without breaking the covalent bonds of the macrocycles. In the bridged catenanes called **pretzelanes**, the two macrocycles are linked with a spacer. In handcuff-shaped catenanes, two connected rings are threaded through the same ring.

See also en.wikipedia.org/wiki/Catenane.

| Catenane | Pretzelane | Handcuff-shaped catenane |

Figure C.5 Families of catenanes.

cathodoluminescence – light emission excited by electron irradiation of matter.
First described by W. Crookes in the middle of the nineteenth century.

Cauchy principal value of the integral $\int_{-\infty}^{\infty} f(x)dx$ is $\lim_{s\to\infty} \int_{-s}^{s} f(x)dx$ provided the limit exists. If the function $f(x)$ is bounded on an interval (a, b) except in the neighborhood of a point c, the Cauchy principal value of $\int_{a}^{b} f(x)dx$ is $\lim_{\delta\to 0}[\int_{a}^{c-\delta} f(x)dx + \int_{c+\delta}^{b} f(x)dx]$ provided the limit exists. It is also known as the principal value.

Cauchy relations – a set of six relationships between the compliance constants of a solid which should be satisfied provided the force between atoms in the solid depends only on the distance between them and acts along the lines joining them, and provided that each atom is a center of symmetry in the lattice.

Cauchy sequence – a sequence with the property that the difference between any two terms is arbitrarily small provided they are both sufficiently far out in the sequence. More precisely it is stated: a sequence $\{a_n\}$ such that for every $\epsilon > 0$ there causality principle

is an integer N with the property that if n and m are both greater than N, then $|a_n - a_m| < \epsilon$.

causality principle asserts that a response of a system follows its cause in time, or that the response at some time depends only on the driving force in the past.

cavitation – a tendency of liquids to form bubbles spontaneously when subjected locally to negative pressure, as they cannot withstand the tension involved.

CCD – an acronym for a **charged-coupled device**.

CCNOT (C^2NOT) gate $\rightarrow$ quantum logic gate.

central limit theorem states that the distribution of sample means taken from a large population approaches a **Gaussian distribution**.

Čerenkov angle $\rightarrow$ Čerenkov radiation.

Čerenkov condition $\rightarrow$ Čerenkov radiation.

Čerenkov radiation – light emitted when relativistic particles pass through non-scintillating matter: a charged particle travelling with a velocity v through a nonconducting medium having an index of refraction $n(f)$ emits electromagnetic radiation at all frequencies f for which the phase velocity c/n is less than v. In principle, a neutral particle having a magnetic moment (e.g. a neutron) will also emit such radiation, although it will be negligibly small because of the smallness of the magnetic moment. The radiation is emitted at one angle, the **Čerenkov angle** $\phi = \arccos(c/nv)$, with respect to the momentum vector of the particle. The light is 100% linearly polarized with the electric vector coplanar with the particle momentum vector. The emitted power is defined as

$$\frac{\mathrm{d}E}{\mathrm{d}z} = \frac{4\pi^2 e^2}{c^2} \int \left[1 - \left(\frac{c}{nv}\right)^2\right] f \, \mathrm{d}f,$$

where the integration is limited to those frequencies for which the **Čerenkov condition** $(c/nv) < 1$ is fulfilled.

The effect is used in high-energy particle physics, as only charged particles that fulfill the Čerenkov condition emit radiation, which then is detected. This is used either to select particles above a certain energy or to discriminate against the heavier particles in a beam of given momentum containing a mixture of particles.

First described in: P. A. Čerenkov, *Visible luminescence from pure liquids subjected to radiation*, Dokl. Akad. Nauk SSSR **2**(8), 451–457 (1934); S. I. Vavilov, *On possible origins of blue luminescence from liquids*, Dokl. Akad. Nauk SSSR **2**(8), 457–461 (1934)–experiment, both in Russian; I. E. Tamm, I. M. Frank, *Coherent* *radiation from a fast electron in a medium*, Dokl. Akad. Nauk SSSR **14**(3), 107–112 (1937)–theory, in Russian.

Recognition: in 1958 P. A. Čerenkov, I. M. Frank, I. E. Tamm received the Nobel Prize in Physics for the discovery and the interpretation of the Čerenkov effect.

See also www.nobel.se/physics/laureates/1958/index.html.

characteristic function – the **Fourier transform** of a probability function.

characteristic temperature T_0 – one of the experimental characteristics of a semi-conductor laser indicating dependence of its threshold current density J_{th} on temperature. In a limited temperature range $J_{th} \sim exp(\frac{T}{T_0})$. The characteristic temperature usually takes values in the range of 100–150 K for AlGaAs lasers and 50–80 K for GaInPAs lasers. A higher characteristic temperature corresponds to a lower temperature sensitivity of the threshold current density. It is worthwhile to note that this sensitivity is considerably reduced in lasers based on low-dimensional structures.

Chargaff's rules $\rightarrow$ **DNA**.

First described in: E. Chargaff, *Chemical specificity of nucleic acids and mechanism of their enzymatic degradation*, Experientia **6**, 201–209 (1950).

charge coupled device (CCD) – an array of integrated metal-oxide-semiconductor capacitors which can accumulate and store charge because of their capacitance. Under the control of an external circuit, each capacitor can transfer its electric charge to one or other of its neighbors. CCDs are used in digital image processing, photometry, and optical spectrometry.

charge density wave – a static modulation of the charge density at atomic sites accommodated by a periodic lattice distortion.

chemical beam epitaxy $\rightarrow$ **molecular beam epitaxy**.

chemical potential – a thermodynamic variable bearing a relation to the flow of matter (or particles) similar to the relation which temperature bears to the flow of heat. Matter (particles) only flows between regions of different chemical potential and then in the direction of decreasing potential. The chemical potential of the *i*-th species is $\mu = (\frac{\partial G}{\partial n_i})_{T,p,n_i} \mu = \frac{\partial G}{\partial n_i}|_{T,p,n_i}$, where G is the Gibbs free energy, n_i is the number of modes of the *i*-th species, T is the absolute temperature, and p is the pressure.

chemical vapor deposition – a thin solid film growth of a material transported from the gas phase onto a suitable solid substrate. The gaseous medium actually contains vapors of the depositing material or a mixture of gaseous reactants capable chemiluminescence

of undergoing chemical conversion at the substrate surface to produce the film growth. The deposition is performed in open flow chambers, where a carrier gas containing the reactive species of the precursors is forced to flow past heated substrates. Chemical composition of the precursors, gas pressure in the reaction chamber and substrate temperature are the most important factors controlling the process to yield monocrystalline (epitaxial), polycrystalline or amorphous films.

chemiluminescence – light generation by electronically excited products of chemical reactions. Such reactions are relatively uncommon as most exothermic chemical processes release energy as heat. The initial step is termed chemiexcitation. Chemical energy is converted into electronic excitation energy. If the excited state species is fluorescent, the process is **direct chemiluminescence**. If the initial excited state species transfer energy to a molecule, which then emits light, the process is **indirect chemiluminescence**.

First described by: E. Wiedemann *Über Fluorescenz und Phosphorescenz*, Ann. Phys. **34**, 446 (1888).

chemisorption – an **adsorption** process characterized by strong chemical bond formation between adsorbed atoms and substrate atoms at the adsorbing surface.

Child–Langmuir law relates the density of the current which can be produced under space charge limited conditions between infinite parallel electrodes to the potential applied between the electrodes. The flow of charged particles is said to be space charge limited if the current density given off by an emitter is not limited by any property of the emitter but rather by the fact that the charge in the space near the emitter reduces the electric field at the emitter to zero. Finally, the current density is proportional to $V^{3/2}$, where V is the potential of the collector respect to the emitter.

First described in: C. D. Child, *Discharge from hot CaO*, Phys. Rev. (Series I) **32**(5), 492–511 (1911); I. Langmuir, *The effect of space charge and residual gases on thermionic currents in high vacuum*, Phys. Rev. **2**(6), 450–486 (1913).

chiral → **chirality**.

chirality – Greek meaning "handedness". It is an indication of a structure asymmetry. The word was first introduced by Lord Kelvin in a *Baltimore Lecture* in 1884 "I call any geometrical figure, or group of points, **chiral**, and say that it has **chirality** if its image in a plane mirror, ideally realized, cannot be brought to coincide with itself". Thus an object or a system is called chiral if it differs from its mirror image, and its mirror image cannot superimpose on the original object. A chiral object and its mirror image are called **enantiomorphs** (Greek "opposite forms") or, when referring to molecules, **enantiomers**. A non-chiral object is called **achiral** (sometimes also **amphichiral**). It can be superimposed on its mirror image. Recently a new definition of **chirality** has been proposed by L. D. Barron "True chirality is shown by systems existing in two enantiomeric states that are interconverted by space inversion but not by time reversal combined with any proper spatial rotation". This definition includes motion; the system breaks parity P but not time reversal T.

First described in: L. Pasteur, *Recherches sur les relations qui peuvent exister entre la forme crystalline et la composition chimique, et le sens de la polarisation rotatoire,* Ann. Chim. Phys. **24**, 442–459 (1848).

More details in: L. D. Barron *Fundamental symmetry aspects of molecular chirality,* in *New Developments in Molecular Chirality,* ed. by P. G. Mezey (Elsevier, Dordrecht, 1991).

chromatin – the complex of **DNA** and **protein** making up **chromosomes**. The functions of chromatin are to package DNA into a smaller volume to fit in the cell, to strengthen the DNA to allow mitosis and meosis, and to serve as a tool to control expression.

First described by: W. Flemming, *Beiträge zur Kenntnis der Zelle und ihre Lebenser-scheinungen,* Arch. Mikr. Anat., **18**, 151–259 (1880).

More details in: A. Wolffe, *Chromatin: Structure and Function* (Academic Press, London, 1998).

chromatography – a technique used to analyze material compositions including separation of the components in the mixture and quantitative measurement of the amount of each component. The mostly employed types of chromatography are:

- gas chromatography (to analyze gases or those liquids and solids, which can be heated to a volatile state without decomposition)
- liquid chromatography (to analyze liquids and solutions of solids when such materials cannot be analyzed by gas chromatography because of their thermal instability)
- ion chromatography (to analyze qualitatively and quantitatively ionic species in complex liquid and solid mixtures)
- gel permeation chromatography (to determine molecular weights and molecular weight distributions of polymer materials by separation of the molecules by size).

First described in: M. S. Tswett, *Adsorptionanalyse und chromatographische Meth-ode–Anwendung auf die Chemie des Chlorophills,* Ber. Dtsch. Bot. Ges. **24**, 316–332 (1906); M. S. Tswett, *Physikalisch-chemische Studien über das Chlorophill. Die Adsorptionen,* Ber. Dtsch. Bot. Ges. **24**, 384–393 (1906).

More details in: V. G. Berezkin, Ed., *Chromatographic adsorption analysis: selected works of M. S. Twett* (Ellis Horwood, New York, 1990).

chromic wheel → **molecular nanomagnet**.

chromosome – a single **DNA** molecule that is the self-replicating genetic structure within the cell which carries the linear nucleotide sequence of genes. In humans (or **eukaryotes**), the DNA is supercoiled, compacted, complexed with accessory proteins, and organized into a number of such structures. Normal human cells contain 46 chromosomes (except the germ cells, egg and sperm): 22 homologous pairs of autosomes and the sex-determining X and Y chromosomes (XX for females and XY for males). **Prokaryotes** carry their entire genome on one circular chromosome of DNA.

Term coined by: W. Waldeyer, *Zur Spermatogenese bei den Saugethieren*, Arch. Mikr. Anat. **32**, 1 (1888).

CIP geometry → **giant magnetoresistance effect**.

circularly polarized light → **polarization of light**.

cis–trans isomers → **isomers**.

cladding layers → **double heterostructure laser**.

Clark rule – a modification of the **Morse rule** relating the equilibrium internuclear distance in a diatomic molecule r and the equilibrium vibrational frequency ω in the form $\omega r^3 (n)^{1/2} = k - k'$, where n is the group number assigned according to the number of shared electrons, k is a constant characteristic of the period, and k' is a correction for singly ionized molecules of that period (zero for neutral molecules).

First described in: C. H. D. Clark, *The relation between vibration frequency and nuclear separation for some simple non-hydride diatomic molecules*, Phil. Mag. **18**(119), 459–470 (1934); C. H. D. Clark, *The application of a modified Morse formula to simple hydride diatomic molecules*, Phil. Mag. **19**(126), 476–485 (1935).

clathrate = **clathrate compound**.

clathrate compound – a chemical substance consisting of a lattice of one type of molecule trapping and containing a second type of molecule. It is formed by inclusion of molecules in cavities existing in crystal lattices or present in large molecules. The constituents are bound in definite ratios, but these are not necessarily integral. The components are not held together by primary valence forces, but instead are the consequence of a tight fit which prevents the smaller partner, the guest, from escaping from the cavity of the host. Consequently, the geometry of the molecules is the decisive factor. A clathrate compound is a weak composite because the captured molecules occupy the space which are left by the other compound.

Inclusion compounds can be subdivided into (1) lattice inclusion compounds (inclusion within a lattice which, as such, is built up from smaller single molecules);

(2) molecular inclusion compounds (inclusion into larger ring molecules with holes); and (3) inclusion compounds of macromolecules.

First described in: P. Pfeiffer, *Organische Molekülverbindungen*, (Ferdinand Enke, Stuttgart, 1927).

Clausius–Mossotti equation connects total polarization of dielectric particles P with dielectric constant ϵ and molecular volume $v(= mol.wt./density)$ as $P = (\epsilon - 1)/(\epsilon + 2)v$. It supposes that particles of a dielectric under the inductive action of an electric field behave as a series of small insulated conductors and can be polarized as a whole.

First described by: O. F. Mossotti, *Memorie di Matematica e di Fisica della Società Italiana delle Scienze Residente in Modena*, **24**(Part 2) 49–74 (1850); R. Clausius, *Die mechanische Behandlung der Electricität*, (Vieweg, Braunschweig 1858) p. 62.

clay–polymer nanocomposite – a **polymer** nanocomposite representing a new class of material alternative to conventional filled polymers. In this new class of material, nanosized inorganic fillers (at least one dimension) are dispersed in polymer matrix offering improvement in performance properties of the polymer. These hybrid compositions exhibit large increases in tensile strength, modulus, and heat distortion temperature as compared with the pristine polymer. The composites also have lower water sensitivity, reduced permeability to gases, and a similar thermal coefficient of expansion. All of these property improvements can be realized without a loss of clarity in the polymer. Typically smectite-type clays are used as fillers.

First described in: A. Usuki, M. Kawasumi, Y. Koyima, A. Okada, T. Karauchi, O. Kamigaito, *Swelling behavior of montmorillonite cation exchanged for ω-amino acids by ε-caprolactam*, J. Mater. Res. **8**(5), 1174–1178 (1993), *Synthesis of nylon 6-clay hybrid*, J. Mater. Res. **8**(5), 1179–1184 (1993).

More details in: *Polymer-Clay Nanocomposites*, edited by T. J. Pinnavaia, G. W. Beall (John Wiley & Sons, Chichester 2000).

Clebsch–Gordan coefficients – the elements of the unitary matrix $\langle m_1 m_2 | jm \rangle$, which are in fact the coefficients of the expansion of the eigenstates $|jm\rangle$ in terms of the eigenstates $|m_1 m_2\rangle$, in accordance with the formula $|jm\rangle = |m_1 m_2\rangle\langle m_1 m_2 | jm\rangle$. These coefficients are called Clebsch-Gordan coefficients. Sometime also termed Clebsch–Gordon, **Wigner**, vector addition or vector coupling **coefficients**. There are many symbols employed in the literature to denote the Clebsch–Gordan coefficients, for which it is possible to obtain explicit formula. The Clebsch–Gordan coefficients play a principal role in the general problem of the combination of angular momenta in quantum mechanics.

First described in: R. F. A. Clebsch, P. A. Gordan, *Theorie der Abelsche Funktionen* (Teubner, Leipzig, 1866).

More details in: A. Messiah, *Clebsch-Gordon Coefficients and 3j Symbols, Appendix C I.*, in: Quantum Mechanics (North Holland, Amsterdam, 1962); L. Cohen, close packed crystal

Tables of the Clebsch-Gordan Coefficients (North American Rockwell Science Center, Thousand Oaks, California, 1974).

close packed crystal – a crystal with atoms arranged so that the volume of the interstitials between the spheres corresponding to the atoms is minimal.

clusters (atomic) – large "molecules" containing typically from 10 to 2000 atoms. They have specific properties (mainly due to their large surface-to-volume ratio), which are size dependent and different from both the atoms and the bulk material.

CMOS – an acronym for a **c**omplementary **m**etal **o**xide **s**emiconductor device. Such a device, typically being an **integrated circuit**, is a combination of p-channel metal oxide semiconductor transistors and n-channel metal oxide semiconductor transistors. It was initiated by technical investigations in the 1940s and started as an industry in the late 1960s. CMOS circuitry dominates the current semiconductor market because of the astonishing power of silicon electronic integration technology. CMOS technology has distinguished itself by the rapid improvements in its products, in a four-decade history, most of the figures of merit of the silicon industry have been marked by exponential growth → **Moore's law**, principally as a result of the ability to exponentially decrease the minimum feature sizes of integrated circuits.
More details in: http://www.sematech.org/public/roadmap/.

CNDO – an acronym for **c**omplete **n**eglect of **d**ifferential **o**verlap → **approximate self-consistent molecular orbital methods.**

CNOT gate → **quantum logic gate.**

coenzyme – a non**protein** portion of an **enzyme** that supports the catalytic process. Coenzymes are often vitamins or metal ions.

confocal fluorescence microscopy → **fluorescence nanoscopy.**
More details in: T. Wilson, C. J. R. Sheppard, *Theory and Practice of Scanning Optical Microscopy* (Academic Press, New York, 1984).

coherence – a quality that expresses the correlation between two stochastic processes or states.

coherent potential approximation is used for calculation of fundamental electronic properties of disordered alloys. It introduces the concept of coherent atom corresponding to an average of disorderly arranged atoms of various species constituting a given alloy. We solve the impurity problem, in which an atom of a given species substitutes a coherent atom in the lattice of the pure crystal of coherent atoms.

The potential of electrons related to the impurity atom can be determined self-consistently in the **local density approximation** if the potential of the coherent atom is known. The latter potential is called the coherent potential. It is determined self-consistently by making use of the potentials of component atoms, which in turn have been determined in terms of the coherent potential by solving the above-mentioned impurity problem.

First described in: P. Soven, *Coherent-potential model of substitutional disordered alloys*, Phys. Rev. **156**(3), 809–813 (1967).

cohesion – a property exhibited by condensed matter: work must be done in order to separate matter into its constituents (free atoms or molecules), which are normally held together by attractive forces. These forces are particularly strong in solids. The work required to dissociate one mole of the substance into its free constituents is called its **cohesive energy**.

collimated light – a beam of light whose rays are parallel and thus has a plane wavefront. A collimated light is usually generated by **laser**s. It has a low beam divergence, so that the beam radius does not undergo significant changes within moderate propagation distances.

collision radiation → **Bremsstrahlung**.

colloid – a substance consisting of a continuous medium and small (**colloidal**) particles dispersed therein.

First described in: M. Faraday, *The Bakerian lecture: experimental relations of gold (and other metals) to light*, Phil. Trans. R. Soc. **147**, 145–181 (1857).

colloidal crystal – a self organized periodic, like in a crystal, arrangement of **colloidal particles**.

More details in: P. Pieranski, *Colloidal crystals*, Contemp. Phys. **24**(1), 25–73 (1983).

colloidal particles – particles smaller than coarse, filterable particles but larger than atoms and small molecules. They may sometimes be large molecules composed of thousands of atoms. There are no generally accepted size limits for colloidal particles, but the upper limit is approximately 200–500 nm, while the smallest particles are measured to be 0.5–2 nm.

colored noise – a noise characterized by a second order correlation function, which is broader than a delta function.

commutator – a notation for operators **A** and **B**, [**A,B**] = **AB** − **BA**. The order of operators is important. They cannot be reordered like numbers. Two operators are said to commute if their commutator is zero. It is possible to measure two physical quantities simultaneously to arbitrary accuracy only if their operators commute.

complementarity – the term used by N. Bohr to describe the fact that in a given experimental arrangement one can measure either the particle aspects or the wave aspects of an electron, or other quantum particle, but not both simultaneously.

First described in: N. Bohr, *Can quantum-mechanical description of physical reality be considered complete?*, Phys. Rev. **48**(8), 696–702 (1935).

complete neglect of differential overlap (CNDO) → approximate self-consistent molecular orbital methods.

compliance constants → elastic moduli of crystals.

compliance matrix → Hooke's law.

Compton effect – electromagnetic radiation scattered by an electron in matter increasing its wavelength so that it becomes

$$\lambda_\theta = \lambda_0 \left[1 + \frac{2h}{mc\lambda_0} \sin^2(\theta/2) \right],$$

where λ_0 is the wavelength of the incident radiation, m is the mass of the electron, and θ is the scattering angle. It is indeed pronounced in the case of X-rays and gamma radiation. The radiation is scattered in some definite direction. The increase of the wavelength ($\Delta\lambda = \lambda_\theta - \lambda_0$) depends only on the angle of scattering θ. The term h/mc is called the **Compton wavelength**.

First described in: A. H. Compton, *A quantum theory of the scattering of X rays by light experiments*, Phys. Rev. **21**(5), 483–502 (1923).

Recognition: in 1927 A. H. Compton received the Nobel Prize in Physics for the discovery of the effect named after him.

See also www.nobel.se/physics/laureates/1927/index.html.

Compton scattering – elastic scattering of photons by electrons.

Compton wavelength → **Compton effect**.

conductance – the real part of the **admittance** of an electric circuit. When the circuit **impedance** contains no reactance, as in a direct current circuit, it is the reciprocal of **resistance**. Conductance is a measure of the ability of the circuit to conduct electricity.

conductivity – the ratio of the electric current density to the electric field in a material.

conduction band – a set of one-electron states most commonly available for conduction of electric current by electrons in solids.

conductor – a substance through which electric currents flow easily. It has resistivity in the range $10^{-6} - 10^{-4}$ Ω cm. A solid is a conductor if there is no energy gap between its **conduction band** and **valence band** and the conduction band is filled with electrons.

constitutive relations → Hooke's law.

confinement factor → optical confinement factor.

continuity equation – an equation which relates the rate of inflow of a quantity to the rate of increase of the amount of that quantity within a given system. It can be written in the most general form for a particle density N as $\frac{dN}{dt} =$ *injection(and/or + production) – absorption – leakage*. Conservation of the particles in three dimensions gives $div\mathbf{J} + \frac{\partial N}{\partial t} = 0$, where $\mathbf{J}$ is the particle density vector. In the one dimensional case it transforms to $\frac{\partial J}{\partial x} + \frac{\partial N}{\partial t} = 0$.

controlled rotation gate (CROT gate) – a logical gate in which the state of one **bit** (the control bit) controls the state of another bit (the target bit). The target bit is rotated through π (that is from state 0 to 1 or vice versa) if and only if the control bit is 1. The transformation matrix of the simplest gate is:

$$\begin{pmatrix} 1 & 0 & 0 & 0 \\ 0 & 1 & 0 & 0 \\ 0 & 0 & 0 & -1 \\ 0 & 0 & 1 & 0 \end{pmatrix}.$$

The CROT gate is equivalent to the standard controlled NOT gate despite the difference in the minus sign placement in their matrix representation.

Cooley–Tukey algorithm – the efficient algorithm for calculations of the discrete Fourier transform → **fast Fourier transform**. It provides these calculations using only $n\log n$ arithmetic operations instead of n^2 operations.

Formally, the discrete **Fourier transform** is a linear transformation mapping any complex vector f of length n to its Fourier transform F. The k-th component of F is $F(k) = \sum_{j=0}^{n-1} f(j)\exp(2\pi ijk/n)$ and the inverse Fourier transform is $f(j) = \frac{1}{n}\sum_{k=0}^{n-1} F(k)\exp(2\pi ijk/n)$. Cooley and Tukey showed how the Fourier transform on the cyclic group $\mathbf{Z}/n\mathbf{Z}$, where $n = pq$ is composite, can be written in terms of Fourier transforms on the subgroup $q\mathbf{Z}/n\mathbf{Z}$ or $\mathbf{Z}/p\mathbf{Z}$. The trick is to change variables so that the one-dimensional formula giving the discrete Fourier tranform is turned into a two-dimensional formula which can be computed in two stages.

Cooley–Tukey algorithm

Define variables j_1, j_2, k_1, k_2 through the equations:

$$j = j(j_1, j_2) = j_1 q + j_2$$

$$0 \leq j_1 < p, 0 \leq j_2 < q$$

$$k = k(k_1, k_2) = k_2 p + k_1$$

$$0 \leq k_1 < p, 0 \leq k_2 < q.$$

Then the Fourier transform can be rewritten as:

$$F(k_1, k_2) \sum_{j_2=0}^{q-1} \exp([2\pi i j_2 (k_2 p + k_1)]/n) \sum_{j_1=0}^{p-1} \exp[2\pi i j_1 (k_1/p)] f(j_1, j_2).$$

Now one can compute F at two stages:
1. For each k_1 and j_2 compute the inner form

$$F_F(k_1, j_2) = \sum_{j_1=0}^{p-1} \exp[2\pi i j_1 (k_1/p)] f(j_1, j_2).$$

This requires at most $p^2 q$ scalar operations.
2. For each k_1 and k_2 compute the outer sum

$$F(k_1, k_2) = \sum_{j_2=0}^{q-1} \exp([2\pi i j_2 (k_2 p + k_1)]/n) F_F(k_1, j_2).$$

This requires an additional $q^2 p$ operations. Thus, instead of $(pq)^2$ operations, the above algorithm uses $(pq)(p + q)$ operations. If n could be factored further, the same trick could be further applied obtaining the transform through only $n \log n$ operations.

The algorithm had a revolutionary effect on many digital and image processing methods and it is still the most widely used method for computing Fourier transforms. It is interesting to note that the idea started from the necessity to develop a method for detecting Soviet Union nuclear tests without visiting the nuclear facilities. One idea was to analyze a large number of seismological time series obtained from off-shore seismometers, which requires fast algorithms for computing discrete Fourier transforms. The algorithm was not patented, and the impressive rapid growth of its applications is testimony to the advantages of open developments.

First described in: J. W. Cooley and J. W. Tukey, *An algorithm for machine calculation of complex Fourier series*, Mathematics of Computations, **19**(2), 297–301 (1965).

Cooper pair – a pair formed by two electrons with differently oriented spins (up and down). As a result the pair has zero spin and energy distribution of such particles follows **Bose-Einstein statistics**. Cooper pairs are used in the explanation of superconductivity.

First described in: L. N. Cooper, *Bound electron pairs in a degenerate Fermi gas*, Phys. Rev. **104**(4), 1189–1190 (1956).

Recognition: in 1972 J. Bardeen, L. N. Cooper and J. R. Schrieffer received the Nobel Prize in Physics for their jointly developed theory of superconductivity, usually called the BCS theory.

See also www.nobel.se/physics/laureates/1972/index.html.

coordination number – the number of nearest neighbors of a certain atom. The polyhedron formed when the nearest neighbors are connected by lines is called its coordination polyhedron.

Corbino disc – a flat doughnut shaped sample (see Figure C.6) that has concentric contacts to the inner and outer cylindrical surfaces.

First described in: O. M. Corbino, *Elektromagnetische Effekte die von der Verzerrung herrühren*, Phys. Z. **12**, 842–845 (1911).

Figure C.6 Corbino disc.

Corbino effect – production of a circumferential electric current in a circular disk by the action of an external axial magnetic field upon a radial current in the disk.

First described in: O. M. Corbino, *Elektromagnetische Effekte die von der Verzerrung herrühren*, Phys. Z. **12**, 842–845 (1911).

Coriolis force – an inertial force arising when an object moves in a rotating coordinate system. As a result, the path of the object appears to deviate. In a moving coordinate system, this deviation makes it look as if a force is acting upon the object, but actually there is no real force acting on the object. The effect is due to rotation (associated with an acceleration) of the coordinate system itself. In a rotating frame of reference (such as the earth), the apparent force is $\mathbf{F} = 2m(\mathbf{v} \times \mathbf{\Omega})$, where m is the mass and $\mathbf{v}$ is the linear velocity of the object, $\mathbf{\Omega}$ is the angular velocity of the coordinate system.

First described in: G. Coriolis, *Mémoire sur l'influence du moment d'inertie du balancier d'une machine a vapeur et de sa vitesse moyenne sur la régularité du mouvement de rotation du va-et-vient du piston communiqué au volant*, J. Ecole Polytechn. **13**, 228–267 (1832).

correlation – the interdependence or association between two variables that are quantitative or qualitative in nature.

correspondence principle → **Bohr correspondence principle**.

Cottrell atmosphere – a cluster of impurity atoms surrounding a dislocation in a crystal.
  First described in: A. H. Cottrell, B. A. Bilby, *Dislocation theory of yielding and strain ageing iron*, Proc. Phys. Soc. Lond. A **62**, 49 (1948).

co-tunneling – a tunneling process through an intermediate virtual state. It becomes relevant when sequential **single-electron tunneling** through quantum dots is suppressed by **Coulomb blockade**. The virtual states originate from quantum fluctuations of the number of electrons there. Co-tunneling is a process of high order tunneling in which only the energy from the initial to the final (after the tunneling events) state needs to be conserved. Elastic and inelastic co-tunneling are distinguished.
  In **elastic co-tunneling** an electron tunnels into the dot and out of the dot through the same intermediate energy state of the dot. For a dot placed between two leads the corresponding current is

$$I = \frac{\hbar \sigma_1 \sigma_2 \Delta}{8 \pi^2 e^2} \left(\frac{1}{E_1} + \frac{1}{E_2} \right) V.$$

Here σ_1 and σ_2 are the barrier conductances in the absence of the tunneling processes, Δ is the energy gap between the states in the dot, and E_1 is the charging energy related to coming of one electron into the dot, and E_2 is the charging energy related to one electron leaving the dot. The current varies linearly with the applied voltage V. At low temperatures it is temperature independent.
  In **inelastic co-tunneling**, an electron that tunnels from one lead into a state in the dot is followed by an electron that tunnels from a different state of the dot into another lead. In this case the current is

$$I = \frac{\hbar \sigma_1 \sigma_2}{6 e^2} \left(\frac{1}{E_1} + \frac{1}{E_2} \right)^2 \left[(k_B T)^2 + \left(\frac{eV}{2\pi} \right)^2 \right] V.$$

It is obvious that the current is temperature dependent and proportional to V^3. In a multidot system with N junctions $I \sim V^{2N-1}$.
  The co-tunneling processes yield the current below the threshold voltage controlled by the Coulomb blockade. This limits the accuracy of the **single-electron turnstile** even under the most favorable conditions.
  First described in: D. V. Averin, A. A. Odintsov, *Macroscopic quantum tunneling of the electric charge in small tunnel junctions*, Phys. Lett. A **140**, 251–257 (1989) – predicted; L. J. Geerligs, D. V. Averin, J. E. Mooij, *Observation of macroscopic quantum tunneling through the Coulomb energy barrier*, Phys. Rev. Lett. **65**(24), 3037–3040 (1990) – experimentally observed.

Coulomb blockade – an interdiction of an electron transfer into a region where it results in a change of the electrostatic energy greater than the thermal energy $k_B T$. Note that if the region is characterized by the capacitance C, its electrostatic energy is increased by $e^2/2C$ with an arrival of one electron. In macroscopic structures, this change in the energy is hardly noticeable, while in **nanostructures**, particularly in quantum dots, the condition $e^2/2C > k_B T$, is easily realized. The change in the electrostatic energy due to transfer of a single electron results in a gap of $e^2/2C$ in the energy spectrum at the **Fermi level** referred to as a **Coulomb gap**. Injection of an electron in a Coulomb blockade regime is inhibited until this energy gap is overcome through an applied bias. The phenomenon clearly manifests itself in **single-electron tunneling**.

First described in: C. J. Gorter, *A possible explanation of the increase of the electrical resistance of thin metal films at low temperatures and small field strengths*, Physica **17**(8), 777–780 (1951) – idea; I. Giaever, H. R. Zeller, *Superconductivity of small tin particles measured by tunneling*, Phys. Rev. Lett. **20**(26), 1504–1507 (1968) – experiment; I. O. Kulik, R. I. Shekhter, *Kinetic phenomena and charge discreteness effects in granulated media*, Zh. Exp. Teor. Fiz. **68**(2), 623–640 (1975) – in Russian, theory.

Recognition: in 1973 I. Giaever received the Nobel Prize in Physics for his experimental discoveries regarding tunneling phenomena in superconductors.

See also www.nobel.se/physics/laureates/1973/index.html.

Coulomb gap → Coulomb blockade.

Coulomb force → Coulomb law (electricity).

Coulomb gauge – the gauge defined by the equation $\nabla \cdot \mathbf{A} = 0$, where $\mathbf{A}$ is the vector potential of the magnetic field.

Coulomb law (mechanics) states that frictional force between two bodies is independent of the velocity of their movement with respect each other → **Amontons' law**.

First described in: C. A. Coulomb, *Théorie des machines simples, en ayant égard au frottement de leurs parties, et la roideur des cordages*, Mem. Math. Phys., 161–342 Paris (1785).

Coulomb law (electricity) states that the electrostatic force of attraction or repulsion exerted by one particle with a charge q_1 on another with a charge q_2, also called the **Coulomb force**, is defined as $F = q_1 q_2/r^2$, where r is the distance between the particles.

First observed by: H. Cavendish in 1773 but not published;

First described by: C. A. Coulomb, *Premier Mémoire sur l'électricité et le magnétisme*, Mem. Acad. Roy. Sci., 569–577 (1785).

Coulomb oscillations → single-electron transistor.

Coulomb potential

Coulomb potential – the potential describing the interaction of two charged particles in the form: $V(r) = \frac{q_1 q_2}{\varepsilon_0 \varepsilon r}$, where q_1 and q_2 are the charges of the interacting particles, ε is the relative permittivity, and r is the distance between the particles.

Coulomb scattering – scattering of charged particles by an atomic nucleus under the influence of the Coulomb forces exerted by the nucleus as a whole, that is the internal nuclear force field is neglected. This means that the charged particles which undergoes scattering does not approach the nucleus sufficiently closely for the nuclear forces acting between the protons and neutrons to have any external effect.

Coulomb staircase – a stair-like relation between electrical parameters of a system in which electronic transport is controlled by the **Coulomb blockade**. The charge accumulated in a **single-electron box** as a function of the gate–voltage and current–voltage characteristics of double barrier single-electron structures are typical examples.

covalent bond – an interatomic bond in which each atom of a bound pair contributes one electron to form a pair of electrons.
   First described in: I. Langmuir, *Isomorphism, isosterism and covalence*, J. Am. Chem. Soc. **41**(10), 1543–1559 (1919).
   Recognition: in 1932 I. Langmuir received the Nobel Prize in Chemistry for his discoveries and investigations in surface chemistry.
   See also www.nobel.se/chemistry/laureates/1932/index.html.

CPHASE gate → **quantum logic gate.**

CPP geometry → **giant magnetoresistance effect.**

Cr-based molecular ring → **molecular nanomagnet.**

CROT gate → **controlled rotation gate.**

crown ethers – heterocyclic chemical compounds that, in their simplest form, are cyclic oligomers of ethylene oxide. Some examples are illustrated in Figure C.7.

12-crown-4   15-crown-5   18-crown-6

Figure C.7 Crown ethers with 4, 5 and 6 repetitions of the ethyleneoxy groups.

The essential repeating unit of any simple crown ether is ethyleneoxy, i.e., $-CH_2CH_2O-$, which repeats twice in dioxane and six times in 18-crown-6. The nine-membered ring 1,4,7-trioxonane (9-crown-3) is often called a crown and can interact with cations. Macrocycles of the $(-CH_2CH_2O-)_n$ type in which $n \geq 4$ are generally referred to as crown ethers rather than by their systematic names. This is because the molecules formed when this special group of heterocycles binds to cations resemble a crown sitting on a head in structure.

The crown ethers are notable for their ability to strongly solvate cations. In other words, the equilibrium is strongly toward the complex. The oxygen atoms are ideally located to coordinate with a cation in the interior of the ring, whereas the exterior of the ring is **hydrophobic**. The result is that the complexed cation is soluble in nonpolar solvents. The size of the interior of the crown ether determines the size of the cation that it can solvate. Therefore, 18-crown-6 has a high affinity for potassium cation, 15-crown-5 for sodium cation and 12-crown-4 for lithium cation.

First described in: C. J. Pedersen, *Cyclic polyethers and their complexes with metal salts,* J. Am. Chem. Soc. **89**(26), 7017–7036 (1967).

Recognition: in 1987 C. J. Pedersen shared with D. J. Cram and J. M. Lehn the Nobel Prize in Chemistry for their development and use of molecules with structure-specific interactions of high selectivity.

See also www.nobel.se/chemistry/laureates/1987/index.html.

cryogenics – production and maintenance of very low temperatures, within a few degrees of absolute zero, and the study of phenomena at these temperatures.

crystal – a three-dimensional body consisting of periodic arrays of ions, atoms or molecules.

crystal point group → **point group**.

cubic lattice – a **Bravais lattice** whose unit cell is formed by perpendicular axes of equal length.

cucurbituril – a cyclic organic macrocyclic compound consisting typically of six units of glycoluril, while the molecules composed of 5, 7, 8, and 10 repeat units have been identified. Its cyclic unit is shown in Figure C.8.

This compound is of fundamental and applied interest because its molecule has roughly the shape of a hollow nanotube. The cavity of cucurbit [6] uril is characterized by the height of approximately 91 nm, the outer diameter is 58 nm and the inner diameter is 39 nm.

First described in: R. Behrend, E. Meyer, F. Rusche, *Über Condensationsprodukte aus Glycoluril und Formaldehyd,* Justus Liebigs Ann. Chem. **339**, 1–37 (1905).

Curie law

Figure C.8 A cyclic unit of cucurbituril–glycoluril.

Curie law governs the variation of magnetic susceptibility with temperature of a paramagnetic material with negligible interaction between the magnetic carriers. It has the form $M = C/T$, where M is the molar susceptibility and C is the Curie constant.

First described in: P. Curie, *Propriétés magnétiques des corps à diverses températures,* Annales de Chimie et de Physique, 7 série, **1**(V), 289–331 (1985).

Curie point (temperature) – the temperature above which particular magnetic properties of the substance disappear. There are three of such points relating to different magnetic materials: i) the ferromagnetic, ii) the paramagnetic, iii) the antiferromagnetic Curie points. The latter is also called the **Néel point (temperature)**.

Curie–Weiss law – the **Curie law** modified to the case when internal interactions play a role. The temperature dependence of the molar susceptibility takes the form $M = C/(T - \Theta)$, where Θ is the **Curie point**.

First described in: P. Weiss, *L'hypothèse du champ moléculaire et la propriété ferromagnétique,* J. Phys. **6**(1), 661–690 (1907).

Curie–Wulff's condition → **Wulff's theorem**.

"current law" → **Kirchhoff laws (for electrical circuits)**.

cyclodextrins – cyclic **oligomer** compounds consisting of six to eight **glucose** units, forming ring-shaped molecules. They are respectively called α- (six-membered ring molecule), β- (seven-membered ring molecule) and γ-cyclodextrin (eight-membered ring molecule).

Topologically, cyclodextrins can be represented as toroids in which primary hydroxyl groups are located "inside" and secondary ones "outside". Because of this arrangement, the interior of the toroids is **hydrophobic** and thus able to host other hydrophobic molecules. They are known to form inclusion complexes with various low-molecular-weight compounds ranging from non-polar **aliphatic** molecules to polar **amines** and acids.

More information in: F. Cramer, *Einschlussverbindungen der Cyclodextrine,* Angew. Chem. **64**, 136 (1952).

cyclopentadienyl complex – a metal complex with one or more **cyclopentadienyl rings**. One-cyclopentadienyl-ring complexes are shown in Figure C.9.

Based on the type of bonding between the metals and the cyclopentadienyl moieties, three categories of the complexes are distinguished: π-complexes, σ-complexes and ionic complexes. These complexes form **metallocenes**.

Me   Me   Me

Figure C.9 Cyclopentadienyl complexes.

cyclopentadienyl ring – the organic complex of composition $(C_5H_5)^-$, in which C atoms are bound up with each other and arranged in the ring structure as in an aromatic ring. It is abbreviated as Cp^-.

cyclotron effect – any free electron of mass m moving in a plane normal to a magnetic field with induction B executes a circular trajectory in this plane with the characteristic frequency $\omega_c = eB/m$, known as the **cyclotron frequency**. The radius of the orbit, called the **cyclotron radius**, is $r_c = (2mE)^{1/2}/(eB)$, where E is the electron kinetic energy. Note that the period of the cyclotron motion is independent of the electron energy, while the radius of the orbit is dependent on the electron energy.

cyclotron frequency → **cyclotron effect**.

cyclotron radius → **cyclotron effect**.

cyclotron resonance – a resonance absorption of energy from an alternating current electric field by electrons in a uniform magnetic field when the frequency of the electric field equals the cyclotron frequency, or the cyclotron frequency corresponding to the electron's effective mass if the electrons are in a solid.

D

From D'Alembert equation to Dzyaloshinskii–Moriya interaction

d'Alembert equation – a system of simultaneous linear homogeneous differential equations with constant coefficients of the form $dy_i/dx = \sum a_{ik} y_k = 0$ ($i = 1, 2, \ldots n$), where $y_i(x)$ are the n functions to be determined and a_{ik} are constants.
First described by: J. d'Alembert in 1747.

d'Alembertian operator is $\partial^2/\partial x^2 + \partial^2/\partial y^2 + \partial^2/\partial z^2 - 1/v^2(\partial^2/\partial t^2)$. It is used in electromagnetic wave theory where the parameter v is the wave velocity in the medium under consideration.

damped wave – a wave whose amplitude drops exponentially with distance because of the energy losses, which are proportional to the square of the amplitude.

Darwin curve – the form of the angular distribution of intensity in the X-ray diffraction pattern from a perfect crystal, as originally calculated by C. G. Darwin.
First described in: C. G. Darwin, *The theory of X-ray diffraction*, Phil. Mag. **27**, 315–333 (1914); C. G. Darwin, *The theory of X-ray diffraction: Part II*, Phil. Mag. **27**, 675–690 (1914).

Davydov splitting – splitting in absorption spectra of molecular crystals due to the interaction between identical molecules having different orientations in the **unit cell** of the crystal.
First described in: A. S. Davydov, *Theory of absorption spectra of molecular systems*, Zh. Exp. Teor. Fiz. **18**(2), 210–218 (1948) – in Russian.

Davisson–Calbick formula gives the focal length for the simplest electrostatic lens composed of a single circular aperture in a conducting plate kept at the potential V and separating two regions of different potentials V_1 and V_2 in the form $f = 4V(V_2 - V_1)$.
First described in: C. J. Davisson, C. J. Calbick, *Electron lenses*, Phys. Rev. **42**(4), 580–580 (1932).

What is What in the Nanoworld. Victor E. Borisenko and Stefano Ossicini
Copyright © 2008 WILEY-VCH Verlag GmbH & Co. KGaA, Weinheim
ISBN: 978-3-527-40783-5

Davisson–Germer experiment – the first demonstration of electron diffraction in which a beam of electrons was directed at the surface of a nickel crystal. The distribution of scattered back electrons was measured with a Faraday cylinder.

First described in: C. J. Davisson, L. H. Germer, *Diffraction of electrons by a crystal of Nickel*, Phys. Rev. **30**, 705–740 (1927).

Recognition: in 1937 C. J Davisson shared with G. P. Thomson the Nobel Prize in Physics for their experimental discovery of diffraction of electrons by crystals.

See also www.nobel.se/physics/laureates/1937/index.html.

de Broglie wavelength appears from de Broglie's statement that any particle, not only photons, with a momentum p should have (in some sense) a wavelength given by the relation $\lambda = h/p$.

First described in: L. de Broglie, *Ondes et quanta*, C. R. Acad. Sci. (Paris) **177**, 507–510 (1923).

Recognition: in 1929 L. de Broglie received the Nobel Prize in Physics for his discovery of the wave nature of electrons.

See also www.nobel.se/physics/laureates/1929/index.html.

Debye effect – selective absorption of electromagnetic waves by a **dielectric** due to molecular dipoles.

First described by: P. Debye in 1933.

Recognition: in 1936 P. Debye received the Nobel Prize in Chemistry for his contributions to our knowledge of molecular structure through his investigations on the dipole moments and on diffraction of X-rays and electrons in gases.

See also www.nobel.se/chemistry/laureates/1936/index.html.

Debye frequency – the characteristic frequency of the lattice vibrations that couple the electrons in the superconducting state.

Debye-Hückel screening – phenomenon in a plasma, where the electric field of a charged particle is shielded by particles having charges of the opposite sign.

First described in: P. Debye, G. Hückel, *Zur Theorie der Elektrolyten. Part II. Law of the limit of electrolytic conduction*, Phys. Z. **24**(10), 305–325 (1923).

Debye–Hückel theory – the theory of the behavior of strong electrolytes according to which each ion is surrounded by an ionic atmosphere of charges of the opposite sign whose behavior retards the movement of ions when an electric current is passed through the medium.

The energy of interaction between the ions and their surrounding ionic atmospheres determine an electrostatic potential energy of the solution, which is assumed to account for the nonideal behavior. The following assumptions are made to calculate this potential energy: i) strong electrolytes are completely dissociated in the concentration range in which the theory is valid, ii) the only forces operative in interionic attraction are the **Coulomb force**s, iii) the dielectric constant of the solution is not essentially different from that of the solvent, iv) the ions can be regarded as point charges, v) the electrostatic potential energy is small compared with the thermal energy.

First described in: P. Debye, E. Hückel, *Zur Theorie der Elektrolyten*, Phys. Zeitschrift **24**(9), 185–206 (1923).

Debye–Hückel screening length – a penetration depth of an impurity charge screened by mobile carriers in semiconductors $L = (e^2 n/\epsilon_0 k_B T)^{-1/2}$, where n is the carrier concentration and T is the temperature. It is valid for lightly doped semiconductors at around room temperatures. The quantity $(L)^{-1}$ is denoted as the **Debye–Hückel screening wave number.**

Debye–Hückel screening wave number → **Debye–Hückel screening length.**

Debye–Jauncey scattering – incoherent background scattering of X-rays from a crystal in directions between those of **Bragg reflections.**

First described in: G. E. M. Jauncey, *Theory of diffuse scattering of X-rays by solids,* Phys. Rev. **37**(10), 1193–1202 (1931).

Debye relaxation – relaxation of a charge **polarization** characterized by a single relaxation time t. It is typical for the relaxation of orientation polarization in a material containing only one type of permanent dipoles. The relaxation current is exponentially decreased with time: $I \sim \exp{(-t/\tau)}$.

Debye screening length = **Debye–Hückel screening length.**

Debye temperature – the characteristic used to describe the slope of the temperature dependence of the heat capacity of matter generally presented as $c_v = f(T/T_\Theta)$. In fact, the Debye temperature T_Θ is dependent on temperature, so the above relation holds linearity in an appropriate power of T/T_Θ in a limited temperature range.

First described in: P. Debye, *Zur Theorie der spezifischen Wärmen*, Ann. Phys. **30**(4), 789–839 (1912).

Debye–Waller factor – a reduction in the intensity of X-rays diffracted at a crystal lattice due to the thermal motion of the atoms in the lattice. It is $D = \exp{[-8\pi^2 u^2 \sin^2{(\theta/2\lambda^2)}]}$, where u^2 is the average value of the square of the displacement of the atom perpendicular to the reflecting plane (a function of T), θ is the **Bragg reflection angle**, and λ is the wavelength of the X-rays.

First described in: I. Waller, *Zur Frage der Einwirkung der Wärmebewegung auf die Interferenz von Röntgenstralen*, Z. Phys. **17**(6), 398–408 (1923).

deca- – a decimal prefix representing 10, abbreviated da.

decoherence (quantum) – a transformation of a pure quantum state into a mixed state due to an interaction of the quantum system with surroundings.

deceleration radiation → **Bremsstrahlung**.

deci- – a decimal prefix representing 10^{-1}, abbreviated d.

deformation potential is introduced in order to describe how **phonons** compress and dilate alternating regions of a solid. A uniform compression or dilatation of the crystal causes the edge of each electronic energy band to move up or down proportionally to the strain. The constant of proportionality is called the deformation potential, Ξ.

The potential energy is $V(z) = \Xi e(z)$, where the longitudinal strain is $e(z) = \partial u / \partial z$, and $u(z)$ is the displacement of an atom at z.

degeneracy means the equality of quantities which are independent and thus in general distinct. In quantum mechanics, it is a number of **orbitals** (energy levels) with the same energy or, identically, the conditions when different wave functions correspond to the same energy.

degenerate states – the states of the same energy. In other words, different states with a common **eigenvalue** are said to be degenerate.

de Haas–van Alphen effect – oscillatory variation of magnetic susceptibility of metals as a function of $1/H$, where H is the strength of a static magnetic field. The effect is observed at sufficiently low temperatures, typically below 20 K. It is a consequence of the quantization of the motion of conduction electrons in a magnetic field. Recording field and temperature dependences of the effect, in particular as a function of the field orientation with respect to the crystal axes, provides finding the electron effective mass, electron relaxation time, and **Fermi level** in the material.

First described by: W. J. de Haas, P. M. van Alphen, *Relation of the susceptibility of diamagnetic metals to the field*, Proc. K. Akad. Amsterdam **33**(7), 680–682 (1930).

More details in: D. Schoenberg, *Magnetic Oscillations in Metals*, (Cambridge University Press, Cambridge, 1984).

Delbrück scattering – a scattering of light produced by a Coulomb field, i.e. by virtual electron–positron pairs.

First described in: L. Meitner, H. Kösters, M. Delbrück, *Über die Streuung kurzwelliger γ-Strahlen*, Z. Phys. **84**, 137–144 (1933).

delta-doped structure – a semiconductor structures with extremely inhomogeneous doping profile characterized by location of all impurities in a very thin inner

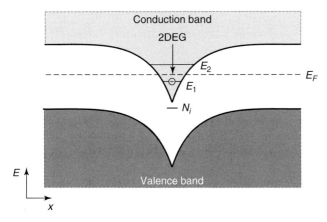

Figure D.1 Energy bands in a delta-doped semiconductor structure.

layer, ideally within one monolayer. The dopant profile resembles the Dirac **delta function**. Electronic bands in such a structure are shown in Figure D.1.

Carriers in the highly doped region are restricted to move far from the ionized impurity because of strong Coulomb interaction with them. The electric field of the ionized impurity atoms is screened by the free carriers they produce. A V-shaped potential well is formed. It can be approximated by

$$U(x) = \frac{me^4}{\epsilon^2 \hbar^2} \left[\left(\frac{15\pi^3}{8\sqrt{2}} N_i^2 a_{B*}^4 \right)^{-1/10} + \sqrt{\frac{2\sqrt{2}}{15\pi} \frac{x}{a_{B*}}} \right]^{-4},$$

where ϵ is the dielectric susceptibility of the material, N_i the sheet concentration of the impurity, and $a_{B*} = \epsilon \hbar^2/me^2$ the effective Bohr radius.

Energy states in the well are quantized according to the quantum confinement regularities. They can accommodate a high density of electrons with a large number of occupied two-dimensional sub-bands.

One of the simplest delta-doped structures consists of a monolayer of silicon deposited onto the surface of monocrystalline GaAs and then buried with more epitaxial GaAs. Donor silicon atoms slightly diffuse away, but remain within some monolayers near the original doping plane. The electron-confining region extends to about 10 nm. The electron density in the 2DEG up to 10^{14} cm^{-2} can be reached, but at the cost of cutting their mobility.

A periodic alignment of n-type and p-type delta-doped layers separated by an intrinsic material is known as a *nipi* structure. When donor and acceptor concentrations in n-type and p-type layers, respectively, are equal the structure in equilibrium has no free carriers to move. Non-equilibrium carriers, generated for example by the light illumination of the structure, appear to be separated by the built-in electric field. Their charge modifies the energy band diagram in the same way as equilibrium carriers do. An identical effect can be also achieved by external electrical biasing of the *n*- and *p*-layers. Both approaches give an opportunity of effective tuning of the energy band diagram.

delta function – a distribution δ such that $\int_{-\infty}^{\infty} f(x)\delta(x-t)dx = f(t)$, also known as a **Dirac delta function**. It is a functional generalization of the **Krönecker delta function**.

demagnetization curve – a functional relationship between the magnetic induction and the strength of the magnetic field when the strength of the field is decreased from the saturation state of the material to the point of zero flux.

Dember effect – an appearance of an electric field in a uniform semiconductor when it is nonuniformly irradiated with light. It is a result of different diffusion lengths of electrons and holes generated by the light.
First described in: H. Dember, *Über eine photoelektromotorische Kraft in Kupfer-oxid-Kristallen*, Phys. Z. **32**, 554–556 (1931) – experiment; J. I. Frenkel, *Possible explanation of superconductivity*, Phys. Rev. **43**(11), 907–912 (1933) – theory.

dendrimer – an artificially manufactured or synthesized branched molecule built up from **monomers**. Dendrimer = dendritic **polymer**, where Greek "dendra" means "tree". A schematic representation of a dendrimer is shown in Figure D.2.
First described in: E. Buhleier, W. Wehner, F. Vögtle, *"Cascade"- and "Nonskid-Chain-like" syntheses of molecular cavity topologies*, Synthesis **1978** 155–158 (1978).

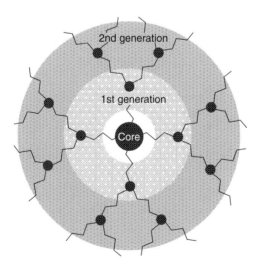

Figure D.2 Schematic structure of a second-generation dendrimer.

dendrite – a tree-like crystal.

Derjaguin–Muller–Toporov force → Derjaguin–Muller–Toporov model.

Derjaguin–Muller–Toporov model approaches the problem of **adhesion** between a sphere and a flat surface considering a **Lennard-Jones potential** acting in the region around the contact area. The friction force becomes $F_f = \tau\pi(R/K)^{2/3}F_{DMT}^{2/3}$ where $F_{DMT} = N + 2\pi R\gamma$ is the so-called **Derjaguin–Muller–Toporov force**. In these formulae N is the applied load, R is the radius of the sphere, and $\gamma = \gamma_1 + \gamma_2 - \gamma_{1,2}$ is the interfacial energy (γ_1 is the surface energy of the sphere, γ_2 is the surface energy of the flat surface, and $\gamma_{1,2}$ is the energy of the interface). For the meaning of K → **friction force theory**. For zero applied load the friction force reduces to $F_{f0} = \tau\pi[(2\pi R^2\gamma)/K]^{2/3}$ and the contact radius is $a_0 = [(2\pi R^2\gamma)/K]^{1/3}$. The model is more suitable for low adhesion and small (tip) radii.

First described in: B. V. Derjaguin, V. M. Muller, Y. P. Toporov, *Effect of contact deformations on the adhesion of particles*, J. Coll. Interface Sci. **67**(2), 314–326 (1975).

density functional theory states that the ground state energy corresponds to the minimum of a functional of the density of electrons, which is a function of position. Then electron distribution in the ground state is shown to be determined by solving the **Schrödinger equation** with an appropriate one-electron potential which takes into account the exchange correlation energy as well as the classical electron–electron and electron–nucleus Coulomb interactions. The exchange-correlation potential, which is a function of the electron density, in principle, can be determined only approximately. The approximation which is widely used is the **local density approximation**. By calculating the one-electron potential self-consistently, one can determine not only the electron structure for a given lattice configuration of nuclei but also the crystal structure itself and lattice spacing on the basis of the energy minimum principle by comparing the total energy of various configurations.

The density functional theory is based on two general theorems, the **Hohenberg–Kohn theorems**:

- any physical property of an interacting electron gas, in its fundamental state, can be written as a unique functional of the electron density $\rho(\mathbf{r})$ and, in particular, its total energy $E[\rho]$ (the total energy of an electron gas in the presence of a background potential may be written as a functional of the charge density)

- $E[\rho]$ reaches its minimum for the true density $\rho(\mathbf{r})$, that is, derived from the Schrödinger equation. The true charge density of the system is that which minimizes this functional subject to it having the correct normalization.

For electrons moving in a potential due to ions (in the CGS system)

$$E[\rho] = -e^2 \int \frac{\rho^-(\mathbf{r})\rho^+(\mathbf{r}')}{|\mathbf{r}-\mathbf{r}'|}d^3\mathbf{r}d^3\mathbf{r}' + \frac{e^2}{2}\int \frac{\rho^-(\mathbf{r})\rho^-(\mathbf{r}')}{|\mathbf{r}-\mathbf{r}'|}d^3\mathbf{r}d^3\mathbf{r}'$$

density of states

$$+\frac{e^2}{2}\int\frac{\rho^+(\mathbf{r})\rho^+(\mathbf{r}')}{|\mathbf{r}-\mathbf{r}'|}d^3rd^3r'+T[\rho(\mathbf{r})]+E_{xc}[\rho(\mathbf{r})],$$

where $\rho^+(\mathbf{r})$ is the number of elementary positive charges per unit volume. The first three terms are due to the classical Coulomb electron–ion, electron–electron and ion–ion interactions, respectively. The fourth term is the kinetic energy of a non-interacting electron system of density $\rho(\mathbf{r})$. The last term is the exchange and correlation energy.

The theory is applicable for simulation of fundamental electronic and related properties of matter.

First described in: P. Hohenberg, W. Kohn, *Inhomogeneous electron gas*, Phys. Rev. **136**(3B), B864–B871 (1964); W. Kohn, L. J. Sham, *Quantum density oscillations in an inhomogeneous electron gas*, Phys. Rev. **137**(6A), A1697–A1705 (1965); W. Kohn, L. J. Sham, *Self consistent equations including exchange and correlation effects*, Phys. Rev. **140**(4A), A1133–A1138 (1965).

More details in: E. K. Gross, F. J. Dobson, M. Petersilka, *Density Functional Theory* (Springer, Heidelberg, 1996).

Recognition: in 1998 W. Kohn received the Nobel Prize in Chemistry for his development of the density functional theory.

See also www.nobel.se/chemistry/laureates/1998/index.html.

density of states – a number of states per energy interval dE per spin orientation per unit volume of real space. Total density of states and projected density of states are distinguished. The **total density of states** is calculated as

$$n(E)=\frac{2}{V_{BZ}}\sum_n\int_{V_{BZ}}\delta(E-E_n(\mathbf{k}))d\mathbf{k},$$

where V_{BZ} is the Brillouin zone volume. The integral is taken over the surface of constant energy E and n is the band index. The **projected density of states** of the atom type t is

$$n_i^t(E)=\frac{2}{V_{BZ}}\sum_n\int_{BZ}Q_i^t(\mathbf{k})\delta(E-E_n(\mathbf{k}))d\mathbf{k},$$

where Q_i^t is the partial charge.

The total densities of states in three-dimensional (3D) and reduced-dimensionality systems are:

$$3D:n_{3D}(E)=\frac{m^*\sqrt{2m^*E}}{\pi^2\hbar^3},$$

$$2D:n_{2D}(E)=\frac{m^*}{\pi\hbar^2l_x}\sum_i\Theta(E-E_i), i=1,2,\ldots,$$

$$1D:n_{1D}(E)=\frac{\sqrt{2m^*}}{\pi\hbar l_xl_y}\sum_{i,j}\sqrt{\frac{1}{(E-E_{i,j})}}, i,j=1,2,\ldots,$$

$$0D : n_{0D}(E) = \frac{2}{l_x l_y l_z} \sum_{i,j,k} \delta(E - E_{i,j,k}), i, j, k = 1, 2, \ldots,$$

where m^* is the effective electron mass, $\Theta(E - E_i)$ is the step function, $\delta(E - E_{i,j,k})$ is the δ function, and l_x, l_y, l_z are the sizes of low dimensionality.

deoxyribonucleic acid $\rightarrow$ DNA.

desorption – removal of a substance from a surface on which it has been adsorbed. The alternative process is **adsorption**.

Destriau effect – **electroluminescence** of zinc sulfide phosphors when excited by an electric field. This effect is the basis for the alternating-current phosphor panel technology. G. Destriau first coined the word "electroluminescence" to refer to the phenomenon he observed.

First described in: G. Destriau, *Recherches sur les scintillations des sulfures de zinc aux rayons*, J. Chem. Phys. **33**, 587–625 (1936).

Deutsch gate $\rightarrow$ quantum logic gate.

Deutsch-Jozsa algorithm – a quantum algorithm for solving the Deutsch-Jozsa problem: the task is to determine whether the Boolean function $f : \{0, 1\}^n \rightarrow \{0, 1\}$ is constant, always returning 0 or always 1 irrespective of the input, or balanced returning 0 for half of the inputs and 1 for the other half, for example, determining whether a coin is fair (heads on one side, tails on the other) or fake (heads or tails on both sides). Functions of this type are often referred to as oracle or black box. The algorithm gives the right answer using just an examination step. A classical algorithm requires in the best case two evaluations, in the worst case $(2^{n-1} + 1)$ evaluations. It is sufficient to require that f is linear and to note that a reversible implementation of f is possible by using at least one extra bit to represent the computation as a permutation of the states of the input bits.

First described in: D. Deutsch, R. Jozsa, *Rapid solutions of problems by quantum computation*, Proc. R. Soc. A **439**, 553–558 (1992).

More details in: A. Galindo, M. A. Martín Delgado, *Information and computation: classical and quantum aspects*, Rev. Mod. Phys. **74**(2), 347–423 (2002); S. Guide, M. Riebe, G. P. T. Lancaster, C. Becher, J. Eschner, H. Häffner, F. Schmidt-Kaler, I. L. Chuang, R. Blatt, *Implementation of the Deutsch-Josza algorithm on a ion-trap quantum computer*, Nature **421**(1) 48–50 (2003).

Recognition: in 2002 D. Deutsch received the International Award on Quantum Communication for his theoretical work on Quantum Computer Science.

diagonal sum rule – a sum of the diagonal elements (the trace) of the matrix of an observable is independent of representation and is just equal to the sum of the characteristic values of the observable weighted by their degeneracies.

diamagnetic – a substance which in an external magnetic field demonstrates magnetization opposite to the direction of the field, so that diamagnetics are repelled by magnets. The magnetization is proportional to the applied field. A substance in a superconducting state exhibits perfect diamagnetism → **magnetism.**

First described in: M. Faraday, *Experimental researches in electricity. - Twentieth Series.* Phil. Trans. **136**, 31–40 (1846); M. Faraday, *Experimental researches in electricity. - Twenty-first Series.* Phil. Trans. **136**, 41–62 (1846).

diamagnetic Faraday effect – the **Faraday effect** at frequencies near an absorption line, which is split due to the splitting of the upper level only.

diamond structure – specific crystalline structure which can be described as two interpenetrating **face-centered cubic (fcc)** lattices that are displaced relative to one another along the main diagonal (atomic arrangement in an fcc lattice can be seen in a **Bravais lattice**). The position of the origin of the second fcc lattice expressed in terms of the basis vector is (1/4,1/4,1/4). The structure takes its name from diamond where carbon atoms are arranged in such a fashion. It supposes three-dimensional covalent bonding in which every atom is surrounded by four nearest neighbors in a tetrahedral configuration. Other elements that crystallize with the same lattice are Si, Ge, and αSn.

diastereoisomers = diastereomers.

diastereomers – stereoisomers that are not **enantiomers** (mirror images of each other). They can have different physical properties and different chemical reactivity. In another definition diastereomers are pairs of **isomers** that have opposite configurations at one or more of the chiral centers but are not mirror images of each other → **isomers.**

Dicke effect – splitting of a **spontaneous emission** of photons by an ensemble of initially excited atoms. If the excited atoms are located in a region smaller than the wavelength of the emitted radiation, they no longer decay independently. Instead, the radiation has a higher intensity and takes place in a shorter time interval than for an ensemble of independent atoms due to the coupling of all atoms to the common radiation field.

In the case of emission from two excited atoms with dipole moments d_1 and d_2 at positions r_1 and r_2 the spontaneous emission rate of photons with wave vector $\mathbf{Q}$ following the **Fermi's golden rule** is $\Gamma_\pm(Q) \sim \sum_Q |g_Q|^2 |1 \pm \exp[i\mathbf{Q}(\mathbf{r}_2 - \mathbf{r}_1)]|^2 \delta(\omega_0 - \omega_q)$, where $Q = \omega_0/c$, ω_0 is the transition frequency between the upper and lower level, and g_Q is matrix element for the mode $\mathbf{Q}$. The two signs "+" and "−" correspond to the two different relative orientations of the dipole moments of the two atoms. The interference of contributions of these atoms leads to splitting of the spontaneous emission of photons into a fast superradiant decay channel described by $\Gamma_+(Q)$ and a slow subradiant decay channel described by $\Gamma_-(Q)$.

From the four possible states in the **Hilbert space** of two two-level systems one can expect to observe their singlet and triplet states. The superradiant decay channel occurs via the triplet and the subradiant decay via the singlet states. In the extreme Dicke limit where the second phase factor is close to unity, $|\exp[i\mathbf{Q}(\mathbf{r}_2 − \mathbf{r}_1)]| \approx 1$, it follows that $\Gamma_-(Q) = 0$ and $\Gamma_+(Q) = 2\Gamma(Q)$, where $\Gamma(Q)$ is the decay rate of one single atom. This limit is theoretically achieved if $|\mathbf{Q}(\mathbf{r}_2 − \mathbf{r}_1)| \ll 1$ for all wave vectors $\mathbf{Q}$, so that the distance between the two atoms is much smaller than the wavelength of the light. In reality, the conditions for which the subradiant rate is zero and the superradiant rate is just twice the rate for an individual atom is never reached.

First described in: R. H. Dicke, *Coherence in spontaneous radiation processes*, Phys. Rev. **93**(1), 99–110 (1954).

Dicke superradiance – a collective decay of an ensemble of excited two-level systems due to spontaneous emission.

First described in: R. H. Dicke, *Coherence in spontaneous radiation processes*, Phys. Rev. **93**(1), 99–110 (1954).

More details in: M. G. Benedict, A. M. Ermolaev, V. A. Malyshev, I. V. Sokolov, E. D. Trifonov, *Super Radiance*, Optics and Optoelectronics Series (Institute of Physics, Bristol, 1996).

dielectric – a substance in which an external moderate electric field once established may be maintained without loss of energy. It is an insulator containing no free charges. Actual dielectrics may be slightly conducting and the line of demarcation between insulators or dielectrics and **semiconductors** or **conductors** is not sharp.

dielectric function – a function $\epsilon(\omega)$ relating the electromagnetic field $\mathbf{E}$ applied to matter and the displacement of charges $\mathbf{D}$ produced in it via $\mathbf{D} = \epsilon(\omega)\mathbf{E}$. It is a complex quantity, so $\epsilon(\omega) = \epsilon_1(\omega) + i\epsilon_2(\omega)$ with the corresponding real $\epsilon_1(\omega)$ and imaginary $\epsilon_2(\omega)$ parts.

The imaginary part of the dielectric function is defined by the quantum-mechanical transition rate W_{ij} in the random phase approximation for transitions from the state i to j via $\epsilon_2 = (n^2/\omega) \sum W_{ij}(\omega)$. The summation runs over all occupied and unoccupied states. For transitions between one-electron Bloch states from the filled valence into empty conduction band states the transition rate within the first-order perturbation theory is $W_{ij}(\omega) = (2\pi/h)|V_{ij}(k)|^2\delta(E_{ij}(k) − \hbar\omega)$, where the interband **matrix element** $|V_{ij}|$ contains the Bloch factors of the valence and conduction band wave functions and the dipole operator of the incident light wave, which is considered as perturbation. The interband matrix element depends on the symmetry of the states participating in the optical transitions and the light polarization. The dipole operator has odd **parity**, therefore V_{ij} vanishes unless the Bloch factors have odd parity. Transitions are termed allowed if V_{ij} is non-zero, otherwise they are termed forbidden. Generally V_{ij} is dependent on the electron wave vector **k**; however, near critical points they are generally independent of **k**.

The real and imaginary parts of the dielectric function are connected by the **Kramers-Kronig dispersion relation**. They are used to calculate the macroscopic optical properties of solids represented by the **refractive index** n^*, **extinction coefficient** k^*, and **absorption coefficient** α:

$$n^* = \left\{ (1/2) \left[\epsilon_1 + (\epsilon_1^2 + \epsilon_2^2)^{1/2} \right] \right\}^{1/2} ,$$

$$k^* = \left\{ (1/2) \left[-\epsilon_1 + (\epsilon_1^2 + \epsilon_2^2)^{1/2} \right] \right\}^{1/2} ,$$

$$\alpha = \frac{4\pi k^*}{\lambda} ,$$

where λ is the wavelength of the light in a vacuum.

differential scanning calorimetry (DSC) – a technique of thermal analysis of matter which involves heating of a sample along with a thermally inert reference material and measuring the relative heat changes that occur in the sample during the heating process. The most common heat change normally observed is the melting transition of a crystalline material, which results in an absorption of heat or energy relative to the inert standard. It is observed as an endothermic peak in the thermogram. The reverse phenomenon, that is crystallization, occurring upon cooling a crystallizable material from the melt manifests itself as a release of heat. It is observed as an exothermic peak. Many other heat-sensitive phenomena can also be characterized by this technique, such as glass and multiple crystal phase transitions, loss of volatiles, melting of blends and alloys, and degradation processes.

diffraction – generation of a weaker wavefront at an opaque edge or around a hole when waves pass through them. These secondary wavefronts will interfere with the primary wavefront as well as with each other to form various diffraction patterns.

diffusion – a directed motion of matter driven by a gradient of the electrochemical potential in a system. It provides homogenization or approach to equilibrium through random atomic motion. Gradients of the electrochemical potential are often related to gradients of particle concentrations and temperature gradients in real systems.

When a gradient of particle concentration N is a driving force for diffusion, the flux of the diffusion particles per unit area is $J = -D\mathrm{grad}N$ (in the one-dimensional case $J = -D(\partial N/\partial x)$), where D is the coefficient of diffusion or particle's diffusivity. Time-dependent variation of the particle concentration is described by the equation $(\partial N/\partial t) = \text{grad}(D\text{grad}N)$ (in the one-dimensional case $(\partial N/\partial t) = (\partial/\partial x)(D\partial N/\partial x)$). The above equations are also referred to as **Fick's laws** of diffusion, the first law and the second law, respectively.

In solids, diffusion motion of atoms needs point lattice defects to be involved in the process. So far the mechanisms of atomic diffusion distinguished in solids are: i) vacancy mechanism (an atom in a substitutional position jumps into a nearest vacancy by exchanging with it), ii) interstitial mechanism (jumps between neighboring interstitial sites in the lattice), iii) place exchange or ring mechanism (the simultaneous movement of two (in place exchange) or more (in ring) neighboring atoms as a result of which they squeeze past each other and exchange sites in the lattice). In the last case, lattice defects are not so important. In practice, superposition of the above mechanisms can take place.

First described mathematically in: Λ. Fick, *Über Diffusion*, Ann. Phys. **94**, 59–86 (1855).

More details in: J. Crank, *Mathematics of Diffusion* (Clarendon Press, Oxford, 1956); B. I. Boltaks, *Diffusion in Semiconductors* (Academic Press, New York, 1963).

diffusion length – an average distance diffused by a particle, i.e. a charge carrier or an atom, from the point at which it is formed or liberated to the point where it is absorbed.

diluted magnetic semiconductors are typically $A^{II}B^{VI}$ and $A^{III}B^{V}$ semiconductors in which element A is partially substituted by magnetic ions - very often by manganese (Mn). In contrast to magnetic semiconductors, such as Eu chalcogenids or $ZnCr_2Se_4$, diluted magnetic semiconductors allow for tuning their magnetic properties by the variation of magnetic ion concentration. $Zn_{1-x}Mn_xSe$, $Zn_{1-x-y}Be_xMn_ySe$, and $Ga_{1-x}Mn_xAs$ are widely used examples of diluted magnetic semiconductors. Magnetic properties of these materials are determined by the manganese atoms. They are controlled by the strong exchange interaction between the hybridized $sp^2 - d$ **orbitals** of the magnetic ions and charge carriers. As a result, giant **Zeeman splitting** is observed in both conduction and valence bands when the material is placed in an external magnetic field. It is illustrated in Figure D.3.

The conduction band of the diluted semiconductors is two-fold spin degenerate, whereas the valence band has four spin degenerate states related to heavy and light holes. In the magnetic field the spin degeneracy is lifted providing vacant places for spin-up ($m_j = +1/2$) and spin-down ($m_j = -1/2$) electrons and both spin-oriented heavy ($m_j = +3/2$ and $m_j = -3/2$) and light holes ($m_j = +1/2$ and $m_j = -1/2$). Manganese incorporates into $A^{II}B^{VI}$ semiconductors isoelectrically, so manganese-containing materials may also be doped by either n- or p-type impurities. In $A^{III}B^{V}$ semiconductors manganese acts as a shallow acceptor precluding the fabrication of *n*-type material.

More details in: A. Twardowski, *Diluted Magnetic Semiconductors* (World Scientific, Singapore, 1996).

dimer

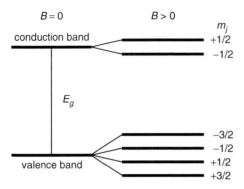

Figure D.3 Splitting of conduction and valence bands of a diluted semiconductor in an external magnetic field (B).

dimer – a molecule composed of two identical sub-units (**monomers**) linked together.

diode – an electronic device permitting electric current to flow in one direction only.

diploid – a **eukaryotic** cell with two sets of **chromosomes**.

dipole – a pair of electric charges, equal in magnitude but opposite in size. The product of the magnitude of the charges and their distance of separation is the **dipole momentum**.
   First described by: P. Debye, *Some Results of a Kinetic Theory of Insulators*, Phys. Z. **13**, 97–100 (1912).
   Recognition: in 1936 P. Debye received the Nobel Prize in Chemistry for his contributions to our knowledge of molecular structure through his investigations on the dipole moments and on diffraction of X-rays and electrons in gases.
   See also www.nobel.se/chemistry/laureates/1936/index.html.

dip-pen nanolithography – the technique providing fabrication of molecular layer patterns with nanometer features with the use of a scanning probe of an **atomic force microscope**. In this technique the water meniscus is formed between the tip of the probe and the substrate. The molecules to be deposited are dissolved in the water thus forming an ink. It acts both as a source of the molecules and as a medium for the molecular transport. The pattern formation is illustrated in Figure D.4.
   First, the ink is deposited onto the tip from an ink-pot. Then, the tip is positioned at the substrate and scanned over it with a combination of contact and non-contact modes. The molecules are deposited only in the regions where the tip is in contact with the substrate and the meniscus is formed. The meniscus formation is dependent on the humidity of the ambient atmosphere and the water reaction of

Ink deposition to the tip   Ink deposition to the substrate

Figure D.4 Pattern formation by dip-pen nanolithography.

the substrate surface (**hydrophilic** or **hydrophobic**). The scan speed determines the amount of ink that is carried out with the tip. At high scan rates the tip can carry less ink. As the molecular transport is essentially diffusion controlled, the number of deposited molecules is proportional to $t^{1/2}$, where t is the time of contact between the tip and the substrate.

The efficiency of the technique depends on the capability of depositing molecules to form self-assembled monolayers. Among already tested compounds, 16-mercapto hexadecanoic acid and 1-octadecane thiol have been shown to meet the requirements. The best results are obtained when the substrate is covered by a nanometer-thick gold film which is patterned first and then used as a mask for dipper profiling of the underlaying region. Surface structures with feature sizes down to 10 nm can be fabricated.

First described in: R. D. Piner, J. Zhu, F. Xu, *"Dip-pen" nanolithography*, Science **283**, 661–663 (1999).

Dirac delta function → **delta function**.

Dirac equation – a relativistic wave equation for a quantum particle in which the wave function has four components corresponding to four internal states specified by a two-valued spin coordinate and an energy, which can have a positive or negative value. It was obtained by employing the following linearized expression for energy of a quantum particle of mass m with momentum $\mathbf{p}$: $E = c\boldsymbol{\alpha}\cdot\mathbf{p} + \beta mc^2$ (instead of $E = (c^2\mathbf{p}^2 + m^2c^4)^{1/2}$ originating from the special theory of relativity). The four symbols $\boldsymbol{\alpha} = (\alpha_1, \alpha_2, \alpha_3)$ and β are mutually anti-commuting operators satisfied by the matrices, i.e.:

$$\alpha_i = \begin{pmatrix} 0 & \sigma_i \\ \sigma_i & 0 \end{pmatrix}, \quad \beta = \begin{pmatrix} 1 & 0 \\ 0 & -1 \end{pmatrix},$$

where σ_i are **Pauli spin matrices**. Thus, $\alpha_i\alpha_j + \alpha_j\alpha_i = 2\delta_{ij}, \beta^2 = 1, \alpha_i\beta + \beta\alpha_i = 0$, where δ_{ij} is the **Krönecker delta function**.

Substituting the expression for energy into the **Schrödinger equation** leads to the Dirac equation:

$$i\hbar\frac{\partial \Psi(r, t)}{\partial t} = (c\boldsymbol{\alpha}\cdot\mathbf{p} + \beta mc^2)\Psi(r, t).$$

Dirac matrix

The wave function $\Psi(r, t)$ is now a four-component entity called a **spinor**. The time-independent Dirac equation for a single particle in the external potential $U(r)$ has the form

$$(c\boldsymbol{\alpha}\cdot\mathbf{p} + U(r) + \beta mc^2)\Psi(r) = (E + mc^2)\Psi(r),$$

where the binding energy E has the same meaning as that in nonrelativistic theory.

In fact, the Dirac equation is a Lorentz invariant relativistic generalization of the Schrödinger (nonrelativistic) equation.

First described in: P. A. M. Dirac, *The quantum theory of the electron* Proc. R. Soc. A **115**, 610–624 (1928).

Recognition: in 1933 P. A. M. Dirac and E. Schrödinger received the Nobel Prize in Physics for the discovery of new productive forms of atomic theory.

See also www.nobel.se/physics/laureates/1933/index.html.

Dirac matrix – any one of four 4×4 matrices designated $\gamma_k (k = 0, 1, 2, 3)$. They satisfy $\gamma_k\gamma_l + \gamma_l\gamma_k = 2\delta_{kl}\mathbf{E}, \gamma_k\gamma_0 + \gamma_0\gamma_k = 0, \gamma_k\gamma_l = \gamma_0^2 = \mathbf{E}$ for $k, l = 1, 2, 3$. Here δ_{kl} is the **Krönecker delta function**, and $\mathbf{E}$ is the single matrix. The matrices are used to operate on the four-component wave function in the **Dirac equation**.

First described in: P. A. M. Dirac, *The quantum theory of the electron*, R. Proc. Soc. A **115**, 610–624 (1928).

Dirac monopole – a quantum of magnetic charge equal to h/e. It is also known as the magnetic monopole.

First described in: P. A. M. Dirac, *The Principles of Quantum Mechanics* (Oxford University Press 1930).

Dirac notation (for quantum states) – any quantum state represented by a vector Ψ in a complex **Hilbert space** may be written as "**ket**"

$$|\Psi_a\rangle \rightarrow \begin{pmatrix} a_0 \\ a_1 \\ a_2 \end{pmatrix}$$

or "**bra**"

$$\langle\Psi_a| \rightarrow (a_0^* a_1^* a_2^*).$$

The state vector should be normalized, so $\sum_i |a_i|^2 = 1$.

Bras and kets are related by Hermitian conjugation: $|\Psi_a\rangle = (\langle\Psi_a|)^\dagger, \langle\Psi_a| = (|\Psi_a\rangle)^\dagger$. The inner product of a bra and a ket $\langle\Psi_a\|\Psi_b\rangle$ denotes $\langle\Psi_a|\Psi_b\rangle$. It is a complex number: $\langle\Psi_a|\Psi_b\rangle = a_0^* b_0 + a_1^* b_1 + a_2^* b_2$. For normalized states $\langle\Psi|\Psi\rangle = 1$. The outer product of a bra and a ket is a linear operator (matrix):

$$|\Psi_a\rangle\langle\Psi_b| = \begin{pmatrix} a_0 b_0^* & a_0 b_1^* & a_0 b_2^* \\ a_1 b_0^* & a_1 b_1^* & a_1 b_2^* \\ a_2 b_0^* & a_2 b_1^* & a_2 b_2^* \end{pmatrix}$$

It is often convenient to work with basis kets and bras representing a state:

$$|\Psi_a\rangle \rightarrow a_0|0\rangle + a_1|1\rangle + a_2|2\rangle$$

with

$$|0\rangle = \begin{pmatrix} 1 \\ 0 \\ 0 \end{pmatrix}, \quad |1\rangle = \begin{pmatrix} 0 \\ 1 \\ 0 \end{pmatrix}, \quad |2\rangle = \begin{pmatrix} 0 \\ 0 \\ 1 \end{pmatrix};$$

$$\langle\Psi_a| \rightarrow a_0^*\langle 0| + a_1^*\langle 1| + a_2^*\langle 2|$$

First described in: P. A. M. Dirac, *The Principles of Quantum Mechanics* (Oxford University Press, 1930).

Dirac wave function – a two-component function of the form

$$\psi = \begin{pmatrix} \phi_1 \\ \phi_2 \end{pmatrix}.$$

It is appropriate for describing spin 1/2 particles and antiparticles.

direct current Josephson effect → Josephson effects.

directed molecular evolution – a strategy within **protein engineering** where mutagenesis is applied to protein generating a greater diversity of mutation. If further diversity is desired a random factor is introduced to make more mutants and a selection regime is used to choose variants with the searched qualities.
More information in: J. A. Brannigan, A. J. Wilkinson, *Protein engineering 20 years on,* Nat. Rev. Molec. Cell Biol. **3**, 964–970 (2002).

dislocation – a region of distorted atom configuration formed between the displaced and normal areas in a crystal when part of the crystal is displaced tangentially.

dispersion curve – an electron energy versus momentum curve usually presented as $E(k)$, where k is the wave vector.

disruptive technologies are those that displace older technologies and enable radically new generations of existing products and processes to take over. The term has been coined by Clayton Christiansen in 1995. Nanotechnology is a disruptive technology.

dissociation – a process causing ingredients of a substance to split into simple groups, which in the case of compounds can be single atoms, groups of atoms or ions.

distributed Bragg reflector – a solid-state structure formed from multiple layers of alternating materials with varying **refractive index** or by periodic variation of some characteristic (such as height) of a dielectric waveguide, resulting in periodic variation in the effective refractive index in the guide. Each layer boundary causes a partial reflection of an optical wave. For waves whose wavelength is close to four times the optical thickness of the layers, the many reflections combine with constructive **interference**. The layers act as a high-quality reflector. The range of wavelengths that are reflected is called the **photonic stopband**. Within this range of wavelengths, light is "forbidden" to propagate in the structure.

Reflectivity of the reflector composed of N repeated pairs of materials with different (low/high) refractive indices n_1 and n_2 can be calculated as

$$R = \left[\frac{n_a(n_2)^{2N} - n_s(n_1)^{2N}}{n_a(n_2)^{2N} + n_s(n_1)^{2N}} \right]^2,$$

where n_a and n_s are the refractive indices of surrounding ambient and the substrate, respectively. The bandwidth of the photonic stopband is $\Delta v = (4v/\pi)\arcsin[(n_2 - n_1)/(n_2 + n_1)]$, where v_0 is the central frequency of the band.

It is evident that an increase in the number of pairs in the reflector increases its mirror reflectivity. An increase in the refractive index contrast between the materials in the pairs brings increases in both the reflectivity and the bandwidth.

distributed-feedback laser – a **laser** with a resonator consisting of a periodic structure, which acts as a **distributed Bragg reflector** in the wavelength range of laser action and containing an optical gain medium. Typically, the periodic structure supposes a phase shift in its middle. This structure is thus basically the direct concatenation of two Bragg gratings with an internal optical gain. It has multiple axial optical modes, but there is typically one mode which is favored in terms of losses. Therefore, single-frequency operation is often easily achieved despite spatial **hole-burning** due to the standing-wave pattern in the gain medium. Most distributed-feedback lasers are either fiber lasers or semiconductor lasers.

DNA – an acronym for a **d**eoxyribo**n**ucleic **a**cid. A molecule of DNA contains six commonly occurring molecular components. They are: two purines–adenine (A) and guanine (G), two pyrimidines–cytosine (C) and thymine (T), sugar–deoxyribose, and phosphoric acid. Their formulae are presented in Figure D.5. The above purines and pyrimidines are called bases. Figure D.6 shows the chemical composition of a fragment of a DNA chain. Its backbone is built from repeating deoxyribose and phosphate groups. Each deoxyribose group is attached to one of four bases (adenine, guanine, cytosine or thymine). The sequence of bases constitutes the genetic code. In space, such chains form a double helix structure, similar to a twisted ladder, schematically shown in Figure D.7. It consists of two strands of DNA twisting around each other. Each strand is composed of a long chain of monomer nucleotides. The nucleotide of DNA consists of a deoxyribose

Purines:

Adenine

Guanine

Pyrimidines:

Cytosine

Thymine

Deoxyribose

Phosphoric acid

Figure D.5 Molecular components of DNA.

sugar molecule to which a phosphate group and one of the four bases are attached. The nucleotides are joined together by covalent bonds between the phosphate of one nucleotide and the sugar of the next, forming a phosphate sugar backbone from which the nitrogenous bases protrude. Hydrogen bonds between the bases couple the two strands together. The chemical bonding is such that adenine base in one chain forms hydrogen bonds with a thymine base in the other chain of the double helix, while guanine and cytosine bases are similarly hydrogen bonded across the axis of the double helix. The base pairs look like the rungs of a helical ladder. This is the **Watson–Crick double-helix model of DNA**. This A–T and G–C bonding, to give what are called **Watson–Crick base pairs**, satisfies **Chargaff's rules** that the molar ratios of A to T and of G to C are very close to unity in all DNA samples, whatever the sequence of bases.

The naturally occurring double helix structure has two preferential forms, A and B. The B form is the main one. It is shown in Figure D.7. Both A and B are right handed double helixes. The double helix is about 2 nm wide, and its repeat

DNA

Figure D.6 Chemical composition of a part of a DNA chain.

consists of about 10.5 nucleotide pairs, each separated by about 0.34 nm. In the A structure, the bases are tilted and the hydrogen bonds are less nearly perpendicular to the axis of the helix. Which form occurs depends not on chemical composition but on the degree of hydration. An artificial left handed form, called Z form, can also exist. It got its name from the zigzag form of the phosphate sugar backbone. It has a repetition period of 4.46 nm containing 12 pairs of bases per one helix turn.

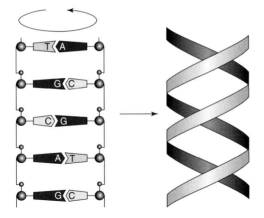

A - adenine   T - thymine   C - cytosine   G - guanine

Figure D.7 Double helix structure of DNA (Figure 1a,b from C. Dekker, M. A. Ratner, *Electronic properties of DNA*, Physics World **14**(8), 29–33 (2001), by permission).

Two pairs of bases are repeated in the Z-DNA in contrast to one pair repetition in A- and B-DNA.

In a living cell, DNA is combined with protein to form chromosomes, which are replicated when a cell divides. DNA controls the functioning of the cell by determining which protein molecules are synthesized there, and the amino acid sequence in the proteins is in turn determined by the sequence of adenine, guanine, cytosine and thymine in the DNA chain. Thus, the sequence of bases in DNA constitutes a code for the amino acid sequence in a protein.

DNA is very important not only as a genetic biomolecule, but also as a molecular template for nanotechnology. Besides, it is a good medium for one-dimensional charge carrier transport predominantly via coherent tunneling and diffusive thermal hopping. That makes DNA molecules attractive for applications in nanoelectronics, molecular electronics and nanomechanics.

First described in: W. T. Astbury, F. O. Bell, *Some recent developments in the X-ray study of proteins and related structures*, Cold Spring Harbor Symposia in Quantum Biology **6**, 109–121 (1938) – DNA composition; J. D. Watson, F. H. C. Crick, *Molecular structure of nucleic acids*, Nature **171**, 737–738 (1953) – DNA double helix structure.

Recognition: in 1962 F. H. C. Crick, J. D. Watson and M. H. F. Wilkins (team leader) received the Nobel Prize in Physiology or Medicine for their discoveries concerning the molecular structure of nucleic acids and its significance for information transfer in living material.

DNA-based molecular electronics – **DNA** molecules seem particularly appealing for **molecular electronics** by virtue of their double-strand **recognition** and their special structuring that suggests their use in **self-assembly**. The idea that double-stranded DNA may function as a conduit for fast electron transport, along the axis of its

DNA probe

base-pair stack, was first advanced in 1962. More and more evidence accumulating from the direct electrical transport measurements shows that it is possible to transport a charge carrier along short single DNA molecules, in bundles, and in a network, although the conductivity is rather poor.

First described in: D. D. Eley, D. I. Spivey, *Semiconductivity of organic substances: nucleic acid in dry state*, Trans. Faraday Soc. **12**, 245 (1962).

More details in: D. Porath, G. Cuniberti, R. Di Felice, *Charge transport in DNA-based devices* in *Long Range Charge Transport in DNA* ed. Gary Schuster, Topics in Current Chemistry (Springer Verlag, Berlin, 2004) pp. 183–227.

DNA probe – a single-stranded piece of **DNA** that binds specifically to a complementary DNA sequence. The probe is labeled, for example with a fluorescent or radioactive tag, in order to detect its incorporation through hybridization with DNA in a sample.

domain wall → **Bloch wall.**

dome – an open crystal form consisting of two faces astride a symmetry plane.

donor (atom) – an impurity atom, typically in semiconductors, which donates electron(s). Donor atoms usually form electron energy levels slightly below the bottom of the conduction band. An electron is readily excited into this band and may then contribute to the electrical conduction of the material.

Doppler effect – a difference between the frequency with which waves are produced by a source and the frequency with which they may be registered by an observer. This difference occurs, in general, when the source, the observer, or both are moving relative to the medium in which the waves are propagated.

First described in: C. Doppler, *Über das farbige Licht der Doppelsterne und einige andere Gestirne der Himmel*, Abhand. Köngl. Böhm. Ges. **2**(5), 465 (1842).

Doppler–Fizeau principle states that the displacement of spectral lines is determined by the distance between, and relative velocity of, the observer and the light source. When the distance decreases, the lines of the spectrum move toward the ultraviolet. When it increases, the lines move toward the red part of the spectrum.

First described by H. Fizeau in 1848.

d orbital → **atomic orbital.**

dot blot – a method for detecting **proteins** by specific binding of an **antibody** or binding molecule to a sample spot on nitrocellulose paper. The bound sample is visualized using an enzymatic or fluorimetric reporter conjugated to the probe.

double heterostructure laser – a semiconductor laser in which active layer of one semiconductor material is sandwiched between two *n*- and *p*-doped layers of another semiconductor that have wider band gaps. A heterostructure combining wider gap AlGaAs and smaller gap GaAs suits this demand well. Energy band diagram of such structure is shown in Figure D.8.

The *n*- and *p*-doped layers, called **cladding layers**, are used to inject carriers into the active region. Electrons and holes injected through the heterojunctions at forward bias appear to be confined to the active region by the potential barriers of the heterojunctions. Their radiative recombination there yields photons with the energy determined by the energy gap of the active layer semiconductor. The double heterostructure forms an efficient waveguide as well because the refractive index of the active layer is higher than that of the cladding layers. Thus, interaction between optical field and injected carriers is facilitated, which is needed for laser action. Polished mirror facets perpendicular to the heterojunction planes are used to form an optical resonator.

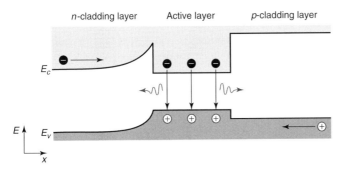

Figure D.8 Energy bands in double heterojunction laser at forward bias.

Most promoted double heterostructure lasers adopt a stripe geometry in which current is injected only within a narrow region beneath a stripe contact in order to keep the threshold current low and to control the optical field distribution in the lateral directions. Their representative examples are schematically shown in Figure D.9. As compared with continuous layer lasers, the threshold current is reduced below 1000 A cm^{-2} roughly in proportion to the area of the contact. Lasing occurs in a limited region of the active layer where a high-density current flows.

Figure D.9 Heterostructure laser in a stripe geometry.

Dresselhaus coupling

In the simplest stripe contact geometry the optical intensity distribution in the lateral direction is determined by the gain profile related to the carrier density distribution. Lasers with such geometry are termed **gain-guided lasers**.

In the stripe-like buried geometry, refractive index variation in the lateral direction takes place. Such lasers are termed **index-guided lasers**. They have the active region surrounded by materials with lower refractive index in both vertical and lateral transverse directions. Thus, waveguiding can occur in both directions.

First described in: Zh. I. Alferov, R. F. Kazarinov, *Semiconductor laser with electric pumping*, Inventor's Certificate no.181737 (1963) – in Russian; H. Kroemer, *A proposed class of heterojunction injection lasers*, Proc. IEEE 51(12), 1782–1783 (1963).

Recognition: in 2000 Zh. I. Alferov and H. Kroemer received the Nobel Prize in Physics for developing semiconductor heterostructures used in high-speed and optoelectronics.

See also www.nobel.se/physics/laureates/2000/index.html.

Dresselhaus coupling – a **spin-orbit coupling** in solids resulting from breaking of the bulk inversion symmetry as takes place in **zinc blende structures**. The related spin-orbit contribution in the **Hamiltonian** is represented by the term $H_D = -\beta(k_x\sigma_x - k_y\sigma_y) + O(k^3)$, where β is the material-dependent parameter specifying the strength of the spin orbit interaction, k is the wave vector component, σ is the **Pauli spin matrix**, and $O(k^3)$ is the qubic component.

First described in: G. Dresselhaus, *Spin–orbit coupling effects in zinc blende structures*, Phys. Rev. **100**(2), 580–586 (1955).

Drude equation explains optical activity of matter containing **chiral** molecules through rotating the plane of polarization of polarized light. A molecule is chiral if it cannot be superimposed on its mirror image. Electrons in chiral molecules are considered to oscillate along a helix, giving the relationship between molecular polarizability α, wavelength of the probing light λ and the relative wavelengths λ_i in the form $\alpha = \sum_i(K_i/(\lambda^2 - \lambda_i^2))$, where K_i are constants proportional to the rotational strength of the transitions resulting in optical activity. Thus the rotation of the plane of polarization of plane-polarized light passing through the optically active media is inversely proportional to the difference between the square of the wavelength of the probing light and the square of the wavelengths relative to the transitions.

First described in: P. Drude, *Lehrbuch der Optik* (Hirzel, Leipzig, 1900).

More details in: S. F. Mason, *Molecular Optical Activity and the Chiral Discriminations* (Cambridge University Press, Cambridge, 1982).

Drude formula gives the frequency dependence of the conductivity of a solid in the form $\sigma(\omega) = \sigma_0(1 + i\omega t)/(1 + \omega^2\tau^2)$, where constant current conductivity $\sigma_0 = e^2 n\tau/m$, n is the density of free electrons, τ is the mean free time between electron scattering events, and m is the electron mass. It was obtained within a quasiclassical approach → **Drude theory** under the assumption that free electrons moving in the solid with friction define the conductivity. It is valid only when the

Fermi electron wavelength $\lambda_F = 2\pi/k_F$ is much smaller than the mean free path $l_F = v_F\tau$, that is $k_F l \gg 1$, $E_F\tau \gg 1$. Accounting for $\lambda_F \sim 0.5$ nm, it means that the quasiclassical theory becomes problematic for $\rho > 10^{-3}$ Ω cm.

First described in: P. Drude, *Zur Elektronentheorie. I*, Ann. Phys. **1**, 566–613 (1900); P. Drude, *Zur Elektronentheorie. II*, Ann. Phys. **3**, 369–402 (1900).

Drude theory was the first attempt to explain electronic properties of metals (later it was generalized for all solids) considering a gas of free electrons moving within a network of positively charged ionic cores. The assumptions applied are: (i) electrons collide and subsequently are scattered only by the ionic core; (ii) electrons do not interact with each other or with ions between the collisions; (iii) collisions are instantaneous and result in a change in the electron velocity; (iv) an electron suffers a collision with the probability per unit time proportional to τ^{-1}, where τ is the relaxation time, that is τ^{-1} is the scattering rate; (v) electrons reach thermal equilibrium with their surroundings only through collisions.

Within the framework of the theory, one may trivially express mobility of charge carriers μ and the conductivity σ of a homogeneous conductor through which the carriers are moving in terms of the electric charge e and the effective mass m of the mobile charge carriers as well as the carrier concentration n: $\mu = e\tau/m$, $\sigma = en\mu = e^2n\tau/m$.

First described in: P. Drude, *Zur Elektronentheorie. I*, Ann. Phys. **1**, 566–613 (1900); *Zur Elektronentheorie. II*, Ann. Phys. **3**, 369–402 (1900).

DSC – an acronym for a **differential scanning calorimetry**.

DSG – an acronym for **dynamic shadowing growth**.

dual beam FIB–SEM system – a dual beam instrument composed of a **focused ion beam (FIB)** system and a **scanning electron microscope (SEM)** as schematically illustrated in Figure D.10. It is a shared experimental facility for sample preparation and fabrication at the nanoscale by ion-mill sculpting or beam-induced deposition

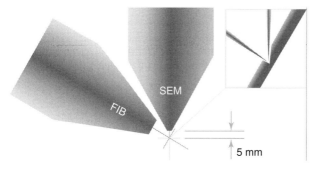

Figure D.10 A dual beam geometry for a combined FIB–SEM sample treatment (courtesy of P. Facci).

and offers high-resolution cross-section imaging. Thus the Dual Beam system combines in a single apparatus an ion column and an electron column working at coincidence. It enables high-resolution sample machining with the FIB and simultaneous high-resolution imaging with the non-destructive SEM probe.

Duane–Hunt law states that the frequency of X-rays resulting from electrons striking a target cannot exceed eV/h, where V is the voltage accelerating the electrons.

First described in: W. Duane, F. L. Hunt, *On X-ray wave-lengths*, Phys. Rev. **6**(2), 166 (1915).

D'yakonov–Perel mechanism – the electron spin relaxation mechanism in semiconductors in which spin splitting of the conduction band via spin–orbit coupling due to the lack of inversion symmetry (as III-V compounds) is the driving force. It is, for example, responsible for the fast spin dephasing in bulk GaAs at elevated temperatures. In bulk materials, the mechanism is usually less pronounced in semiconductors with a larger direct band gap because of the smaller influence of the **valence band** on the **conduction band**.

The related **spin relaxation time** τ can be been expressed in the semiphenomenological form as $1/\tau = A(\alpha^2/\hbar^2 E_g)\tau_p T^3$ for bulk semiconductors and $1/\tau = (\alpha^2 < p_z^2 >^2 /2\hbar^2 m^2 E_g)\tau_p T$ for **quantum well** structures. Here α describes conduction band spin splitting due to lack of inversion symmetry (for example $\alpha = 0.07$ for GaAs), E_g is the band gap, T is the absolute temperature, A is a numerical coefficient depending on the orbital scattering mechanism, $< p_z^2 >$ is the average square of the momentum in the quantum well growth direction, and τ_p is the average momentum relaxation time, being a phenomenological parameter.

Such spin relaxation increases both in (100) oriented quantum wells and in quantum wires because of the further decrease in the symmetry and increase in the momentum related to quantum confinement there. One requirement for room temperature spin transport in III-V semiconductors is therefore the suppression of the D'yakonov–Perel mechanism. This is possible since the electron interaction resulting in the spin relaxation depends on the semiconductor material and the direction of electron momentum and spin in the host crystal. This mechanism has been demonstrated to be suppressed by choosing a quantum well with special crystal axes. For instance, in an undoped (110) oriented GaAs quantum well a decrease of nearly one order of magnitude in the spin relaxation time with increasing temperature, reaching spin relaxation times as long as 2 ns at room temperature is observed. This is much longer than in (100) oriented quantum wells. An even stronger increase of nearly two orders of magnitude of the spin relaxation time with temperature is observed in ZnSe and ZnCdSe quantum wells. The slower spin dephasing at higher temperatures results from the reduced electron-hole exchange interaction, which is a very efficient source of spin relaxation due to thermal ionization of **excitons**.

First described in: M. I. D'yakonov, V. I. Perel, *On spin orientation of electrons in interband absorption of light in semiconductors*, Zh. Eksp. Teor. Fiz. **60**(5), 1954–1965 (1971) – in Russian.

More details in: I. Žutić, J. Fabian, S. Das Sarma, *Spintronics: Fundamentals and applications*, Rev. Mod. Phys. **76**(2), 323–410 (2004).

dye – an organic molecule whose absorption bands are in the visible spectral region.

dye-sensitized solar cell – a device, also called a **Grätzel solar cell**, in which in contrast to the conventional silicon systems, where the semiconductor assumes both the task of light absorption and that of charge carrier transport, the two functions are separated. Light is absorbed by a sensitizer, which is anchored to the surface of a wide band gap **oxide** semiconductor. Charge separation takes place at the interface via photo-induced electron injection from the **dye** into the **conduction band** of the solid. Carriers are transported in the conduction band of the semiconductor to the charge collector. The use of sensitizers having a broad **absorption** band in conjunction with oxide films of nanocrystalline morphology permits one to harvest a large fraction of sunlight. Efficiencies of ~11% have been reached. On similar principles are also based the so-called **McFarland and Tang solar cells**.

First described in: B. O'Regan, M. Grätzel, *A low-cost, high-efficiency solar cell based on dye-sensitized colloidal TiO$_2$ films*, Nature **353**, 737–740 (1991); M. Grätzel, *Photoelectrochemical cells*, Nature **414**, 338–344 (2001).

More details in: M. Grätzel, *Conversion of sunlight to electric power by nanocrystalline dye-sensitized solar cells*, J. Photochem. Photobiol. A: Chem. **164**, 3–14 (2004); G. E. Tulloch, *Light and energy – Dye solar cells for the 21st century*, J. Photochem. Photobiol. A: Chem. **164**, 209–219 (2004).

dynamic force microscopy – an **atomic force microscopy** in a dynamic mode, that is when the tip is oscillating perpendicular to the sample surface. Two modes of operation are usually employed: the amplitude modulation mode (also called tapping mode) and the self-excitation mode. The first separates the purely attractive force interaction regime from the attractive-repulsive regime. It is primarily used for imaging in air and liquid. The second dominates for imaging in ultrahigh vacuum to get real atomic precision.

More details in: A. Schirmeisen, B. Anczykowski, H. Fuchs, *Dynamic force microscopy*, in: Handbook of Nanotechnology, edited by B. Bhushan (Springer, Berlin, 2004), pp. 449–473.

dynamic redox system – a compound that can be reversibly converted into the corresponding charged species upon electron transfer but accompanied by drastic structural changes and/or covalent bond making/breaking. Although transformation between the neutral and charged species proceeds quantitatively in most cases, their electrochemical processes are irreversible owing to the chemical reactions followed by the electron transfer. This situation endows the dynamic **redox** pairs with very high bistability, which is a prerequisite in order to construct the molecular response systems.

dynamic shadowing growth (DSG)

More details in: T. Suzuki, H. Higuchi, T. Tsuji. J. I. Nishida, Y. Yamashita, T. Miyashi, *Dynamic redox systems: Towards realization of the unimolecular memory*, in: *Chemistry of Nanomolecular Systems. Towards the realization of Nanomolecular Devices*, edited by T. Nakamura, T. Matsumoto, H. Tada, K. I. Sugiura (Springer, Berlin Heidelberg, 2003).

dynamic shadowing growth (DSG) – a technique that involves physical vapor deposition in which the angle of the vapor-beam direction can be controlled so that different shapes of material can be deposited on a substrate. It is used to fabricate catalytic nanomotors that mimic the way biological motors work.

First described in: Y. He, J. Wu, Y. Zhao, *Designing catalytic nanomotors by dynamic shadowing growth*, Nano Lett. **7**(5), 1369–1375 (2007).

dynamics – a study of the behavior of objects in motion. It can be distinguished in this way from statics, which deals with objects at rest or in a state of uniform motion, and also from **kinematics**, which studies the geometry of motion only without any reference to the forces causing it.

Dzyaloshinskii–Moriya interaction – a combination of superexchange and **spin–orbital interactions** in atomic systems. It originates from symmetry effects. The related antisymmetric spin coupling from purely symmetry grounds and the symmetric pseudodipolar magnetic interaction are estimated to be proportional to $(\Delta g/g)J$ and $(\Delta g/g)^2 J$, respectively, where g is the gyromagnetic ratio and Δg is its deviation from the value for a free electron, J is the isotropic superexchange energy. For a particular spin system the allowed components of the interaction are determined by the symmetry of the spin complex. Corrections to the energy spectrum due to the interaction are usually small being proportional to D^2/Δ, where D is the magnitude of the interaction and Δ is the separation of the **singlet** and **triplet** energy levels.

This interaction is considered to be a consequence of the spin–orbital coupling in the presence of relativistic effects under peculiar structure symmetry in some crystals. Its effects can be explicitly exhibited only when an external magnetic field is applied.

First described in: I. E. Dzyaloshinskii, *A thermodynamic theory of "weak" ferromagnetism of antiferromagnetics*, J. Phys. Chem. Solids **4**(4), 241–255 (1958); T. Moriya, *Anisotropic superexchange interaction and weak ferromagnetism*, Phys. Rev. **120**(1), 91–98 (1960).

E

From (e,2e) reaction to Eyring equation

(e,2e) reaction → (e,2e) spectroscopy.

(e,2e) spectroscopy – very detailed analysis of electronic structures of atoms, molecules and solids, including ionization energy, electron momentum distribution and wavefunction mapping, can be obtained by means of the **(e,2e) reaction**. This is a scattering process in which a high-energy electron collides inelastically with an atomic, molecular or solid target. As a consequence, a bound target electron is ejected, and the outgoing electrons (one scattered and one knocked out) are detected. This process is very rich in information, since one can observe kinetic energies and momenta of three electrons (one primary and two secondary particles), and thus deduce some fundamental properties of the electronic structure of the target. The energies and momenta of the three electrons and of the target are interrelated by conservations laws.

More details in: M. De Crescenzi, M. N. Piancastelli, *Electron Scattering and Related Spectroscopies* (World Scientific Publishing, Singapore, 1996).

E91 protocol – the protocol for quantum cryptographic key distribution developed by A. K. **E**kert in **1991** (the acronym uses the bold characters). Within this scheme the generalized **Bell's inequality** safeguards confidentiality in the transmission of pairs of spin-1/2 particles entangled like **Einstein–Podolsky–Rosen pairs**.

The idea consists of replacing the quantum channel carrying two **qubits** from Alice (conventional name of a sender) to Bob (conventional name of a receiver) by a channel carrying two qubits from a common source, one qubit to Alice and one to Bob. The source generates a random sequence of correlated particle pairs with one member of each pair (one qubit) being sent to each party. Spin-(1/2) particles or pairs of so-called Einstein–Podolsky–Rosen photons whose polarizations are able to be detected by the parties can be used for that.

Alice and Bob both measure their particle in two bases (two different orientations of the analyzer detecting spin orientation or photon polarization), again chosen independently and randomly. The source then announces the bases via an open channel. Alice and Bob divide results of their measurements into two separate groups: a first group for which they used different bases and a second group for

What is What in the Nanoworld. Victor E. Borisenko and Stefano Ossicini
Copyright © 2008 WILEY-VCH Verlag GmbH & Co. KGaA, Weinheim
ISBN: 978-3-527-40783-5

which they used the same bases. Subsequently, they reveal publicly the results they obtained within the first group only. Comparing the results allows them to know whether there is an eavesdropper on their communication channel. The eavesdropper would have to detect a particle to read the signal, and retransmit it in order for his presence to remain unknown. However, the act of detection of one particle of a pair destroys its quantum correlation with the other, and the two parties can easily verify whether this has been done, without revealing the results of their own measurements, by communication over an open channel. After Alice and Bob become sure that the particles (qubits) they received are not disturbed by an eavesdropper, they can be converted into a secret string of bits representing the key. This secret key may be then used in a conventional cryptographic communication between them.

First described in: A. K. Ekert, *Quantum cryptography based on Bell's theorem*, Phys. Rev. Lett. **67**(6), 661–663 (1991).

ebit – an acronym for an **e**ntangled **bit**. It is defined as an amount of **quantum entanglement** in a two-**qubit** state maximally entangled, or any other pure bipartite state for which entropy of entanglement is 1. One ebit represents a computing resource made up of a shared **Einstein–Podolsky–Rosen pair**.

First described in: C. H. Bennett, D. P. Di Vincenzo, J. Smolin, W. K. Wootters, *Mixed-state entanglement and quantum error correction*, Phys. Rev. A **54**(5), 3824–3851 (1996).

edge states appear at a hard edge of a wire when an external magnetic field is applied perpendicular to the current flow in it. The classical behavior of electrons near the wire edge is shown in Figure E.1.

Electrons deep inside the wire execute circular **cyclotron orbits** with no net drift. Near the edge, the orbits are interrupted when the electrons hits the boundary. The result is a skipping orbit that bounces along the boundary. The electron thus acquires a net drift velocity, which is higher for orbits closer to the edge. The states on opposite edges travel in opposite directions. The states squeeze against the side of the wire with higher energy than those in the middle.

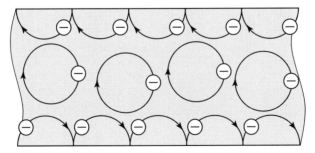

Figure E.1 Orbits of electrons in a wire placed in a magnetic field perpendicular to the electron flow in the wire.

eddy currents = Foucault currents – currents induced in a conducting medium subjected to a varying magnetic field. Such currents result in energy dissipation, termed eddy current loss, and a reduction of the apparent magnetic permeability of the conducting medium. In accordance with **Lenz's law**, the eddy currents circulate in such direction as to tend to prevent variation of the magnetic field.

First described by J. B. L. Foucault in 1855.

EDXS – an acronym for **energy dispersive X-ray spectroscopy.**

EELS – an acronym for **electron energy loss spectroscopy.**

effective mass (of charge carriers) is a characteristic of an extremum point at an electron energy band represented by the **dispersion curve** $F(k)$. A Taylor expansion of $E(k)$ around the extremum at $k = k_0$ and ignoring the high-order terms gives:

$$E(k) = E_0 + \frac{\hbar^2 (k - k_0)^2}{2m^*},$$

with

$$m^* = \hbar^2 \left(\frac{d^2 E}{dk^2} |_{k=k_0} \right)^{-1}$$

referred to as the effective mass of charge carriers occupying the band.

The effective mass is evidently inversely proportional to the curvature of $E(k)$ at $k = k_0$. Unlike the gravitational mass, the effective mass can be positive, negative, zero, or even infinite. An infinite effective mass corresponds to an electron localized at a lattice site. In this case, the electron bound to the nucleus assumes the total mass of the crystal. A negative m^* corresponds to a concave downward band and means that the electron moves in the opposite direction to the applied force. If the band structure of a solid is anisotropic, the effective mass becomes a tensor, characterizing by its m_x^*, m_y^*, m_z^* components. The average effective mass in this case can be calculated as $m^* = (m_x^* m_y^* m_z^*)^{1/3}$.

Four types of critical points, namely M_0, M_1, M_2, and M_3, are distinguished as a function of the sign of the effective mass components in a particular point of the **Brillouin zone** with a zero energy-gradient on k. In the M_0 critical point all three components are positive. In the M_1 and M_2 one and two components, respectively, are negative. In the M_3 point all three components are negative.

effective mass approximation supposes that the structure of energy bands $E(k)$ near extremum points, which are minima in conduction bands and maxima in valence bands, has a parabolic character well described within an assumption that they are formed by electrons or holes with an effective mass m^*, i.e. $E(k) = \hbar^2 k^2 / 2m^*$.

effective Rabi frequency → **Rabi frequency.**

effusion – a molecular flow of gases through an aperture when the molecules do not collide with each other as they escape through the aperture.

effusion (Knudsen) cell – the unit used in **molecular beam epitaxy** machines to create an atomic or molecular beam. It is schematically shown in Figure E.2.

An appropriate solid source material is loaded inside a cylindrical cell with a very small orifice and heated until it vaporizes. As the vapor escapes from the cell through the small nozzle, its atoms or molecules form a well-collimated beam, since the ultrahigh vacuum environment outside the cell allows them to travel without scattering.

Figure E.2 Effusion (Knudsen) cell.

Efimov quantum states – a universal set of bound **trimer** states appearing for three identical **bosons** with a resonant two-body interaction. These states should even exist in the absence of a corresponding two-body bound state.

First described in: V. Efimov, *Energy levels arising from resonant two-body forces in a three-body system*, Phys. Lett. B **33**(8), 563–564 (1970); V. Efimov, *Weakly bound states of three resonantly interacting particles*, Sov. J. Nucl. Phys. **12**, 589–595 (1971) – theoretical prediction.

EFM – an acronym for **electrostatic force microscopy**.

Ehrenfest adiabatic law states that if the **Hamiltonian** of a system undergoes an infinitely slow change and if the system is initially in an eigenstate of the Hamiltonian, then at the end of the change it will be in the eigenstate of the new Hamiltonian that derives from the original state by continuity, provided certain conditions are met.

First described in: P. Ehrenfest, *Adiabatische Invarianten und Quantentheorie*, Ann. Phys. **51**, 327–352 (1916).

Ehrenfest theorem states that the motion of a quantum mechanical **wave packet** will be identical to that of the classical particle it represents, provided that any potentials acting upon it do not change appreciably over the dimensions of the packet. In this interpretation, the position and momentum of the particle are replaced by their quantum mechanical expectation values.

First described in: P. Ehrenfest, *Bemerkung über die angenäherte Gültigkeit der klassischen Mechanik innerhalb der Quantenmechanik,* Z. Phys. **45**(7/8), 455–457 (1927).

Ehrlich–Schwoebel effect – an asymmetry in the adatom attachment and detachment to and from atomic steps on crystal surfaces. The steps capture atoms diffusing on the surface with certain probabilities and, in addition, the capture probability depends on the direction from which adsorbed atoms approach the step.

First described in: G. Ehrlich, F. G. Hudda, *Atomic view of surface self-diffusion: tungsten on tungsten,* J. Chem. Phys. **44**(3), 1039–1049 (1966); R. L. Schwoebel, E. J. Shipsey, *Step motion on crystal surfaces,* J. Appl. Phys. **37**(10), 3682–3686 (1966).

eigenfunction – was initially introduced in relation to the mathematics of function spaces or **Hilbert spaces**. The concept is a natural generalization of the notion of an eigenvector of a linear operator in a finite dimensionally linear vector space.

The eigenfunction, or characteristic function, is related to a linear operator defined on a given vector space of functions on some specified domain. If a function f, being not the null function of the functional space, is mapped by the operator **O** into a multiple of itself, that is $\mathbf{O}f = af$, where a is a scalar, f is said to be an eigenfunction of **O** belonging to the **eigenvalue** a.

Eigenfunctions find wide physical applications in classical field theories but they appear to be best known for their extensive use in Schrödinger's formulation of quantum mechanics.

eigenstate – a state of a quantum system for which the observables in some commuting, and hence simultaneously measurable, set have definite values, in the orthodox interpretation of the theory. It is a **wave function** in quantum mechanics.

eigenvalue = energy (in quantum mechanics).

Einstein–Bohr equation defines the frequency of the radiation emitted or absorbed by a system when it undergoes a transition from one energy state to another as $\nu = \Delta E/h$, where ΔE is the energy difference between the states.

First described in: A. Einstein, *Über einen die Erzeugung und Verwandlung des Lichtes betreffenden heuristischen Gesichtspunkt,* Ann. Phys. **17**, 132–148 (1905); N. Bohr, *On the constitution of atoms and molecules,* Phil. Mag. **26**, 1–25 (1913).

Einstein–de Haas effect – axial rotation of a rod of a magnetic material, which occurs when the rod experiences a magnetic field applied parallel to its axis. Sometimes it also called the **Richardson effect** after Richardson who first predicted such behavior of magnetic materials.

First described in: A. Einstein, W. J. de Haas, *Experimenteller Nachweis der Ampereschen Molekularströme,* Dtsch. Phys. Ges., Verh. **17**(8), 152–170 (1915).

Einstein frequency

Einstein frequency – a single frequency with which each atom vibrates independently of the other atoms in the model of lattice vibrations. It is equal to the frequency observed in infrared absorption studies.

Einstein photochemical equivalence law → **Stark–Einstein law.**

Einstein photoelectric law states that the energy of an electron emitted from a system in the **photoelectric effect** is $(h\nu - W)$, where ν is the frequency of the incident radiation, W is the energy needed to remove the electron from the system. If $h\nu < W$, no electrons are emitted.

First described in: A. Einstein, *Über einen die Erzeugung und Verwandlung des Lichtes betreffenden heuristischen Gesichtspunkt*, Ann. Phys. **17**, 132–148 (1905).

Recognition: in 1921 A. Einstein received the Nobel Prize in Physics for his services to Theoretical Physics, and especially for his discovery of the law of the photoelectric effect.

See also www.nobel.se/physics/laureates/1921/index.html.

Einstein–Podolsky–Rosen pair – nonlocal pairwise entangled quantum particles → **quantum entanglement**. There is a strong correlation between characteristics of presently noninteracting members of a pair that have interacted in the past. Such nonlocal correlations occur only when the quantum state of the entire system is entangled.

First described in: A. Einstein, B. Podolsky, N. Rosen, *Can quantum mechanical description of physical reality be considered complete?* Phys. Rev. **47**(5), 777–780 (1935).

Einstein–Podolsky–Rosen paradox assumes as a necessary property for a physical theory to be complete that every element of the physical reality must have a counterpart in the physical theory. Originally it was formulated as "If, without in any way disturbing a system, we can predict with certainty (that is with probability equal to unity) the value of a physical quantity, then there exists an element of physical reality corresponding to this physical quantity". It is a sufficient condition for an element of physical reality. It means that if two quantum particles are in an entangled state, then a measurement of one can affect the state of the other instantaneously, even if they are separated.

First described in: A. Einstein, B. Podolsky, N. Rosen, *Can quantum mechanical description of physical reality be considered complete?* Phys. Rev. **47**(5), 777–780 (1935).

Einstein relationship = **Einstein–Smoluchowski equation.**

Einstein's coefficients A and B denote rates of **absorption** and **stimulated emission** per unit electromagnetic energy density (within the frequency interval between ν and $\nu + \Delta\nu$) due to the transition of an electron between the level i and another level j – the coefficient B_{ij}, and the rate for **spontaneous emission** of radiation due to transition from the level i to the level j – the coefficient A_{ij}. For two nondegenerate levels in a medium with a refractive index n the coefficients are related as: $B_{ij} = B_{ji}$, $A_{ij} = 8\pi h\nu^3 n^3 c^{-3} B_{ij}$. The total emission rate of radiation from the level i to the level j for a system in thermal equilibrium at a given temperature is $R_{ij} = A_{ij} + B_{ij}\rho_e(\nu) = A_{ij}(1 + N_p)$, where $\rho_e(\nu)$ is the photon energy density and N_p is the photon occupation number.

First described in: A. Einstein, *Strahlungs-Emission and -Absorption nach der Quantentheorie*, Verh. Deutsch. Phys. Ges. **18**, 318–323 (1916); A. Einstein, *Zur Quantentheorie der Strahlung*, Phys. Z. **18**, 121–128 (1917).

Einstein shift – a shift toward longer wavelengths of spectral lines emitted by atoms in strong gravitational fields.

Einstein–Smoluchowski equation states that the relation between diffusivity D of a particle and its mobility μ at the absolute temperature T is $D/\mu = k_B T/e$.

First described in: A. Einstein, *Über die von der molekularkinetischen Theorie der Wärme geforderte Bewegung von in ruhenden Flüssigkeiten suspendierten Teilchen*, Ann. Phys. **17**, 549–560 (1905); M. Smoluchowski, *Molekular-kinetische Theorie der Opaleszenz von Gasen im kritischen Zustande, sowie einiger verwandter Erscheinungen*, Ann. Phys. **25**, 205–226 (1908).

elastic moduli of crystals connect stresses and strains in a crystal. The coefficients of proportionality between the stresses and the strains, that is when the stresses are expressed in terms of the strains, are known as the crystal elastic **stiffness constants**. Another set of coefficients, which are crystalline **compliance constants**, is obtained when the strains are written in terms of the stresses.

elastic scattering – a scattering process in which the total energy and momentum of interacting (colliding) particles are conserved. The alternative is **inelastic scattering**.

electret – a dielectric body which retains an electric momentum after an externally applied electric field has been reduced to zero.

electrochromism – a color change of matter caused by an electric current.

electroluminescence – light emission from a solid excited by injection of electrons and holes via an external electric current. G. Destriau first coined the word "electroluminescence" to refer to the light emission he observed from zinc sulfide phosphors when excited by an electric field → **Destriau effect**.

First described in: H. J. Round, *A note on carborundum*, Electr. World **19** (February 9), 309 (1907).

electrolyte – a substance in which electric charges are transported by ions. Those ions bearing a negative charge called **anions**, while whose carrying a positive charge are called **cations**.

electromigration – net motion of atoms caused by the passage of an electric current through matter.

First described in: M. Gerardin, *De l'action de la pile sur lessels de potasse et de soude et sur les alliages soumis à la fusion ignée*, C. R. Acad. Sci. **53**, 727 (1861).

electron – a stable negatively charged elementary particle having a mass of 9.10939×10^{-31} kg, a charge of 1.602177×10^{-19} C, and a spin of 1/2.

Term coined by: G. J. Stoney, *Of the "Electron," or Atom of Electricity*, Phil. Mag. **38**, 418–420 (1894).

First described in: J. J. Thomson, *Cathode rays*, Phil. Mag. 44(269), 293–316 (1897); J. J. Thomson, *On the mass of ions in gases at low pressures*, Phil. Mag. 48(295), 547–567 (1899).

Recognition: in 1906 J. J. Thomson received the Nobel Prize in Physics in recognition of the great merits of his theoretical and experimental investigations on the conduction of electricity by gases.

See also www.nobel.se/physics/laureates/1906/index.html.

electron affinity – an energy distance between the bottom of the conduction band of a semiconductor or isolator and the energy level in a vacuum.

electron affinity rule → **Anderson's rule.**

electron-beam lithography – a technique for patterning integrated circuits and semiconductor devices by the use of an electron beam. A typical electron beam lithography machine includes a vacuum column with an electron source, accelerating electrodes, magnetic lenses, and a steering system. The electron beam formation system produces a flux of electrons accelerated to energies in the range 20–100 keV and focused into a spot of 0.5–1.5 nm in diameter. A conventional **scanning electron microscope** or scanning transmission electron microscope providing these facilities is often used for electron beam nanolithography. Moreover, conducting probes of a **scanning tunneling microscope** or **atomic force microscope** can be employed for low-energy electron exposure as well. The fundamental resolution limit is given by **Heisenberg's uncertainty principle**. Patterning below 10 nm is achievable.

The electron beam is scanned over the resist-covered surface exposing certain regions and avoiding others under the control of a computer pattern generator. The most frequently used organic positive resist is PMMA (poly(methylmethacrylate)), a long-chain polymer. Where the electron beam hits the resist the chain length is shortened and the polymer can then be easily dissolved away in an appropriate developer. The threshold of sensitivity to the electron exposure is around 5×10^{-4} C cm^{-2}.

Among organic negative resists, **calixarene** (MC6AOAc–hexaacetate *p*-methyl calixarene) and α-methylstyrene best meet the requirements of nanolithography, showing high durability against plasma etching. Calixarene has a cyclic structure in

116

the form of a ring-shaped molecule about 1 nm in diameter. Its main component is a phenol derivative, which seems to have high durability and stability, originating from the strong chemical coupling of the benzene ring. Calixarene is almost 20 times less sensitive to electron exposure than PMMA. Meanwhile, Cl methylated calixarene, which is one of the calixarene derivatives where Cl atoms replace methyl groups of the calixarene, has practically the same sensitivity. The small size of the calixarene molecule and high molecule uniformity result in surface smoothness of the resist film and in ultrahigh resolution.

PMMA, calixarene and α-methylstyrene resist films with a typical thickness of 30–50 nm provide formation of nanopatterns with a resolution of 6–10 nm. Other polymer electron beam resists, such as positive resist ZEP, developed by Nippon Zeopn Company and negative resist SAL601 from Shipley Company have similar characteristics. The resolution limit of organic resists is set by low-energy secondary electrons (up to 50 eV), which are generated by a primary beam on impact with these materials. The secondary electrons expose the resist material around the beam to a distance of about 3–5 nm thus giving a writing resolution limit close to 10 nm.

Besides organic materials, inorganic compounds, such as SiO_2, AlF_3-doped LiF and NaCl, have shown good prospects as resist materials for electron beam nanopattern fabrication. They are promising to achieve sub-5-nm resolution level but their sensitivity, characterized by the threshold of sensitivity to the electron exposure above 0.1 C cm^{-2}, remains too small.

Direct nanopatterning of SiO_2 by an electron beam is indeed attractive for device fabrication. It can be realized with the use of oxide films less than 1 nm thick. Irradiation of such ultrathin films is performed by a focused electron beam at room temperature. The irradiated local regions are then decomposed and evaporated by annealing at 720–750°C in high vacuum. This method is attractive for *in situ* fabrication of nanostructures, because all processes of the SiO_2 mask formation, and subsequent deposition of other materials can be performed in an ultra high vacuum chamber without taking the sample out into air.

In conventional electron beam lithography, with electrons having energy in the range of 20–50 keV, the major resolution restrictions come from secondary electrons and electrons backscattered from the substrate. Furthermore, penetrating into the substrate, such energetic electrons generate radiation damages there. These effects can be significantly suppressed by using low-energy electron beams in the range of 2–10 keV. It is achieved at the expense of the decrease in the resist sensitivity and related increase of the exposed dose.

Once the resist mask is fabricated, two principle approaches can be used in order to transfer the resist pattern into features of metal, dielectric or semiconductor layers on the substrate. It can be done by etching to remove material in the windows of the resist mask or by material deposition onto developed resist mask.

The first approach is typical for conventional semiconductor technology. So far the etching rate of PMMA and calixarene is comparable to that of silicon and other electronic materials, and their durability is sufficient to fabricate dielectric and semiconductor nanostructures by plasma dry etching.

electron-beam lithography

Figure E.3 Nanostructure fabrication by lift-off process with the use of positive electron resist (PMMA): (a) electron beam exposure of the resist film, (b) development of the resist, (c) metal deposition, (d) lift-off of the resist with the metal on its surface.

The second approach is also, known, as a **lift-off process**. It is mainly involved in fabrication of metal nanosize elements. The main steps of the process are shown in Figure E.3.

The resist film, usually spin-on deposited onto the substrate, is first exposed by a single pass of the electron beam. Then it is developed to form the windows in the film. Metal is evaporated onto the patterned surface in a direction normal to it in order to coat the resist and the substrate surface in the windows. After this the substrate is immersed in a powerful solvent, which dissolves the unexposed resist (for example, acetone for PMMA) with the metal deposited on top of it. The undesired metal is removed, leaving behind a copy of the elements that were initially patterned in the resist. The metal islands left can be used as an element of nanoelectronic devices or as a mask for etching semiconductor or dielectric films beneath it.

Electron beam lithography is being constantly developed. Nevertheless, its readiness for insertion into pilot production lines is less than satisfactory. The low speed of electron-beam wafer processing remains a major concern. Several advanced lithography schemes, based on the use of low-energy electrons are being developed. One of them proposes the application of arrays of microcolumn electron-beam guns. Many guns working in parallel significantly increase the throughput of the lithographic system. Further developments in applications of low-energy electrons for the fabrication of nanostructures are also expected using rapidly improving proximal probe techniques.

electronegativity – a qualitative measure of "power of an atom in a molecule to attract electrons to itself ". A difference in electronegativity between atoms A and B is regarded as a measure of the degree of the electron transfer from atom A to atom B on forming the chemical bond between them. More electronegative atoms attract electrons more efficiently.

electron energy eigenstate = energy level.

electron energy loss spectroscopy (EELS) – energy-resolved registration of electrons passed through matter. When an electron beam is incident into a specimen, some electrons are inelastically scattered and lose a part of their energy. The energy loss is a fingerprint of a particular chemical element and particular atomic bonding. Thus, elemental composition and atomic bonding states in solids can be determined. It is usually realized with a spectroscope attached under the sample holder in an electron microscope.

More details in: M. De Crescenzi, M. N. Piancastelli, *Electron Scattering and Related Spectroscopies* (World Scientific Publishing, Singapore, 1996).

electronic band structure – a diagram presenting variation of the electron energy E in a solid as a function of the wave vector k. Since $E(k)$ is periodic in crystalline solids, the presentation is usually done within the first **Brillouin zone**.

electronic polarization $\rightarrow$ **polarization of matter**.

electron magnetic resonance (EMR) = electron paramagnetic resonance.

electron microscopy – an analytical imaging technique, whereby a beam of accelerated (up to 10–200 keV) electrons is directed onto a sample for analysis, and reflected or transmitted electrons are collected over a narrow solid angle and focused by electronic or magnetic lenses onto the image plane, thus providing an enlarged version of the sample structure. Scanning of the probing electron beam is employed in order to enlarge the analyzing area.

Among the variety of the techniques, **scanning electron microscopy (SEM)** and **transmission electron microscopy (TEM)** are distinguished in principle. SEM produces images by detecting scattered or secondary electrons which are emitted from the surface of a sample due to excitation by the primary electron beam. TEM forms images by detecting electrons that are transmitted through the sample. Generally, the space resolution of TEM (reaching atomic level) is about an order of magnitude better than that of SEM. TEM analysis usually needs sample thinning in order to make the sample transparent for probing electrons. SEM requiers no special sample preparation, except deposition of thin conductive film when the sample itself does not have appropriate conductivity for leaking the charge brought about by probing electrons.

First described in: M. Knoll, E. Ruska *Beitrag zur geometrischen Elektronenoptik I und II*, Ann. Phys. **12**, 607–640, 641–661 (1932).

Recognition: in 1986 E. Ruska received the Nobel Prize in physics (shared with G. Binning and H. Rohrer) for his fundamental work in electron optics, and for the design of the first electron microscope.

See also www.nobel.se/physics/laureates/1986/index.html.

electron-optical configuration for nanodiffraction – an **electron microscope** technique able to produce a **diffraction** pattern from an area of material just 10 nm in diameter with unprecedented precision. The configuration utilizes one of the round lenses in an objective prefield **aberration** corrector. The method allows one to tune carefully the size of the diffracting area to the features of interest in a particular material, excluding the surrounding material.

First described in: C. Dwyer, A. I. Kirkland, P. Hartel, H. Müller, M. Haider, *Electron nanodiffraction using sharply focused parallel probes*, Appl. Phys. Lett. **90**, 151104 (2007).

electron paramagnetic resonance (EPR) – a process of resonant absorption of microwave radiation by paramagnetic ions or molecules with at least one unpaired electron spin in the presence of a static magnetic field. It arises from the magnetic moment of unpaired electrons in a **paramagnetic** substance or in a paramagnetic center in a **diamagnetic** substance. Such resonance is used in a radiofrequency spectroscopy in which transitions among the energy levels of a system of weakly coupled electronic magnetic dipole moments in a magnetic field are detected.

It is also known as **electron spin resonance (ESR)** or **electron magnetic resonance (EMR)**.

First described in: E. K. Zavoisky, *The paramagnetic absorption of a solution in parallel fields*, Zh. Exp. Teor. Fiz. **15**(6), 253–257 (1945) – in Russian.

More details in: J. E. Wertz, J. R. Bolton, *Electron Spin Resonance: Elementary Theory and Practical Applications*, (Chapman and Hall, New York, 1986).

electron spectroscopy for chemical analysis (ESCA) = X-ray photoelectron spectroscopy (XPS).

electron spin resonance (ESR) = electron paramagnetic resonance.

electrooptical Kerr effect → **Kerr effects**.

electrophoresis – the electrokinetic phenomenon consisting in motion of charged particles or agglomerates through a static liquid under the influence of an applied electric field.

More details in: A. T. Andrews, *Electrophoresis* (Clarendon Press, Oxford, 1986).

electrostatic force microscopy – detection of electrostatic forces with an **atomic force microscope**. In this technique tip and sample are regarded as two electrodes of a capacitor. If they are electrically connected and have different **work functions**, electrons will flow between tip and sample until their Fermi levels are equalized. As a result, an electric field and, consequently, an attractive electrostatic force exists between them at zero bias. This contact potential difference can be balanced by applying an appropriate bias voltage. It has been demonstrated that individual doping atoms in semiconducting materials can be detected by electrostatic interactions due to the local variation of the surface potential around them. Meanwhile, in practical applications, the technique is usually used to visualize the distribution of electric potential, charge, contact potential difference and their related characteristics of solid surfaces.

First described in: Y. Martin, D. V. Abraham, H. K. Wickramasinghe, *High-resolution capacitance measurement and potentiometry by force microscopy*, Appl. Phys. Lett. **52**(13), 1103–1105 (1988).

More details in: Roadmap of Scanning Probe Microscopy, edited by S. Morita (Springer Verlag, Berlin Heidelberg, 2007).

electrostriction – elastic deformation of a dielectric caused by the force produced, when the dielectric is placed in an inhomogeneous electric field.

Elliott–Yafet mechanism – the electron spin relaxation mechanism in solids, which results from spin–orbit scattering during collisions with phonons or impurities. The related spin relaxation time τ_s in semiconductors can be determined analytically using approximations of the band and phonon structures, as most important states are usually around high-symmetry points. The spin relaxation of conduction electrons with energy E_k in this case is described by

$$\frac{1}{\tau_s(E_k)} = A \left(\frac{\Delta_{so}}{E_g + \Delta_{so}} \right)^2 \left(\frac{E_k}{E_g} \right)^2 \frac{1}{\tau_p(E_k)},$$

where $\tau_p(E_k)$ is the momentum scattering time at energy E_k, E_g is the energy gap, and Δ_{so} is the spin–orbit splitting of the valence band. The numerical factor A, which is of order 1, depends on the dominant scattering mechanism (charge or neutral impurity, phonon, electron-hole).

The mechanism is important for small-gap semiconductors with large spin-orbit splitting. The temperature dependence of $1/\tau_s$ for metals and degenerate semiconductors follows the dependence of $1/\tau_p$. In metals this means a constant at low T and a linear increase at large T. For nondegenerate semiconductors,

$1/\tau_s(T) \sim T^2/\tau_p(T)$. In the important case of scattering by charged impurities $(\tau_p(T) \sim T^{3/2}), 1/\tau_s(T) \sim T^{1/2}$.

First described in: R. J. Elliott, *Theory of the effect of spin–orbit coupling on magnetic resonance in some semiconductors*, Phys. Rev. **96**(2), 266–279 (1954); Y. Yafet, *g-factors and spin lattice relaxation of conduction electrons*, in: Solid State Physics, vol. **14**, edited by F. Seitz, D. Turnbull (Academic Press, New York, 1963), p. 1–98.

More details in: I. Žutić, J. Fabian, S. Das Sarma, *Spintronics: fundamentals and applications*, Rev. Mod. Phys. **76**(2), 323–410 (2004).

ellipsometry – a procedure for determining the nature of a specular surface from the change of the polarization state of the polarized light beam when it reflects from this surface.

First described in: P. Drude, *Zur Elektronentheorie der Metalle*, Ann. Phys. **36**, 566–613 (1889).

elliptical polarization of light → polarization of light.

embedded-atom method – the semiempirical technique for description of an atomic interaction in solids based on the **local density approximation**. The energy of each atom in a monoatomic solid is given by $E(r_0) = F(n_1\rho^\alpha(r_0)) + 0.5n_1 V(r_0)$ where r_0 is the first-neighbor distance, n_1 is the number of first neighbors, ρ^α is the spherically averaged atomic electron density at a distance r from the nucleus, V is the energy of atom-atom pair interaction and F is the embedding function. The method is used for molecular dynamics simulation of deformed crystals.

First described in: M. S. Daw, M. J. Baskes, *Applications of the embedded-atom method to covalent materials*, Phys. Rev. B **29**(11), 6443–6449 (1984).

embossing – one of the approaches used for a pattern fabrication by **imprinting**. The idea of the embossing process is that a reusable mold is stamped at high pressure into a polymer film on a heated substrate, and then removed. It is schematically illustrated in Figure E.4.

The mold is made so that it has desired topographic features raised from the surface. It is coated with a very thin film of mold release compound in order to protect its surface and prevent sticking during embossing. The substrate to be patterned is coated with a thin film of thermoplastic polymer. The substrate is heated above the glass transition temperature of the polymer to make it viscoelastic. The mold is then pressed into the polymer. The heating time and the pressure hold time last typically a few minutes each. After that the system is cooled back down below the glass transition temperature. The pattern freezes into the polymer film. After removal of the imprinter, the trenches formed are cleaned with either an oxygen plasma or solvents to remove any residual polymer from their bottom. The patterned polymer film is used as a mask for subsequent etching or lift-off process.

The most frequently used polymer is poly(methylmethacrelate) (PMMA), a well-known resist for electron beam lithography. Its good imprint properties

Figure E.4 Imprint process by embossing: (a) substrate with deposited resist film and master prior to imprinting, (b) imprinting, (c) patterned resist mask after imprinting with a residual polymer in the windows, (d) etching of the residual polymer.

are determined by an appropriate viscoelasticity of this material. PMMA has a glass transition temperature of about 105°C, thus providing good imprints in the temperature range of 190–200°C. Demounting and separation of the mold and substrate take place when both are at about 50°C. The ultimate resolution is of the order of 10 nm. It is limited by the radius of gyration of the polymer.

First described in: S. Y. Chou, P. R. Krauss, P. J. Renstrom, *Imprint of sub-25 nm vias and trenches in polymers*, Appl. Phys. Lett. **67**(21), 3114–3116 (1995).

EMR – an acronym for an **electron magnetic resonance = electron paramagnetic resonance**.

emulsion – a system containing two substantially immiscible liquids, one of which (the "internal" or "disperse" phase) is subdivided as fine droplets within the other (the "external" or "continuous" phase).

enantiomers → **chirality** and **isomers**.

enantiomorphs → **chirality**.

energy dispersive X-ray spectroscopy (EDXS) – registration of characteristic X-ray radiation excited by energetic electron bombardment of matter. The characteristic radiation arises when the incident electron transfers to an inner orbital electron engineering

the energy in excess of its binding energy, the orbital electron is ejected from the atom and the X-ray photon is generated when the vacancy is filled by an electron from the deeper inner orbital. As each chemical element has a unique electronic structure, the series of X-ray photons generated are characteristic of the particular element. The intensity of the X-rays is proportional to the number of atoms generating the rays. Thus, the energy dispersive analysis of the characteristic X-ray radiation is used for qualitative and quantitative study of matter.

engineering – an application of scientific principles to the exploitation of natural resources, to the design and production of commodities and to the provision and maintenance of utilities.

entanglement → **quantum entanglement**.

entanglement monotone – a minimum number of nonincreasing parameters characterizing quantum nonlocal resources of a **qubit** system.
First described in: G. Vidal, *Entanglement of pure states for a single copy*, Phys. Rev. Lett. **83**(5), 1046–1049 (1999).

entangled state – a state of a composite system whose parts can be spatially delocalized. Such states can have intrinsically nonlocal correlations. For more details → **quantum entanglement**.

enthalpy – a quantity conveniently expressing results of thermodynamic processes occurring under certain special conditions. It has the dimensions of energy and is also called heat content. A value of enthalpy for any system in a definite state depends only upon the state and not upon its prehistory, that is upon the way by which the system has evolved into this state.

entropy – a measure of the capacity of a system to undergo a spontaneous change. The term derives from the Greek expression for "transformation". Two consistent definitions of entropy are possible: thermodynamic and statistical. According to the thermodynamic definition, the entropy S defined by its change $dS = \delta Q_{rev}/T$. That is, the entropy change in an infinitesimal process is equal to the heat absorbed divided by the absolute temperature at which the heat is absorbed. The heat absorbed must be that characteristic of a reversible process, if entropy is to be a state function, that is $\Delta S = S_2 - S_1$. From the statistical point of view $S = k_B \log \Omega + \text{constant}$, where Ω is the probability defined as the number of configuration states available to the system.
First described in: R. Clausius *Abhandlungen über die mechanische Wärmetheorie* (Friedrich Vieweg und Sohn, Braunschweig, 1864).

envelope function method is based on a plane wave expansion

$$\psi(\mathbf{r}) = \sum_{kG} \tilde{\psi}(\mathbf{k} + \mathbf{G}) \exp(i(\mathbf{k} + \mathbf{G}) \cdot \mathbf{r}) = \sum_{kG} \tilde{\psi}(\mathbf{k}) \exp(i(\mathbf{k} + \mathbf{G}) \cdot \mathbf{r}),$$

where the $\mathbf{G}$ are reciprocal lattice vectors of an underlying **Bravais lattice** and the $\mathbf{k}$ are wavevectors confined to a primitive cell of the reciprocal lattice usually taken as the first **Brillouin zone**. $\tilde{\psi}(\mathbf{k} + \mathbf{G})$ is the Fourier transform of ψ and $\tilde{\psi}(\mathbf{k})$ is an alternative notion that emphasizes the decomposition of the wavevector $\mathbf{k} + \mathbf{G}$ into a reciprocal lattice vector and a wavevector in the primitive cell. A complete set of functions, $U_n(\mathbf{r})$, periodic in the Bravais lattice, is introduced

$$U_n(\mathbf{r}) = \sum_G U_{nG} \exp(i\mathbf{G} \cdot \mathbf{r})$$

usually chosen to be orthonormal so that expressing the plane waves in terms of them one has:

$$\exp(i\mathbf{G} \cdot \mathbf{r}) = \sum_n U_{nG}^* U_n(\mathbf{r}).$$

Substitution for $\exp(i\mathbf{G} \cdot \mathbf{r})$ finally gives

$$\psi(\mathbf{r}) = \sum_n F_n(\mathbf{r}) U_n(\mathbf{r}),$$

with the envelope functions, $F_n(\mathbf{r})$, given by

$$F_n(\mathbf{r}) = \sum_{kG} U_{nG}^* \tilde{\psi}(\mathbf{k} + \mathbf{G}) \exp(i\mathbf{k} \cdot \mathbf{r}) = \sum_{kG} U_{nG}^* \tilde{\psi}_G(\mathbf{k}) \exp(i\mathbf{k} \cdot \mathbf{r}),$$

which describes the slowly varying mesoscopic part of the locally highly oscillating microscopic wave functions. This expansion was introduced by Luttinger and Kohn to derive the **effective mass** equation for periodic structures in the presence of a slow perturbing potential. Thus it has sometimes been called the effective mass representation. Later Bastard introduced a prescriptive envelope function method in the case of **superlattices**, in which the real space equations satisfied by the envelope functions were equivalent to the $\mathbf{k} \cdot \mathbf{p}$ method, but the band edges were allowed to be functions of position, implicitly assuming that the equations would be valid at any atomically abrupt interfaces. Only the envelope of the nanostructure wavefunction is described, regardless of the atomic details. Despite the numerous assumptions involved, the envelope function approximation has had great success, mainly due to a fair compromise between the simplicity of the method and the reliability of the results. A method for deriving an exact envelope functions equation starting from the **Schrödinger** equation appropriate for the calculation of electronic states and photonic modes in nanostructures has been introduced by M. G. Burt.

First described in: J. M. Luttinger, W. Kohn, *Motion of Electrons and Holes in Perturbed Periodic Field*, Phys. Rev. **97**(4), 869–833 (1955) – introduction of the method of developing an "effective mass" equation for electrons moving in a perturbed periodic structure. G. Bastard, *Superlattice band structure in the envelope function approximation*, Phys. Rev. B **24**(10) 5693–5697 (1981) – application to superlattices. M. G. Burt, *An exact formulation of the envelope function method for the determination of electronic states in semiconductor microstructures*, Semicond. Sci. Techn. **3**(8), 739–753 (1998) – application to nanostructures.

More details in: M. G. Burt, *Fundamentals of envelope function theory for electronic states and photonic modes in nanostructures*, J. Phys: Condens. Matter **11**(9), R53–R83 (1999); A. Di Carlo, *Microscopic theory of nanostructured semiconductor devices: beyond the envelope-function approximation*, Semicond. Sci. Technol. **18**(1), R1–R31 (2003).

enzyme – a **protein** molecule partly associated with a nonprotein portion, the **coenzyme** or **prosthetic group**. It is a biocatalyst controlling all reactions within living organisms.

More details in: H. Bisswanger, *Proteins and enzymes*, in: *Encyclopedia of Applied Physics*, Vol. 15, edited by G. L. Trigg (Wiley-VCH, Weinheim, 1996).

epitaxy – a crystal growth onto a monocrystalline substrate. The term is derived from the Greek words *epi*, meaning "on", and *taxis*, meaning "orderly arrangement". For epitaxy techniques → **solid phase epitaxy, liquid phase epitaxy, chemical vapor deposition** and **metal-organic chemical vapor deposition, molecular beam epitaxy, chemical beam epitaxy**.

More details in: M. A. Herman, *Epitaxy* (Springer Verlag, Berlin, 2004).

EPR – an acronym for an **electron paramagnetic resonance**.

equation of state – a mathematical expression which defines the physical state of a homogeneous substance, that is gas, liquid or solid, by relating its volume to pressure and absolute temperature for a given mass of the material.

equivalent circuit – a representation of a real circuit or transmission system by a network of lumped elements.

ergodic hypothesis – the assumption that in the course of time a system will take up the totality of microstates compatible with the macroscopic conditions. It has been shown that this assumption cannot be justified mathematically and the original hypothesis has been replaced by a less strict formulation, the so called quasi ergodic hypothesis. This only requires that the phase curve of the system, that is the time succession of the system state represented in the phase space, should approximate as closely as may be desired to the energy surface in the phase space.

First described in: J. C. Maxwell *On the dynamical theory of gases*, Phil. Trans. R. Soc. Lond. **157**, 49 (1867); L. Boltzmann, *Studien über das Gleichgewicht der lebendigen Kraft zwischen bewegten materiellen Punkten*, Wiener Berichte **58**, 517 (1868).

ergodicity – property of a system that ensures equality of statistical ensemble averages and time averages, along the trajectory of the system through phase space.

Ericson fluctuations – reproducible fluctuations of nuclear reaction cross sections as a function of the nuclear reaction energy. They occur in the "continuum" regime when a large number of compound nuclear states overlap, owing to the short lifetime of the compound nucleus.
First described in: T. Ericson, *Fluctuations of nuclear cross sections in the "continuum" region*, Phys. Rev. Lett. **5**(9), 430–431 (1960).

error function $\text{erf}(z) = 2/\sqrt{\pi} \int_0^z \exp(-y^2/2)\mathrm{d}y$, the complementary error function is $\text{erfc}(z) = 1 - \text{erf}(z)$.

ESCA – an acronym for **electron spectroscopy for chemical analysis**.

ESR – an acronym for **electron spin resonance = electron paramagnetic resonance**.

esters – organic compounds that are alcohol derivatives of carboxylic acids. Their general formula is RCOOR′, where R may be a hydrogen, alkyl group ($-C_nH_{2n+1}$), or aromatic ring, and R′ may be an alkyl group or aromatic ring but not a hydrogen.

Ettingshausen coefficient → Ettingshausen effect.

Ettingshausen effect – if a conductor carries a current density **J** in a transverse magnetic field **H**, a temperature gradient appears in a direction normal to both. Assuming the heat current to be zero $\nabla_t T = PJx\mathbf{H}$, where P is the **Ettingshausen coefficient**.
First described in: A. von Ettingshausen, W. Nernst, *Über das Auftreten electromotorischer Kräfte in Metallplatten, welche von einem Wärmestrome durchflossen werden und sich im magnetischen Felde befinden*, Ann. Phys. **265**(10), 343 (1886).

Ettingshausen–Nernst effect → Nernst effect.

eukaryote – an organism possessing a nucleus with a double layer of membrane and other membrane-bound organelles. It includes such unicellular or multicellular members as all members of the protist, fungi, plant, and animal kingdoms. The name came from the Greek root "karyon", meaning "nut", combined with the prefix "eu-" meaning "good" or "true".

Euler equations of motion describe motion of a mechanical system consisting of connected particles. Basically they are forms of the two vector equations $\mathrm{d}\mathbf{p}/\mathrm{d}t = \mathbf{R}$ and $\mathrm{d}\mathbf{h}/\mathrm{d}t = \mathbf{G}$, which are referred to as an inertial frame of reference with a stationary origin. The vectors **p** and **h** are the linear and angular momenta of the system referred to this frame, **R** is the vector sum of the forces applied to the system, the vector **G** is the sum of the vector moments of this applied forces about the origin. If the origin always coincides with the center of mass of the system, the second equation is unchanged, although dh/dt is now referred to a frame with moving origin. This means that one may use velocities relative to the center of mass instead of actual velocities. This is a principle of independence of translation and rotation.

When the mechanical system is a rigid body, the expression for **h** involves the moments and products of inertia and, in general, dh/dt will involve the rates of change of these coefficients of inertia owing to the motion of the body relative to the frame of reference. This difficulty is avoided by using axes which are fixed in the body.

More details in: L. Euler, *Opera omnia* (Teubner und Fussli, Leipzig, 1911).

Euler law states that the friction force between two bodies is proportional to the loading force → **Amontons' law**.

Euler–Maclaurin formula gives a connection between the integral and the sum for a smooth function $f(x)$ defined for all real numbers x between 0 and n, where n is the natural number, in the form $\int_0^n f(x)dx = f(0)/2 + f(1) + \cdots + f(n-1) + f(n)/2$.
First described in: L. Euler, Comment. Acad. Sci. Imp. Petrop. **6**, 68 (1738).

Euler's formula (topology) gives a relation between the numbers of vertices V, of edges E, and faces F of a convex polyhedron, that is a polyhedron any two points of which can be joined by a straight line entirely contained in the polyhedron. It states that $V - E + F = 2$.
First described in: L. Euler, *Elementa doctrinae solidorum*, Novi. Comm. Acad. Sci. Imp. Petrop. 4, 109 (1752).

even function – a function $f(x)$ with the property that $f(x) = f(-x)$.

Ewald sphere – a sphere superimposed on the reciprocal lattice of a crystal used to determine the directions in which an X-ray or other beam will be reflected by a crystal lattice.
First described in: P. P. Ewald, *Zur Theorie der Interferenzen der Röntgenstrahlen in Kristallen*, Phys. Z. **14**, 465–472 (1913).
More details in: J. B. Pendry, *Low Energy Electron Diffraction* (Academic Press, London, 1974).

EXAFS – an acronym for an **extended X-ray absorption fine structure**.

exchange interaction – a quantum mechanical effect which increases or decreases the energy of two or more electrons when their **wave functions** overlap. Arising from the **Pauli exclusion principle**, this energy change is the result of the identity of particles, exchange symmetry, and the electrostatic force. Exchange interaction is the mechanism by which the electron spins in many magnetic materials are lined up parallel or opposed.

First described in: W. Heisenberg, *Mehrkörperproblem und Resonanz in der Quantenmechanik*, Z. Phys. **38**(6–7), 411–426 (1926); P. A. M. Dirac, *On the theory of quantum mechanics*, Proc. R. Soc. Lond. A **112**(762), 661–677 (1926).

excimer – a combination of two atoms or molecules that survives only in an **excited state** and which dissociate as soon as the excitation energy has been discarded. The term is a contraction of "excited dimer". Excimers are used for light generation in lasers → **excimer laser**.

excimer laser – a rare-gas halide or rare-gas metal vapor laser emitting in the ultraviolet (from 126 to 558 nm) that operates on electronic transitions of molecules, up to that point diatomic, whose ground state is essentially repulsive. Excitation may be by done e-beam or electric discharge. Lasing gases include ArCl, ArF, KrCl, KrF, XeCl and XeF.

excited state – a high energy state of some entire system.

exciton – an electron–hole pair in a bound state in a hydrogen–atom–like fashion. Exciton formation processes in photon irradiated semiconductors are illustrated in Figure E.5. If the energy of the photon is larger than the band gap of the irradiated semiconductor (high energy excitation regime), then a free electron is created in the conduction band and an empty state, that is a hole, is left in the valence band. After relaxation they occupy the lowest energy states in the conduction and valence bands, respectively. The attractive force acting between the electron and hole leads to a reduction in their total energy, by an amount E_{X0}, and makes them bound. The exciton is formed. A photon with the energy just below the band gap can be resonantly absorbed, thus crating the exciton directly.

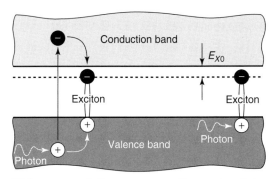

High energy excitation   Resonant excitation

Figure E.5 Non-resonant and resonant exciton formation in semiconductors.

exciton Bohr radius

As the hole mass is generally much greater than the electron mass, the exciton can be considered like a two-body system resembling a hydrogen atom with the negatively charged electron orbiting the positive hole. Similarly to the hydrogen atom, the exciton is characterized by the **exciton Bohr radius** $a_B = \epsilon \hbar^2 / \mu e^2 = \epsilon m_0 / \mu (x0.053)$ nm where μ is the electron hole reduced mass $1/\mu = 1/m_e^* + 1/m_h^*$; $\epsilon = \epsilon_r \epsilon_0$ is the permittivity of the material. It is the natural physical separation in a crystal between an electron in the conduction band and the hole it leaves behind in the valence band. The size of this radius controls how large a crystal must be before its energy bands can be treated as continuous. Therefore, the exciton Bohr radius can rightly be said to define whether a crystal can be called a semiconductor quantum dot, or simply a bulk semiconductor.

The exciton binding energy is $E_{X0} = -\mu e^4 / (32\pi^2 \hbar^2 \epsilon^2)$. The reduced electron–hole mass is smaller than the electron rest mass m_0, and the dielectric constant ϵ is several times bigger than that of vacuum. This is why the exciton Bohr radius is significantly larger and the exciton energy is significantly smaller than the relevant values for the hydrogen atom. Absolute values of a_B for most semiconductors range in the interval 1–10 nm, and the energy takes the values from 1 meV to 100 meV. The lifetime of excitons is of the order of hundreds of ps to ns.

The concentration of excitons n_{exc} and of the free electrons and holes $n = n_e = n_h$ are related via the ionization equilibrium equation known as the **Saha equation**:

$$n_{exc} = n^2 \left(\frac{2\pi \hbar^2}{k_B T} \frac{m_e^* + m_h^*}{m_e^* m_h^*} \right)^{3/2} e^{(E_{X0}/k_B T)}.$$

For $k_B T \gg E_{X0}$ most of the excitons are ionized and the properties of the electron subsystem of the crystal are determined by free electrons and holes. At $k_B T \leq E_{X0}$ a significant part of electron–hole pairs exists in the bound state.

Weakly bound excitons, typical for semiconductors, are also called **Mott-Wannier** excitons, while tightly bound excitons, usually found in solid inert gases, are called **Frenkel excitons**.

First described in: J. I. Frenkel, *Transformation of light into heat in solids*, Phys. Rev. **37**(1), 17–44 (1931).

exciton Bohr radius → exciton.

exclusion principle → Pauli exclusion principle.

exon – a segment of a **gene** present in mature mRNA transcripts that specify the amino acid sequence of a polypeptide during translation. Exons of a gene are linked together by mRNA splicing.

expectation value = mean value.

ex situ – Latin meaning "away from". The phrase "ex situ analysis of experimental samples" is used to show that the samples are analyzed at the place different from that where they have been fabricated. An alternative is *in situ* (Latin "at the spot") analysis supposing that the samples are analyzed in the same place where they have been fabricated.

extended Hückel method (theory) supposes molecular orbital treatment in which the interactions between electrons in different orbitals are ignored when the electronic bands in solids are simulated.

First described in: R. Hoffmann, *An extended Hückel theory*, J. Chem Phys. **39**(6), 1397–1412 (1963).

Recognition: in 1981 R. Hoffmann and K. Fukui received the Nobel Prize in Chemistry for their theories, developed independently, concerning the course of chemical reactions.

See also www.nobel.se/chemistry/laureates/1981/index.html.

extended state – a quantum mechanical state of a single electron in a random potential that extends throughout the whole sample.

extended X-ray absorption fine structure (EXAFS) spectroscopy – when an X-ray passes through a material, it attenuates progressively and undergoes some discontinuities for energies corresponding to electronic transitions of core electron towards unoccupied states above the **Fermi level**. In the case of a crystalline solid, the atom excited by the X-ray emits a photoelectron described by a spherical outgoing wave centered on the atomic target. The emitted photoelectron is retrodiffused by the electrons of the nearest neighbor atoms. This retrodiffusion is represented by spherical waves centered on the sites of neighboring atoms. A portion of these waves interferes with the first outgoing wave and gives rise to a modulation of the absorption coefficient. EXAFS spectroscopy reflects this local order, and the amplitude of the oscillations depends on the coordination number of neighbors N_j of type j located at a distance r_j from the absorbing central atoms. The absorption spectrum is composed by three regions:

1. an edge onset corresponding to the transition of a core electron to the Fermi level

2. a region close to the edge (**XANES – X-ray absorption near-edge structure**) where the emitted photoelectron carries little energy and because of the long mean free path the multiple scattering effects become important and reflect the geometry and the bond direction

3. a region showing oscillations of weak intensity, which is the EXAFS region, characterized basically by a single-scattering regime.

The absolute intensity of these fine structures is referred to the absorption coefficient of the isolated atom and in general is of the order of 10–20% of the total absorption.

external quantum efficiency

More details in: B. K. Teo, D. C. Joy, *EXAFS Spectroscopy and Related Techniques* (Plenum Press, New York, 1981); M. De Crescenzi, M. N. Piancastelli, *Electron Scattering and Related Spectroscopies* (World Scientific Publishing, Singapore, 1996).

external quantum efficiency → **quantum efficiency**.

extinction – complete absorption of plane-polarized light by a polarizer whose axis is perpendicular to the plane of polarization.

extinction coefficient – a property of a material characterizing its ability to attenuate light → **dielectric function** and **refractive index**.

Eyring equation gives the specific reaction rate for a chemical reaction in the form $V = C^*(k_B T/h)p$, where C^* is the concentration of the activated complex formed by the collision of reactant molecules possessing the required energy, $k_B T/h$ is the classical vibration frequency of a bond which is broken when the complex dissociates, and p is the probability that this dissociation will lead to products.

First described in: H. Eyring, *The activated complex in chemical reactions*, J. Chem. Phys. **3**(2), 107–115 (1935).

F

From Fabry–Pérot resonator to FWHM (full width at half maximum)

Fabry–Pérot resonator consists of two parallel plates with their reflecting surfaces facing each other and spaced at a distance equal to $\lambda n/2$ ($n = 1, 2, \ldots$), where λ is the wavelength of the light desired to be in resonance conditions in it.

First described in: C. Fabry, A. Pérot, *Mesure de petites épaisseurs en valeur absolue*, Compt. Rend. Acad. Sci. **123**, 802 (1896).

face centered cubic (fcc) structure – a cubic lattice with lattice points on the faces of the unit cell in addition to the eight corner points. It is schematically shown in Figure F.1.

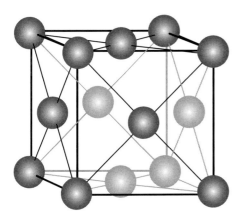

Figure F.1 Face centered cubic structure.

Falicov–Kimball model considers spinless electrons interacting with classical particles (ions) in a lattice. In fact, it is the **Hubbard model** including a few extra complications in order to investigate metal-insulator phase transitions in rare-earth materials and transition-metal compounds. These transitions are supposed to be regulated by the interactions between electrons in two electronic states: non-localized states (itinerant electrons) and states that are localized around the sites corresponding to the metallic ions of the crystal (static electrons).

What is What in the Nanoworld. Victor E. Borisenko and Stefano Ossicini
Copyright © 2008 WILEY-VCH Verlag GmbH & Co. KGaA, Weinheim
ISBN: 978-3-527-40783-5

Fang–Howard wave function

The two species of particles are supposed to play a role in a crystal lattice within a **tight-binding approach**. The first species consists of spinless quantum fermions referred to as "electrons". The second species consists of localized holes or electrons considered as classical particles. Electrons hop between nearest-neighbor sites but classical particles do not. Both species obey **Fermi–Dirac statistics**. The **Pauli exclusion principle** prevents more than one particle of a given species to occupy the same site. Interactions are on-site and thus involve particles of different species. They can be repulsive or attractive.

The model can be viewed as the simplest model where quantum particles interact with classical fields.

First described in: L. M. Falicov, J. C. Kimball, *Simple model for semiconductor-metal transitions: SmB_6 and transition-metal oxides*, Phys. Rev. Lett. **22**(19), 997–999 (1969).

More details in: //arxiv.org/PS_cache/cond-mat/pdf/9811/9811299v1.pdf.

Fang–Howard wave function – $\psi(x) = x \exp[(1/2)bx]$. It vanishes at $x = 0$ and decays roughly like an exponent as $x \to 0$.

First described in: F. F. Fang, W. E. Howard, *Negative field-effect mobility on (100) Si surfaces*, Phys. Rev. Lett. **16**(18) 797–799 (1966).

Fano factor – the dimensionless characteristic of a short noise in an electronic device. It is defined as $\gamma = S_I/(2eI)$, with S_I being the spectral density of current fluctuations at low frequency, I the current flowing in the device, and e the elementary quantum of charge determining the current. In the absence of correlation between current pulses $j = 1$. This case corresponds to full shot noise. Deviations from this ideal situation are a signature of existing correlations between different pulses. A negative correlation corresponds to $\gamma < 1$ and a suppressed shot noise. In case of $\gamma > 1$ the correlation is positive and the shot noise is enhanced.

Fano interference $\rightarrow$ Fano resonance.

Fano resonance – arises from coupling (**Fano interference**) of a discrete energy state degenerate with a continuum. The mixing of a configuration belonging to a discrete spectrum with continuous spectrum configurations gives rise to the phenomenon of **autoionization**. Autoionized levels manifest themselves in continuous absorption spectra as asymmetric peaks because, in mixing of configurations to form a stationary state of energy E, the coefficients vary sharply when E passes through an autoionized level. It has been shown that the shapes of the absorption lines in the ionization continuum of atomic and molecular spectra are represented by the formula $\sigma(\epsilon) = \sigma_a[(q + \epsilon)^2/(1 + \epsilon^2)] + \sigma_b$, where $\epsilon = [(E - E_r)/(\Gamma/2)]$ indicates the deviation of the incident photon energy E from the idealized resonance energy E_r, which pertains to a discrete autoionizing level of the atom. This deviation is expressed in a scale whose units are the half-width $\Gamma/2$ of the line; $\sigma(\epsilon)$ is the absorption cross section for photons of energy E, and σ_a and σ_b are two portions of the cross section corresponding to transitions to states of the continuum that do and do not interact with the discrete auto-ionizing level, and q is a parameter (**Fano parameter**) which characterizes the line profile as shown in Figure F.2.

The Fano resonance is a universal phenomenon observed in **rare gas** spectra, impurity ions in semiconductors, electron-phonon coupling, photodissociation, bulk GaAs in a magnetic field, **superlattices** in electric field, and so on. It can be semiclassically understood in a coupled oscillator model. The discrete state drives the continuum oscillators, with eigenstates consisting of a mixture of the discrete state and the adjacent continuum. The continuum oscillators on either side of the discrete state move with opposite phase, according to whether the driving frequency is above or below their resonant frequency. Thus, on one side they interfere constructively with the discrete state and on the other destructively, leading to the characteristic asymmetric Fano resonance line shape.

First described in: U. Fano, *Sullo spettro di assorbimento dei gas nobili presso il limite dello spettro d'arco*, Nuovo Cimento **12**(2), 156 (1935); U. Fano, *Effects of configuration interaction on intensities and phase shifts*, Phys. Rev. **124**(6), 1866–1878 (1961).

More details in: U. Fano, J. W. Cooper, *Spectral distribution of atomic oscillator strengths*, Rev. Mod. Phys. **40**(3), 441–507 (1968).

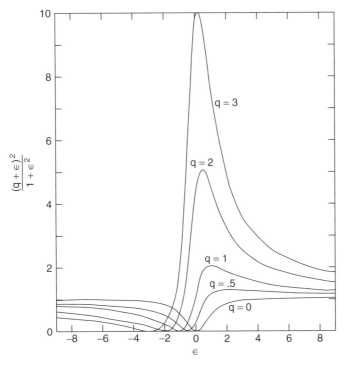

Figure F.2 Natural line shapes for different values of q.
(Reverse the scale of abscissas for negative values of q.)

Fano parameter

Fano parameter → Fano resonance.

Faraday cell – a magneto optical device consisting of a fused silica or optical glass core inserted into a solenoid. When a linearly polarized light beam passes through the device, the direction of its polarization is rotated by an angle proportional to the magnetic flux density along the propagation direction.

Faraday configuration – the term used to indicate a particular orientation of a **quantum film** structure with respect to the magnetic field applied, that is when the field direction is perpendicular to the film plane. The alternative orientation is labeled as a **Voigt configuration**.

Faraday effect – a rotation of the plane of polarization of polarized light when it passes through a homogeneous medium along a magnetic field. The magnitude and direction of this magnetic rotation depend on the strength and direction of the field, but do not depend on the direction of the light propagation through the field. Mathematically it can be described by the relationship $\theta = VHL$, where θ is the amount of rotation, H is the strength of the magnetic field, and L is the path length of the light through the sample. The proportionality constant V is known as the **Verdet constant**. It is approximately proportional to the reciprocal of the square of the wavelength of the light. This constant can be either positive or negative. A positive value refers to a substance in which the direction of rotation is the same as the direction of the current producing the magnetic field. A negative one represents the opposite situation.

First described by: M. Faraday on September 13, 1845 and *published in* M. Faraday, *Experimental researches in electricity. – Nineteenth Series.* Phil. Trans. **136**, 1–20 (1846); M. E. Verdet, *Recherches sur les propriétés optiques développées dans les corps transparents par l'action du magnétisme,* Ann. Chim. **41**, 370–412 (1854).

Faraday law (of electrolysis) governs electrolysis within the following statements: i) the amount of chemical action produced by a current is proportional to the quantity of electricity passed, ii) the masses of different substances deposited or dissolved by the same quantity of electricity are in the same ratios as those of their chemical equivalents.

First described in: M. Faraday, *Experimental researches in electricity. – Seventh series,* Phil. Trans. **124**, 77–122 (1834).

Faraday law (of electromagnetic induction) states that any change of a magnetic flux Φ in an electric circuit induces the electromotive force E proportional to the rate of the flux change, so $E \propto (d\Phi/dt)$.

First described by: M. Faraday in 1831 and *published in* M. Faraday, *Experimental researches in electricity. – Third series,* Phil. Trans. **123**, 23–54 (1834).

fast Fourier transform (FFT) – a powerful tool used in many engineering and physical disciplines. It was developed to compute the discrete **Fourier transform** and its inverse (formally a discrete Fourier transform is a linear transformation mapping any complex vector of length N to its Fourier transform) of a sequence of N numbers in a faster and efficient manner. It is of great importance in several applications, from digital signal processing to solving partial differential equations to developing algorithms for quickly multiplying large integers. In particular, the **Cooley–Tukey algorithm**, being one of the fast Fourier transform algorithms, reduces the number of computations from $O(N^2)$ to $O(N \log N)$. It is interesting to note that an algorithm similar to the FFT has been recently attributed to Carl Friedrich Gauss.

More details in: M. T. Heidemann, D. H. Johnson, C. Sidney Burrus, *Gauss and the History of Fast Fourier Transform*, IEEE Acoustic, Speech, and Signal Processing Magazine, **1**(10), 14–21 (1984).

Fechner ratio – the differential luminance threshold divided by the luminance.
First described in: G. T. Fechner, *Vorschule der Ästhetik* (1876).

Feher effect – a dynamical nuclear polarization arising as a result of hyperfine coupling between the electron and nuclear spins together with different temperatures representing electron velocity and electron spin populations.
First described in: G. Feher, *Nuclear polarization via 'hot' conduction electrons*, Phys. Rev. Lett. **3**(3), 135–137 (1959).

Felgett advantage – the signal-to-noise ratio obtained with a **Fourier transform** spectrometer when detector noise prevails, which exceeds the signal-to-noise ratio of a scanning spectrometer in the same time by a factor proportional to the square root of the number of spectral elements studied. The increase is due to the ability of a Fourier transform spectrometer to observe all spectral elements simultaneously.
First described in: P. Felgett, *A propos de la théorie du spectrométric interférentiel multiplex*, J. Phys. Radium **19**, 187–222 (1954).

femto- – a decimal prefix representing 10^{-15}, abbreviated f.

Fermat's principle (in optics) states that the actual path between two points taken by a beam of light is the one which is traversed in the least time. In this original form however, the Fermat's principle is somewhat incomplete. An improved form is: a light ray, in going between two points, must traverse an optical path length which is stationary with respect to variations of the path. In such a formulation, the paths may be maxima, minima, or saddle points.
First described by P. Fermat in 1635. See in P. Fermat, *Oeuvres* (Gauthier-Villars, Paris, 1891).

Fermi–Dirac distribution (statistics)

Fermi–Dirac distribution (statistics) gives the probability that in an ideal electron gas in thermal equilibrium at a temperature T an electron state with an energy E will be occupied

$$f(E) = \frac{1}{\exp\left[(E - \mu)/k_B T\right] + 1},$$

where μ is the chemical potential. Differently from the **Bose–Einstein statistics** the Fermi–Dirac distribution also takes into account the **Pauli exclusion principle** – there can be no more than one particle in each quantum state. At absolute zero the chemical potential is equal to the **Fermi energy** E_F. The Fermi energy is often used instead of the chemical potential in the Fermi–Dirac distribution, but one must know that E_F is a temperature dependent quantity.

First described in: E. Fermi, *Zur Quantelung des idealen einatomigen Gases*, Z. Phys. **36**(11/12), 902–912 (1926); P. A. M. Dirac, *Theory of quantum mechanics*, Proc. R. Soc. Ser. A. **112**, 661–677 (1926).

Recognition: in 1933 P. A. M. Dirac and E. Schrödinger received the Nobel Prize in Physics for the discovery of new productive forms of atomic theory.

See also www.nobel.se/physics/laureates/1933/index.html.

Fermi energy – the energy of the topmost filled level in the ground state of an electron system at absolute zero (for more details → **Fermi–Dirac distribution**).

Fermi gas – a system of noninteracting **fermions**.

Fermi level = **Fermi energy**.

Fermi liquid – a system of interacting **fermions**. The low-energy excitations (or **quasiparticles**) of this system act almost like completely free electrons, moving entirely independently of one another.

fermions – quantum particles whose energy distribution obeys the **Pauli exclusion principle** and **Fermi–Dirac statistics**.

Fermi's golden rule – if an electron (or hole) in a state $|i\rangle$ of energy E_i experiences a time-dependent perturbation $\mathbf{H}$ which could scatter (transfer) it into any one of the final states $|f\rangle$ of energy E_f, then the lifetime of the carrier in the state $|i\rangle$ is given by

$$\frac{1}{\tau_i} = \frac{2\pi}{h} \sum_f |\langle f|\mathbf{H}|i\rangle|^2 \delta(E_f - E_i).$$

Thus, the rule describes the lifetime of a particle in a particular state with respect to scattering by a time varying potential.

The formula was firstly proposed by P. A. M. Dirac. More than 20 years later, Fermi, in his Chicago lectures, called the formula a golden rule.

First described in: P. A. M. Dirac, *The quantum theory of emission and absorption of radiation*, Proc. R. Soc. (London) A **114**, 243–265 (1927); E. Fermi, *Nuclear Physics* (University of Chicago Press, 1950).

Fermi sphere – the **Fermi surface** of an assembly of **fermions** in the approximation that they are free particles.

Fermi surface – a constant energy surface in the space containing the wave vectors of states, that is k space, defined by the **Fermi energy**.

Fermi–Thomas approximation supposes that if the total potential V of electrons in a solid does not vary greatly over the distance r corresponding to the electron wavelength, the local Fermi kinetic energy E_F must also vary in order to compensate the variation of the potential, so that $V + E_F =$ constant.

Fermi wave vector – the momentum corresponding to the **Fermi energy**.

Fermi wavelength – $\lambda_F = 2\pi/k_F$, where k_F is the **Fermi wave vector**.

ferric wheel $\rightarrow$ **molecular nanomagnet**.

ferrimagnetic – a substance in which atomic magnetic momenta are arranged in such a way that a resulting momentum occurs as a consequence of the tendency for antiparallel orientation of the magnetic momenta of certain neighboring atoms $\rightarrow$ **magnetism**.

ferrite – a nonmetallic magnetic compound containing Fe_2O_3 as a major component. Examples: $MeFe_2O_4$ (Me = Mn, Fe, Co, Ni, Zn, Cd, Mg,) – spinel ferrites; $R_3Fe_5O_{12}$ (R- rare earth metal or yttrium) – garnets.

ferritin – a self-assembled multi-subunit protein involved in biological functions such as iron storage and heme production. It is about 12 nm in diameter. This protein is constructed from 24 nearly identical subunits forming a spherical shell that encloses an internal cage 8 nm in diameter. The cavity contains up to 4500 iron atoms in the form of a poorly crystalline Fe(III) oxy-hydroxide mineral, ferrihydrite. The size of the ferrihydrite core is constrained by the 8-nm inner cavity of the protein.

More details in: E. L. Mayers, S. Mann, *Mineralization in nanostructured biocompartments: biomimetic ferritins for high density data storage*, in: Nanobiotechnology. *Concepts, Applications and Perspectives*, edited by C. M. Niemeyer and C. A. Mirkin (Wiley-VCH, Weinheim, 2004).

ferrocene – the prototypical **metallocene** compound $Fe(C_5H_5)_2$ consisting of two **cyclopentadienyl** rings bound on opposite sides of a central iron atom and forming a sandwich structure schematically shown in Figure F.3.

It has two five-membered carbon rings which are parallel, with the iron ion sandwiched between them. The bonding is between π orbitals on the rings and d orbitals on the Fe^{2+} ion. That provides an electron configuration like that of an inert gas, making it chemically stable.

The compound demonstrates electrochemically reversible one-electron **redox** behavior. The ferrocenium ion, which is formed by oxidation of ferrocene is also stable in the air and easy to handle.

Ferrocene is used as a combustion control additive in fuels, an antiknock agent in gasoline, and for heat stabilization in greases and plastics. Its redox responsivity is of interest for the design of molecular devices.

First described in: T. J. Kealy, P. L. Pauson, *A new type of organo-iron compound*, Nature **168**, 1039–1040 (1951).

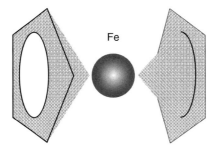

Figure F.3 Ferrocene molecule.

ferroelectric – a dielectric for which a nonlinear hysteretic relationship exists between **polarization** and applied electric field over certain temperature ranges. It can be considered as the electric analog of a **ferromagnetic**. All ferroelectrics are **piezoelectric** and **pyroelectric**, but they additionally possess a reversible, nonvolatile macroscopic spontaneous electric dipole moment in the absence of an external electric field. Ferroelectric crystals can be seen as an assembly of batteries with a particular orientation, which remains stable unless an external electric field is applied to change its direction. Their polar state is a consequence of the structural transition from a high-temperature, high-symmetry paraelectric phase to a low-temperature, low-symmetry ferroelectric phase. These materials also behave as high-dielectric-constant insulators useful for capacitors and energy storage materials.

First described in: J. Valasek, *Piezo-electric and allied phenomena in Rochelle salt*, Phys. Rev. **17**(4), 475–481 (1921).

ferromagnetic – a substance exhibiting: (i) high value of magnetization in weak magnetic fields, coming to saturation with an increase of the field strength, (ii) abnormally high permeability that depends on the field strength and prior magnetic history (hysteresis), (iii) residual magnetism and spontaneous magnetization. All these properties disappear above a certain temperature called the **Curie point** and the material behave like a **paramagnetic** → **magnetism**.

ferromagnetic resonance – a resonant absorption of energy in **ferromagnetics** from an external high frequency circularly polarized magnetic field when its frequency is equal to the internal frequency of precession. Note that if total magnetic momentum of a specimen is displaced from its equilibrium position, it precesses about the direction of the uni-directional magnetic field applied.

First described by V. K. Arkadiev in 1912 – prediction; J. H. E. Griffiths, *Anomalous high frequency resistance of ferromagnetic metals*, Nature **158**, 670–671 (1946) – experimental observation.

Feshbach resonance – a resonance of a many-body system, with more than one degree of freedom, which would turn into a bound state if coupling between some degrees of freedom and the degrees of freedom associated to the fragmentation were set to zero.

For instance, in a system containing two neutral atoms the electrostatic interaction between the electrons and nuclei creates a potential well that is attractive at large distances and repulsive at short distances. In Feshbach resonance a change of magnetic field mimics a change in the depth of the well. Crucially, a bound state of two atoms can exist when the depth of the well is large. Moreover, if the atoms occupying this bound state are **fermions**, the pair has integer spin and is therefore a **boson**. By cooling the system to the critical temperature where **Bose–Einstein condensation** takes place, the composite bosons can then undergo a phase transition to a superfluid state.

The phenomenon is important in the study of **Fermi gases**, as these resonances allow for the creation of Bose–Einstein condensates. In such systems the resonance occurs when the energy of a bound state of an interatomic potential is equal to the kinetic energy of a colliding pair of atoms, which have hyperfine structure coupled via Coulomb or **exchange interaction**s. This condition is rare, but can be satisfied in ultracold alkali atoms.

First described in: H. Feshbach, *Unified theory of nuclear reactions*, Ann. Phys. **5**(4), 357–390 (1958).

Feynman diagram – an intuitive picture of a term in perturbation expansion of a **scattering matrix** element or other physical quantity associated with interactions of particles: in the diagrams each line represents a particle, each vertex an interaction.

First described in: R. P. Feynman, *Space-time approach to quantum electricity*, Phys. Rev. **76**(6), 769–787 (1949).

Feynman–Hellman theorem states that the derivative of the energy E_n of a state n with respect to some parameter q is equal to the expectation value of the derivative

Feynman integral

of the **Hamiltonian H** in that state:

$$\frac{\partial E_n}{\partial q} = \left\langle \frac{\partial H}{\partial q} \right\rangle_n = \left\langle n \left| \frac{\partial H}{\partial q} \right| n \right\rangle.$$

First described in: H. Hellman, *Einführung in die Quanten Chemie* (Leipzig, 1937); R. P. Feynman, *Forces in molecules*, Phys. Rev. **56**(4), 340–343 (1939).

Feynman integral – a term in perturbation expansion of a **scattering matrix** element. It is an integral over the **Minkowski space** of various particles or over the corresponding momentum space of the product of propagators of these particles and quantities representing interactions between the particles.
First described in: R. P. Feynman, *Space-time approach to non-relativistic quantum mechanics*, Rev. Mod. Phys. **20**(2), 367–387 (1948).

Feynman propagator – a factor of the form $(p + m)/(p^2 - m^2 + i\epsilon)$ in a transition amplitude corresponding to a line that connects two vertices in a **Feynman diagram**, and that represents a virtual particle.
First described in: R. P. Feynman, *Space-time approach to non-relativistic quantum mechanics*, Rev. Mod. Phys. **20**(2), 367–387 (1948).

Fibonacci sequence – the sequence of integers produced by adding the two previous terms: 1, 2, 3, 5, 8, 13, 21,... One of the fascinating illustrations of the sequence in nature is that it predicts the number of rabbits there will be after a certain number of months starting from a single breeding couple. The ratio of successive terms tends to the golden ratio, which is about 1.618 – a number that appears throughout nature in different geometrical guises such as seashells, pine cones, and cauliflowers.
First described in: L. P. Fibonacci. *Liber Abbaci* (1202).

fibrous protein → **protein**.

Fick's laws → **diffusion**.
First described in: A. Fick, *Über Diffusion*, Ann. Phys. **94**, 59–86 (1855).

field emission – emission of electrons from a solid surface caused by an applied electric field.

field ion microscopy (FIM) – a projection-type microscopy in which individual atoms protruding on the surface of a needle-shaped specimen are imaged by preferential ionization of imaging gas, that is usually He or Ne, at the protruding site.
The sample to be analyzed is prepared in the form of a sharp tip. A positive potential is applied to the tip such that a very large electric field is present at the tip. The ambient gas surrounding the tip, that is the imaging gas, is usually He

or Ne at a reduced pressure. The gas atoms are ionized in collisions with an atom protruding on the surface. Then they are accelerated away from the tip where they strike a fluorescent screen. The net effect of many gas atoms is to create a pattern on the fluorescent screen showing spots of light which correspond to individual atoms on the tip surface.

First described in: E. W. Müller, *Das Feldionenmikroskop*, Z. Phys. **131**(1), 136–142 (1951).

figure of merit – a performance rating that governs the choice of a device for a particular situation. For instance, the dimensionless thermoelectric characteristic (ZT) of a material is used as a figure of merit to evaluate the material's suitability for thermoelectric power generation. It is defined as $ZT = (S^2 T)/(\kappa \rho)$, where S is the **Seebeck coefficient**, κ is the total thermal conductivity, and ρ is the electrical resistivity.

FIM – an acronym for a **field ion microscopy**.

fine structure constant – the dimensionless constant $\alpha = e^2/(hc)$ used in scattering theory.

In quantum electrodynamics it represents the strength of the interaction between **electron**s and **photon**s. This value is $\alpha = 0.007297352533$, and for the often used inverse value $\alpha^{-1} = 137.03599977$.

Finnis–Sinclair potential – the interatomic short-range potential based on the second moment approximation within the **tight-binding approach**. It has the form

$$V = 0.5 \sum_{ij} V(r_{ij}) - A \sum_{i} \sqrt{\rho_i},$$

where $V(r_{ij})$ is the pair potential, r_{ij} is the distance between atoms i and j, and A is the fitting parameter. The total electron density ρ_i at an atom position i is calculated as a linear superposition of electron-density contributions from neighboring atoms

$$\rho_i = \sum_{i \neq j} \phi(r_{ij}),$$

where the electron density contribution from an atom j to an atom i is $\phi(r_{ij}) = (r_{ij} - d)^2$ for $r_{ij} \leq d$ and $\phi(r_{ij}) = 0$ for $r_{ij} > d$ with d to be the fitted cutoff distance.

This potential is often used in the embedded atom method.

First described in: M. W. Finnis, J. E. Sinclair, *A simple empirical N-body potential for transition metals*, Phil. Mag. A **50**(1), 45–55 (1984).

More details in: G. A. Mansoori, *Principles of Nanotechnology: Molecular-based Study of Condensed Matter in Small Systems* (World Scientific, Singapore, 2005).

Firsov potential

Firsov potential – the semiclassical interatomic pair potential in the form:

$$V(r) = \frac{Z_1 Z_2 e^2}{r} \chi \left[(Z_1^{1/2} + Z_2^{1/2})^{2/3} \frac{r_0}{a} \right],$$

where

$$\chi(x) = \chi \left[(Z_1^{1/2} + Z_2^{1/2})^{2/3} \frac{r_0}{a} \right]$$

is the screening function with the **Firsov scale factor** $V(r) = A/r^2$, with $A = (3.05 \times 10^{-16} Z_1 Z_2)/(Z_1^{1/2} + Z_2^{1/2})^{2/3}$. Here r is the interatomic distance, Z_1 and Z_2 are the atomic numbers of the interacting atoms, and r_0 and a are fitting parameters.

First described in: O. B. Firsov, *Calculation of the interaction potential of atoms*, Zh. Exp. Teor. Fiz. **33**(3), 696–699 (1957) – in Russian.

Firsov scale factor $\rightarrow$ **Firsov potential**.

FISH – an acronym for **fluorescence *in situ* hybridization**.

FLAPW method = **full potential augmented plane wave (FLAPW) method**.

Fleming tube – the first vacuum tube diode, consisting of a heated filament and cold metallic electrode placed in an evacuated glass tube.

First described by J. A. Fleming and patented on November 16, 1904.

flicker noise – the noise observed on the direct current (I) flowing through vacuum tubes and semiconductor devices. Its spectrum is $S(f) = KI^n/f^a$, where K is a constant dependent on the material, $n \approx 2$, and a varies between 0.5 and 2. This noise is also called **1/f noise** or **pink noise**. Its major cause in semiconductor devices is traced to surface properties of the material. The generation and recombination of charge carriers in surface energy states are the most influencing factors.

First described in: J. B. Johnson, *The Schottky effect in low-frequency circuits*, Phys. Rev. **26**(1), 71–85 (1925).

fluctuation–dissipation theorem states a general relationship between the response of a given system to an external disturbance and the internal fluctuation of the system in the absence of the disturbance. Such a response is characterized by a response function or equivalently by an admittance or an impedance. The internal fluctuation is described by a correlation function of relevant physical quantities of the system fluctuating in thermal equilibrium, or equivalently by their fluctuation spectra. The most striking feature of this theorem is that it relates, in a fundamental manner, the fluctuations of a physical quantity pertaining to the equilibrium state of the given systems to a dissipative process which, in practice, is realized only when the system is subjected to an external force that drives it away from equilibrium.

This means that the response of a given system to an external perturbation (nonequilibrium property) can be expressed in terms of fluctuation properties of the system in thermal equilibrium (equilibrium property).

Probably one of the first forms of the fluctuation–dissipation theorem was given by the **Einstein relationship** that relates the viscous friction of a Brownian particle to the diffusion constant of the particle in the form $D = k_B T/m\gamma$, where D is the diffusion constant and $m\gamma$ is the friction constant. The Einstein relation can be also written as: $\mu = 1/m\gamma = D/k_B T = (1/k_B T) \int_0^\infty < u(t_0)u(t_0 + t) > dt$, which means that the mobility μ, the inverse of the friction constant, is related to the fluctuation of the velocity $u(t)$ of the **Brownian motion**; this is in fact the fluctuation–dissipation theorem.

The theorem can be used in two ways. From the known characteristics of the admittance or the impedance one can predict the characteristics of the fluctuation or the noise intrinsic to the system. Alternatively, from the analysis of thermal fluctuations of the system one can derive the admittance or the impedance. A general proof of the fluctuation–dissipation theorem has been given, in modern times, by the linear response theory.

First described: It is difficult to say definitely when the theorem was formulated for the first time. It appeared several times in different fields of applications. For two applications in the senses outlined above see H. Nyquist, *Thermal agitation of electrical charge in conductors*, Phys. Rev. **32**(1), 110–113 (1928), L. Onsager, *Reciprocal relations in irreversible processes. I*, Phys. Rev. **37**(4), 405–426 (1931), for a proof within the linear response theory see H. B. Callen, T. A. Welton, *Irreversibility and generalized noise*, Phys. Rev. **83**(1), 34–40 (1951).

More details in: R. Kubo, *The fluctuation–dissipation theorem*, Rep. Prog. Phys. **29**, 255–284 (1966).

Recognition: in 1968 L. Onsager received the Nobel Prize in Chemistry for the discovery of the reciprocal relations bearing his name, which are fundamentals for the thermodynamics of irreversible processes.

See also www.nobel.se/chemistry/laureates/1968/index.html.

fluorescence – classically the emission of radiation of a characteristic wavelength by a substance having been excited, which occurs no longer than 10^{-8} s after removal of the excitation. Longer light emission is called **phosphorescence**. In today words, it is the light emitted as a result of spin-allowed radiative deactivation of **excited states**.

First described in: G. Stokes, *On the change of refrangibility of light*, Proc. Roy. Soc. London **6**, 195–199 (1852).

fluorescence *in situ* hybridization (FISH) – an analytical technique employing fluorescent molecular tags to detect probes hybridized to **chromosomes** or chromatin. It is used for genetic mapping and detecting chromosomal abnormalities.

fluorescence microscopy

fluorescence microscopy – an optical imaging technique in which an analyzed specimen is irradiated at a wavelength exciting either natural or artificially introduced fluorescent molecules known as fluorophores. The image is then studied through a filter that transmits at the fluorescence wavelength, but absorbs light at the excitation wavelength.

fluorescence nanoscopy – a set of techniques employed to overcome the so-called **diffraction** or **Abbe's resolution limit**.

In **confocal fluorescence microscopy**, the excitation light wave formed by the lens to a spherical cap produces a three-dimensional diffraction spot, generating fluorescence in the focal region. A point-like detector registers fluorescence mostly from the main maximum, thus providing a slightly improved resolution (>200 nm in the focal plane and to >450 nm along the optical axis) with respect to conventional **fluorescence microscopy**.

In **structured illumination microscopy (SIM)** a diffraction grating is placed in the path of the excitation light generating a diffraction pattern on the sample. This fine pattern of light interacts with fine patterns in the sample and creates an interference pattern whose periodicity is much greater than that of the original pattern. Thus, fine patterns that were previously below the **Abbe's resolution limit** can now be visualized.

By combining the wavefront caps of two opposing lenses the emitted light can be harvested by a greater portion of the sphere surrounding the sample, this so called **FourPi (4Pi) microscopy** forms a narrower spot along the optical axis and hence an improved resolution along the optical axis (80 to 150 nm).

Stimulated emission depletion (STED) microscopy make use of a regularly focused excitation beam that is superimposed by a doughnut-shaped STED beam that instantly quenches excited molecules at the periphery of the excitation spot, thus confining fluorescence emission to the doughnut zero. Saturated quenching results in a fluorescent spot far below diffraction whose scanning across the sample yields a sub-diffraction-resolution image.

In **incoherent interference illumination (I^3M) microscopy** the light, coming from two objectives, interferes constructively at the focal plane allowing an stronger concentration of the effective excitation. If the emitted light through the two objectives is collected and constructively combined on a single detector one obtains a sharper pattern. This last technique is called **image interference (I^2M) microscopy**. These two last techniques can be combined creating **image interference (I^2M) + incoherent interference illumination (I^3M) microscopy (I^5M)**. The resolution reached is of the same order as that in **FourPi (4Pi) microscopy**.

Other techniques use bright and dark molecular states to break the diffraction limit.

Parallelized **STED, ground-state depletion (GSD)** and **saturated pattern excitation spectroscopy (SPEM)** (similar to **SSIM**) all use reversible saturable or photoswitchable transitions between two molecular states. In particular they use photophysical transitions. One limitation here is the finite number of cycles between the two states. These techniques are able to resolve ensembles of molecules.

Instead if one uses photochemical transitions in which atoms are relocated or bonds are formed or broken one has **photoactivable localization microscopy (PALM)** and **stochastic optical reconstruction microscopy (STORM)** that are able to resolve single molecules.

The acronym **RESOLFT** (i.e. **reversible saturable optically linear fluorescence transitions**) is now used for all these techniques.

More details in: W. A. Wells, *Man the nanoscopes*, J. Cell. Biol. **164**(3), 337–340 (2004); S. W. Hell, *Far-field optical nanoscopy*, Science **316**, 1153–1158 (2007).

fluxon – a quantum of magnetic flux $\Phi_0 = h/2e = 2.07 \times 10^{-15}$ Wb.

Foch space – the space in the quantum field theory where the number of particles is generally not fixed. It is sometimes convenient to represent the state of a system by an infinite set of wave functions, each of which refers to a fixed number of particles.

Fock exchange potential → Hartree–Fock approximation.

focused ion beam (FIB) system – a system for sample preparation and fabrication at the nanoscale by ion-mill sculpting or beam-induced deposition. Moreover, exposing sensitive surfaces to a focused ion beam makes it possible to fabricate and write dimensions into resist materials well below 20 nm, which in fact is a FIB lithography technique. A beam of singly charged ions (Ga^+) is usually used for this. Typical FIB instruments provide a focal spot size below 5 nm at a current of a few pA. Practical resolution is a few tens of nanometers. The proximity effect of FIB lithography is much weaker than that of **electron beam lithography** because of the shorter range of the exposing particles (ions), smaller lateral spread, weaker secondary electron generation, and lower energy of secondary electrons. A serial writing process employing scanning of the ion beam or a substrate is used to expose the whole substrate surface. This causes the throughput of the technique to be quite limited and not effective in mass production.

Fogden–White theory generalizes the **friction force theory (Hertz theory)** and **Johnson–Kendall–Roberts theory** for contact elasticity by introducing a condensable vapor atmosphere, in which capillary condensation is well known to significantly affect the force of **adhesion** between the solids.

First described in: A. Fogden, L. White, *Contact elasticity in the presence of capillary condensation: I. The nonadhesive Hertz problem*, J. Coll. Inter. Sci. **138**(2), 414–430 (1989).

Foiles potential describes the pair interaction between two different atoms approximated by the geometric mean of the pair interaction of the individual atoms. For atoms A and B separated by the distance r it has the form $V(r) = Z_A(r)Z_B(r)/r$,

Fokker–Planck equation

where $Z(r) = Z_0 \exp(-ar)$. The value of Z_0 is given by the number of outer electrons of the atom and a is the fitting parameter.

First described in: S. M. Foiles, *Calculation of the surface segregation of Ni-Cu alloys with the use of embedded-atom method*, Phys. Rev. B **32**(12), 7685–7689 (1985).

Fokker–Planck equation – an equation for the distribution function $f(r, p, t)$ of particles in a gas or a liquid, analogous to the **Boltzmann distribution** but applying where forces are long-range and collisions are not binary:

$$\frac{\partial f}{\partial t} + \frac{p}{M}\frac{\partial f}{\partial r} - \frac{\partial V}{\partial r}\frac{\partial f}{\partial p} = \frac{\partial}{\partial p}\xi\left(\frac{p}{M}f + k_B T\frac{\partial f}{\partial p}\right),$$

where r is the coordinate, p is the momentum, M is the mass of the gas particles; t is time and T is temperature; V is the potential from an external force acting on the particles and ξ is the friction coefficient. It describes the evolution of the distribution function in the presence of diffusion and driving force.

First described in: A. Einstein, *Über die von der molekularkinetischen Theorie der Wärme geforderte Bewegung von in ruhenden Flüssigkeiten suspendierten Teilchen*, Ann. Phys. **17**, 549–560 (1905), then by A. Fokker, *Die mittlere Energie rotierender elektrischer Dipole im Strahlungsfeld*, Ann. Phys. **43** 810–820 (1914) and by M. Smoluchowski, *Über Brownsche Molekularbewegung unter Einwirkung äusserer Kräfte und deren Zusammenhang mit der verallgemeinerten Diffusionsgleichung*, Ann. Phys. **48**, 1103–1112 (1915), and in the final form in: M. Planck, *Über einen Satz der statistischen Dynamik und seine Erweiterung in der Quantentheorie*, Sitzungsber. Preuss. Akad. Wiss. Berlin **24**, 324–341 (1917).

folded spectrum method supposes, instead of seeking solutions of the time-independent **Schrödinger equation**, that solutions are sought to the alternative expression: $(\mathbf{H} - E_{ref})^2 \psi_{n,k} = (E_{n, k} - E_{ref})^2 \psi_{n,k}$, where the reference energy E_{ref} can be chosen to lie within the fundamental gap and hence the **valence** and **conduction band** edge states are transformed from being arbitrary high-energy states to being the lowest states. The technique serves to minimize the expectation value $\langle\psi|(\mathbf{H} - E_{ref})^2|\psi\rangle$, where the standard empirical pseudopotential operator can be employed.

First described in: L. W. Wang, A. Zunger, *Solving Schrödinger's equation around a desired energy: application to quantum dots*, J. Chem. Phys. **100**(3), 2394–2397 (1994).

f orbital → atomic orbital.

Ford–Kac–Mazur formalism was originally proposed to study Brownian motion in coupled oscillators. It was later extended to a general analysis of a quantum particle coupled to a quantum mechanical heat bath. Within this formalism the bath is considered as a collection of oscillators which are initially in equilibrium. Its degrees of freedom are then eliminated, leading to quantum Langevin equations for the remaining degrees of freedom of the whole system. Thus, such a bath can be viewed as a source of noise and dissipation in the system. The formalism is used for simulation of classical heat transport in disordered harmonic chains as well as for a detailed study of quantum transport in disordered electronic and phononic systems.

First described in: W. Ford, M. Kac, P. Mazur, *Statistical mechanics of assemblies of coupled oscillators*, J. Math. Phys. **6**(4), 504–515 (1965).

Förster energy transfer – a long-range Coulomb-induced transfer of optically excited excitons.

First described in: Th. Förster, *Transfer mechanisms of electronic excitation*, Disc. Faraday Soc. **27**, 7–17 (1959).

More details in: Th. Förster, *Delocalized excitation and excitation transfer*, in: Modern Quantum Chemistry, IIIB, edited by O. Sinanoglu (Academic Press, New York, 1965), pp. 93–137.

Foucault currents → **eddy currents**.

FourPi (4Pi) microscopy → **fluorescence nanoscopy**.
First described in: S. W. Hell, E. H. Stelzer, *Properties of a 4PI-confocal fluorescence microscope*, J. Opt. Soc. Am. A **93**, 277–282 (1992).

fountain effect occurs when two containers of superfluid helium are connected by a capillary tube and one of them is heated, so that helium flows through the tube in the direction of higher temperature.

Fourier analysis – a mathematical tool by which discontinuous and periodic functions may be represented in terms of continuous functions. If $f(x)$ is single valued in the range of $a < x < b$, and is sectionally continuous, that is it contains only a finite number of discontinuities in the range which are finite in magnitude, then it is possible to expand $f(x)$ in an infinite set of functions orthogonal in this range. By suitable scaling the range can be made $-\pi < x < \pi$, and it is then convenient to choose the circular functions for the expansion in the form $f(x) = A_0 + \sum_{k=1}^{\infty}(A_k \cos kx + B_k \sin kx)$, which is termed the **Fourier expansion** or **Fourier series**. The coefficients are $A_0 = 1/2\pi \int_{-\pi}^{\pi} f(x)dx$, $A_k = 1/\pi \int_{-\pi}^{\pi} f(x)[\cos kx]dx$, $B_k = 1/\pi \int_{-\pi}^{\pi} f(x)[\sin kx]dx$. If $f(x)$ is an even function, the B_k are zero. If $f(x)$ is an odd function, the A_k are zero.

Instead of using sine and cosine functions for the expansion, one can employ the exponential function with the imaginary argument: $f(x) = \sum_{-\infty}^{\infty} C_k \exp(ikx)$ with $C_k = 1/2\pi \int_{-\pi}^{\pi} f(x) \exp(-ikx)dx$. A Fourier series may be integrated term by term, but it cannot generally be differentiated.

First described in: J. Fourier, *La Thèorie Analytique de la Chaleur* (Didot, Paris, 1822).

Fourier equation – the differential equation giving the temperature T at any place and at any time when heat transport is by the conduction mechanism only. In the

Fourier law

absence of internal heat sources and for a nonhomogeneous material it has the form $\rho C(\partial T/\partial t) = \nabla \cdot (K\nabla T)$, where ρ is the material density, C is the specific heat of the material, t is the time, and K is the thermal conductivity of the material.

Fourier law states that the rate of heat flow by conduction across an infinitesimally small area S is proportional to the temperature gradient normal to the area: $dQ/dT = -KSdT/dr$, where Q is the quantity of heat, t is the time, K is the thermal conductivity of the material at the place in question, r is the distance measured normal to S in the direction of decreasing temperature.

First described in: J. Fourier, *La Theorie Analytique de la Chaleur* (Didot, Paris, 1822).

Fourier series → **Fourier analysis**.

Fourier transform for a function $f(x)$ is given by

$$F(\xi) = 1/\sqrt{2\pi} \int_{-\infty}^{\infty} f(x) \exp(i\xi x)dx.$$

If the Fourier transform $F(\xi)$ of a function is known, the function itself can be found by means of the Fourier inversion theorem

$$f(x) = 1/\sqrt{2\pi} \int_{-\infty}^{\infty} F(\xi) \exp(-i\xi x)dx.$$

First described in: J. Fourier, *La Théorie Analytique de la Chaleur* (Didot, Paris, 1822).

Fowler–Nordheim tunneling describes charge carrier transport through insulators. The current density is $J = CE^2 \exp(-E_0/E)$, where C and E_0 are constants in terms of effective mass and barrier height, and E is the electric field in the insulator (V/m).

First described in: R. H. Fowler, L. Nordheim, *Electron emission in intense electric fields*, Proc. R. Soc. A **119**, 173–181 (1928).

fractal – a rough or fragmented geometric shape that can be subdivided into parts, each of which is (at least approximately) a reduced-size copy of the whole. As a geometric object, a fractal is characterized by the following features: (i) it has a fine structure at arbitrarily small scales; (ii) it is too irregular to be easily described in traditional Euclidean geometric language; (iii) it is self similar, at least approximately; (iv) it has a Hausdorf dimension which is greater than its topological dimension, while this requirement is not met by space-filling curves such as the Hilbert curve; (v) it has a simple and recursive definition.

First described in: B. Mandelbrot, *Les objets fractals: forme, hasard et dimension*, (Flammarion, Paris, 1975).

More details in: K. Falconer, *Fractal Geometry: Mathematical Foundations and Applications* (John Wiley and Sons, Chichester, 2003).

fractional quantum Hall effect → **Hall effect**.

Franck–Condon excited state – an excitation of an electron in the potential adapted to the ground state.

Franck–Condon principle states that as nuclei, being heavy particles, are so much more massive than electrons, an electronic transition occurs faster than the nuclei can respond. It also states that for the transition probability not to be negligibly small it is necessary: (i) that there should be an overlap between the classically allowed regions of coordinates for the states participating in the transition, and (ii) that the classical velocities of the heavy particles within the region of the overlap should not be very different for these states. This principle explains the fact that in solids a luminescent center emits light at a longer wavelength than it absorbs, and that in order to impel an electron optically from an impurity level into the conduction band the necessary minimum energy is greater than the energy difference between the bottom of the conduction band and the impurity level.

First described in: J. Franck, *Elementary processes of photochemical reactions*, Trans. Faraday Soc. **21**(3), 536–542 (1925); E. Condon, *A theory of intensity distribution in band systems*, Phys. Rev. **28**(6), 1182–1201 (1926).

Franck–Hertz experiment first demonstrated that intrinsic energy of atoms could not be changed instantaneously. It was done by recording the kinetic energy lost by electrons in inelastic collisions with atoms.

First described in: J. Franck, G. Hertz, *Über Zusammenstösse zwischen Elektronen und den Molekülen des Quecksilberdampfes und die Ionisierungsspannung desselben,* Verh. Deutsche Phys. Ges. **15**(20), 929–934 (1914).

Recognition: in 1925 J. Franck and G. Hertz received the Nobel Prize in Physics for their discovery of the laws governing the impact of an electron upon an atom.

See also www.nobel.se/physics/laureates/1925/index.html.

Frank partial dislocation – a partial dislocation whose **Burger's vector** is not parallel to the fault plane, so that it can only diffuse and not glide, in contrast to a **Shockley partial dislocation**.

First described in: F. C. Frank, W. T. Read, *Multiplication processes for slow moving dislocations,* Phys. Rev. **79**(4), 722–723 (1950).

Frank–Read source – a self-regenerating configuration of dislocations from which, under the action of applied stress, successive concentric loops of dislocations can be thrown off in a glide plane. It provids a mechanism for continuing slip in a crystal and an increase in its dislocation content, without essential limit, through plastic deformation.

Frank–Van der Merwe growth mode (of thin films)

First described in: F. C. Frank, W. T. Read, *Multiplication processes for slow moving dislocations,* Phys. Rev. **79**(4), 722–723 (1950).

Frank–Van der Merwe growth mode (of thin films) – layer-by-layer growth of thin films during their deposition as illustrated in Figure F.4.

It arises when deposited atoms are more strongly attracted to the substrate than they are to themselves.

First described in: F. C. Frank, J. H. van der Merwe, *One-dimensional dislocation. II. Misfitting monolayers and oriented overgrowth,* Proc. R. Soc. A **198**, 216–225 (1949).

Deposited material

Figure F.4 Frank–Van der Merwe growth mode during thin film deposition.

Franz–Keldysh effect – an effect of an electric field on the optical absorption edge in semiconductors. It is illustrated in Figure F.5, where band structure of a semiconductor under external biasing is sketched. An optically induced transition of an electron from the valence into conduction band in an undisturbed semiconductor can occur only if two states have an energy separation $\Delta E \leq h\nu$, where ν is the light frequency. Moreover, the states have to overlap in space. Interband light absorption in a semiconductor is impossible if $h\nu$ is smaller than the band gap E_g because there are no available states for corresponding electron transitions.

Electric biasing of a semiconductor changes the states in both valence and conduction bands provided that the states are present at all energies. Their overlap in space depends on the difference in energies. It is strong if $\Delta E > E_g$, when the oscillating part of the wave functions overlap. Meanwhile, only the tails of the wave functions overlap if $\Delta E < E_g$, and this decays rapidly with an increase of $E_g - \Delta E$. Thus the absorption edge gains a tail tunneling into the previously forbidden gap, as well as due to changes in the wave function. Electron transitions in such conditions are considered as photo-assisted field-induced tunneling across

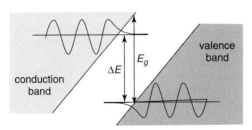

Figure F.5 Interband electron transitions determining light absorption in an electrically biased semiconductor

the band gap. They result in the shift of the absorption edge to a lower energy range. The related oscillations in imaginary part of the **dielectric function** above the band gap is known as **Franz–Keldysh oscillations**.

First described in: W. Franz, *Einfluss eines elektrischen Felds auf eine optische Absorptionskante*, Z. Naturforsch. A**13**(6), 484–489 (1958) – in German; and independently in: L. V. Keldysh, *Effect of a strong electric field on the optical properties of non conducting crystals*, Zh. Eksp. Teor. Fiz. **34**(5), 1138–1141 (1958) – in Russian.

Franz–Keldysh oscillations → **Franz–Keldysh effect**.

Fraunhofer diffraction – diffraction of radiation passed through an aperture and registered at a distance from it so large that waves from the virtual sources at the aperture arrive in phase at a point on the normal to the screen used for registration.

Fraunhofer lines – a set of dark lines appearing against the bright background of the continuous solar spectrum. The lines are mainly produced by absorption of light by the cooler gases in the Sun's outer atmosphere at frequencies corresponding to the atomic transition frequencies in particular chemical elements composing these gases. Some of the lines are the result of the absorption by oxygen in the Earth's atmosphere. Characteristics of the major Fraunhofer lines are presented in the Table F.1.

First described by W. H. Wollaston in 1802, then systematically studied by J. von Fraunhofer in 1814 and *published in*: J. Fraunhofer, Gilberts Ann. **56**, 264 (1817).

Table F.1 Major Fraunhofer lines.

Designation	Element	Wavelength (Wl) nm	Designation	Element	Wl nm
y	O_2	898.765	c	Fe	495.761
Z	O_2	822.696	F	H β	486.134
A	O_2	759.370	d	Fe	466.814
B	O_2	686.719	e	Fe	438.355
C	H α	656.281	G'	H γ	434.047
a	O_2	627.661	G	Fe	430.790
D_1	Na	589.592	G	Ca	430.774
D_2	Na	588.995	h	H δ	410.175
D_3 (or d)	He	587.5618	H	Ca^+	396.847
e	Hg	546.073	K	Ca^+	393.368
E_2	Fe	527.039	L	Fe	382.044
b_1	Mg	518.362	N	Fe	358.121
b_2	Mg	517.270	P	Ti^+	336.112
b_3	Fe	516.891	T	Fe	302.108
b_4	Fe	516.751	t	Ni	299.444
b_4	Mg	516.733			

Fredkin gate

Fredkin gate → quantum logic gate.

free electron laser – a **laser** producing **stimulated emission** by passing a beam of free electrons (not bound to an atom or molecule) through an undulator or "wiggler." The undulator creates a magnetic field of alternating polarity (in another version it guides the electrons along a helical path), causing the electrons to "wiggle" and thus release radiation. It covers a wide range of wavelengths – from the soft X-ray region to millimeters.

free energy – a state function of a system under consideration. There are two commonly used notations for it. The **Helmholtz free energy** is $F = U - TS$, where U is the internal energy, S is the **entropy**, T is the temperature. The **Gibbs free energy** is $G = H - TS = U + PV - TS = F + PV$, where H is the **enthalpy** of the system, P is the pressure, and V is the system volume. These functions serve immediately to give the maximum work available from the system in an isothermal process.

Frenkel defect – a crystal defect consisting of a lattice vacancy and a host interstitial atom. It is also known as a **Frenkel pair**.
First described in: J. Frenkel, *Über die Wärmebewegung in festen und flüssigen Körpern*, Z. Phys., **35**(8–9), 652–669 (1926).

Frenkel excitation – an excitation of electron-hole pairs with a distance between electron and hole, called the **exciton Bohr radius**, that is smaller than the length of the lattice **unit cell**. Such excited pairs are called **Frenkel excitons**. An alternative case is represented by **Wannier excitation** and **Wannier excitons**.
First described in: J. Frenkel, *On the transformation of light into heat in solids. I*, Phys. Rev. **37**(1), 17–44; J. Frenkel, *On the transformation of light into heat in solids. II*, Phys. Rev. **37**(10) 1276–1294 (1931).

Frenkel excitons → Frenkel excitation.

Frenkel pair → Frenkel defect.

Frenkel–Poole emission – an electric field assisted thermal excitation of trapped electrons into the conduction band in solids. It is also known as **Frenken–Poole law** or field-induced emission. Insulators and semiconductors display in high electric fields an increase in their electrical conductivity, which finally leads to a breakdown. The dependence of the conductivity on the electric field was represented by Poole in an exponential way (**Poole law**). The Frenkel–Poole law differs from the Poole law by the substitution for the electric field of the square root of the electric field itself. Finally, the current related to the Frenkel–Poole emission is defined as

$$J = CE \exp \left[-\frac{e}{k_B T} \left(\phi_b - \sqrt{\frac{eE}{\pi \epsilon_i}} \right) \right],$$

where C is a constant in terms of the trapping density in the insulator (semiconductor), E is the strength of the electric field, ϕ_b the barrier height, and ϵ_i the dynamic permittivity.

First described in: H. H. Poole, *On the dielectric constant and electrical conductivity of mica in intense field*, Phil. Mag. **32**, 112–129 (1916); J. Frenkel, *On the theory of electrical breakdown of dielectrics and semiconductors*, Technical Physics USSR **5**, 685–687 (1938); J. Frenkel, *On pre-breakdown phenomena in insulators and electronic semi-conductors*, Phys. Rev. **54**(8), 647–648 (1938).

Frenkel–Poole law → Frenkel–Poole emission.

Fresnel diffraction – the diffracted field is studied at a distance from the aperture, which is large compared with the wavelength of the diffracting radiation and the aperture dimensions, but yet not so large that the phase difference between the waves from the aperture sources can be neglected, even along the normal to the screen used for registration. It contrasts with **Fraunhofer diffraction**.

Fresnel law states the conditions of interference of polarized light: (i) two rays of light emanating from a common polarized beam and polarized in the same plane interfere in the same manner as ordinary light, (ii) two rays of light emanating from a common polarized beam and polarized at right angles to each other will only interfere if they are brought into the same plane of polarization, (iii) two rays of light polarized at right angles and emanating from ordinary light will not interfere if brought into the same plane of polarization.

First described by A. J. Fresnel in 1823.

friction force theory (Hertz theory) describes the perfectly elastic, nonadhering case of the contact between two spheres of radii R_1, R_2, **Young's moduli** E_1 and E_2, Poisson ratio ν_1 and ν_2, and external applied load N. The friction force becomes $F_f = \tau \pi (R^*/K)^{2/3} N^{2/3}$, where τ is the **shear modulus**, $R^* = R_1 R_2 / (R_1 + R_2)$, $1/K = (3/4)(\frac{1-\nu_1^2}{E_1} + \frac{1-\nu_2^2}{E_2})$. Here the **real contact area** A_R and the applied load N are related through the formula $A_R = \pi (R^*/K)^{2/3} N^{2/3}$. When $R_2 = \infty$ the contact is between a sphere and a flat surface and $R^* = R_1$.

This theory does not take into account **adhesion** force at the interface.

First described in: H. Hertz, *Über die Berührung fester elastischer Körper*, J. Reine Angew. Math **92**, 156–171 (1881).

friction force microscopy – an **atomic force microscopy** measuring forces along the scanning direction. It is used for atomic scale and micro scale studies of friction and lubrication.

Friedel law

More details in: B. Bhushan, *Micro/nanotribology and materials characterization studies using scanning probe microscopy,* in: Handbook of Nanotechnology, edited by B. Bhushan (Springer, Berlin, 2004), pp. 497–541.

Friedel law states that X-ray or electron diffraction measurements cannot determine whether or not a crystal has a center of symmetry.
First described in: G. Friedel *Sur les symmétries cristallines que peut révéler la diffraction des rayons X,* C. R. **157**, 1533–1536 (1913).

Friedel oscillations – the oscillatory behavior of the screened potential of a point charge at large distances r as $\cos(2k_F r)/(k_F r)^3$, where k_F is the **Fermi wave vector**. The phenomenon originates from the singularity of the dielectric function at $q = 2k_F$. These oscillations, present in a variety of phenomena, depending on the context in which they are present, are also called **Ruderman–Kittel oscillations**.
First described in: J. Friedel, *The distribution of electrons around impurities in univalent metals,* Phil. Mag. **43**(2), 153 (1952); M. A. Ruderman, C. Kittel, *Indirect exchange coupling of nuclear magnetic moments by conduction,* Phys. Rev. **99**(1), 96–102 (1954).

Fröhlich interaction → **phonon**.

f-sum rule → **Kuhn–Thomas–Reiche sum rule**.

Fulcher bands of hydrogen – a system of bands of molecular hydrogen, which are preferentially excited by a low-voltage discharge. They consist of a number of regularly spaced lines in the red and green spectral range.
First described in: G. S. Fulcher, *Spectra of low-potential discharges in air and hydrogen,* Astrophys. J. **37**, 60 (1913).

fullerene – a carbon nanostructure in which 60 or 70 carbon atoms are bound together in a closed hollow cage having the form of a ball. The designation fullerenes is taken from the name of an American architect, R. Buckminster-Fuller, who had designed a dome having the form of a football for the 1967 Montreal World Exhibition.
The prototype fullerene is C_{60}, a molecule formed out of 60 carbon atoms distributed on a sphere with a diameter of 0.7 nm. These carbon atoms are assembled in the form of a truncated dodecahedron with a carbon atom sitting at each corner of the polyhedron. It is shown in Figure F.6.
The carbon atoms outline 20 hexagons and 12 pentagons, the latter giving rise to the curvature and thus enabling the closed quasi spherical structure of the molecule. Valence electrons of the carbon atoms are predominantly sp^2 hybridized. There is also some, typical for a diamond admixture, sp^3 hybridization due to the finite curvature of the molecule.

Figure F.6 Structure of C_{60} fullerene.

First described in: E. Osawa, Kagaku (Kyoto) **25**, 854–863 (1970) – in Japanese, prediction of the truncated icosahedron isomer of C_{60}; D. A. Bochvar, E. G. Gal'perin, *About hypothetical systems: carbododekaedr, s-icosahedron and carbo-s-icosahedron*, Dokl. Akad. Nauk SSSR **209**, 610–612 (1973) – in Russian, Proc. Acad. Sci. USSR **209**, 239–241 (1973) – in English, first theoretical simulation of C_{60} cluster; H. W. Kroto, R. F. Curl, R. E. Smalley, J. R. Heath, *C-60 buckminsterfullerene*, Nature **318**(6042), 162–163 (1985) – fabrication and beginning of experimental study of C_{60} fullerenes.

Recognition: in 1996 H. W. Kroto, R. F. Curl and R. E. Smalley received the Nobel Prize in Chemistry for their discovery of fullerenes.

See also www.nobel.se/physics/laureates/1996/index.html.

fullerite – a solid-state compact manifestation of **fullerene**s and related carbon nanostructures in which these carbon nanounits form a crystal-type lattice.

Pure solid C_{60} fullerite has an **FCC** crystal structure with relatively small **van der Waals** cohesive energy. The fullerite does not appear to melt, but rather sublimes at low pressure. Below the structural phase transition point ($\sim$257 K) the fullerite structure transforms into a simple cubic one. In this structure, molecules with different orientations make up a larger unit cell than the FCC cell: the corner and face-centre molecules become inequivalent, that is there are four C_{60} molecules in the crystal basis.

Fullerites are harder than diamond. Ultrahard fullerites have a hardness value of 310 ± 40 GPa, while diamond is characterized by the maximum hardness of about 230 GPa.

Pure C_{60} fullerite is a semiconductor with a bandgap of 1.6 eV, a value reasonably close to the **HOMO–LUMO** gap of the free molecule.

full potential augmented-plane-wave (FLAPW) method – one of the most powerful schemes for electronic structure calculations of solids within the **density functional theory**. It is based on the **augmented plane wave method** extended by the assumption of no shape approximations for the charge density and the potential. Both the functionalization (of surfaces)

charge density and the effective one-electron potential are represented by the same analytical expansion, that is by a Fourier representation in the interstitial region, in spherical harmonics inside the spheres, and for surface and thin film calculations a two-dimensional Fourier series in the vacuum region. Thus, the charge density $\rho(\mathbf{r})$ is given (with a similar representation for the effective potential) by

$$\rho(\mathbf{r}) = \begin{cases} \sum_i \rho_j \exp i\mathbf{G_j} \cdot \mathbf{r} & \text{for } \mathbf{r} \text{ interstitial} \\ \sum_{lm} \rho_{lm}^{\alpha}(r_{\alpha}) Y_{lm}(r_{\alpha}) & \text{for } \mathbf{r} \text{ inside the sphere } \alpha \\ \sum_q \rho_q(z) \exp i\mathbf{K_q^{\|}} \cdot \mathbf{r} & \text{for } \mathbf{r} \text{ in vacuum} \end{cases}$$

where $\mathbf{G}_j$ denote reciprocal lattice vectors in the bulk, $\mathbf{K}_q$ reciprocal lattice vectors in vacuum and Y_{lm} are the spherical harmonics. The coefficients in the equations are used in each iteration to create the **Coulomb potential** via the solution of the **Poisson equation** and to construct the exchange-correlation potential. These coefficients for the input and output densities are also used to perform the self-consistency procedure mixing at each iteration for input and output density.

First described in: E. Wimmer, H. Krakauer, M. Weinert, A. J. Freeman, *Full-potential self-consisted linearized-augmented-plane-wave method for calculating the electronic structure of molecules and surfaces: O$_2$ molecule*, Phys. Rev. B **24**(2), 864–875 (1981).

More details in: E. Wimmer, A. J. Freeman, *Fundamentals of the electronic structure of surfaces*, in: Handbook of Surface Science, Vol. 2, edited by K. Horn, M. Scheffler, (Elsevier Science B.V., 2000), pp. 1–92.

functionalization (of surfaces) – design, tailoring and preparation of surfaces that are able to define, induce and control distinct architecture and outgrowth. In particular in the case of biofunctional surfaces one must be able to match the sophisticated (bio)**recognition** ability of biological systems on the nanoscale as well as on larger length scales.

FWHM – an acronym for a **f**ull **w**idth at **h**alf **m**aximum, which is a parameter used to characterize optical spectral lines.

G

From gain-guided laser to gyromagnetic frequency

gain-guided laser → double heterostructure laser.

galvanoluminescence – emission of light produced by the passage of an electrical current through an appropriate electrolyte in which an electrode, made of certain metals such as aluminum or tantalum, has been immersed.

gamma function $\Gamma(x) = \int_0^\infty y^{x-1} \exp(-y) dy$, where x is a real number greater than zero. It can be also expressed as a contour integral: $\Gamma(x) = 1/(\exp(i2\pi x) - 1) \int_C y^{x-1} \exp(-y) dy$, where the contour C denotes a path which starts at $+\infty$ on the real axis, passes round the origin in the positive direction and ends at $+\infty$ on the real axis. The initial and final values of the argument are taken to be 0 and 2π respectively.

gas immersion laser doping (GILD) – an introduction of doping atoms contained in a gas ambient, for example BCl_3, into a semiconductor, for example silicon, by means of laser beam irradiation of the semiconductor making its surface amorphous. Doped layers of less than 20 nm thickness with a sharp doping impurity profiles are formed.

First described in: G. Kerrien, J. Boulmer, D. Débarre, D. Bouchier, A. Grouillet, D. Lenoble, *Ultra-shallow, super-doped and box-like junctions realized by laser-induced doping*, Appl. Surf. Sci. **186**(1–4), 45–51 (2002).

Gaussian distribution – the function describing how a certain property is distributed among members of a group, also called the frequency distribution, for the case when deviations of the members of the set from some assigned values are the algebraic sum of a very large number of small deviations. It is of the form $f(x) = 1/(\sqrt{2\pi}\sigma) \exp\{-1/2[(x - \bar{x})/\sigma]\}$, where $\bar{x} = 1/n \sum_{k=1}^{n} x_k$ is the mean value of x and $\sigma^2 = 1/n \sum_{k=1}^{n}(x_k - \bar{x})^2$ is the mean square deviation of x. This distribution is also called the **normal distribution**.

First described by K. F. Gauss in 1809.

What is What in the Nanoworld. Victor E. Borisenko and Stefano Ossicini
Copyright © 2008 WILEY-VCH Verlag GmbH & Co. KGaA, Weinheim
ISBN: 978-3-527-40783-5

Gauss theorem

Gauss theorem states that the total electric flux from a charge q is $\int_S D dS = \int_V \nabla D dv = 4\pi \int_V \rho dv = 4\pi q$, where D is the flux of electric induction, ρ is the bulk density of the charge, and S is any surface surrounding the volume V, which contains this charge. In the magnetic case, since $\nabla B = 0$ one has $\int_S B dS = 0$.
First described by K. F. Gauss in 1830.

gel $\rightarrow$ **sol–gel technology.**

gel point indicates the stage at which a liquid begins to exhibit increased viscosity and elastic properties.

gene – a constituent part or zone of a **chromosome**. It represents a **nucleotide** sequence of **DNA** that codes for a protein. It determines a trait in an organism.

gene mapping – a linear map determining the relative position of **genes** along a **chromosome** or plasmid. Distances are established by linkage analysis and measured in linkage units.

generalized gradient approximation (GGA) assumes that the exchange-correlation energy in solids is a function of the electron concentration n and its gradients ∇n, that is $E_{xc} = \int n(\mathbf{r})\epsilon_{xc}(n(\mathbf{r}), \nabla n(\mathbf{r}))d\mathbf{r}$ where ϵ_{xc} is the exchange-correlation energy per electron for the electron gas at a given point in space $\mathbf{r}$. The combined exchange-correlation function is typically split up into two additive terms ϵ_x and ϵ_c for exchange and correlation ones, respectively.
First described in: D. C. Langreth, J. P. Perdew, *Theory of nonuniform electronic systems. I. Analysis of the gradient approximation and a generalization that works*, Phys. Rev. B **21** (12), 5469–5493 (1980).
More details in: J. P. Perdew, *Density functional approximation for the correlation energy of the inhomogeneous electron gas*, Phys. Rev. B **33** (12), 8822–8824 (1986); J. P. Perdew, Y. Wang, *Accurate and simple density functional for the electronic exchange energy: generalized gradient approximation*, Phys. Rev. B **33** (12), 8800–8802 (1986).

genetics – a field of science studying properties of single **genes** or groups of genes.

genome – a complete **DNA** sequence of one set of **chromosomes** in an organism. The chromosome set is species specific for the number genes and linkage groups carried in genomic DNA.

genomics – a field of science studying global properties of a **genome** (DNA sequence) of organisms.

g-factor – a ratio of the magnetic momentum of a particle to its mechanical momentum. It is defined by $g\mu_B \equiv -\gamma\hbar$, where γ is the gyromagnetic ratio (the

ratio of the magnetic moment to the angular momentum). For an electron spin
$g = 2.0023$.

GGA – an acronym for a **generalized gradient approximation**, which is an approach
used for calculation of electronic band structure of solids.

GHZ theorem $\rightarrow$ **Greenberger–Horne–Zeilinger theorem**.

giant magnetoresistance effect – dramatic change in the resistance of layered magnetic thin-film structures composed of alternating layers of a nonmagnetic material
between differently magnetized **ferromagnetic** materials when placed in a magnetic
field. The effect is observed when the resistance of the structure is measured by the
current passing either parallel to the layers (Current In Plane – **CIP** geometry) or
perpendicular to the layer plane (Current Perpendicular to Plane – **CPP** geometry).

A thin-film structure with a parallel geometry of the observation of the giant
magnetoresistance effect is schematically shown in Figure G.1.

Figure G.1 Giant magnetoresistance effect measured with a
current flowing in the plane of a layered magnetic material
sandwich: antialigned magnetic moments – high resistance,
aligned magnetic moments – low resistance.

The as-deposited ferromagnetic layers have opposite magnetizations that can be
achieved by their deposition in the magnetic fields of opposite orientations. In the
absence of the magnetic field the resistance measured by current flowing in the
plane of the layers is the greatest when the magnetic moments in the alternating
layers are oppositely aligned. The spin exclusion results in high interface scattering
and forces the current to flow within narrowed pathways.

The resistance is the smallest when the magnetic moments become oriented
all in one direction in the external magnetic field. The magnetic field required
to achieve all magnetization states to be parallel (lowest resistance) is commonly
termed the saturation field. The resistance decrease can reach a few hundreds
% at low temperatures. The most pronounced effect is observed in Fe/Cr and
Co/Cu multilayer structures. It increases with the number of layers and reaches its
maximum for approximately 100 periods at layer thicknesses of a few nanometers.

Measurements of magnetoresistance in a perpendicular geometry yield a larger
effect because of the elimination of the shunting current passing through intermediate nonmagnetic layers separating ferromagnetic layers. In this case all the

carriers must undergo spin scattering at every interface when traversing the layered structure. However, the low resistance of all-metal structures requires the application of **nanolithography** techniques to fabricate vertical elements with very small cross-section in order to get detectable resistance variation. Figure G.2 presents schematically the main features of the vertical transport. When the magnetizations of the two ferromagnetics are antialigned, the spin polarized carriers coming into the nonmagnetic layer from one ferromagnetic layer cannot be accommodated in the other ferromagnetic layer. They are scattered at the interface giving rise to the high resistance. In contrast, aligned magnetizations of both ferromagnetics ensure identity in the spin polarization of injected electrons and electron states in the next ferromagnetic layer. Thus, the interface scattering is minimized, which correspond to the lowest vertical resistance of the structure.

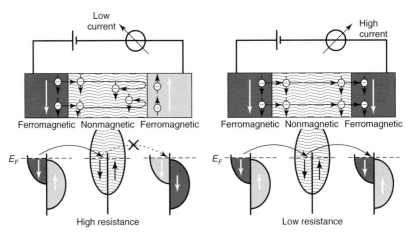

Figure G.2 Spin-polarized transport across a layered ferromagnetic /nonmagnetic /ferromagnetic structure: antialigned magnetic moments – high resistance, aligned magnetic moments – low resistance.

The spin relaxation length has proved to be much longer than the typical thickness of the layers, that is around 10 nm. An electron can pass through many layers before its spin orientation is changed. Within this length, each magnetic interface can act as a spin filter. The more scattering interfaces an electron interacts with, the stronger the filtering effect. This explains the increase in the giant magnetoresistance effect with the number of layers.

The interfacial spin scattering itself ultimately derives from lattice matching of the contacting materials and from the degree to which their conduction bands at the **Fermi level** are matched at the interface. For example, the interfaces in the popular Fe/Cr structure are formed by two **bcc** lattice matched metals and the d conduction band of chromium (paramagnetic) closely matches the minority (spin down) d conduction band of iron. There is no close match to the conjugate majority (spin up) d conduction band of iron. This suggests severe discrimination between spin up and spin down electrons at the interface. As a result, the profound giant magnetoresistance effect is observed.

A thin-film structure consisting of two ferromagnetic layers with different magnetic properties is called a **spin valve**. The most commonly used method of attaining anti-parallel orientation of two magnetic films in a spin valve structure is to deposit two ferromagnetic materials that respond differently to magnetic fields, for instance cobalt and permalloy ($Ni_{80}Fe_{20}$). The coercivity of permalloy is smaller than that of cobalt. Thus, if both a permalloy film and a cobalt film are initially saturated in the same direction (low resistance state) and then a reversed magnetic field is applied having a value larger than the coercivity of the permalloy film but less than the coercivity of the cobalt film, the anti-parallel magnetic orientations (high resistance state) can be attained.

An improvement in making the two magnetic layers with different magnetic behavior is the use of an **antiferromagnetic** in contact with one of the ferromagnetic films to effectively "pin" the magnetization in the ferromagnetic layer. Under proper deposition and annealing conditions, the antiferromagnetic and the ferromagnetic layers couple at their interface. This coupling pins the ferromagnetic layer to fields of typically about 10^4 Am^{-1}, and even if much higher fields are applied, the original state of the pinned layer is restored upon relaxation of the field.

A further refinement of antiferromagnetic pinning involves the strong anti-parallel coupling between two magnetic films with a thin interlayer of certain metals, such as ruthenium. The pair of ferromagnetic films can be coupled antiparallel with an equivalent field of about 10^5 Am^{-1}, well above the normal applied fields used in most devices. This structure is commonly called a synthetic antiferromagnetic. When one of the ferromagnetic layers in the synthetic antiferromagnetic is then pinned on an outside surface with an antiferromagnetic layer, the resulting structure is very stable to quite large fields and to temperatures approaching the **Néel temperature** of the antiferromagnetic. This allows for wide margins for magnetic fields where the structure can be in the high-resistance state.

There is another variant of the spin valve structure, that is the so-called **pseudo-spin valve**. It has two magnetic layers that have mismatched properties so that one tends to switch at lower fields than the other. No pinning layer is used and the two magnetic layers may be of the same composition, but different thickness, if the composite film layers are etched into small areas. In this case, the small lateral size of the structure contributes demagnetizing fields, which causes the thinner of the two film layers to switch at lower magnetic fields than the thicker film layer. Without switching the hard layer, the magnetization of the soft layer can be manipulated to be parallel or antiparallel to the hard layer. The resistance is the lowest at the fields where magnetizations of both layers are aligned with each other.

Spin valve and pseudo-spin valve structures usually have magnetoresistance values at room temperature of between 5% and 10% and saturation fields of between 800 and 8000 Am^{-1}.

First described in: M. N. Baibich, J. M. Broto, A. Fert, F. N. Van Dau, F. Petroff, *Giant magnetoresistance of (001)Fe/(001)Cr magnetic superlattices*, Phys. Rev. Lett. **61**(21), 2472–2475 (1988); G. Binasch, P. Grünberg, F. Saurenbach, and W. Zinn,

giant magnetoresistance nonvolatile memory

Enhanced magnetoresistance in layered magnetic structures with antiferromagnetic interlayer exchange, Phys. Rev. B **39**(7), 4828–4830 (1989).

Recognition: in 2007 A. Fert and P. Grünberg received the Nobel Prize in Physics for the discovery of giant magnetoresistance.

See also www.nobel.se/physics/laureates/2007/index.html.

giant magnetoresistance nonvolatile memory – a thin-film memory device that employs the giant magnetoresistance effect. Giant magnetoresistance elements are fabricated in arrays in order to get a net of elements functioning as a memory. The principle is illustrated in Figure G.3.

Figure G.3 Fragment of random access memory constructed of giant magnetoresistance elements connected in series.

The elements are essentially spin valve structures that are arranged in series connected by conducting spacers to form a sense line. The sense line stores the information and has a resistance that is the sum of the resistances of its elements. Current is run through the sense line and amplifiers at the ends of the lines detect the changes in the total resistance. The magnetic fields needed to manipulate the magnetization of the elements are provided by additional lithographically defined wires passing above and below the elements. These wires cross the sense line in an *xy* grid pattern with intersection at each of the giant magnetoresistance information storage elements. The wire passing parallel to the sense line acts as a world writing line and the wire crossing the sense line perpendicularly acts as a bit writing line. All the lines are electrically isolated. When current pulses are run through world and bit lines, they generate magnetic fields controlling the resistance of giant magnetoresistance elements.

A typical addressing scheme uses pulses in the world and bit lines that are half-select. This means that the field associated with a world-line pulse is half that needed to reverse the magnetization of the giant magnetoresistance element. Where any two lines in the *xy* grid overlap, the two half-pulses can generate a combined field that is sufficient to selectively reverse a soft layer or, at higher current levels, sufficient to reverse a hard layer also. Typically one pulse rotates the layer 90°. Through this *xy* grid, any element of an array can be addressed either to store information or to interrogate the element.

The exact information storage and addressing schemes may be highly varied. One scheme may store information in the soft layer and use "destroy" and restore procedure for interrogation. Alternatively, another scheme could construct the individual giant magnetoresistance elements so that high-current pulses are used to store information in the hard layer. Low-current pulses can then be used to "wiggle" the soft layer to interrogate the element by sensing the change in resistance, without destroying and restoring the information. There are many additional variations on these schemes, and the exact scheme used is often proprietary and dependent on the specific requirements of the memory application.

giant magnetoresistance read head – a thin-film read head that employs the **giant magnetoresistance effect**. It is also called a **spin valve** read head. The head reads the magnetic bits that are stored on the surface of discs or tapes in the form of differently oriented magnetic domains as small as 10–100 nm. Where the heads of two of the oppositely magnetized domains meet, uncompensated positive poles generate a magnetic field directed out of the media perpendicular to the domain surface that is a positive domain wall. In the place where the tails of two domains meet, the walls contain uncompensated negative poles that generate a sink for magnetic lines of flux returning back into the media, that is a negative domain wall. The head senses the changes in the direction of the magnetic field at the domain walls.

Figure G.4 Schematic representation of a giant magnetore-sistance read head that passes over recording media containing magnetized regions.

The principles of the read head and its operation are illustrated in Figure G.4. The sensing element of the head is a typical spin valve consisting of two layers: one with easily reversed magnetization and the other with a fixed (or difficult-to-reverse) magnetization.

Magnetic momentum in the magnetically soft layer is parallel to the plane of the media carrying magnetic domains in the absence of any applied fields. The magnetic momentum in the magnetically hard layer is perpendicular to the media plane. When the head passes over a positive domain wall, the magnetic field existing there pushes the magnetization of the easily reversed layer up. When the head passes over a negative domain wall, the magnetic momentum is pulled down. As soon as magnetization direction in the soft layer responds to the fields that emanate from the media by rotating either up or down, the measured change in the resistance is sensed by the current passing through the spin valve structure. Spin valve read heads enable very high aerial packing densities for hard drives, up to 25 Gbits per square inch.

More information in: Magnetic Multilayers and Giant Magnetoresistance: Fundamentals and Industrial Applications, edited by U. Hartmann, Springer Series in Surface Sciences, Vol. **37** (Springer, Berlin, 2000).

Gibbs distribution (statistics) asserts that the thermal equilibrium probability of finding an N particle system in an N particle stationary state of energy E at the temperature T is proportional to $\exp(-E/k_B T)$. It is generally valid for classical or quantum systems, whether or not they are weakly interacting. The **Maxwell–Boltzmann, Fermi–Dirac, Bose–Einstein distributions**, characterizing the occupation of single particle levels for weakly interacting particles or excitations, all follow from the Gibbs distribution in the appropriate special cases.

First described in: J. W. Gibbs, *On the equilibrium of heterogeneous substances*, Trans. Conn. Acad. **3**, 108–248 (1875); 227–229 (1876); 343–524 (1877).

Gibbs–Duhem equation connects the variations of the chemical potentials μ_i in a mixture containing n_i moles of species i, at given temperature and pressure in the form $\sum_i n_i d\mu_i = \sum_i x_i d\mu_i = 0$, where the mole fraction $x_i = n_i / \sum_i n_i$.

First described in: J. W. Gibbs, *On the equilibrium of heterogeneous substances*, Trans. Conn. Acad. **3**, 108–248 (1875), 227–229 (1876), and 343–524 (1877); P. Duhem, *Traité d'énergétique ou de thermodynamique générale*, (Gauthier-Villars, Paris, 1911).

Gibbs free energy → **free energy**.

First described in: J. W. Gibbs, *On the equilibrium of heterogeneous substances*, Trans. Conn. Acad. **3**, 108–248 (1875), 227–229 (1876), and 343–524 (1877).

Gibbs theorem states that the **entropy** change is zero for the process in which a perfect gas A occupying a volume v and a perfect gas B occupying separately also volume v are mixed isothermally so as to form a perfect gas mixture occupying a total volume v.

First described in: J. W. Gibbs, *On the equilibrium of heterogeneous substances*, Trans. Conn. Acad. **3**, 108–248 (1875), 227–229 (1876), and 343–524 (1877).

giga- – a decimal prefix representing 10^9, abbreviated G.

GILD – an acronym for **gas immersion laser doping**.

Ginzburg–Landau equations – a pair of coupled differential equations relating a wave function Ψ representing superconducting electrons, a vector potential $\mathbf{A}$ at a point $\mathbf{r}$ representing a microscopic field in this point and the supercurrent $\mathbf{J}$:

$$\frac{1}{2m}\left(-i\hbar\nabla - \frac{e}{c}\mathbf{A}\right)^2 \Psi + \frac{\partial F}{\partial \Psi^*} = 0,$$

$$J = -\frac{ie\hbar}{mc}(\Psi^*\nabla\Psi - \Psi\nabla\Psi^*) - \frac{e^2}{mc}\Psi^*\Psi A,$$

where the free energy of the superconductor *F* can be represented by the series $F = F_0 + \alpha|\Psi|^2 + \beta/2|\Psi|^4$ with phenomenological parameters α and β.

Although the equations have some limitations, they are very useful for calculations of important characteristic length scales, for instance the penetration depth of a magnetic field, for different superconductors. In general, they provide the basis for most understanding of spatially varying superconducting states in solids.

First described in: V. L. Ginzburg, L. D. Landau, *To the theory of superconductivity,* Zh. Exp. Teor. Fiz. **20**(12), 1064–1082 (1950) – in Russian.

Recognition: in 2003 V. L. Ginzburg received the Nobel Prize in Physics for pioneering contributions to the theory of superconductors and superfluids.

See also www.nobel.se/physics/laureates/2003/index.html.

glass – a noncrystalline, inorganic mixture of various metallic oxides fused by heating with glassifiers such as silica, or boric or phosphoric oxides. Most glasses are transparent in the visible spectrum and up to about 2.5 μm in the infrared, but some are opaque such as natural obsidian. These are, nevertheless, useful as mirror blanks. Traces of some elements such as cobalt, copper and gold are capable of producing a strong coloration in glass. Laser glass contains a small amount of didymium oxide. Opal glass is opaque and white, with the property of diffusing light. Some opal glass has a thin layer of opal material flashed onto the surface of ordinary glass. Tempered glass has a high degree of internal strain, caused by rapid cooling, which gives it increased mechanical strength.

globular protein → **protein.**

glycolipids → **lipids.**

glycoluril → **cucurbituril.**

GNR – an acronym for a **graphene nanoribbon.**

Goldschmidt law states that crystal structure is determined by the ratios of the numbers of the constituents, the ratios of their size and their polarization properties.

First described in: V. M. Goldschmidt, T. Barth, G. Lunde, W. Zachariasen, *Geochemische Verteilungsgesetze: VII: Die Gesetze der Krystallochemie,* Skr. Norsk. Vid. Akademie, Oslo, Mat. Nat. Kl. **2** (1926), and *Geochemische Verteilungsgesetze: VIII: Untersuchungen über Bau und Eigenschaften von Krystallen,* Skr. Norsk. Vid. Akademie, Oslo, Mat. Nat. Kl. **8** (1926).

g permanence rule states that the sum of **g-factors** is the same for strong and weak magnetic fields for the same value of the magnetic quantum numbers.

Goodenough–Kanamori rules

First described in: W. Pauli, *Über die Gesetzmässigkeiten des anomalen Zeemaneffektes*, Z. Phys. **16**(3), 155–164 (1923).

Goodenough–Kanamori rules provide a description of the exchange interactions between magnetic atoms, and hence the magnetic ordering pattern, in terms of the relative orientation of valence orbitals on neighboring lattice sites. They state that if there is a large overlap between partly occupied orbitals at two magnetic ions, the superexchange interaction between them is strongly **antiferromagnetic** because of the **Pauli principle**, whereas overlap between partly occupied and unoccupied orbitals gives weakly **ferromagnetic** interaction due to **Hund rules**.

First described in: J. B. Goodenough, *Theory of the role of covalence in the perovskite-type manganites [La, M(II)]MnO₃*, Phys. Rev. **100**(2), 564–573 (1955); J. Kanamori, *Superexchange interaction and symmetry properties of electron orbitals*, J. Phys. Chem. Solids **10**(2–3), 87–98 (1959).

More details in: J. B. Goodenough, *Magnetism and the Chemical Bond* (Interscience, New York, 1963).

Grätzel solar cells = dye-sensitized solar cells.

graphene – the name given to a single layer of carbon atoms densely packed into a benzene-ring structure, thus being a two-dimensional honeycomb net made up of sp^2 bonded carbon atoms (for a picture of graphene → **carbon nanotube**). It is widely used to describe properties of many carbon-based materials, including graphite, large **fullerene**s, and carbon nanotubes. Graphene was isolated in 2004. Starting with a graphite crystal, layers of carbon atoms were peeled off one by one with adhesive tape until a single-layer flake was left. Graphene's conductivity never falls below a minimum value corresponding to the quantum unit of conductance, even when concentrations of charge carriers tend to zero. The **quantum Hall effect** in graphene occurs at half-integer filling factors. Electron transport is essentially governed by **Dirac**'s relativistic **equation**, and also **Berry phase** has been observed.

First described in: K. S. Novoselov, A. K. Geim, S. V. Morozov, D. Jiang, Y. Zhang, S. V. Dubonos, I. V. Grigorieva, A. A. Firsov, *Electric field effect in atomically thin carbon films*, Science **306**, 666–669 (2004); K. S. Novoselov, D. Jiang, F. Schedin, T. J. Booth, V. V. Khotkevich, S. V. Morozov, and A. K. Geim, *Two-dimensional atomic crystals*, PNAS, **102**(30), 10451–10453 (2005); K. S. Novoselov, A. K. Geim, S. V. Morozov, D. Jiang, M. I. Katsnelson, I. V. Grigorieva, S. V. Dubonos, A. A. Firsov, *Two-dimensional gas of massless Dirac fermions in graphene*, Nature **438**, 197–200 (2005); Y. Zhang, Y.-W. Tan, H. L. Stormer, P. Kim, *Experimental observation of the quantum Hall effect and Berry's phase in graphene*, Nature **438**, 201–205 (2005).

More details in: C. L. Kane, *Erasing electron mass*, Nature **438**, 168–169 (2005).

graphene nanoribbon (GNR) – the structure fabricated when **graphene** is etched or patterned along one specific direction, which in fact is a strip of graphene. It is a quasi-one-dimensional structure. The structure of graphene nanoribbons is closely related to that of **carbon nanotubes**, since they can be viewed as unrolled single-wall carbon nanotubes. Thus also in the case of graphene nanoribbons it is possible to describe the whole structure using only two indices. There are **armchair** and zigzag graphene nanoribbons. The main topological differences with respect to nanotube and graphene is related to the finite width, which implies two different atomic sites: bulk sites, with three neighboring carbon atoms, and edge sites, with only two C–C bonds. The GNRs with armchair-shaped edges are classified by the number of dimer lines across the ribbon width, and the ribbons with zigzag-shaped edges are classified by the number of zigzag chains across the ribbon width. These simple features of GNR have a great impact on both symmetry and electronic properties, for example the energy gap scales inversely with the ribbon width.

First described in: M. Fujita, K. Wakabayashi, K. Nakada, K. Kusakabe, *Peculiar localized state at zigzag graphite edge*, J. Phys. Soc. Jpn **65**(7), 1920–1923 (1996); M. Y. Han, B. Özyilmaz, Y. Zhang, P. Kim, *Energy band-gap engineering of graphene nanoribbons*, Phys. Rev. Lett. **98**(20), 206805 (2007).

Greenberg–Horne–Zeilinger states – multiparticle quantum states which have perfect correlations and such that these correlations are incompatible with a classical conception of locality: $|\Psi^{\pm}\rangle = 1/\sqrt{2}(|00\ldots0\rangle \pm |11\ldots1\rangle)$.

First described in: D. M. Greenberger, M. A. Horne, A. Shimony, A. Zeilinger, *Bell's theorem without inequalities*, Am. J. Phys. **58**(12), 1131–1143 (1990).

Greenberger–Horne–Zeilinger theorem, being an extension of **Bell's theorem** to three or more particles, shows that assumptions of the **Einstein–Podolsky–Rosen paradox** lead to predictions which are exactly opposite of those of quantum theory.

First described in: D. M. Greenberger, M. A. Horne, A. Zeilinger, *Going beyond Bell's theorem*, in: *Bell's Theorem, Quantum Theory, and Conceptions of the Universe*, edited by M. Kafatos (Kluver, Dordrecht, 1989), pp. 73–76.

Green's function arises in the solution w of the differential equation $\mathbf{L}w = f(\mathbf{x})$ satisfying certain boundary conditions. The operator $\mathbf{L}$ is a linear differential operator and $f(\mathbf{x})$ is a prescribed function of the n variables $(x_1, x_2, \ldots, x_n) = \mathbf{x}$ defined uniquely through a region Ω on whose boundary the prescribed boundary conditions are formulated. If one can find a function $G(\xi, \mathbf{x})$ such that this solution can be written in the form $w(\mathbf{x}) = \int_{\Omega} G(\xi, \mathbf{x}) f(\xi) d\xi$, the function $G(\xi, \mathbf{x})$ is called the Green's function corresponding to the operator $\mathbf{L}$ and the prescribed boundary conditions.

Introducing the n dimensional **Dirac delta function** as $\delta(\mathbf{x} - \xi) = \delta(x_1 - \xi_1) \ldots \delta(x_n - \xi_n)$ makes obvious that $G(\xi, \mathbf{x})$ is the appropriate solution of the differential equation $\mathbf{L}w = \delta(\mathbf{x} - \xi)$.

The name "Green's function" seems to have been applied to such a function for the first time by H. Burkhardt in 1894, because it is in some ways analogous to the function introduced in 1828 by George Green in his memoir on electricity and magnetism.

Green's theorem

First described in: G. Green, *An essay on the application of mathematical analysis to the theories of electricity and magnetism* (1828). Reissued by Lord Kelvin in J. Reine Angew. Math. **39**, 73–89 (1850), **44**, 356–374 (1852), **47**, 161–221 (1854).

Green's theorem states that if $\mathbf{A} = (P, Q, R)$ is a vector each of whose components is a continuous function of x, y, z with continuous derivatives at each point of a simple volume V enclosed by a surface S then

$$\int_V \int \int \left(\partial P/\partial x + \partial Q/\partial y + \partial R/\partial z \right) \mathrm{d}x\mathrm{d}y\mathrm{d}z = \int_S \int (lP + mQ + nR)\mathrm{d}S,$$

where $(l, m, n) = \mathbf{n}$ are the direction cosine of the outward drawn normal to S.

First described in: G. Green, *An essay on the application of mathematical analysis to the theories of electricity and magnetism* (1828). Reissued by Lord Kelvin in J. Reine Angew. Math. **39**, 73–89 (1850), **44**, 356–374 (1852), **47**, 161–221 (1854).

Greenwood–Williamson theory reconciles **Amonton's law** and the **Bowden–Tabor law**. Taking two rough surfaces in contact through a large number of asperities **(multi-asperity contact model)** the **real contact area** depends linearly on the applied load.

This theory does not contemplate **adhesion** forces.

First described in: J. A. Greenwood, J. B. P. Williamson, *Contact of nominally flat surfaces*, Proc. R. Soc. London A, **295**, 300–319 (1966).

Grimm–Sommerfeld rule states that the absolute valency of an atom is numerically equal to the number of electrons engaged in attaching the other atoms.

First described in: H. G. Grimm, A. Sommerfeld, *Über den Zusammenhang des Abschlusses der Elektronengruppen im Atom mit den chemischen Valenzzahlen*, Z. Phys. **36**(1), 36–59 (1926).

ground state – the state of an entire system which is of minimum energy, in which, therefore, each electron is represented by an energy **eigenstate** corresponding to the lowest available energy level.

ground state depletion (GSD) microscopy → **fluorescence nanoscopy**.

First described in: S. W. Hell, M. Kroug, *Ground-state depletion fluorescence microscopy, a concept for breaking the diffraction resolution limit*, Appl. Phys. B **60**, 495–497 (1995).

group velocity → **wave packet**.

Grover algorithm – a quantum mechanical algorithm that would search an unsorted database of N elements in a time that scales roughly $N^{1/2}$ faster than any possible classical search algorithm. It supposes the following steps.

Step 1. Initialize the system to the superposition:

$$(1/N^{1/2}, 1/N^{1/2}, \ldots, 1/N^{1/2});$$

that is, there is the same amplitude to be in each of the N states. This super position can be obtained in M (log N) steps.

Step 2. Repeat the following unitary operations $M(N^{1/2})$ times (the precise number of repetitions is important).

2 a. Let the system be in any state S: in case $C(S) = 1$, rotate the phase by π radians; in case $C(S) = 0$, leave the system unaltered.

2 b. Apply the diffusion transform D, which is defined by the matrix $\mathbf{D}$ as follows: $D_{ij} = 2/N$ if $i \neq j$ and otherwise $D_{ij} = -1 + 2/N$ ($\mathbf{D}$ can be implemented as a product of three elementary matrices).

Step 3. Measure the resulting state. This will be the state S_v (that is the desired state that satisfies the condition $C(S_v) = 1$) with a probability of at least 0.5.

Note that the steps 1 and 2 are a sequence of unitary operations. The step 2 a is a phase rotation. In an implementation it would involve a portion of the quantum system sensing the state and then deciding whether or not to rotate the phase. It would do it in such a way so that no trace of the state of the system would be left after this operation, so as to ensure that paths leading to the same final state were indistinguishable and could interfere.

By having the input and output in quantum superpositions of states, one can find an object in $M(N^{1/2})$ quantum mechanical steps instead of $M(N)$ classical steps. This discovery started a flurry of theoretical activity searching for quantum computer applications.

First described in: L. K. Grover, *Quantum mechanics helps in searching for a needle in a haystack*, Phys. Rev. Lett. **78**(2), 325–328 (1997).

Grüneisen equation of state → **Grüneisen rules**.

Grüneisen rules are based on the **Grüneisen equation of state** for solids $PV + G(V) = \gamma E$, where P is the pressure, V is the molar volume, $G(V)$ depends on the interatomic potential energy, E is the energy of the atomic vibrations and γ is a dimensionless quantity. The **first rule** governs the change of the volume of a solid as $\triangle V = \gamma K_0 E$, where K_0 is the isothermal compressibility of the solid at absolute zero temperature. The **second rule** connects the volume coefficient of thermal expansion β and the molar specific heat of the solid C_v in the form $\beta = \gamma K_0 C_v / V$.

First described in: E. Grüneisen, *Theorie des festen Zustandes einatomiger Elemente*, Ann. Phys. **39**, 257–306 (1912), and in *Handbuch der Physik*, **10** (Springer, Berlin, 1926).

GSD – an acronym for **ground state depletion (GSD) microscopy**.

Gudden–Pohl effect – a light flash that occurs when an electrical field is applied to a phosphor already excited by ultraviolet radiation.

Gunn effect

First described in: B. Gudden; R. W. Pohl, *Über Ausleuchtung der Phosphoreszenz durch elektrische Felder,* Z. Phys. **2**(2), 192–196 (1920).

Gunn effect – generation of coherent microwaves when a constant electric field that exceed a critical threshold value of several thousand volts per centimeter is applied across a randomly oriented, short, *n*-type sample of GaAs or InP. The frequency of oscillations is approximately equal to the reciprocal of the carrier transit time across the length of the sample. The mechanism of observed oscillations is field-induced transfer of conduction band electrons from a low-energy, high-mobility valley to higher-energy, low-mobility satellite valleys.

First described in: J. B. Gunn, *Microwave oscillations of current in III-V semiconductors,* Solid State Commun. **1**(4), 88–91 (1963).

Gurevich effect – in electrical conductors with a pronounced phonon electron interaction, phonons carrying a thermal current in the presence of a temperature gradient tend to drag the electrons with them from hot to cold.

First described in: L. Gurevich, *Thermoelectric properties of conductors,* J. Phys. **9**, 477–488 (1945); ibid. **10**, 67–80 (1945).

GW approximation (GWA) is the approximation in which the self-energy of an electronic system of solids is calculated using **Green's function** G and screened Coulomb interaction W. The GWA may be regarded as a generalization of the **Hartree-Fock approximation** (HFA) but with a dynamically screened Coulomb interaction. The non-local exchange potential in the HFA is given by $\Sigma^x(\mathbf{r}, \mathbf{r}') = \sum_{kn}^{occ} \psi_{kn}(\mathbf{r}) \psi_{kn}^*(\mathbf{r}') v(\mathbf{r} - \mathbf{r}')$. In the Green's function theory, the exchange potential is written as $\Sigma^x(\mathbf{r}, \mathbf{r}', t - t') = iG(\mathbf{r}, \mathbf{r}', t - t') v(\mathbf{r} - \mathbf{r}') \delta(t - t')$, which when Fourier transformed yields the previous equation. The GWA corresponds to replacing the bare Coulomb interaction by a screened interaction $W : \Sigma(1, 2) = iG(1, 2) W(1, 2)$. This is physically well motivated especially in metals where the HFA leads to unphysical results such as zero density of states at the Fermi level, due to the lack of screening.

The GWA has been successful in predicting the quasiparticle systems such as free-electron like metals and semiconductors as well as Mott insulators and d- and f-band metals. It was also found that GWA gives a rather good description of the band gap of strongly correlated systems like NiO.

First described in: L. Hedin, *New method for calculating the one-particle Green's function with application to the electron-gas problem,* Phys. Rev. **139**(3A), A796–A823 (1965).

More details in: F. Aryasetiawan, O. Gunnarsson, *The GW method,* Rep. Prog. Phys. **61**(3), 237–312 (1998).

gyromagnetic effect – the rotation induced in a body by a change of its magnetization or magnetization resulting from a rotation.

gyromagnetic frequency – the term denoting the frequency of precession of a charged revolving particle of mass m placed in a magnetic field of strength H. This frequency is $qHp/(2mch)$, where q is the charge and p is the angular momentum of the particle.

From habit plane to hyperelastic scattering

habit plane – the crystallographic plane or system of planes along which certain phenomena such as twinning occur.

Hadamard gate → **Hadamard transformation** and **quantum logic gate**.

Hadamard operator produces self inverse operation forming superposition of states with equal coefficients → **Hadamard transformation**.

Hadamard product defines the product of two power series as the power series $P \times Q = \sum_{n=0}^{\infty} a_n b_n x^n$, where $P = \sum_{n=0}^{\infty} a_n x^n$ and $Q = \sum_{n=0}^{\infty} b_n x^n$.

Hadamard transformation – a unitary, orthogonal, real transformation. The basis functions could have only the value 1 and -1. In the matrix form we have

$$H = \frac{1}{\sqrt{2}} \begin{pmatrix} 1 & 1 \\ 1 & -1 \end{pmatrix}.$$

Thus, one has, for example

$$H|0\rangle = \frac{1}{\sqrt{2}}|0\rangle + \frac{1}{\sqrt{2}}|1\rangle$$

$$H|1\rangle = \frac{1}{\sqrt{2}}|0\rangle - \frac{1}{\sqrt{2}}|1\rangle.$$

The Hadamard transformation matrix (called also **Hadamard gate**) is a special case of a generalized **qubit** rotation.
First described in: J. Hadamard, Bull. Sci. Math., **17**, 240–248 (1893).

Hagen–Rubens relation – an equation that links the reflectivity of a solid to the frequency of radiation and the conductivity of the solid: $R_{opt} = 1 - (f/\sigma)^{1/2}$, where f is the frequency of light, and σ is the electrical conductivity.

What is What in the Nanoworld. Victor E. Borisenko and Stefano Ossicini
Copyright © 2008 WILEY-VCH Verlag GmbH & Co. KGaA, Weinheim
ISBN: 978-3-527-40783-5

It applies at wavelengths long enough that the product of the frequency and the relaxation time is much less that unity.

First described in: E. Hagen, H. Rubens, *Über Beziehungen des Reflexions- und Emissionsvermögens der Metalle zu ihrem elektrischen Leitvermögen*, Ann. Phys. **11**(8b), 873–901 (1903).

half-metallic material – a material which can be considered in its electronic properties as a hybrid between a metal and a semiconductor or dielectric. Its main particular feature is that electrons at the Fermi level show complete one-direction spin polarization. For that reason half-metallic alloys and compounds have well pronounced **ferromagnetic** properties, making them prime targets for spintronic devices.

Typical examples of such materials are **diluted magnetic semiconductors** from $A^{II}B^{VI}$ and $A^{III}B^{V}$ groups, ternary compounds with the general formula AB_2O_4 (such as $Fe_3O_4 = FeFe_2O_4$), **Heusler alloys** with the general formula A_2MnB (such as Co_2MnSi), half Heusler alloys $AMnB$ (such as $NiMnSb$), and the binary compound CrO_2. See Figure H.1.

First described in: R. A. de Groot, F. M. Mueller, P. G. van Engen, K. H. J. Buschow, *New class of materials: half-metallic ferromagnets*, Phys. Rev. Lett. **50**(25), 2024–2007 (1983) – coined the name of the half-metallicity phenomenon.

More details in: Half-metallic Alloys-Fundamentals and Applications, Lecture Notes in Physics, edited by P. H. Dederichs, I. Galanakis (Springer Verlag, Berlin, 2005).

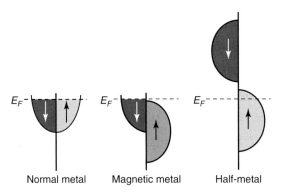

E_F E_F E_F

Normal metal   Magnetic metal   Half-metal

Figure H.1 Normal metal, magnetic metal and half-metal.

halide – a chemical compound of an element wtith a halogen. .

Hall effect – in a conducting strip placed in magnetic field perpendicular to the direction of the current flow in the strip (see Figure H.2) the resistance across the current path linearly depends on the strength of the magnetic field. The electrons experience a **Lorentz force** that is perpendicular to both the magnetic field and their initial direction. They are pushed toward one side of the specimen (depending

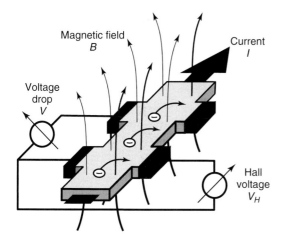

Figure H.2 Sample geometry for measurements of the Hall effect.

on the direction of the magnetic field), giving rise to charge accumulation on that side as compared with the other. The voltage drop V measured along the current characterizes the material resistance $R = V/I$. The voltage V_H induced by the magnetic field across the current path is called **Hall voltage**. The related **Hall resistance** is defined as $R_H = V_H/I$.

In the classical Hall effect $R_H = B/(en)$, where B is the magnetic induction, e is the electron charge and n is the electron density per unit area in the strip. Most remarkably, the Hall resistance does not depend on the shape of the specimen. It increases linearly with the magnetic field, while longitudinal resistance R is predicted to be roughly independent of the magnetic field, as illustrated in Figure H.3 left. Because of its independence from the sample geometry, the classical Hall effect has become a standard technique for determination of the type, density and mobility of free carriers in metals and semiconductors.

When the Hall effect is measured at low temperatures in a sample containing two-dimensional electron gas (2DEG) in which electrons are confined to move only within the plane, the Hall resistance is found to deviate from classical behavior. At sufficiently high fields, a series of flat steps appear in the graph of the Hall voltage versus magnetic field (see Figure H.3 right), which is referred to as the **quantum Hall effect (QHE)**. At the plateaus of the Hall voltage the longitudinal voltage becomes zero. The Hall resistance at the position of the plateaus of the step is quantized to a few parts per billion to $R_H = h/(ie^2)$, where h is the Planck constant and i is an integer. Consequently this effect is called the **integer quantum Hall effect (IQHE)**. The effect is definitely material independent. The quantum of the resistance, h/e^2, as measured reproducibly with high precision via the integer quantum Hall effect, has become the resistance standard. The effect can be explained in terms of occupation of **Landau levels**.

Electrons moving in a perpendicular magnetic field are forced into circular orbits, following the **Lorentz force**. They perform a cyclotron motion with the angular frequency $\omega_c = eB/m$, called the **cyclotron frequency**, where m is the electron mass.

Hall effect

Figure H.3 Variation of the longitudinal voltage drop and the Hall voltage as a function of the magnetic field in classic and quantum Hall effect.

So far, available energy levels for electrons become quantized. These quantized energies are known as Landau levels. They are given by $E_i = (i + 1/2)\hbar\omega_c$ with $i = 1, 2, \ldots$. In an ideal perfect system containing a 2DEG these levels have a form of δ-function, as illustrated in Figure H.4. The gap between nearest levels is determined by the cyclotron energy $\hbar\omega_c$. The levels are broadened with temperature rise. Clearly one needs $k_B T \ll \hbar\omega_c$ to observe well-resolved Landau levels.

Electrons can only reside at the energies of Landau levels, but not in the gap between them. The existence of the gaps is crucial for the occurrence of the quantum Hall effect. Here 2DEG differs decisively from electrons in three dimensions. Motion in the third dimension, along the magnetic field, can add any amount of energy to the energy of the Landau levels, thus filling the gaps. Therefore, there are no energy gaps in three dimensions and the quantum Hall effect causes the background to disappear.

Two years after the discovery of the integer quantum Hall effect it was found that quantization i could take fractional values, like 1/3, 2/3, 2/5, 3/5 and so on. In general, $i = p/q$, where p and q are integers with q odd. The phenomenon got the name of **fractional quantum Hall effect (FQHE)**. As an explanation of the fractional quantum Hall effect it was supposed that electrons in a 2DEG placed in a strong magnetic field condense into an exotic new collective state, a quantum

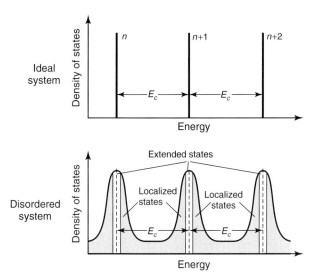

Figure H.4 Occupation of Landau levels by electrons in 2DEG placed in a magnetic field at absolute zero temperature. $E_c = \hbar\omega_c$.

liquid, similar to the way in which collective states form in superfluid helium. A quantum of magnetic flux and an electron exist as a **quasiparticle** that carries a fractional charge. Such particles follow neither **Fermi–Dirac** nor **Bose–Einstein statistics**. Instead they obey special so-called fractional statistics.

First described in: E. H. Hall, *On a new action of the magnet on electric currents*, Am. J. Math. **2**, 287–292 (1879) – the classical effect; K. von Klitzing, G. Dorda, M. Pepper, *New method for high-accuracy determination of the fine-structure constant based on quantized Hall resistance*, Phys. Rev. Lett. **45**(6), 494–497 (1980) – the integer quantum Hall effect; D. C. Tsui, H. L. Störmer, A. C. Gossard, *Two-dimensional magnetotransport in the extreme quantum limit*, Phys. Rev. Lett. **48**(22), 1559–1562 (1982) and R. B. Laughlin, *Anomalous quantum Hall effect: an incompressible quantum fluid with fractionally charged excitations*, Phys. Rev. Lett. **50**(18), 1395–1398 (1983) – the fractional quantum Hall effect.

Recognition: in 1985 K. von Klitzing received the Nobel Prize in Physics for the discovery of the quantized Hall effect; in 1998 R. B. Laughlin, H. L. Störmer and D. C. Tsui received the Nobel Prize in Physics for their discovery of a new form of quantum fluid with fractionally charged excitations.

See also www.nobel.se/physics/laureates/1985/index.html.

See also www.nobel.se/physics/laureates/1998/index.html.

Hall resistance → **Hall effect.**

Hall voltage → **Hall effect.**

Hallwachs effect

Hallwachs effect – the ability of ultraviolet radiation to discharge a negatively charged body in a vacuum.

First described in: W. Hallwachs, *Über den Einfluss des Lichtes auf electrostatisch geladene Körper*, Ann. Phys. **33**, 301–344 (1888).

halogens – F, Cl, Br, I, At.

Hamilton equations of motion form a set of first order symmetrical equations describing the motion of a classical dynamical system:

$$\dot{q} = \frac{\partial H}{\partial p_i},$$

$$\dot{p} = -\frac{\partial H}{\partial q_i},$$

with $i = 1, 2, \ldots$, where q_i are generalized coordinates of the system, p_i is the momentum conjugate to q_i, and **H** is the **Hamiltonian**. They are also termed canonical equations of motion. The Hamiltonian determines the time evolution of the system.

First described in: W. R. Hamilton, *Second essay on a general method in dynamics*, Phil. Trans. R. Soc. **125**, 95–144 (1835).

Hamiltonian operator (or, shortly, Hamiltonian) – a differential **operator** typically used to represent energy relations governing behavior of quantum particles. It is denoted

$$\mathbf{H} = -\frac{\hbar^2}{2m} \nabla^2 + V(\mathbf{r})$$

and includes the kinetic energy term supposing that the particle has a mass m and the potential energy term $V(\mathbf{r})$. The operator takes the second derivative of a function (after multiplication by $-\hbar^2/2m$) and adds it to the result of multiplying this function by $V(\mathbf{r})$.

First described in: W. R. Hamilton, *Theory of systems of rays*, Trans. R. Irish Acad. **15**, 69–174 (1828); *Supplement to an essay on the theory of systems of rays*, Trans. R. Irish Acad. **16**, 1–61 (1830); *Second supplement to an essay on the theory of systems of rays*, Trans. R. Irish Acad. **16**, 93–125 (1831).

Hamilton–Jacobi equation –

$$\mathbf{H} \left(q_1, \ldots, q_n, \frac{\partial \phi}{\partial q_1}, \ldots, \frac{\partial \phi}{\partial q_n}, t \right) + \frac{\partial \phi}{\partial t} = 0,$$

where $q_1, \ldots, q_n$ are generalized coordinates, t is the time coordinate, **H** is the Hamiltonian, and ϕ is a function that generates a transformation by means of which the generalized coordinates and momenta may be expressed in terms of new generalized coordinates and momenta, which are constants of motion.

First described in: W. R. Hamilton, *On a General Method of Expressing the Paths of Light, and of the Planets, by the Coefficients of a Characteristic Function* (Dublin, 1833); C. G. Jacobi, *Opuscula mathematica* (Berlin, 1846–1857).

Hanle effect – magnetic field-induced depolarization of the **luminescence** in solids related to dephasing of electron spins as a result of their precessing along the magnetic field. Linear and nonlinear effects are distinguished.

The **linear Hanle effect** supposes a change in the degree of polarization of the resonant luminescence of an atomic or molecular sample in the presence of a magnetic field. The **nonlinear Hanle effect** is a change in the nonlinear absorption coefficient of an atomic or molecular sample as a function of the applied magnetic field. Since it is related to the removal of degeneracy of sublevels by the external field, it is also known as stimulated zero-field level crossing.

First described in: W. Hanle, *Über magnetische Beeinflussung der Polarisation der Resonanzfluoreszenz*, Z. Phys. **30**, 93–105 (1924).

More details in: G. Moruzzi, F. Strumia, *The Hanle Effect and Level-Crossing Spectroscopy* (Plenum Press, New York, 1991).

haploid – a cell or individual with a genetic complement containing one copy of each nuclear chromosome.

hard-sphere potential – the classical interatomic pair potential in the form:

$$V(r) = \begin{cases} 0 & \text{if } r > r_0 \\ \infty & \text{if } r \leq r_0 \end{cases}$$

where r_0 is the radius of the atom.

Hardy–Schulze rule states that the ion which is effective in causing precipitation in a sol is that whose charge is opposite to that on the colloidal particle, and its effect is greatly increased with increasing valence of the ion.

First described in: H. Schulze, *Schwefelarsen in wässriger Lösung*, Journal für Praktische Chemie **25**, 431–452 (1882); W. B. Hardy, *A preliminary investigation of the conditions which determine the stability of irreversible hydrosols*, Proc. R. Soc. Lond. **66**, 110–125 (1900) and J. Phys. Chem. 4(4), 235–253 (1900).

harmonic function – a function which, both itself and its first derivative in a given domain, are continuous functions. It must also satisfy the **Laplace equation**. Special cases are spherical harmonic functions, which are homogeneous in variables x, y, z, for example **Legendre polynomials**.

harmonic oscillation – the simplest form of oscillation of a particle about a point arising when the force which tends to pull the particle back to this point is proportional to the displacement of the particle from the point. The displacement of the particle at any time t is $A\cos(\omega t + \phi)$, where A is the amplitude, ω is the frequency and ϕ is the phase of the motion.

Hartree approximation – the independent-particle approximation wherein the total wavefunction is approximated by a product of orthonormal molecular orbitals used to simplify finding a solution of the **Schrödinger equation** for many-electron systems. It is assumed that each electron moves independently within its own orbital and sees only the average field generated by all the other electrons. The Hartree wavefunction (for an N-electron system) is $\Psi = \psi_1(\mathbf{r}_1)\psi_2(\mathbf{r}_2)\ldots\psi_N(\mathbf{r}_N)$, where $\psi_N(\mathbf{r}_N)$ are orthonormal one-electron wave functions consisting of a spatial orbital and one of two spin functions representing spin up and spin down states.

First described in: D. R. Hartree, *Wave mechanics of an atom with a non Coulomb central field. Part I. Theory and methods*, Proc. Camb. Phil. Soc. **24**, 89–110 (1928); D. R. Hartree, *Wave mechanics of an atom with a non Coulomb central field. Part II. Some results and discussion*, Proc. Camb. Phil. Soc. **24**, 111–132 (1928).

Hartree–Fock approximation considers interaction of electrons in a many-electron system in terms of a potential. Interaction between particles is included by allowing each particle to be governed by an effective potential due to all other particles. The effective potential for one electron then has two additional terms, usually called the **Hartree potential** and the **Fock exchange potential**. The first, which is the electrostatic potential due to the charge distribution of the other electrons, is easily to understand. The second is a nonlocal potential due to the exchange interaction of the electron with the others.

Mathematically, the approximation consists in finding the best set of one-electron orbitals in the **Slater determinant** representing the ground state. According to the **variational principle**, the best set is the one which yields the lowest total energy for the interacting system. The resultant orbitals are given by a set of modified one-particle **Schrödinger equations**:

$$\left[-\frac{\hbar^2}{2m}\nabla^2 + V(\mathbf{r}) + \int d\mathbf{r}' \frac{e^2}{|\mathbf{r}-\mathbf{r}'|}n(\mathbf{r}')\right]\psi_i(\mathbf{r})$$

$$-\int d\mathbf{r}' \frac{e^2}{|\mathbf{r}-\mathbf{r}'|}n(\mathbf{r},\mathbf{r}')\psi_i(\mathbf{r}') = E\psi_i(\mathbf{r})$$

referred to as the **Hartree–Fock equation**. It supposes that each electron moves under the combined influence of the external potential $V(\mathbf{r})$, the Hartree potential and the Fock exchange potential. The Hartree potential is expressed here as the electrostatic potential due to the electron density distribution $n(\mathbf{r})$ via the Coulomb interaction $e^2/|\mathbf{r}-\mathbf{r}'|$. The nonlocal Fock exchange potential is due to the one particle density matrix $n(\mathbf{r},\mathbf{r}')$ as a consequence of the **Pauli principle**. Both the density and density matrix are given by the N lowest energy orbitals of the above equation. The interaction of an orbital with itself in the Hartree and Fock terms cancels. The many-electron wave function Ψ is approximated as a product of single-particle functions ψ_i, *i.e.* $\Psi(\mathbf{r}_1, \mathbf{r}_2, \ldots) = \psi_1(\mathbf{r}_1) \cdot \psi_2(\mathbf{r}_2) \cdot \ldots \cdot \psi_N(\mathbf{r}_N)$. In addition to the ground state, the equation yields orbitals which can be used to construct excited states. The excitation energy over the ground state is simply the sum and difference of the single-particle energies of the orbitals added to or taken from the ground state. Note that the Hartree-Fock approximation misses so called correlation effects in many-electron systems.

First described in: D. R. Hartree, *Wave mechanics of an atom with a non Coulomb central field. Part I. Theory and methods*, Proc. Camb. Phil. Soc. **24**, 89–110 (1928); D. R. Hartree, *Wave mechanics of an atom with a non Coulomb central field. Part II. Some results and discussion*, Proc. Camb. Phil. Soc. **24**, 111–132 (1928); V. Fock, *Approximate method of solution of the problem of many bodies in quantum mechanics*, Z. Phys. **61**(1/2), 126–148 (1930).

Hartree–Fock equation → Hartree–Fock approximation.

Hartree method – an iterative variational method of finding an approximate wave function for a system of many electrons. Within the method one attempts to find a product of single-particle wave functions each one of which is a solution of the **Schrödinger equation** with the field deduced from the charge density distribution due to all the other electrons.

First described in: D. R. Hartree, *Wave mechanics of an atom with a non Coulomb central field. Part I. Theory and methods*, Proc. Camb. Phil. Soc. **24**, 89–110 (1928).

Hartree potential → Hartree–Fock approximation.

Hartree wave function → Hartree approximation.

Heaviside step function – zero for a negative argument and one for a positive argument.

First described by Oliver Heaviside in 1880.

hecto- – a decimal prefix representing 10^2, abbreviated h.

Heggie potential – a semiclassical interatomic many-body potential based on the proximity cell (the **Wigner–Seitz primitive cell**). It introduces three internal degrees of freedom per atom, representing the magnitude and direction of the p-orbital that is not involved in sp-hybridization. One of the variants of the potential has the form:

$$V(r_{ij}) = \left(\frac{1}{2} t^{-4} + \frac{1}{4} |\mathbf{p}'_{ij}|^{-4} + \frac{1}{4} |\mathbf{p}''_{ij}| \right)^{-1/4}$$

Heisenberg equation of motion

$$t = \left[1 - f(\triangle_{ij}, \triangle_l, \triangle_u)\right] \left[\mathbf{p}_i \mathbf{p}_j |\mathbf{p}'_{ij}| |\mathbf{p}''_{ij}| (2 - |\mathbf{p}'_{ij}|)(2 - |\mathbf{p}''_{ij}|)\right]^4,$$

where $\mathbf{p}_i$ is the direction of the unhybridized p orbital on atom i being between 0 (sp^3 hybridization, for example diamond) and 1 (sp^2 hybridization, for example graphite), vectors $\mathbf{p}'_{ij}$ and $\mathbf{p}''_{ij}$ represent the projection of $\mathbf{p}_i$ onto the bond face and $\mathbf{p}_j$ onto the latter direction, respectively, as follows: $\mathbf{p}'_{ij} = \mathbf{p}_i - (\mathbf{p}_i \cdot \hat{\mathbf{r}}_{ij})\hat{\mathbf{r}}_{ij}$, $\mathbf{p}''_{ij} = (\mathbf{p}_i \cdot \hat{\mathbf{p}}'_{ij})\hat{\mathbf{p}}'_{ij}$; $\mathbf{r}_{ij} = \mathbf{r}_j - \mathbf{r}_i$ is the distance between atoms i and j.

First described in: M. I. Heggie, *Semiclassical interatomic potential for carbon and its application to the self-interstitial in graphite*, J. Phys.: Condens. Matter **3**(18), 3065–3078 (1991).

Heisenberg equation of motion gives the rate of change of an **operator** corresponding to a physical quantity within the **Heisenberg representation**.

First described in: W. Heisenberg, *Über quantentheoretische Umdeutung kinematischer und mechanischer Beziehungen*, Z. Phys. **33**(1), 879–893 (1925).

Heisenberg exchange interaction → **exchange interaction**.

First described in: W. Heisenberg, *Mehrkörperproblem und Resonanz in der Quantenmechanik*, Z. Phys. **38**(67), 411–426 (1926); P. A. M. Dirac, *On the theory of quantum mechanics*, Proc. R. Soc. Lond., Series A **112**, 661–677 (1926).

Heisenberg force – a force between two nucleons derivable from a potential with an **operator**, which exchanges both the positions and the **spins** of the particles.

First described in: W. Heisenberg, *Über den Bau der Atomkerne. I*, Z. Phys. **77**(1–2), 1–11 (1932); W. Heisenberg, *Über den Bau der Atomkerne. II*, Z. Phys. **78**(3–4), 156–164 (1932); W. Heisenberg, *Über den Bau der Atomkerne. III*, Z. Phys. **80**(9–10), 587–596 (1933).

Heisenberg representation – a mode of a description of a system in which stationary vectors represent dynamic states and **operators**, which evolve in the course of time, represent physical quantities.

First described in: W. Heisenberg, *Über quantentheoretische Umdeutung kinematischer und mechanischer Beziehungen*, Z. Phys. **33**(1), 879–893 (1925); M. Born, P. Jordan, *Zur Quantenmechanik*, Z. Phys. **34**(1), 858–888 (1925); M. Born, W. Heisenberg, P. Jordan, *Zur Quantenmechanik II*, Z. Phys. **35**(8–9), 557–615 (1925).

Heisenberg's uncertainty principle states that it is impossible to specify simultaneously, with arbitrary precision, both the momentum (p) and the position (r) of a quantum particle. The quantitative version of the position-momentum uncertainty relation is $\triangle p \triangle r \geq \hbar/2$ or equivalently for time (t) and energy (E) → $\triangle t \triangle E \geq \hbar/2$.

First described in: W. Heisenberg, *Über den anschaulichen Inhalt der quantentheoretischen Kinematik und Mechanik*, Z. Phys. **43**(3/4), 172–198 (1927).

Recognition: in 1932 W. Heisenberg received the Nobel Prize in Physics for the creation of quantum mechanics, the application of which has, among other things, led to the discovery of the allotropic forms of hydrogen.

See also www.nobel.se/physics/laureates/1932/index.html.

Heitler−London method – a way to describe molecules and molecular states supposing that each molecule is composed of atoms and their electronic structure is represented by atomic orbitals of the composing atoms, which later becomes the superposition of atomic orbitals. The method is an alternative to the molecular orbital method → **Hund−Mulliken theory**. Both are an approximation to the complete solution. Consider two atoms brought together starting from a large distance. One can consider and distinguish three regions. The first region is that located at long distances, where the atoms maintain their character but are polarized by each other. In the second, intermediate region, there is a "real" molecule and the atoms lose part of their identity. In the third region, at short distances, the atoms begin to behave like a single united atom. The molecular description in the Heitler−London method tries to describe the molecular states starting from the third region, while the Hund−Mulliken theory starts from the first region.

First described in: W. Heitler, F. London, *Wechselwirkung neutraler Atome and homopolare Bindung nach der Quantenmechanik*, Z. Phys. **44**, 455–472 (1927).

More details in: J. C. Slater, *Molecular orbital and Heitler−London Methods*, J. Chem. Phys. **43**(10), S11–S17 (1965); C. A. Coulson, *Valence* (Oxford University Press, Oxford, 1961).

helium I – the phase of liquid ^{4}He which is stable at temperatures above the **lambda point** (2.178 K) and has the properties of a normal liquid, except low density.

helium II – the phase of liquid ^{4}He which is stable at temperature between absolute zero and the **lambda point** (2.178 K) and demonstrates such exceptional properties as vanishing viscosity called **superfluidity** and extremely high heat conductivity. It also shows the **fountain effect**.

First described by P. L. Kapitza in 1937 – experiment; L. Landau in 1941 – theory.

Recognition: in 1962 L. D. Landau received the Nobel Prize in Physics for his pioneering theories for condensed matter, especially liquid helium.

See also www.nobel.se/physics/laureates/1932/index.html.

helium neon laser – a type of **ion laser** with a mixture of ionized helium and neon gases as the active medium. It generates light peaking at 633 nm (red light).

Hellman−Feynman theorem states that within the **Born−Oppenheimer approximation** the forces on nuclei in molecules or solids are those which would arise electrostatically if electron probability density were treated as a static distribution of negative electric charge.

Helmholtz double layer

First described in: H. Hellmann, *Zur Rolle der kinetischen Elektronenenergie für die zwischenatomaren Kräfte*, Z. Phys. **85**, 180 (1933); R. P. Feynman, *Forces in molecules*, Phys. Rev. **56**(4), 340–343 (1939).

Helmholtz double layer represents an electrical double layer of positive and negative charges one molecule thick which occurs at a surface where two different materials are in contact.

Helmholtz equation – an elliptical partial differential equation of the form $\nabla^2\phi + k^2\phi = 0$, where ϕ is a function, ∇^2 is the Laplace operator, k is a number. When $k = 0$, the equation reduces to the Laplace equation. With imaginary k, the equation becomes the space part of the diffusion equation.

First described in: H. Helmholtz, *Die Thermodynamik chemischer Vorgänge*, Sitzungsber. Berlin Akad. Wissen. **1**, 21, 825 (1882); H. Helmholtz, *Versuch an Chlorzinkelementen*, Sitzungsber. Berlin Akad. Wissen. **2**, 958, 1895 (1883); H. Helmholtz, *Folgerungen die galvanische Polarisation betreffend*, Sitzungsber. Berlin Akad. Wissen. **3**, 92 (1883).

Helmholtz free energy → **free energy**.

First described in: H. Helmholtz, *Über die Erhaltung der Kraft*, (Berlin 1847).

Hermann–Mauguin symbols (notations) – symbols representing the 32 symmetry classes of crystals. They consist of series numbers giving the multiplicity of symmetry axes in descending order with other symbols indicating inversion axes and mirror planes. They are widely used in crystallography. This type of symbol, first proposed by C. Hermann, was greatly simplified and owes its present form to the efforts of C. Mauguin.

First described in: C. Hermann, *Zur systematischen Strukturtheorie I. Eine neue Raumgruppensymbolik*, Z. Kristallogr. **68**, 257–287 (1928); C. Hermann, *Zur systematischen Strukturtheorie. II.*, Z. Kristallogr. **68**, 377–401 (1928); C. Hermann, *Zur systematischen Strukturtheorie. III. Ketten- und Netzgruppen*, Z. Kristallogr. **69**, 259–270 (1929); C. Hermann, *Zur systematischen Strukturtheorie. IV. Untergruppen*, Z. Kristallogr. **69** 533–555 (1929); C. Mauguin, *Sur le symbolisme des groupes de répétition ou de symétrie des assemblages cristallins*, Z. Kristallogr. **76**, 542–558 (1931).

More details in: F. A. Cotton, *Chemical Applications of Group Theory*, 3rd edn. (Wiley, New York, 1990).

Hermite equation – a differential equation of the form $y'' - 2xy' + 2uy = 0$, where u is a constant, but not necessary an integer. Introducing the operator $\mathbf{D} = d/dx$ one can rewrite this as $(\mathbf{D}^2 - 2x\mathbf{D} + 2u)y = 0$.

Hermite polynomial has the form

$$H_n(x) = (2x)^n - n\frac{n-1}{1!}(2x)^{n-2} + \frac{(n-1)(n-2)(n-3)}{2!}(2x)^{n-4} \dots,$$

where n is its degree. The first Hermite polynomials are: $H_0(x) = 1, H_1(x) = 2x, H_2(x) = 4x^2 - 2, H_3(x) = 8x^3 - 12x, H_4(x) = 16x^4 - 48x^2 + 12$. They are solutions of the **Hermite equation** for $n = u$.

First described in: C. Hermite, *Sur un nouveau développement en série des fonctions,* Compt. Rend. **58** 93–100 (1864).

Hermitian conjugate operator $\rightarrow$ adjoint operator.

Hermitian operator – a linear **operator A** acting on vectors in a **Hilbert space** such that if x and y are in the range of **A** then the inner products $(\mathbf{A}x, y)$ and $(x, \mathbf{A}y)$ are equal, or in the case of a function $\psi(r)$ one has $\int \psi^*(r)\mathbf{A}\psi(r) = \int (\mathbf{A}\psi(r))^* \psi(r)$.

Herring's γ plot $\rightarrow$ Wulff's theorem.

First described in: C. Herring, *Some theorems on the free energies of crystal surfaces,* Phys. Rev. **82**(1), 87–93 (1951).

Hertzian vector – a single vector describing an electromagnetic wave from which both the electrical and magnetic intensities can be obtained by derivation. The vector is represented by the equation $\mathbf{Z} = \int A dt$, where A is the vector potential. Thus, the intensity of the electric field is

$$E = \nabla(\nabla \mathbf{Z}) - \frac{1}{c^2}\frac{\partial^2 \mathbf{Z}}{\partial t^2}$$

and the intensity of the magnetic field is

$$H = \frac{1}{c}\nabla \times \frac{\partial \mathbf{Z}}{\partial t}.$$

Hess's law states that in a chemical reaction the evolved or absorbed heat is the same whether the reaction takes place one step or several steps.

First described in: G. H. Hess, *Recherches thermochimiques,* Bull. Sci. Acad. Imp. Sci. St Petersburg **8**, 257–272 (1840).

heterojunction – an interface between two dissimilar materials.

heterojunction solar cell makes use of **heterojunction**s formed from different $A^{III}B^V$ compounds or their alloys to produce the rectifying junction. In these cells, the total current is a sum of the currents generated in the different materials but voltage is controlled by the lowest of the two band gaps. Also double- and triple-junction devices $\rightarrow$ **multijunction solar cell**s have been developed that allow the monolithic stacking of multiple cells on top of one another, so each efficiently converts a relatively narrow range of photon energies suited to its band gap.

heterostructure

First described in: Zh. I. Alferov, V. M. Andrew, M. B. Kagan, I. I. Protasov, V. G. Trosfim, Sov. Phys. Semicond. **4**, 2047 (1971).

Recognition: in 2000 Zh. I. Alferov received (shared with H. Kroemer and J. S. Kilby) the Nobel Prize in Physics for developing semiconductor heterostructures used in high-speed- and optoelectronics.

See also www.nobel.se/physics/laureates/2000/index.html.

heterostructure – structure comprising more than one kind of material. Heterostructures are formed from multiple **heterojunctions**.

heterostructure field effect transistor (HFET) → **high electron mobility transistor.**

Heusler alloy – an alloy made of metals that are only partially magnetically aligned in their pure state but have all their spins aligned at room temperature in the alloy form.

First described in: F. Heusler, Verh. Dtsch. Ges. **5**, 219 (1903).

hexagonal close-packed (hcp) structure is represented by a **unit cell** consisting of three layers of atoms as schematically shown in Figure H.5.

The top and bottom layers contain six atoms at the corners of a hexagon and one atom at the center of each hexagon. The middle layer contains three atoms nestled between the atoms of the top and bottom layers. This structure is adopted by many metals.

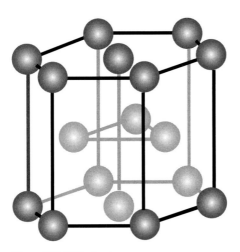

Figure H.5 HCP structure.

hexagonal lattice – a particular type of **Bravais lattice** whose unit cell is a right prism with a hexagonal base and whose lattice points are located at the vertices of the unit cell and at the center of the hexagon.

Heyd–Scuseria–Ernzerhof (HSE) functional – the hybrid functional employing screened short-range **Hartree–Fock** exchange instead of the full exact exchange. The exchange-correlation energy is written as

$$E_{xc}^{HSE} = aE_x^{HF,SR} + (1-a)E_x^{PBE,SR} + E_x^{PBE,LR} + E_c^{PBE},$$

where $E_x^{HF,SR}$ is the short-range Hartree–Fock exchange, $E_x^{PBE,SR}$ and $E_x^{PBE,LR}$ are the short-range and long-range components of the **Perdew–Burke–Ernzerhof** (PBE) exchange functional, respectively, and $a = 1/4$ is the Hartree–Fock exchange mixing parameter usually determined via perturbation theory. The form of the functional can be viewed as an adiabatic connection for the short-range portions of exchange, whereas long range exchange and correlation are treated with the PBE functional.

First described in: J. Heyd, G. E. Scuseria, M. Ernzerhof, *Hybrid functionals based on a screened Coulomb potential*, J. Chem. Phys. **118**(15), 8207–8215 (2003).

hidden variable – a hypothetical variable, in addition to the quantum state, that specifies a definite value for the observable of the quantum system.

high electron mobility transistor (HEMT) – a transistor exploiting the high in-plane mobility of electrons near a modulation-doped heterojunction. Synonyms are **heterostructure field effect transistor (HFET)**, **modulation doped field effect transistor (MODFET)**, **two-dimensional electron gas field effect transistor (TEGFET)**.

high-T_c superconductors $\rightarrow$ **superconductivity**.

Hilbert space – a many-dimensional normed linear vector space with an inner product that makes it complete. Points of real Hilbert space are sequences of numbers $x_1, x_2, \ldots$ such that the infinite sum $\sum x_i^2$ is finite. Elements of the basis are the interiors of balls in which the distance between $p = (x_1, x_2, \ldots)$ and $q = (y_1, y_2, \ldots)$ is the square root of $\sum(y_i - x_i^2)$. Hilbert space is an abstract mathematical tool for calculating the evolution in time of the energy levels of systems. This evolution occurs in ordinary space–time scale. It is widely used in quantum mechanics to define an observable value to be an operator on a Hilbert space.

More details in: D. Hilbert, *Grundlagen der Geometrie* (Teubner, Leipzig-Berlin, 1930).

Hilbert transform of a function $f(x)$ is $g(\xi) = 1/\pi \int_{-\infty}^{\infty} f(x)/(\xi - x)dx$, where the integral is understood as a **Cauchy principle value**. If the transform of a function is known, then the inverse formula gives $f(x) = -1/\pi \int_{-\infty}^{\infty} g(\xi)/(\xi - x)d\xi$.

Hohenberg–Kohn theorems relate to any system consisting of electrons moving under the influence of an external potential. Theorem 1: The external potential, and hence the total energy, is a unique functional of the electron density. Theorem 2: The ground-state energy can be obtained variationally: the density that minimizes the total energy is the exact ground-state density.

It is necessary to note that an electronic system with a given **Hamiltonian** has a ground-state energy as well as a ground-state **wave function**, which is completely determined by the minimization of the total energy as a functional of the wave function. Moreover, as the external potential together with the number of electrons completely determines the Hamiltonian, these two quantities determine all properties of the ground state.

The theorems provide the formal justification of the **Thomas-Fermi theory**, and the foundation of the **density functional theory**. They effectively reduce the $3N$-dimensional minimization problem of finding the wave function to a three-dimensional minimization problem of determining the density.

First described in: P. Hohenberg, W. Kohn, *Inhomogeneous electron gas*, Phys. Rev. **136**(3B), B864–B871 (1964).

Recognition: in 1998 W. Kohn received the Nobel Prize in Chemistry for his development of the density functional theory.

See also www.nobel.se/chemistry/laureates/1998/index.html.

Hofstadter butterfly – a diagram presenting energy levels in a two-dimensional crystal as a function of the magnetic field applied. In a sense it looks like a butterfly.

First described in: D. R. Hofstadter, *Energy levels and wave functions of Bloch electrons in rational and irrational magnetic fields*, Phys. Rev. B **14**(6), 2239–2249 (1976).

hole-burning $\rightarrow$ **spectral hole-burning**.

hologram – an interference pattern that is recorded on a high-resolution plate when two interfering beams formed by a coherent beam from a laser and light are scattered by an object. If after processing, the plate is viewed correctly by monochromatic light, a three-dimensional image of the object is seen. The name, coined by D. Gabor in 1947, derived from the greek word "holos" (the whole) because the interference pattern contains all the information.

First described in: D. Gabor, *A new microscopic principle*, Nature **161**, 777–778 (1948); D. Gabor, *Microscopy by reconstructed wavefronts*, Proc. R. Soc. (Lond.) A **197**, 454–487 (1949).

Recognition: in 1971 D. Gabor received the Nobel Prize in Physics for his invention and development of the holographic method.

See also www.nobel.se/physics/laureates/1971/index.html.

HOMO – an acronym for a **h**ighest **o**ccupied **m**olecular **o**rbital.

homopolar bond is formed between atoms, which share a pair of electrons with opposite spins.

Hooke's law originally stated that the power of any springy body is in proportion to the extension. In the modern corrected form it states that the extension Δl of an elastic spring and its tension F, but not power, are linearly related to each other: $F = k\Delta l$, where k is the spring constant. The law holds up to a limit, called the elastic limit, after which springs suffer plastic deformation up to the limit of plasticity, after which they break down. Later Cauchy generalized Hooke's law to three-dimensional elastic bodies and stated that the 6 components of stress are linearly related to the 6 components of strain. In the matrix form it is $\epsilon = \mathbf{S} \times \mathbf{s}$ and $\mathbf{s} = \mathbf{C} \times \epsilon$, where ϵ is the 6 components stress vector, $\mathbf{s}$ is the 6 components strain vector, $\mathbf{S}$ is the compliance matrix (36 components), and $\mathbf{C}$ is the **stiffness matrix** (36 components). It is evident that $\mathbf{S} = \mathbf{C}^{-1}$. In general, the above stress-strain relationships are known as **constitutive relations**.

First described in: R. Hooke, *A description of helioscopes and some other instruments,* Phil. Trans. **10**, 440–442 (1675).

hopping conductivity – the charge transport mechanism where by conductivity is determined by electrons hopping via **localized states** without activation to the conduction band. If it is energetically favorable for localized electrons to hop beyond the nearest-neighbor impurity centers, so that the average hopping length exceeds the average distance between them, variable range hopping conductivity sets in the system. The average hopping length increases with the temperature decrease for disordered systems in which electronic states are localized close to the **Fermi level**. The temperature dependence of the hopping conductivity is described by the **Mott law**: $\ln \sigma \sim T^{-1/4}$, which in the general form looks like $\ln \sigma \sim T^{-1/n}$, where the parameter n depends on the dimensionality of the system $d (n = d + 1)$. The Mott law is derived assuming that the density of states near the Fermi level of the system is constant.

Accounting for the existence of a parabolic gap of width ΔE_C in the density of states near the Fermi level of the system, due to electron–electron Coulomb inter-action, comes to the **Shklovskii-Efros law** of temperature dependence of hopping conductivity: $\ln \sigma \sim T^{-1/2}$. It is observed at low temperatures when $\Delta E_C > k_B T$. The power of temperature does not depend on the dimensionality of the system.

A crossover between the Mott and Shklovskii-Efros hopping regimes can be observed as a function of temperature.

First described in: N. F. Mott, *Conduction in non crystalline materials. III. Localized states in a pseudogap and near extremities of conduction and valence bands,* Phil. Mag. **19**(160), 835–852 (1969).

Recognition: in 1977 P. W. Anderson, N. F. Mott and J. H. van Vleck received the Nobel Prize in Physics for their fundamental theoretical investigations of the electronic structure of magnetic and disordered systems.

See also www.nobel.se/physics/laureates/1977/index.html.

hot carrier – a charge carrier, which may be an electron or a hole, that has relatively high energy with respect to the carriers normally found in the material.

hot-carrier solar cell – a **solar cell** designed to collect irradiation-induced electron–hole pairs in a semicondutor before they thermalize to respective band edges. This requires both an absorber with carrier cooling properties and a collection of carriers over a limited range of energies, such that cold carriers in the external contacts do not cool the hot carriers to be extracted.

More details in: R. T. Ross, A. J. Nozik, *Efficiency of hot-carrier solar energy converters,* J. Appl. Phys. **53**(5), 3813–3818 (1982).

HSE functional = **Heyd–Scuseria–Ernzerhof functional**.

Hubbard Hamiltonian – a common model phenomenological **Hamiltonian** used to describe properties of strong electron correlation in solids. In this model, which is in fact the **Hubbard model**, strongly correlated electrons move in a lattice. The related Hamiltonian has two terms: the first term describes a band of tight-binding electrons, the second term involves a repulsive electron–electron interaction. Thus this Hamiltonian studies the effects induced by the competition between localization and correlations.

The interaction between electrons only occurs if they are located on the same site. Only one orbital per site is considered. Thus the Hamiltonian is given by $\mathbf{H} = \sum_{i,j,\sigma} t_{ij} a_{i\sigma}^{+} a_{j\sigma} + U \sum_{j} n_{i\uparrow} n_{i\downarrow}$, where $n_{i\sigma} = a_{i\sigma}^{+} a_{i\sigma}$ is the occupation number operator, $a_{i\sigma}^{+}$ creates and $a_{i\sigma}$ annihilates an electron with spin σ on the i-th site of the lattice, U represents the effective on-site interactions, and t_{ij} is the tight binding hopping integral between the i-th and j-th sites. Strong correlations occur when the interaction U is large in comparison with the band width W of the system, that is for large U/t, called the interaction strength. The model successfully describes the itinerant magnetism and the metal–insulator transition → **Mott–Hubbard transition**. Despite the simplicity of the Hamiltonian, no exact solution has been found, except in the one-dimensional case. Nevertheless, it has been demonstrated that the Hubbard model has a non-trivial limit in higher dimensions.

First described in: J. Hubbard, *Electron correlations in narrow energy bands. I,* Proc. R. Soc. A **276**, 238–257 (1963); J. Hubbard, *Electron correlations in narrow energy bands. II,* Proc. R. Soc. A **277**, 237–259; J. Hubbard, *Electron correlations in narrow energy bands. III,* Proc. R. Soc. A **281**, 401–419 (1964); J. Hubbard, *Electron correlations in narrow energy bands. IV,* Proc. R. Soc. A **285**, 542–560 (1965).

Hubbard model → **Hubbard Hamiltonian**.

Hückel approximation is used to calculate molecular orbitals. The following assumptions are made:

1. If atoms i and j are not directly bonded, then H_{ij} (bond energy between atoms i and j) equals zero.

2. If all atoms and bond lengths are identical in a molecule, then all H_{ij} values for directly bonded atoms are identical.

3. If all atoms are identical, all H_{ii} integrals are equal and will be symbolized by α.

4. The atomic orbitals are normalized, therefore $\int \phi_i \phi_j d\tau = 1$.

5. The atomic orbitals are mutually orthogonal, therefore $\int \phi_i \phi_j d\tau = 0$.

The form of the Hückel secular equations is:

$$
\begin{array}{ccc}
c_{11}(\alpha - E) + & \ldots + & c_{1n}\beta_{1n} = 0 \\
\vdots & & \vdots \\
c_{n1}\beta_{n1} + & \ldots + & (\alpha - E) = 0
\end{array}
$$

and the Hückel secular determinant is:

$$
\begin{vmatrix}
(\alpha - E) & \ldots & \beta_{1n} \\
\vdots & & \vdots \\
\beta_{n1} & \ldots & (\alpha - E)
\end{vmatrix} = 0
$$

It should be remembered that the first assumption states that non-directly bonded atoms have $H_{ij} = 0$, ($\beta = 0$). The number of atomic orbitals equals the number of secular equations and the number of molecular orbitals.

First described in: E. Hückel, *Quantentheoretische Beiträge zum Benzolproblem*, Z. Phys. **70**, 204–286 (1931).

Hund-Mulliken theory supposes that in the case of an atom an atomic orbital is best taken to be an eigenfunction of the one-electron **Schrödinger equation** based on the nuclear attraction for the electron and on the average repulsion of all other electrons. The molecular orbital is defined in the same way. Now its one-electron Schrödinger equation is based on the attraction of two or more nuclei plus the averaged repulsion of all other electrons. It is also known as **molecular orbital theory**.

As in the theory of atoms, it is possible to use the **aufbau principle** (the building-up principle) in order to feed electrons into molecular orbitals. For example, in the case of a diatomic molecule the binding electrons occupy a quantum state specific to the extended chemical bond, thus Hund introduced the Greek symbols Σ, Π and Δ for designing the molecular electronic states, Mulliken used instead the symbols S, P and D. Differently from the valence-bond method or **Heitler–London method**, the molecular orbital method considers a molecule as a self-sufficient unit and not as a mere composite of atoms.

First described in: F. Hund, *Zur Deutung der Molekelspektren I*, Z. Phys. **36**, 657 (1926); F. Hund, *Zur Deutung der Molekelspektren II*, Z. Phys. **37**, 742 (1927); F. Hund, *Zur Deutung der Molekelspektren III*, Z. Phys. **42**, 93 (1927); F. Hund, *Zur Deutung der Molekelspektren IV*, Z. Phys. **43**, 805 (1927); R. S. Mulliken, *The assignment of quantum numbers for electrons in molecules. I*, Phys. Rev. **32**(2) 186–222 (1928); R. S. Mulliken, *The assignment of quantum numbers for electrons in molecules.*

Hund rules

II. Correlation of molecular and atomic electron states, Phys. Rev. **32**(5) 761–772 (1928); R. S. Mulliken, *The assignment of quantum numbers for electrons in molecules. III. Diatomic hydrides*, Phys. Rev. **33**(5) 730–747 (1929).

More details in: R. S. Mulliken, *Molecular scientists and molecular science: some reminiscences*, J. Chem. Phys. **43**(10), S2–S11 (1965); J. C. Slater, *Molecular orbital and Heitler–London methods*, J. Chem. Phys. 43(10), S11–S17 (1965).

Recognition: in 1966 R. S. Mulliken received the Nobel Prize in Chemistry for his fundamental work concerning chemical bonds and the electronic structure of molecules by the molecular orbital method.

See also www.nobel.se/chemistry/laureates/1966/index.html.

Hund rules give the order in the energy of atomic states formed by equivalent electrons: (i) among the terms given by equivalent electrons, the ones with greatest multiplicity have the least energy, and of these the one with greatest orbital angular momentum is lowest; (ii) the state with a multiplet with lowest energies is given by that state in which the total angular momentum is the least possible if the shell is less than half filled, and the greatest possible if the shell is more than half filled.

First described in: F. Hund, *Deutung von Spektra*, Z. Phys. **33**(5/6), 345–371 (1925).

Huntington potential – the interatomic pair potential in the form of **Born–Mayer potential**:

$$V(r) = A \exp\left(a\frac{r_0 - r}{r_0} \right),$$

where r is the interatomic distance, r_0 is the equilibrium distance between atoms in the perfect crystal, and A and a are fitted parameters defined from experimental data.

First described in: H. B. Huntington, *Mobility of interstitial atoms in a face-centered metal*, Phys. Rev. **91**(5), 1092–1098 (1953).

Huygens principle states that every point on a propagating wavefront serves as the source of spherical secondary wavelets, such that the wavefront at some later time is the envelope of these wavelets. If the propagating wave has a certain frequency and is transmitted through the medium at a certain speed, then the secondary wavelets will have the same frequency and speed.

First described in: C. Huygens, *Traité de la lumière* (Leyden, 1690).

hybridization – a process by which a number of constituent **orbitals** of an atom are linearly combined (hybridized) to form an equal number of equivalent and directed orbitals (hybrids).

hydride – a chemical compound of an element with hydrogen. Hydrides suitable for technological applications in fabrication of semiconductor electron devices are listed in Table H.1.

Table H.1 Hydrides for chemical vapor deposition of semiconductors.

Element	Compound		Melting point (°C)	Boiling point (°C)	Decomposition temperature (°C)
	Formula	Name			
Si	SiH_4	Silane	−185	−111.9	450
	Si_2H_6	Disilane	−132.5	−14.5	
	Si_3H_8	Trisilane	−117.4	52.9	
	Si_4H_{10}	Tetrasilane	−108	84.3	
Ge	GeH_4	Germane	−165	−88.5	350
	Ge_2H_6	Digermane	−109	29	215
	Ge_3H_8	Trigermane	−105.6	110.5	195
P	PH_3	Phosphine	−133	−87.8	
As	AsH_3	Arsine	−116	−62.5	
S	H_2S	Hydrogen sulfide	−85.5	−59.6	
Se	H_2Se	Hydrogen selenide	−65.7	−41.3	

hydrocarbons – compounds that are composed entirely of carbon and hydrogen atoms bonded to each other by covalent bonds. Saturated and unsaturated hydrocarbons exist.

Saturated hydrocarbons, called **alkanes**, contain single bonds. The alkanes, also known as parafins, have straight or branched-chain hydrocarbon compounds with only single covalent bonds between the carbon atoms. The general molecular formula for alkanes is C_nH_{2n+2}, where n is the number of carbon atoms in the molecule. Each carbon atom is connected to four other atoms by four single covalent bonds. These bonds are separated by angles of 109.5° (the angle between lines drawn from the center of a regular tetrahedron to its corners). Alkane molecules contain only carbon–carbon and carbon–hydrogen bonds, which are symmetrically directed toward the corners of a tetrahedron. Therefore, alkane molecules are essentially nonpolar.

Alkanes that have lost one hydrogen atom are called **alkyl groups**. These groups have the general formula $-C_nH_{2n+1}$. The missing hydrogen atom may be detached from any carbon in the alkane. The name of the group is formed from the name of the corresponding alkane by replacing *-ane* with the *-yl* ending, for example methane (CH_4) → methyl ($-CH_3$).

Unsaturated hydrocarbons consist of three families of compounds that contain fewer hydrogen atoms than the alkane with the corresponding number of carbon atoms and contain multiple bonds between carbon atoms. These include: **alkenes** with carbon–carbon double bonds; **alkynes** with carbon–carbon triple bonds; and **aromatic compounds** with benzene rings, also called **aromatic rings**, that are arranged in a six-membered ring with one hydrogen atom bonded to each carbon atom and three carbon–carbon double bonds, as illustrated in Figure H.6.

hydrogen bond

Figure H.6 Aromatic ring.

hydrogen bond – an association linkage between hydrogen atoms and the lone pairs of either nitrogen, oxygen or fluorine atoms, which are the most electronegative atoms in the periodic table, on a different molecule. This binding is mainly electrostatic, as the small radius of the positively charged hydrogen atom makes possible a very close approach to the electronegative atom. Hydrogen bonding is the strongest type of intermolecular force because the molecules that they exist between are very polar.

Hydrogen bonds can be formed within different parts of a single molecule (intramolecularly, as in proteins and nucleic acids) or between molecules (intermolecularly, as in water).

Figure H.7 shows some examples of hydrogen bonds.

More details in: Y. Maréchal, *The Hydrogen Bond and the Water Molecule: The Physics and Chemistry of Water, Aqueous and Bio-Media,* (Elsevier Science, Amsterdam, 2006).

Water   Formic acid

Figure H.7 Hydrogen bonding (indicated by dashed lines) in different compounds.

hydrolysis – chemical reactions in which water as a reactant produces a bond rupture in compounds.

hydronium ion → **protonated water cluster.**

hydrophilic – literally, from the Greek, water loving. A hydrophilic compound is one that "loves" water and easily dissolves in it. Having lots of potential for **hydrogen bond**ing or having a charge will make a compound hydrophilic. Most inorganic salts and some organic molecules including ethanol and diethyl ether are hydrophilic. Opposite of **hydrophobic**.

hydrophobic – literally, from the Greek, water fearing. A compound is hydrophobic if it "hates" water and does not dissolve in it. Little **hydrogen bond**ing capacity and no charge makes a molecule hydrophobic. Most organic molecules are hydrophobic. Opposite of **hydrophilic**.

hydroxide – an inorganic chemical compound having one or more hydroxyl (–OH) groups.

hyperelastic scattering – a scattering process in which an excited heavy particle, like an atom, collides with a light particle, like a free electron, and the latter increases its kinetic energy by nearly the whole of the excitation energy leaving the heavy particle in its normal (unexcited) state.

From ideality factor to isotropy (of matter)

ideality factor – the parameter n in the semiconductor diode current–voltage relationship $j = j_0[exp(eV/nk_BT) - 1]$, where j_0 is the saturation current density and V is the applied voltage. It is equal to 1 for an ideal diode when carrier diffusion dominates the current flow, while it can approach 2 for real structures when recombination current dominates.

I^{2}M – an acronym for **image interference microscopy**.

I^{3}M – an acronym for a **incoherent interference illumination microscopy**.

I^{5}M – an acronym for a **image interference + incoherent interference illumination microscopy**.

image force – a force on a charge due to the charge or polarization that it induces on surrounding solids.

image interference microscopy (I^{2}M) → **fluorescence nanoscopy**.
   First described in: M. G. L. Gustafsson, *Extended resolution fluorescence microscopy*, Curr. Opin. Struct. Biol. **9**, 627–634 (1999).

image interference (I^{2}M) + incoherent interference illumination (I^{3}M) microscopy (I^{5}M) → **fluorescence nanoscopy**.
   First described in: M. G. L. Gustafsson, D. A. Agard, J. W. Sedat, *I^{5}M: 3D wide-field light microscopy with better than 100 nm axial resolution*, J. Microsc. **195**, 10–16 (1999).

imide – a chemical compound containing the $>$NH group attached to two **carbonyl** (or equivalent) groups.

impact – a collision between two bodies where relatively large contact forces exist during a very short interval of time.

What is What in the Nanoworld. Victor E. Borisenko and Stefano Ossicini
Copyright © 2008 WILEY-VCH Verlag GmbH & Co. KGaA, Weinheim
ISBN: 978-3-527-40783-5

impedance – the ratio of voltage to current amplitudes in a two-terminal electric network operating with alternating current. It is a complex characteristic including the real part called resistance and the imaginary part called **reactance**.

implicit function theorem states that if a function has nonzero slope at a point, then it can always be approximated by a linear function at this point.

imprinting – a lithography technique used for nanopattern fabrication. There are two approaches within it. The first is based on the transfer of a monolayer of self-assembled molecules from an elastomeric stamp to a substrate and is referred to as ink contact printing or simply **inking**. The second approach uses molding of a thin polymer film of the mask material by a stamp under controlled pressure at elevated temperature and is referred to as **embossing**. No kind of irradiation is involved in the pattern formation. Therefore, both techniques avoid shortcomings associated with wave diffraction and scattering.

The common feature of the imprint approaches is that they in fact are replication techniques. The patterns envisaged must be available as an original: the master or stamp. In analogy with conventional lithography, there is a possibility of positive and negative patterning. The originals must be fabricated following state of the art in **nanolithography** using, for example, electron beam lithography or scanning probes. For the imprint technology, where the stamp is the functional equivalent to the photo-mask in conventional projection lithography, a stepper can be used for the manufacturing process.

impurity photovoltaic effect solar cell – a **solar cell** based on the idea of exploiting two-step carrier generation via impurity states within the band gap to utilize sub-band gap **photon**s and therefore enhance solar cell performance. Three transitions available in that system, in principle, enable better matching to the solar spectrum. High efficiencies have been predicted for these solar cells. The challenge is to find a suitable host wide-band-gap semiconductor combined with a sufficiently radiatively efficient impurity.

First described in: M. A. Green, A. S. Brown, *Impurity photovoltaic effect: fundamental energy conversion efficiency limit,* J. Appl. Phys. **92**, 1329–1326 (2002); G. Beaucarne, A. S. Brown, M. J. Keevers, R. Corkish, M. A. Green, *The impurity photovoltaic (IPV) effect in wide-bandgap semiconductors: an opportunity for very-high-efficiency solar cells?,* Prog. Phot. Res. Appl. **10**, 345–353 (2002).

incoherent interference illumination microscopy (I³M) → fluorescence nanoscopy.
First described in: M. G. L. Gustafsson, *Extended resolution fluorescence microscopy,* Curr. Opin. Struct. Biol. **9**, 627–634 (1999).

incoherent light – electromagnetic radiation in the optical range consisting of waves not all of the same phase and possibly wavelength.

incoherent scattering – scattering of particles or waves in which the scattering centers act independently of one another, so that there are no definite phase relationships among the different parts of the scattered beam.

independent electron approximation is used to solve the **Schrödinger equation** for electrons in solids. It removes the complications of the electron–electron interactions and replaces them with a time-averaged potential.

index guided lasers → **double heterostructure laser**.

inductance – the property of an electric circuit or of two neighboring circuits whereby, by electromagnetic induction, an electromagnetic force is generated in one circuit by a change of current in itself or in the other circuit.

inelastic collision – a collision process in which the total kinetic energy of the colliding particles is not the same as before. Part of the kinetic energy is converted into another type of energy, for example into heat, excitation, and so on.

inelastic electron tunneling spectroscopy consists of recording and analysis of tunnel currents and voltages across a metal–insulator–adsorbate–metal structure in order to extract vibrational and electronic spectroscopic information about the metals (**magnon**s and **phonon**s), the insulator, and the adsorbate.

The advantages of the method include: (i) ultra-high sensitivity–fewer than 10^{13} adsorbed molecules are required to provide a spectrum, (ii) overtone and combination bands are exceptionally weak thus making the identification of fundamentals in this method easier than in IR or **Raman** spectroscopy, (iii) optically forbidden transitions may be observed as strong bands, (iv) oxide bands are generally less intense than adsorbate bands, which makes it possible to obtain adsorbate spectra in the "IR opaque" regions of the oxide spectrum. Broad spectral range extending from 50 to 19 000 cm^{-1} with resolution better than 5 cm^{-1} is typically covered by the method. It is successfully used in studies of catalysis, biology, trace impurity detection, and electronic excitations.

More details in: K. W. Hipps, U. Mazur, *Inelastic electron tunneling spectroscopy*, in: *Handbook of Vibrational Spectroscopy*, edited by J. M. Chalmers, P. R. Griffiths (John Wiley and Sons, Chichester, 2002).

inelastic scattering – a scattering process in which the total momentum of interacting (colliding) particles is conserved but their total kinetic energy is not. Transformation of one type of energy into another type, for example kinetic energy of motion into heat or excitation energy, takes place. The alternative is **elastic scattering**.

inert gases – He, Ne, Ar, Kr, Xe, Rn, also called **rare gases**.

inhibitor

inhibitor – a substance whose addition to a reacting system results in a decrease in the rate of reaction.

injection laser → **semiconductor injection laser.**

inking – one of the approaches used for pattern fabrication by imprinting. The process is shown schematically in Figure I.1.

An elastomer stamp with a patterned surface is covered with ink and pressed onto the substrate. The ink composition is chosen to be able to form a **self-assembled** monolayer after coming into contact with the substrate. This monolayer then serves as a mask for further processing either by etching or surface reaction. The latter is particularly attractive since functionalization is intrinsic to the inking process.

The elastomeric stamp is normally made of polydi-methylsiloxane (PDMS). Thiols and alkanethiolates are suitable to be used as inks. Some of the technological challenges for the application of inking are, the alignment and diffusion of the self-assembled monolayer and the deformation of the stamp.

Mechanical stresses introduced into the PDMS stamp by handling, gravity or forces generated on the stamp/substrate interface during the propagation of the contact front negatively affect the alignment. Furthermore, because of large thermal expansion in the elastomer (for PDMS this is of the order of $10^4 \ K^{-1}$), even small thermal fluctuations in the environment make it difficult to control the effective stamp dimension. A solution is to put a very thin pattern film ($<10 \ \mu m$ thick PDMS) on a rigid carrier, like a silicon wafer, which reduces the distortion effects acting on the stamp.

Diffusion of the ink occurs during the printing stage, that is the period of actual contact between stamp and surface. Diffusion phenomena constitute a complex interplay between gas diffusion, movement of ink molecules and lateral movement of chemisorbed ink. Reduced diffusion leads to better resolution and hence to smaller printed feature sizes. The straightforward approach is printing with heavy inks. Long-chain thiols can be extended only to a certain limit, since further elongation leads to less ordered, and thus less etch-resistant monolayers. With thiols, the resolution is about 100 nm.

(a)   (b)

Figure I.1 Inking process steps: (a) stamp covered with monomer (top) and substrate (bottom) prior to imprinting, (b) self-assembled monolayer pattern left on the substrate after the stamp removal.

First described in: A. Kumar, G. M. Whitesides, *Features of gold having micrometer to centimetre dimensions can be formed through a combination of stamping with an elastomeric stamp and an alkanethiol 'ink' followed by chemical etching,* Appl. Phys. Lett. **63**(14), 2002–2004 (1993).

in situ – Latin meaning "at the spot", "within". The phrase *"in situ* analysis of experimental samples" is often used to show that the samples are analyzed at the place where they have been fabricated, for instance in the same vacuum chamber. An alternative is *ex situ* (Latin "away from") analysis indicating that the samples are analyzed in the place different from that where they have been fabricated.

insulator – a substance highly resistant to a flow of electric currents. It has a resistivity more than 10^{10} Ω cm and high dielectric strength. A solid is an insulator if there is an energy gap between its **conduction band** and **valence band** and the conduction band is not filled with electrons.

integrated circuit – an electronic solid state circuit composed of semiconductor transistors and diodes, resistors and capacitors, other electronic components and their interconnects fabricated in a single technological process into/onto a single substrate. Monocrystalline silicon is currently a widely used material for integrated circuits. The invention of the integrated circuit started the revolution in electronics and information technologies.

First described by J. S. Kilby in 1958.

More details in: J. S. Kilby, *Invention of the integrated circuits,* IEEE Trans. Electron. Dev. **23**(7), 648–654 (1976).

Recognition: in 2000 J. S. Kilby shared the Nobel Prize in Physics with Z. I. Alferov and H. Kroemer for his part in the invention of the integrated circuit.

See also www.nobel.se/physics/laureates/2000/index.html.

interface – a shared boundary between two materials.

interference – variation with distance or time of the amplitude of a wave, which results from the superposition (algebraic or vector addition) of two or more waves having the same or nearly the same frequencies.

intermediate band solar cell – a **solar cell** employing the fact that when this type of cell absorb sunlight, electrons are excited not only between the **valence** and **conduction band**s of the semiconductor material constituting the cell, but also between these bands and one or more additional bands that lie within the normally forbidden band gap of the material. Therefore, this device has the potential to reach high conversion efficiencies by increasing the photogenerated current, due to these multiple step excitations, without reducing the open-circuit voltage.

internal photoelectric effect

More details in: A. S. Brown, M. A. Green, *Intermediate band solar cell with many bands: Ideal performance,* J. Appl. Phys. **94**(9), 6150–6158 (2003) – theory; A. Luque, A. Marti, N. Lopez, E. Antolín, E. Canovas, C. Stanley, C. Farmer, L. J. Caballero, L. Cuadra, J. L. Balenzategui, *Experimental analysis of the quasi-Fermi level split in quantum dot intermediate-band solar cells,* Appl. Phys. Lett. **87**, 083505 (2005) – experiment.

internal photoelectric effect – excitation of electrons from the **valence band** of a semiconductor to its **conduction band** as a result of light absorption.

internal quantum efficiency → **quantum efficiency**.

interstitial – a position in a crystal lattice between regular lattice positions of atoms. The term is used for an atom occupying an interstitial position.

intrinsic semiconductor – if the concentration of charge carriers in a semiconductor is a characteristic of the material itself rather than dependent upon an impurity content or structural defects of the crystal, then the semiconductor is defined an intrinsic semiconductor.

intron – a **nucleotide** sequence intervening between **exons** (coding regions) that is excised from a gene transcript during **RNA** processing.

invariance – the property of a physical quantity or physical law of being unchanged by a certain transformations or operations such as reflection of spatial coordinates, time reversal, charge conjugation, rotations or **Lorentz transformations**.

inversion symmetry – the principle that laws of physics are unchanged by the operation of inversion.

ionic bond – a type of chemical bonding of atoms in which one or more electrons are transferred completely from one atom to another, thus converting the neutral atoms into electrically charged ions.

ionic crystal – a solid formed by **ionic bonds** between constituents.

ionization energy (of an atom) – the minimum energy required to ionize the atom starting from its ground state.

ion laser – a **laser** in which the transition involved in stimulated emission of radiation takes place between two energy levels in an ionized gas. One of the inert gases (argon, helium, neon, or krypton) is used as the active medium. The device is schematically shown in Figure I.2.

The gases are electrically excited in a container called a plasma tube, which typically consists of an alumina or ceramic envelope that is vacuum sealed at each end by either two Brewster windows or one Brewster window and one sealed cavity mirror. The optical cavity is defined by a 100% reflecting mirror and a partially transmissive output coupling mirror. It provides moderate to high continuous-wave output of typically 1 mW to 10 W. For single-frequency operation, the high reflector is replaced by a Brewster prism, and an etalon is inserted.

Figure I.2 Ion laser.

irradiance – an energy per unit time crossing a unit area oriented normal to the propagation direction. The term is usually applied to characterize light beams.

isodesmic structure – a crystal structure formed by ions in which there are no distinct groups formed within the structure, that is where no bond is stronger than any other → **anisodesmic** and **mesodesmic structures**.

isoelectronic atoms – atoms with the same number of valence electrons.

isoelectronic principle states that any two or more molecules which are isoelectronic will have similar molecular orbitals. Molecules possessing the same number of electrons distributed over a similar molecular framework are isoelectronic.

isomers – molecules with the same chemical formula but possessing different properties. There are two main forms of isomers: structural (constitutional) isomers and spatial isomers – **stereoisomers**. The tree of isomeric forms is presented in Figure I.3. In **structural isomers**, the atoms and functional groups are joined together in different ways, as in the example of propyl alcohol. This group includes **chain isomers** in which hydrocarbon chains have variable amounts of branching, **position isomers** which deal with the position of a functional group on a chain, and **functional group isomers** in which one functional group is split up into different smaller ones.

In **stereoisomers**, the bond structure is the same but the geometrical positioning of atoms and functional groups in space differs. This class includes **enantiomers** where different isomers are mirror-images of each other, and **diastereomers** when

isomorphic solids

they are not. Diastereomers are again subdivided into **conformers** (conformal isomers) when isomers can interconvert by chemical bond rotations and **cis-trans isomers** when this is not possible. Note that although conformers can be referred to as having a diastereomeric relationship, the isomers over all are not diastereomers, since bonds in conformers can be rotated to make them mirror images. Such conformers are referred to as **rotamers**.

First described by J. J. Berzelius and F. Woehler in 1827.

Figure I.3 Isomeric forms of molecules (from Wikipedia).

isomorphic solids – solids which crystallize with the same symmetry, with similar facial development, and have closely similar values of the interfacial angles. Such solids often form mixed compounds with one another.

isotope – one of two or more nuclides having the same atomic number, hence constituting the same element, but differing in mass number because of the difference in the content of neutrons in their nuclei.

isotropy (of matter) – the same physical properties of a medium found in all directions. The alternative is **anisotropy**.

From Jahn–Teller effect to Joule's law of electric heating

Jahn–Teller effect → Jahn–Teller theorem.

Jahn–Teller theorem states that if a particular symmetric configuration of a non-linear molecule causes it to have an orbitally degenerate ground state, then this configuration is unstable with respect to one of lower symmetry which does not give rise to an orbitally degenerate ground state. Thus, a system having a degenerate ground state will spontaneously evolve to lower its symmetry unless the degeneracy is simply a spin degeneracy.

The tendency of molecular systems to distort when electronic degeneracy is present is referred to be the **Jahn–Teller effect**.

First described in: H. A. Jahn, E. Teller, *Stability of polyatomic molecules in degenerate electronic states. Part I. Orbital degeneracy*, Proc. R. Soc. A **161**, 220–235 (1937).

Jastrow correlation factor → Slater–Jastrow wave function.

Jaynes–Cummings model describes a system of a two-level atom interacting with a quantized mode of an optical cavity, with or without the presence of light. The **Hamiltonian** describing the interacting system is $H = H_{field} + H_{atom} + H_{JC}$, where H_{field} is the free field Hamiltonian, H_{atom} is the atomic excitation Hamiltonian and the Jaynes-Cummings interaction Hamiltonian assuming the zero field energy at zero is $H_{JC} = g(A_c\sigma_+ e^{i\delta t} + A_a\sigma_- e^{-i\delta t})$ with $g = d(\omega/\hbar V\epsilon_0)^{1/2}$ and $\delta = |\omega - \nu|$, where d is the atomic transition moment, V is the mode volume of the cavity, ω is the atomic transition frequency, and ν is the radiation field angular frequency. Here A_c and A_a are the Bosonic creation and annihilation operators, respectively. The operators $\sigma_+ = \sigma_x + i\sigma_y$ and $\sigma_- = \sigma_x - i\sigma_y$ are the level raising and lowering operators or the atomic spin-flip operators in terms of **Pauli matrices**.

The model serves to find out how quantization of the radiation field affects the predictions for the evolution of the state of a two-level system in comparison with semi-classical theory of light–atom interaction.

First described in: E. T. Jaynes, F. W. Cummings, *Comparison of quantum and semiclassical radiation theories with application to the beam maser*, Proc. IEEE 51(1), 89–109 (1963).

What is What in the Nanoworld. Victor E. Borisenko and Stefano Ossicini
Copyright © 2008 WILEY-VCH Verlag GmbH & Co. KGaA, Weinheim
ISBN: 978-3-527-40783-5

Johnsen–Rahbek effect

More details in: C. C. Gerry, P. L. Knight, *Introductory Quantum Optics* (Cambridge University Press, Cambridge, 2005).

Johnsen–Rahbek effect is observed when a semiconductor is placed between two electrodes. The frictional force between the electrodes and the semiconductor is increased if an electric bias is applied to the electrodes. The effect is particularly strong in the case of suspensions of fine semiconductive powders of high dielectric constant in low-viscosity oils. It results in a marked increase in the viscosity of the liquid.

First described in: A. Johnsen, K. Rahbek, *A physical phenomenon and its applications to telegraphy, telephony, etc.*, J. Sci. Inst. Elect. Eng. **61**, 713–725 (1923).

Johnson and Lark–Horowitz formula defines the resistivity ρ of a metal or degenerate semiconductor as $\rho \sim N^{1/3}$, where N is the density of impurities.

First described in: V. A. Johnson, K. Lark–Horowitz, *Electronic mobility in germanium*, Phys. Rev. **79**(2), 409–410 (1950).

Johnson–Kendall–Roberts theory takes into account **adhesion** forces starting from the **friction force theory (Hertz theory)**. Considering the contact between a sphere and flat surfaces and taking into account the interfacial energies one obtains for the friction force $F_f = \tau \pi (R/K)^{2/3} F_{\text{JKR}}^{2/3}$, where τ is the **shear modulus** and $F_{\text{JKR}} = N + 3\pi R\gamma + \sqrt{6\pi R\gamma N + (3\pi R\gamma)^2}$ defines the **Johnson–Kendall–Roberts force**. In this force N is the applied load, R is the radius of the sphere, and $\gamma = \gamma_1 + \gamma_2 - \gamma_{1,2}$ is the interfacial energy (γ_1 is the surface energy of the sphere, γ_2 is the surface energy of the flat surface, and $\gamma_{1,2}$ is the energy of the interface). For the meaning of $K \rightarrow$ **friction force theory**. Practically, the effect of considering adhesion is to replace the external applied load N with a total load that includes the adhesive force. The contact area and the friction force can be different from zero even if the external applied load is zero, because of the presence of adhesion. For zero applied load, the friction force reduces to $F_{f0} = \tau \pi [(6\pi R^2 \gamma)/K]^{2/3}$ and the contact radius becomes $a_0 = [(6\pi R^2 \gamma)/K]^{1/3}$. The adhesion expressed in this theory can be measured by **atomic force microscopy (AFM)** withdrawing the probe until the contact is broken. The negative attractive force applied to the probe at this point is called the **pull-off force** $F_{\text{JKR}}^{PO} = -(3/2)\pi R\gamma$.

This theory is more useful for highly adhesive systems with low stiffness and large (tip) radii.

First described in: K. L. Johnson, K. Kendall, A. D. Roberts, *Surface energy and the contact of elastic solids*, Proc. R. Soc. Lond. A **324**, 301–313 (1971).

Johnson–Kendall–Roberts force → **Johnson–Kendall–Roberts theory**.

Johnson–Nyquist noise – equilibrium fluctuations of the electric potential inside a conductor occurring without any applied voltage. It is a result of the random thermal motion of the charge carriers. It is to be distinguished from **shot noise**, which describes the additional current fluctuations that occur when a voltage is applied and a macroscopic current starts to flow.

First described in: J. Johnson, *Thermal agitation of electricity in conductors*, Phys. Rev. **32**(1), 97–109 (1928) – experiment; H. Nyquist, *Thermal agitation of electric charge in conductors*, Phys. Rev. **32**(1), 110–114 (1928) – theory.

Josephson effects arise in tunneling of superconducting electron pairs from a superconductor through a layer of an insulator into another superconductor.

Direct current Josephson effect: a tunnel junction between superconductors shows a zero-voltage supercurrent due to the tunneling of condensed pairs; a direct current flows across the junction in the absence of any electric or magnetic field.

Alternating current Josephson effect: if a voltage difference V was maintained across the junction, the current would be an alternating current of frequency $2eV/h$. A constant voltage applied across the junction causes high frequency current oscillations across the junction.

Macroscopic long-range quantum interference: a constant magnetic field applied through a superconducting circuit containing two junctions causes the maximum supercurrent to show interference effects as a function of magnetic field intensity. This effect is used in sensitive magnetometers.

First described in: B. D. Josephson, *Possible new effects in superconductive tunneling*, Phys. Lett. **1**(7), 251–253 (1962).

More details: A. Barone, G. Paterno, *Physics and Applications of Josephson Effects* (John Wiley, New York, 1982).

Recognition: in 1973 B. D. Josephson received the Nobel Prize in Physics for his theoretical predictions of the properties of a supercurrent through a tunnel barrier, in particular those phenomena which are generally known as the Josephson effects.

See also www.nobel.se/physics/laureates/1973/index.html.

Joule's law of electric heating – the heating effect of an electric current I passing through a resistance R for a time t is proportional to I^2Rt.

First described in: J. P. Joule, *On the heat evolved by metallic conductors of electricity*, Phil. Mag. **19**, 260–265 (1841).

From Kane model to Kuhn–Thomas–Reiche sum rule

Kane model is required if materials with narrow direct energy gaps at the first **Brillouin zone** center are considered. The energy of the Γ valley is assumed to be non-parabolic, spherical, and of the form $\hbar^2 k^2/2m^* = E(1 + \alpha E)$, where $\mathbf{k}$ denotes the wavevector, E represents the energy, m^* is the effective mass, and $\alpha = (1 - m^*/m_0)^2/E_g$ is the non-parabolicity coefficient with E_g to be the energy gap.

First described in: E. O. Kane, *Band structure of indium antimonide*, J. Phys. Chem. Solids **1**, 249–261 (1957).

Keesom potential – the intermolecular potential accounting for a contribution from a dipole interaction in the **Lennard–Jones potential** in the form:

$$V(r_{ij}) = \epsilon[(r_0/r_{ij})^{12} - 2(r_0/r_{ij})^6] - \frac{\mu_i^2 \mu_j^2}{3k_B T r_{ij}^2},$$

where r_{ij} is the distance between molecules i and j, μ_i and μ_j being the dipole moments of these molecules. Here the first term is the classical non-polar **Lennard–Jones potential** while the second one is responsible for the dipole interaction. The potential is useful for phase equilibria simulations, in particular in vapor-liquid systems.

First described in: W. H. Keesom, *Die van der Waalschen Kohäsionskräfte*, Phys. Z. **22**, 129–141 (1921).

More details in: J. Israelachvili, *Intermolecular and Surface Forces*, 2nd Edn. (Academic Press, London, 1991).

Keldish theory describes the process of multiphoton ionization in which an atom is ionized as a consequence of the rapid absorption of a sufficient number of photons. The ionization rate is predicted to be dependent primarily upon the ratio of the mean binding electric field to the peak strength of the incident electromagnetic field and upon the ratio of the binding energy to the energy of photons in the field.

First described in: L. V. Keldysh, *Ionization in the field of a strong electromagnetic wave*, Sov. Phys. JETP **20**(5), 1018–1027 (1965).

What is What in the Nanoworld. Victor E. Borisenko and Stefano Ossicini
Copyright © 2008 WILEY-VCH Verlag GmbH & Co. KGaA, Weinheim
ISBN: 978-3-527-40783-5

Kelvin equation

Kelvin equation gives the increase in vapor pressure of a substance at a temperature T, which accompanies an increase in the curvature of its surface: $p_s/p_f = (2\sigma v)/(rR_0 T)$, where p_s and p_f are vapor pressures at spherical and flat surfaces respectively, σ is the surface tension, v is the molecular volume of the condensed phase, and r is the mean surface curvature. The equation predicts the greater rate of evaporation of a small liquid droplet than that of a larger one and the greater solubility of small solid particles than that of larger particles.

First described in: W. Thomson (Lord Kelvin), *On the equilibrium of vapor at a curved surface of liquid*, Phil. Mag. **42**, 448–452 (1871).

Kelvin problem is formulated as What shape would equal "volume bubbles need to take on in order to yield the minimum surface area for a foam?". Lord Kelvin thought (1887) that it was a body-centered cubic structure filled with distorted 14 sided cells, while later (1994) D. Weaire and R. Phelan showed that a lower surface area could be produced in a foam containing two kinds of bubbles: one with 14 sides and the other with 12 sides.

First described in: W. Thomson (Lord Kelvin), *On the division of space with minimum partitional area*, Phil. Mag. **24**(151), 503 (1887); D. Weaire, R. Phelan, *A counterexample to Kelvin's conjecture on minimal surfaces*, Phil. Mag. Lett. **69**, 107–110 (1994).

Kelvin relations – interrelations of the coefficients describing thermoelectric effects in a loop composed of two materials, say A and B, at absolute temperature T: $\alpha_{AB} T = \pi_{AB}$ and $T d\alpha_{AB}/dT = \tau_A - \tau_B$, where α_{AB} is the **Seebeck coefficient**, π_{AB} is the **Peltier coefficient**, τ_A and τ_B are the **Thomson coefficients**.

First described in: W. Thomson (Lord Kelvin), *On a mechanical theory of thermoelectric currents*, Proc. R. Soc. Edinburgh: 91–98 (1851).

Kennard packet indicates a wave packet for which the product of the root mean square deviations of position and momentum from their mean values is as small as possible, being equal to $h/4\pi$.

First described in: E. H. Kennard, *Wave mechanics of deflected electrons*, Phys. Rev. **36**(11), 1667–1668 (1930).

Kerr effects: Electrooptical Kerr effect – a transparent isotropic medium becomes doubly refracting when it is placed in an electric field. Its optical properties are similar to those of a birefringent crystal with its optical axis in the direction of the electric field. **Magnetooptical Kerr effect** – changes produced in the optical properties of a reflecting surface of a **ferromagnetic** when it is magnetized. This applies especially to the elliptical polarization of reflected light, when the ordinary rules of metallic reflection would give only plane polarized light.

First described in: J. Kerr, *A new relation between electricity and light: dielectrified media birefringent*, Phil. Mag. **50**, 337–348 and 446–458 (1875); J. Kerr, *On rotation of the plane of polarization by reflection from the pole of a magnet*, Phil. Mag. **3**(Ser. 5), 321–343 (1877); J. Kerr, *On reflection of polarized light from the equatorial surface* *of a magnet*, Phil. Mag. **5**(Ser. 5), 157–161 (1878); J. Kerr, *Measurements and law in electro-optics*, Phil. Mag. **9**(Ser. 5), 157 (1880).

ket → **Dirac notation**.

ketones – organic compounds that have an **alkyl group** ($-C_nH_{2n+1}$) or **aromatic ring** bonded to a **carbonyl group** ($>C=O$). These may be RCOR′ or ArCOR or ArCOAr compounds with R and R′ representing alkyl groups ($-C_nH_{2n+1}$) and Ar representing *aromatic ring*.

Kikoin–Noskov effect – appearance of an electric field in a light-irradiated semiconductor placed in a magnetic field. This electric field is perpendicular to the magnetic field and to the direction of diffusion of the light- generated carriers from the illuminated surface to the bulk.
   First described in: I. K. Kikoin, M. M. Noskov, *Hall effect internal photoelectric effect in cuprous oxide*, Phys. Z. Sowjetunion **4**(3), 531–550 (1933).

kilo- – a decimal prefix representing 10^3, abbreviated k.

kinematics is the study of the motion of material particles without reference to the forces acting on them.

kinetics – from Greek meaning "motion". The term is used to describe how a process or chemical reaction develop in a time scale.

Kirchhoff equation – the rate of change with temperature of the heat ΔH of a process (such as a chemical reaction) carried out at a constant pressure is given by

$$\left(\frac{\partial \Delta H}{\partial T}\right)_P = \Delta\left(\frac{\partial H}{\partial T}\right)_P = \Delta C_P,$$

where ΔC_P is the change in the heat capacity at constant pressure for the same process. Similarly for a process carried out at a constant volume.

Kirchhoff law (for radiation) states that the ratio of the emissive power for thermal radiation of a given wavelength is the same for all bodies at the same temperature and is equal to the emissive power of a black body at that temperature.
   First described in: G. R. Kirchhoff, *Über einen Satz der mechanischen Wärmetheorie und einige Anwendungen desselben*, Ann. Phys. **103**, 177–205 (1858).

Kirchhoff laws (for electrical circuits) refer to the voltage and electric currents in an electric network. The "**current law**" states that the sum of all currents flowing toward a node of the network must be zero $\sum_n I_n = 0$. The "**voltage law**" states that when traversing the network along a closed loop the sum of all voltages encountered must be zero $\sum_n V_n = 0$.

First described in: G. R. Kirchhoff, *Über den Durchgang eines elektrischen Stromes durch eine Ebene, insbesondere durch eine kreisförmige*, Ann. Phys. **64**, 497–514 (1845); G. R. Kirchhoff, *Über die Auflösung der Gleichungen, auf welche man bei Untersuchung der linearen Vertheilung galvanischer Ströme geführt wird*, Ann. Phys. **72**, 497–508 (1847).

Kirkendall effect – a movement of the interface between different phases due to high temperature interdiffusion of the constituents of these phases. It is observed in the migration of atomic markers that occurs when the markers are placed at the interface between an alloy and a metal. The markers usually move toward the alloy region because of an appropriate concentration gradient of the metal. The phenomenon supports the idea that atomic diffusion occurs mainly through a vacancy exchange mechanism.

The effect has important practical consequences. One of these is the prevention or suppression of voids formed at the boundary interface in various kinds of alloy-to-metal bonding. These voids are referred to as **Kirkendall voids**.

First described in: A. D. Smigelskas, E. O. Kirkendall, *Zinc diffusion in alpha brass*, Trans. AIME **171**, 130–142 (1947).

Kirkendall voids → **Kirkendall effect**.

KKR method → **Korringa–Kohn–Rostoker method**.

Klein–Gordon equation is $d^2 \psi/dt^2 = -(c^2 \mathbf{p}^2 + m^2 \mathbf{c}^4)\psi$, where $\mathbf{p}$ is the momentum operator of the particle with rest mass m. It accounts for the special theory of relativity. It is used in the theory of quantized fields of particles with spin 0 or 1. This equation was contemporarily formulated by many authors.

First described in: E. Schrödinger, *Quantisierung als Eigenwertproblem. (Vierte Mitteilung)*, Ann. Phys. **81**, 109–139 (1926); J. Kudar, *Zur vierdimensionalen Formulierung der undulatorischen Mechanik*, Ann. Phys. **81**, 632–636 (1926); O. Klein, *Quantentheorie und fünfdimensionale Relativitätstheorie*, Z. Phys. **37**(12), 895–906 (1926); V. Fock, *Zur Schrödingerschen Wellenmechanik*, Z. Phys. **38**(3), 242–250 (1926); V. Fock, *Über die invariante Form der Wellen- und der Bewegungsgleichungen für einen geladenen Massenpunkt*, Z. Phys. **39**(2–3), 226–232 (1926); W. Gordon, *Der Comptoneffekt nach der Schrödingerschen Theorie*, Z. Phys. **40**(1–2), 117–133 (1926).

More details in: H. Kragh, *Equation with the many fathers. The Klein–Gordon equation in 1926*, Am. J. Phys. **52**(11), 1024–1033 (1984).

Klein–Nishina formula gives the differential cross section $d\phi$ for **Compton scattering** into a small solid angle $d\Omega$ of an *x*- or gamma-photon by an unbound electron:

$$d\phi = \left(\frac{r_0 \nu}{2\nu_0}\right)^2 \left(\frac{\nu_0}{\nu} + \frac{\nu}{\nu_0} - 2 + 4\cos^2\theta\right) d\Omega,$$

where v and v_0 are the frequencies of incident and scattered photons, θ is the angle between their polarization vectors, and $r_0 = e^2/mc^2$ the classical radius of an electron.

First described in: O. Klein, I. Nishina, *Scattering of radiation by free electrons on the new relativistic quantum dynamic of Dirac*, Z. Phys. **52**(11/12), 853–868 (1929).

Knight shift – a fractional increase in the frequency for **nuclear magnetic resonance** of a given nuclide in a metal subjected to an external magnetic field relative to that for the same nuclide in a nonmetallic compound in the same field, caused by the orientation of the conduction electrons in the metal.

First described in: C. H. Townes, C. Herring, C. Knight, *The effect of electronic paramagnetism on nuclear magnetic resonance frequencies in metals*, Phys. Rev. **77**(6), 852–853 (1950).

Knudsen cell → **effusion (Knudsen) cell**.

Knudsen cosine rule states that individual gaseous molecules impinging on an irregular solid surface are reflected in a direction totally independent of the direction of incidence. The probability ds that a molecule leaves the surface in the solid angle $d\omega$ forming an angle θ with the normal to the surface is defined as $\pi ds = d\omega \cos \theta$.

Kohlrausch laws – (i) the "**law of the independent migration of ions**" states that every ion at infinite dilution contributes a definite amount toward the equivalent conductance of an electrolyte, irrespective of the nature of the other ions with which it is associated; (ii) "**Kohlrausch's square root law**" states that the equivalent conductance of a strong electrolyte in a very dilute solution gives a straight line when plotted against the square root of concentration.

First described in: F. Kohlrausch, *Über das Leistungvermögen einiger Electrolyte in äusserst verdünnter wässriger Lösung*, Ann. Phys. **26**, 161–225 (1885).

Kohlrausch's square root law → **Kohlrausch laws**.
First described in: F. Kohlrausch, *Über die elastische Nachwirkung bei der Torsion*, Ann. Phys. **119**, 337–367 (1863).

Kohn–Sham equations promote practical applicability of the **density functional theory** (DFT) to map the many-particle problem to an independent particle problem in an affective potential. In DFT the total energy is described by:

$$E[n] = T_s[n] + E_H[n] + E_{ext}[n] + E_{xc}[n],$$

where $T_s[n]$ is the kinetic energy of noninteracting electrons, that is independent particles and can be given explicitly by the single-electron orbitals $\psi_i(\mathbf{r})$; $E_H[n]$ is the Hartree energy describing the Coulomb interaction of the electrons in their

Kondo cloud

electrostatic potential; $E_{ext}[n]$ is the energy caused by the external potential; $E_{xc}[n]$ is the exchange-correlation energy functional. The final result of applying the **variation principle** leads to the Kohn–Sham equations:

$$\left(-\frac{\hbar^2}{2m}\nabla^2 + V_H(\mathbf{r}) + V_{ext}(\mathbf{r}) + V_{xc}(\mathbf{r}) \right) \psi_i(\mathbf{r}) = \epsilon_i \psi_i(\mathbf{r}),$$

where ϵ_i are Lagrange parameters, which can be often interpreted as excitation energies.

This form of equations is equivalent to a one-particle **Schrödinger equation** in an effective potential consisting of the Hartree potential V_H:

$$V_H(\mathbf{r}) = e^2 \int \frac{n(\mathbf{r}')}{|\mathbf{r}-\mathbf{r}'| d^3 r},$$

the external potential V_{ext} and the exchange-correlation potential V_{xc}:

$$V_{xc}(\mathbf{r}) = \frac{\delta E_{xc}[\mathbf{r}]}{\delta n(\mathbf{r})}.$$

First described in: W. Kohn, L. J. Sham, *Self-consistent equations including exchange and correlation effects*, Phys. Rev. A **140**(4), 1133–1137 (1965).

Kondo cloud → **Kondo effect.**

Kondo effect – an increase in the electrical resistance of a bulk sample when approaching absolute zero temperature related to nonzero total spin of all electrons in it. The effect is observed in bulk metals containing a small fraction of magnetic impurities (like Fe, Co, Ni) and in quantum dots; however, in this last case it is the conduction that shows an increase. The temperature T_K at which the resistance starts to increase again is called the **Kondo temperature**.

An explanation of the Kondo effect can be given within the model of a magnetic impurity proposed by P. W. Anderson (1961). It is illustrated by the energy diagrams presented in Figure K.1.

The magnetic impurity atom in a nonmagnetic metal is represented by a quantum well that has only one electron energy level E_0 below the Fermi energy of the host metal. This level is occupied by one electron with a particular spin, for instance the electron has its spin up as shown in the diagram. The impurity atom is surrounded by a sea of electrons from the host metal atoms in which all the states with energies below the Fermi level are occupied, while the high energy states are empty. An electric bias applied to the sample makes the occupied levels in the sea of electrons slightly different at the sides of the well.

Adding another electron to the well is prohibited by the Coulomb interaction characterized by the energy barrier U. Removal of the electron from the impurity atom would need at least E_0 to be added to the system. However, the **Heisenberg's**

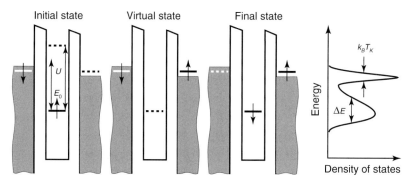

Figure K.1 The Anderson's model of a magnetic impurity atom in a metallic electrically biased sample.

uncertainty principle allows the electron to leave the level in the well for a short time around h/E_0. The electron may tunnel out of the well to briefly occupy a classically forbidden virtual state outside the well, and then be replaced by an electron from the outside sea of electrons. The newcomer may have an oppositely directed spin, and it flips the spin of the impurity. As a result, a difference between the initial and final spin states of the impurity appears.

The spin exchange qualitatively modifies the energy dependence of the density of states in the system. Many such processes occurring together are known as **Kondo resonance**. It provides a new state, called the **Kondo state**, with exactly the same energy as the Fermi level. This state is always "on resonance" as it is attached to the Fermi energy. Since the Kondo state is created by the exchange process between the electron localized near the magnetic impurity atom and the free electrons of the host metal, the Kondo effect is a typical many-body phenomenon. The electrons that have previously interacted with the same magnetic impurity form so-called **Kondo clouds**. Since each of these electrons contains information about the same impurity, they inevitably carry information about each other. So, the electrons in the clouds are mutually correlated.

The increase in the resistance with a temperature drop in the low temperature range is the first hint of the existence of the Kondo state. This state increases scattering of electrons with energies close to the Fermi level. The hybridization of conduction electrons with the localized spin of a magnetic impurity atom in a metal leads to an enhancement of resistivity at low temperature. The total resistivity as a function of temperature has the form $\rho = AT^5 - B\ln T + C$, where A, B, C are the constants dependent on concentration of magnetic ions, exchange energy, and strength of the exchange scattering. The Kondo temperature is:

$$T_K = \frac{\sqrt{\Delta EU}}{2} \exp \frac{\pi E_0(E_0 + U)}{\Delta EU},$$

where ΔE is the width of the impurity energy level broadened by electron tunneling from it. The Kondo temperature varies as the one-fifth power of the concentration of the magnetic impurity. In bulk metallic systems it ranges from 1 to 100 K.

Kondo effect

For a Kondo system, the ratio between its actual resistance R and its resistance at absolute zero R_0 depends only on temperature as $R/R_0 = f(T/T_K)$. All materials that contain spin 1/2 impurities can be described by the same temperature-dependent function $f(T/T_K)$. Thus, the system can be completely characterized by the Kondo temperature instead of the set containing U, E_0 and ΔE.

Quantum dots are another class of systems, whose transport properties can also be affected by the Kondo effect, but which offer a much higher degree of control over the system parameters with respect to magnetic impurities in bulk metals. A quantum dot confining a well-defined number of electrons can act as a magnetic impurity. The total spin is zero or an integer for an even number of electrons and an half integer for a odd number of electrons. The latter case is a canonical example to observe the Kondo effect. All electrons can be ignored, except for the one with the highest energy; that is, the case of a single, isolated spin, s = 1/2.

A dot supplied with gates can be switched from a Kondo system to a non-Kondo system as the number of electrons in the dot is changed from odd to even. Such a structure is shown in Figure K.2.

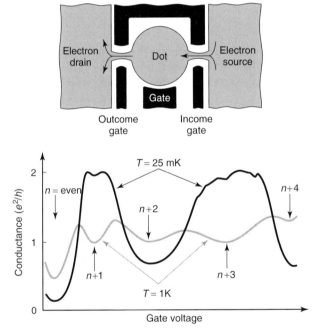

Figure K.2 An external bias-operated quantum dot and its conductance as a function of the gate voltage changing the number n of confined electrons at two different temperatures.

The dot is connected with two reservoirs of electrons by tunneling channels with transparencies controlled by the bias applied to the income and outcome gates. Coupling of the dot with the electron source and drain defines the energy

218

broadening ΔE of the state in the dot. The number of electrons in the dot and their energy levels are tuned by the voltage applied to the central gate. The Kondo temperature can be varied by means of the gate voltage as a single particle energy state nears the **Fermi energy**.

One of the main distinctions between a quantum dot and a bulk metal with a magnetic impurity relates to the nature of the electron states there. In a metal, the electron states are plane waves. Their scattering at the impurity atoms results in mixing of the electron waves with different momenta. This change of the momentum increases the material resistance.

In a quantum dot, all electrons have to travel through the dot, as there is no electrical pass around it. In this case the Kondo resonance makes it easier for states belonging to the two opposite electrodes to mix. Such mixing increases the conductance–the resistance is decreased. So far the Kondo effect in quantum dots leads to the opposite behavior of the resistance to that in a bulk metal. Figure K.2 shows variation of the dot conductance as a function of the gate voltage tuning the number of electrons confined in the dot at two different low temperatures. When an even number of electrons is trapped, the conductance decreases as the temperature is lowered. Such behavior indicates that there is no Kondo effect in this case. The opposite temperature dependence is observed for an odd number of electrons when the Kondo effect takes place.

Like the resistance of a bulk sample in the Kondo regime, the conductance of a quantum dot depends only on the ratio T/T_K. At the lowest temperature the conductance approaches the quantum limit of conductance $2e^2/h$. The fact that the conductance reaches this value implies that electrons are transmitted through the dot perfectly. The Kondo effect allows a dot to become completely transparent.

The Kondo effect can also be observed in quantum dots with an even number of electrons, but the sample has to be placed in a magnetic field in order to get appropriate spin splitting and energy level occupation.

First described in: J. Kondo, *Resonance minimum in dilute magnetic alloys*, Prog. Theor. Phys. **32**(1), 37–49 (1964).

Kondo resonance → **Kondo effect**.

Kondo temperature → **Kondo effect**.

Koopmans theorem states that the ionization potentials for closed-shell molecules may be approximated by the negative of orbital energies obtained from a Hartree–Fock calculation. This provides a physical interpretation of the computed orbital energies. Thus, the lowest ionization potential of the neutral species can be estimated from the orbital energy of the highest occupied molecular orbital (**HOMO**). This is the so-called vertical ionization potential. "Vertical" because the underlying assumption is that no geometrical or electronic structural changes occur during the ionization process. Hence, in this model the wavefunction for the cation is identical to that of the neutral atom.

Kopp law

First described in: T. Koopmans, *Über die Zuordnung von Wellenfunktionen und Eigenwerten zu den einzelnen Electronen eines Atoms*, Physica **1**, 104–113 (1934).

Kopp law states that for solids, the molar heat capacity of a compound at room temperature and pressure approximately equals the sum of the heat capacities of the constituent elements.

First described in: H. Kopp, *Untersuchungen über die spezifische Wärme der starren und tropfbar-flüssigen Körper*, Ann. Chem. Pharm. Suppl. **3**, 289 (1864).

Korringa–Kohn–Rostoker (KKR) method – a method of band structure calculations of solids using standard **Green's functions** to transform the **Schrödinger equation** into the equivalent integral equation.

Within the method, the wave function $\psi(\mathbf{k},\mathbf{x})$ is calculated via the Green's function $G(\mathbf{k},\mathbf{x}\text{-}\mathbf{x}';E)$ satisfying the free-particle Schrödinger equation

$$\left(\frac{\hbar^2}{2m}\nabla^2 + E\right)G(\mathbf{k},\mathbf{x}\text{-}\mathbf{x}';E) = \delta^3(\mathbf{x}\text{-}\mathbf{x}')$$

describing the response to a delta function $\delta^3(\mathbf{x}\text{-}\mathbf{x}')$. The final integral equation for the wave function takes the form

$$\psi(\mathbf{k},\mathbf{x}) = \int d^3x' G(\mathbf{k},\mathbf{x}\text{-}\mathbf{x}';E)V(\mathbf{x}')\psi(\mathbf{k},\mathbf{x}'),$$

where $V(\mathbf{x})$ is the crystal potential.

First described in: J. Korringa, *On the calculation of the energy of a Bloch wave in a metal*, Physica **13**, 392–400 (1947); W. Kohn, N. Rostoker, *Solution of the Schrödinger equation in periodic lattices with an application to metallic lithium*, Phys. Rev. **94**(5), 1111–1120 (1954).

Kossel crystal – a crystal in which repletion takes place in steps. Its widespread graphic representation in two dimensions is shown in Figure K.3.

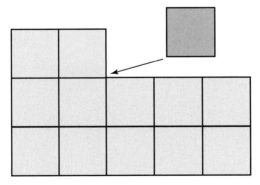

Figure K.3 Two-dimensional Kossel crystal.

It typically has a simple cubic lattice and central symmetry of the additive interaction forces between its building elements.

First described in: W. Kossel, *Zur Theorie des Kristallwachstums*, Nachr. Akad. Wiss. Göttingen, math. phys. Kl. **27**, 135–143 (1927).

Kossel effect arises when characteristic X-rays generated by atoms in a single crystal produces series of cones of reflected X-rays.

Kossel–Stranski surface – a crystal surface built by step-by-step repetitions. It is typical for a **Kossel crystal**.

First described in: W. Kossel, *Molecular forces in crystal growth*, Phys. Z. **29**, 553–555 (1928); I. N. Stranski, *Zur Theorie des Kristallwachstums*, Z. Phys. Chem. **136**(3/4), 259–278 (1928).

Kramers theorem states that in the absence of an external magnetic field the stationary states of an atomic system are always degenerate when the system contains an odd number of electrons, the degree of degeneracy being an even number. If the number of electrons is even, the energy will not be influenced in the first approximation by a magnetic field.

First described in: H. A. Kramers, *General theory of paramagnetic rotation in crystals*, Proc. K. Akad. Amsterdam **33**(9), 959–972 (1930).

Kramers–Kronig dispersion relation connects real $\epsilon_1(\omega)$ and imaginary $\epsilon_2(\omega)$ parts of the **dielectric function** via

$$\epsilon_1(\omega) = 1 + (2/\pi)P \int \epsilon_2(\omega')\omega'd\omega'/(\omega'^2 - \omega^2),$$

where P means principal value of the following integral. The dispersion relation is based on the causality principle.

First described in: R. L. Kronig, *Theory of dispersion of X-rays*, J. Opt. Soc. Am. **12**(June), 547–557 (1926); H. A. Kramers, *Estratto degli Atti del Congresso Internazionale dei Fisici, Como*, vol. **2** (Bologna, 1927) p. 545.

Kronecker delta function denoted as $\delta_{m,n}$ has the value 1 when $m = n$ and 0 otherwise. It is also known as **Kronecker symbol**.

First described by L. Kronecker in 1866.

Kronecker product – a special tensor product of two matrices of arbitrary size. If $\mathbf{A}$ is an m-by-n matrix and $\mathbf{B}$ is a p-by-q matrix, then their Kronecker product is the

Kronecker symbol

mp-by-nq block matrix:

$$A \otimes B \quad \begin{bmatrix} a_{11}b_{11} & \cdots & a_{11}b_{1q} & \cdots & a_{1n}b_{11} & \cdots & a_{1n}b_{1q} \\ a_{11}b_{21} & \cdots & a_{11}b_{2q} & \cdots & a_{1n}b_{21} & \cdots & a_{1n}b_{2q} \\ \cdots & \cdots & \cdots & \cdots & \cdots & \cdots & \cdots \\ a_{11}b_{1p} & \cdots & a_{11}b_{pq} & \cdots & a_{1n}b_{p1} & \cdots & a_{1n}b_{pq} \\ \cdots & \cdots & \cdots & \cdots & \cdots & \cdots & \cdots \\ a_{m1}b_{11} & \cdots & a_{m1}b_{1q} & \cdots & a_{mn}b_{11} & \cdots & a_{mn}b_{1q} \\ a_{m1}b_{21} & \cdots & a_{m1}b_{2q} & \cdots & a_{mn}b_{21} & \cdots & a_{mn}b_{2q} \\ \cdots & \cdots & \cdots & \cdots & \cdots & \cdots & \cdots \\ a_{m1}b_{p1} & \cdots & a_{m1}b_{pq} & \cdots & a_{mn}b_{p1} & \cdots & a_{mn}b_{pq} \end{bmatrix},$$

Kronecker symbol → **Kronecker delta function.**

Kronig–Penney model – an idealized one-dimensional model for a crystal in which the potential energy of an electron is an infinite sequence of periodically spaced square wells. Such a rectangular in space potential is called the **Kronig–Penney potential**.
First described in: R. de L. Kronig, W. G. Penney, *Quantum mechanics of electrons in crystal lattices*, Proc. R. Soc. A **130**, 499–513 (1931).

Kronig–Penney potential → **Kronig–Penney model.**

Kubo formalism – the technique allowing formal calculations of many transport coefficients without the use of a kinetic equation. It supposes that a system is described by a Hamiltonian and that its equilibrium thermodynamic properties are determined by the equilibrium density matrix $\rho^{(0)}$. This system is now subject to an external influence described by $AF(t)$, where A is an arbitrary **operator** and F an arbitrary C-number function of time. The temporal behavior of the system is governed by the density matrix $\rho(t)$, which satisfies the **von Neumann equation** $i\hbar \partial \rho / \partial t = [H_t, \rho]$ with $H_t \equiv H + AF(t)$. So far everything is traditional and exact. The fundamental approximation made in calculating $\triangle\rho \equiv \rho(t) - \rho^{(0)}$ is to solve the von Neumann equation neglecting quadratic terms in either $\triangle\rho$ or A. This allows one to solve the equation for $\triangle\rho$ in that approximation. The average of a quantity B in this new time-dependent state $\rho_0 + \triangle\rho$ assumes a very elegant form

$$\langle B(t) \rangle = B_0 + Tr(B\triangle\rho) = B_0 + \int_0^\infty dt' \phi_{BA}(t - t')F(t'),$$

where B_0 is the equilibrium value of B and ϕ_{BA} is the response function

$$\phi_{BA}(\tau) = \frac{1}{i\hbar} Tr\rho^{(0)}\left[A, B(\tau)\right]$$

with

$$B(\tau) \equiv \exp\left(\frac{1}{i\hbar}H\tau\right)B\exp\left(-\frac{1}{i\hbar}B\tau\right).$$

Kuhn–Thomas–Reiche sum rule (or f-sum rule)

This formalism is a characteristic of linear response. If an electric current caused by the electric field A and $F(t)$ varies like $\exp(i\omega t)$, the above formulae yield, after considerable manipulation, the frequency-dependent conductivity tensor:

$$\sigma_{\mu\nu} = \frac{1}{k_B T} \int_0^\infty dt \exp(i\omega t) \langle j_\nu j_\mu(t) \rangle.$$

First described in: R. Kubo, *Statistical-mechanical theory of irreversible processes. I. General theory and simple applications to magnetic and conduction problems*, J. Phys. Soc. Japan **12**(6), 570–586 (1957).

Kubo oscillator – an oscillator with a random frequency.

Kuhn–Thomas–Reiche sum rule (or f-sum rule) states that for an atom whose electrons are undergoing all possible transitions from or to a given level (usually labeled level 2), if the f's are the corresponding oscillator strength, then $\sum_1 f_{21} + \sum_3 f_{23} = n$, where 1 indicates any level below 2, and 3 any level above 2, including the continuum, and n is the number of optical electrons. The absorption and emission oscillator strengths are related by $\omega_1 f_{12} = -\omega f_{21}$, where ω is the statistical weight.

First described in: W. Kuhn, *Über die Gesamtstärke der von einem Zustande ausgehenden Absorptionslinien*, Z. Phys. **33**(1), 408–410 (1925); W. Thomas, *Über die Zahl der Dispersionselektronen, die einem stationären Zustande zugeordnet sind (Vorläufige Mitteilung)*, Naturwissen. **13**, 627 (1925); F. Reiche, W. Thomas, *Über die Zahl der Dispersionselektronen, die einem stationären Zustand zugeordnet sind*, Z. Phys. **34**(1), 510–525 (1925).

L

From lab-on-a-chip to Lyman series

lab-on-a-chip – a combination of fluidic elements, sensor(s), detection and signal processing units integrated on a chip in order to perform a complete sequence of chemical reactions or biomedical analysis. Sample preparation, appropriate reactions, separation of the products and their identification are usually realized by such a device.

Lagrange equation of motion has the form

$$\frac{d}{dt}\left(\frac{\partial \mathbf{L}}{\partial v_i}\right) - \frac{\partial \mathbf{L}}{\partial r_i} = 0, i = 1, 2, \ldots, n,$$

where $\mathbf{L}$ is the Lagrangian. Given a set of n independent generalized coordinates r_i and velocities v_i that describe the state of a conservative system (one in which all forces derive from some potential energy function U), so that $\mathbf{L} = \mathbf{L}(r_i, v_i, t)$.

First described in: J. Lagrange, *Mécanique Analytique* (1788).

Lagrangian function (or, shortly, Lagrangian) – the difference between the kinetic and potential energies of a particle or system. The function enables the equation of motion of classical mechanics and Hamilton's principle to be written in simple form. For systems involving relativistic velocities or in the preserve of a magnetic field the function may have a more complex form. See also **Lagrange equation of motion**.

lambda point – the temperature equal to 2.178 K at which the transformation between liquids **helium I** and **helium II** takes place.

Lambert law states that illumination of a surface by a light ray varies as the cosine of the angle of incidence between the normal to the surface and the incident ray.

First described in: J. H. Lambert, *Photometria* (Augsburg, 1760).

Lambert surface – an ideal perfectly diffusing surface for which the intensity of reflected radiation is independent of direction.

What is What in the Nanoworld. Victor E. Borisenko and Stefano Ossicini
Copyright © 2008 WILEY-VCH Verlag GmbH & Co. KGaA, Weinheim
ISBN: 978-3-527-40783-5

Lamb shift

Lamb shift – a small shift in the energy levels of a hydrogen atom and of hydrogen-like ions from those predicted by the Dirac electron theory, in accord with principles of quantum electrodynamics.

First described in: W. E. Lamb Jr., R. C. Reserford, *Fine structure of the hydrogen atom by a microwave method*, Phys. Rev. **72**(3), 241–243 (1947).

Recognition: in 1955 W. E. Lamb received the Nobel Prize in Physics for his discoveries concerning the fine structure of the hydrogen spectrum.

See also www.nobel.se/physics/laureates/1955/index.html.

Landau damping – damping of an excitation in a system without collisions of the particles constituting this system when wave-particle interaction is involved.

First described in: L. D. Landau, *On the vibrations of the electronic plasma*, Zh. Eksp. Teor. Fiz., **16**, 574 (1946) – in Russian; J. Phys. USSR, **10**, 25 (1946).

Landauer–Büttiker formalism considers electron transport in low-dimensional structures in terms of transmitted and reflected electron waves. Within this point of view, the conductance is determined by the number of one-dimensional channels available for charge carriers injected from phase randomizing contacts and by the transmission properties of the structure for each channel. It can be illustrated by the schematic multi-terminal device shown in Figure L.1.

The device has i terminals, which connect it with reservoirs of electrons. Each of the reservoirs is characterized by some chemical potential μ_i. The net current in the i-th terminal is $I_i = (2e/h)[(1 - R_i)\mu_i - \sum_{i \neq j} T_{ij}\mu_j]$. It accounts for the fact that the current injected in the i-th terminal is reduced by the reflected back component, described by the reflection coefficient R_i, and by that transmitted from other reservoirs, described by the transmission coefficient T_{ij} representing the transmission probability from the j-th to the i-th reservoir. The factor 2 is included in order to represent electron spin degeneracy.

For a multi-mode regime of the i-th terminal, when there are N_i channels at the **Fermi energy**, the current has been obtained to be $I_i = (2e/h)[(N_i - R_i)\mu_i - \sum_{i \neq j} T_{ij}\mu_j]$. This is the **Landauer–Büttiker formula** for the conductance of a

Figure L.1 Multi-terminal low-dimensional structure illustrating the Landauer–Buttiker formalism.

device with many terminals. The reservoirs with electrons are assumed to feed all the channels equally up to the respective Fermi energy in the given terminal. Electrons reflected and transmitted in different m and n modes have to be accounted for by summation: $R_i = \sum_{mn}^{N_i} R_{i,mn}$, $T_{ij} = \sum_{mn}^{N_i} T_{ij,mn}$. Conservation of the current requires that this must be equal to the total current injected from terminal i that leaves the device through other terminals. This resulted in the sum rule for the reflection ant transmission coefficients: $R_i + \sum_{i \neq j} T_{ij,mn} = N_i$.

In many ways, the Landauer–Büttiker formula is the Ohm's law for low-dimensional structures. It is useful in explanation of a number of experimental observations in qualitative details. It works well in open structures where the picture of noninteracting electron waves is valid and interactions contribute only through finite dephasing times. In closed quantum dots, where charging energy is important, it usually does not hold except in special cases.

First described in: R. Landauer, *Spatial variation of currents and fields due to localized scatters in metallic conduction*, IBM J. Res. Dev. **1**(6), 223–231 (1957); M. Büttiker, *Four-terminal phase-coherent conductance*, Phys. Rev. Lett. **57**(14), 1761–1764 (1986).

More details in: S. Datta, *Electronic Transport in Mesoscopic Systems* (Cambridge University Press, Cambridge, 1995).

Landauer–Büttiker formula → **Landauer–Büttiker formalism.**

Landauer formula defines the resistance of single-mode one-dimensional conducting channel to be:

$$r = \left(\frac{h}{2e^2} \right) \frac{R}{1 - R},$$

where R is the reflection probability of electron waves.

First described in: R. Landauer, *Spatial variation of currents and fields due to localized scatters in metallic conduction*, IBM J. Res. Dev. **1**(6), 223–231 (1957).

Landau fluctuations – variations in the losses of energy of different particles in a thin detector resulting from random variations in the number of collisions and in the energies lost in each collision of the particle.

Landau levels – quantized orbits of free charge carriers in a crystal placed in a magnetic field.

Electrons moving in a perpendicular magnetic field B are forced onto circular orbits, following the **Lorentz force**. They perform a cyclotron motion with the angular frequency $\omega_c = eB/m$, called the **cyclotron frequency**, where m is the electron mass. So far, available energy levels for electrons become quantized. These quantized energies are Landau levels. They are given by $E_i = (i + 1/2)\hbar\omega_c$ with $i = 1, 2, \ldots$.

In an ideal perfect system containing a two-dimensional electron gas these levels have a form of a δ-function. The gap between nearest levels is determined by the cyclotron energy $\hbar\omega_c$. The levels are broadened by a temperature rise. Clearly one

Landé equation

needs $k_B T \ll \hbar \omega_c$ to observe well-resolved Landau levels. Electrons can only reside at the energies of Landau levels, but not in the gap between them.

Landé equation gives the **g-factor** for free atoms in the form

$$g = 1 + \frac{J(J+1) + S(S+1) - L(L+1)}{2J(J+1)},$$

where J, S, L refer to the total, spin and orbital angular momentum quantum numbers, respectively.

First described in: A. Landé, *Über den anomalen Zeemaneffekt (Teil I)*, Z. Phys. **5**(4), 231–241 (1921); A. Landé, *Über den anomalen Zeemaneffekt (Teil II)*, Z. Phys. **7**(1), 398–405 (1921).

Landé g-factor determines the total magnetic moment M of an atom in spin–orbit coupling in the direction of the total mechanical moment via $M = -g_L(e/2mc)J$, where g_L is the Landé g-factor, m is the electron mass, and J is the resultant angular momentum. In fact, the Landé g-factor explicitly determines the splitting energy of each magnetic substance under a given magnetic field. See also **g-factor** and **Zeeman effect**.

First described in: A. Landé, *Über den anomalen Zeemaneffekt (Teil I)*, Z. Phys. **5**(4), 231–241 (1921); A. Landé, *Über den anomalen Zeemaneffekt (Teil II)*, Z. Phys. **7**(1), 398–405 (1921).

Landé interval rule states that when the spin-orbit interaction is weak enough to be treated as a perturbation, an energy level having definite spin angular momentum and orbital angular momentum is split into levels of differing total angular momentum, so that the interval between successive levels is proportional to the larger of their total angular momentum values.

First described in: A. Landé, *Termstruktur und Zeemaneffekt der Multipletts I*, Z. Phys. **15**(1), 189–205 (1923); A. Landé *Termstruktur und Zeemaneffekt der Multipletts II*, Z. Phys. **19**(1), 112–123 (1923).

Landé γ-permanence rule states that the sum of the energy shifts produced by the spin–orbit interaction, over the series of states having the same spin and orbital angular momentum quantum numbers (or the same total angular momentum quantum numbers for individual electrons but different total angular momenta, and having the same total magnetic quantum number, is independent of the strength of an applied magnetic field.

Langevin–Debye formula gives the total polarizability of a substance per molecule in the form $\alpha = \alpha_0 + (p^2/3k_B T)$, where α_0 is the deformation polarizability being the sum of electronic and ionic contributions, and p is the dipole moment of the molecule. First described by P. Langevin in 1905 for magnetic susceptibility of a **paramagnetic** and in the final form by P. Debye in 1912.

First described in: P. Langevin, *Sur la theory du magnetism*, J. Phys. Radium, **4**, 678–693 (1905); P. Langevin, *Magnetism et theory des electrons*, Ann. Chim. Phys. **5**, 70 (1905); P. Debye, *Einige Resultate einer kinetischen Theorie der Isolatoren*, Phys. Z. **13**, 97 (1912).

Recognition: in 1936 P. Debye received the Nobel Prize in Chemistry for his contributions to our knowledge of molecular structure through his investigations on the dipole moments and on diffraction of X-rays and electrons in gases.

See also www.nobel.se/chemistry/laureates/1936/index.html.

Langevin function is $L(a) = \coth a - 1/a$. For $a \ll 1, L(a) = a/3$. It occurs in the expressions for the paramagnetic susceptibility of a classical collection of magnetic dipoles and for the polarizability of molecules having a permanent electric dipole moment.

Langmuir–Blodgett film – an organic film that is deposited onto a solid substrate when the substrate is dipped into a liquid, usually aqueous, that contains an organic film on the surface. The substrate adsorbs a layer of the film as it is lowered into the solution and a second layer as it is withdrawn. This technique of film deposition is called the **Langmuir–Blodgett method**. The organic molecules used for that contain two types of functional groups at their ends. One end has the hydrocarbon chain including an acid or alcohol group soluble in water. It makes this end **hydrophilic**, which means that it "likes" water. The other end contains insoluble hydrocarbon groups making it **hydrophobic**, which means that it is "afraid of" water. Such molecules, called amphiphilic, form a film on the water surface with the hydrophilic ends in the water side and the hydrophobic ends in the air side. This is a **Langmuir film**. A classical example of an amphiphile is stearic acid ($C_{17}H_{35}CO_2H$) where the long hydrocarbon tail ($C_{17}H_{35}$) is hydrophobic, and carboxyl acid head group ($-CO_2H$) is hydrophilic.

There can be three types of deposition of a Langmuir film onto a solid substrate, as illustrated in Figure L.2.

Figure L.2 Types of deposition of Langmuir–Blodgett films.

Langmuir–Blodgett method

When a substrate is moved through the monolayer at the water/air interface, the monolayer can be transferred during immersion (dipping) or emersion (withdrawal). If the substrate surface is hydrophobic, a monolayer will be transferred in the immersion via interaction of the hydrophobic tails with the surface. On the other hand, if the substrate surface is hydrophilic, the monolayer will be transferred during withdrawal when hydrophilic head groups bond to the surface. Thus, a hydrophilic substrate appears to get a hydrophobic surface after the first immersion and withdrawal. The second monolayer will be transferred in the immersion. The immersion-withdrawal cycle can be repeated to add monolayers in a head to head, tail to tail configuration. This type of deposition is denoted as **Y-type**. This is the most usual mode of multilayer formation for amphiphilic molecules with a very hydrophilic head groups (like $-COOH$, $-PO_3H_2$) and where the tail is an alkyl chain. The deposited multilayer film is centrosymmetric.

When a hydrophobic surface, like clean silicon, is passed from air to water, the hydrophobic tails will stick to the surface. If now one moves the substrate to part of the bath where water is not covered by the foreign molecules and withdraws it there, a sequence of head to tail layers can be formed on the substrate. This is denoted as **X-type** deposition. The third possible deposition is **Z-type**, when the film composed of head to tail layers is formed only during withdrawal. Both X- and Z-types can be realized when the head group is not so hydrophilic (like $-COOMe$) or when the alkyl chain is terminated by a weak polar group (like $-NO_2$). In these cases, the interaction between adjacent monolayers is hydrophilic hydrophobic. Therefore the multilayers formed are less stable than those formed via Y-type deposition. X- and Z-deposited multilayer films are noncentrosymmetric.

A unique feature of Langmuir–Blodgett films is the ability to form a well-ordered structure on noncrystalline substrates.

First described in: K. B. Blodgett, *Films built by depositing successive monomolecular layers on a solid surface*, J. Am. Chem. Soc. **57**(6), 1007–1022 (1935); I. Langmuir, K. B. Blodgett, *Methods of investigation of monomolecular films*, Kolloid Z. **73**, 257–263 (1935).

More details in: A. Ulman, *An Introduction to Ultrathin Organic Films. From Langmuir–Blodgett to Self-Assembly* (Academic Press, Boston, 1991).

Langmuir–Blodgett method → **Langmuir–Blodgett film**.

Langmuir effect – ionization of atoms of low ionization potential that come into a contact with a hot metal with a high work function.

First described by I. Langmuir in 1924.

Recognition: in 1932 I. Langmuir received the Nobel Prize in Chemistry for his discoveries and investigations in surface chemistry.

See also www.nobel.se/chemistry/laureates/1932/index.html.

Langmuir film → **Langmuir–Blodgett film**.

lanthanides → rare-earth elements.

Laplace equation – the basic equation of potential theory:

$$\frac{\partial^2 \phi}{\partial x^2} + \frac{\partial^2 \phi}{\partial y^2} + \frac{\partial^2 \phi}{\partial z^2} = 0$$

or in terms of the **Laplace operator** ∇^2 it is $\nabla^2 \phi = 0$. A function ϕ satisfying the Laplace equation is said to be harmonic.

First described in: P. S. Laplace, Mem. Acad. R. Sci. 249–267 (1787–1789).

Laplace operator (or, shortly, Laplacian) – the linear operator of the form

$$\nabla^2 = \frac{\partial^2}{\partial x^2} + \frac{\partial^2}{\partial y^2} + \frac{\partial^2}{\partial z^2} = 0.$$

Laplace transform for a function $f(x)$ defined for all real positive x is the function $L(f) = G$ defined by $L(f(s)) = G(s) = \int_0^\infty f(x) \exp(-\delta x) \mathrm{d}x$. If $f(x)$ is summable over every finite interval and there is a constant c such that $\int_0^\infty |f(x)| \exp(-\delta |x|) \mathrm{d}x$ exists, then the Laplace transform exists whenever δ is any complex number $s = s + it$, such that $s = c$. The Laplace transform is related to the Fourier transform by a simple change of variables: if $\delta = -i\omega$, the Laplace transform $G(s) = (2\pi)/1/2F(\omega)$, where $F(\omega)$ is the Fourier transform of the function $f(x)$.

Laplacian = Laplace operator.

Laporte selection rule states that an electric dipole transition can occur only between states of opposite parity.

First described in: O. Laporte, G. Wentzel, *"Dashed" and displaced terms in spectra*, Z. Phys. **31**(1–4), 335–338 (1925); O. Laporte, W. F. Meggers, *Some rules of spectral structure*. J. Opt. Soc. Am. **11**(5), 459 (1925).

LAPW – an acronym for a linearized augmented plane wave used in the **linearized augmented plane wave (LAPW)** method of calculation of electronic band structure of solids.

Larmor frequency → Larmor theorem.

Larmor theorem states that for an atom in a magnetic field H the motion of the electrons is, to the first order in H, the same as a possible motion in the absence of H except for the superposition of a common precession of angular frequency $\omega_L = -eH/(2mc)$, where m is the electron mass. This frequency of precession is called the **Larmor frequency**.

First described in: J. Larmor, *On the theory of the magnetic influence on spectra and on the radiation from moving ions*, Phil. Mag. **44**, 503–512 (1897).

laser – an acronym for a device producing light amplification by stimulated emission of radiation. An operational principle of a laser, as well as of a **maser**, employs the **stimulated emission** of radiation in matter.

A laser consists of a gain medium inside an optical cavity, with a means to supply energy to the gain medium. The gain medium is a material (gas, liquid, solid or free electrons) with appropriate optical properties. An optical cavity in its simplest form consists of two mirrors arranged such that light bounces back and forth, each time passing through the gain medium. Typically, one of the two mirrors, the output coupler, is partially transparent in order to let the light go out of the device. The outgoing light is very intense and coherent (composed of one frequency). It can be easily focused.

laser ablation – instant evaporation of materials by an intense laser beam. As a technological tool it is used to evaporate complicated compounds from a target and deposit them onto a substrate. The composition of the target material is thus reproduced on the substrate.

latent heat – the amount of heat absorbed or evolved by 1 mole or a unit mass of a substance during a change of state, such as fusion, sublimation or vaporization, at constant pressure and temperature.

lateral force microscopy (LFM) – detection of the torsion of the scanning probe of **AFM** (lateral force) which is correlated to local frictional forces. It is an extension of a contact mode AFM, whereby an additional detected parameter is the torsion of the cantilever, which changes depending on the friction force. The LFM can provide material-sensitive contrast since different components of a composite material exert different friction forces on the scanning tip. The technique is also useful for obtaining edge-enhanced images.

Figure L.3 illustrates two sources of lateral deflections of the cantilever, these being changes in the surface friction and changes in the slope of surface structures. In the first case, the tip may experience greater friction as it traverses some areas, causing the cantilever to twist more strongly. In the second case, the cantilever may twist when it encounters a steep slope. To separate one effect from the other, LFM and AFM images should be collected simultaneously.

Figure L.3 Lateral deflection of the scanning cantilever as a result changes in surface friction and slope of the surface structures.

lattice (of a crystal) – a parallel net-like arrangement of points (atoms) with the special property that the environment about any particular point (atom) is in every way the same as that about any other point (atom) of the lattice.

Laue equations specify a set of conditions, which govern the diffraction of an X-ray beam by a crystal. They are also known as Laue conditions. These are illustrated in Figure L.4.

The point O is the origin of the unit cell in a crystal structure and A, B, C are equivalent points along the three axes OX, OY, OZ of the unit cell, so that $OA = a$, $OB = b$, $OC = c$, where a, b, c are the unit cell edges. Consider that a plane wave of monochromatic X-rays from a distant point P is falling on the assembly of points and let us deduce the conditions necessary to give a diffracted wave in the direction of Q. The direction cosines of OP are supposed to be $\alpha_1, \beta_1, \gamma_1$ and of $OQ - \alpha_2, \beta_2, \gamma_2$.

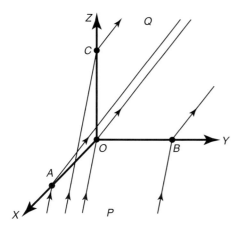

Figure L.4 Scheme of X-ray scattering by equivalent points of a crystal.

The rays with wavelength λ scattered from the points O and A, O and B, O and C will be in phase if $a(\alpha_1 - \alpha_2) = h\lambda, b(\beta_1 - \beta_2) = k\lambda, c(\gamma_1 - \gamma_2) = l\gamma$, where h, k, l are integers. These three equations are called the Laue equations and together constitute the conditions for a diffracted beam in the direction Q. In general, that is for any arbitrary set of incident direction cosines, all three will not be satisfied, since any direction of diffraction is completely specified by two of its direction cosines. This means that only for certain specified incident directions will any diffracted beam be formed when the incident X-rays are monochromatic.

The Laue conditions for reinforcement of the rays scattered from equivalent points of a crystal are identical with the **Bragg equation** governing X-ray diffraction by crystal planes.

First described in: W. Friedrich, P. Knipping, M. Laue, *Interferenzerscheinungen bei Röntgenstrahlen*, Sitzungsber. Bayer. Akad. Wiss. (Math. Phys. Klasse), **303** (1912).

law of rational intercepts

Recognition: in 1914 M. von Laue received the Nobel Prize in Physics for his discovery of the diffraction of Röntgen rays by crystals.

See also www.nobel.se/physics/laureates/1914/index.html.

law of rational intercepts → **Miller law**.

law of the independent migration of ions → **Kohlrausch laws**.
First described in: F. Kohlrausch, *Über das Leistungvermögen einiger Electrolyte in äusserst verdünnter wässriger Lösung*, Ann. Phys. **26**, 161–225 (1885).

LCAO – an acronym for **linear combination of atomic orbitals**, which is a method of calculation of the electronic band structure of solids.

LDA – an acronym for a **local density approximation**, which is an approach used for calculation of the electronic band structure of solids.

LDA+U means **local density approximation** (LDA) with an orbital-dependent correction of the LDA potential. The main idea of the method is to separate electrons into two subsystems – localized d or f electrons for which the Coulomb d–d or f–f interactions should be taken into account in model **Hamiltonian** (through one-site Coulomb interaction U) and delocalized s and p electrons which could be described by using an orbital-independent one-electron potential (LDA). The orbital-dependent potential is written as $V_i(\mathbf{r}) = V_{LDA}(\mathbf{r}) + U(1/2 + n_i)$, where n_i indicates the occupation of the i-th orbital. The LDA + U may be regarded as a generalization of the **Hartree–Fock approximation** but with an orbital-dependent screened Coulomb interaction and, at least for localized states such as d–and f–orbitals of transition metal or rare-earth metal ions, the LDA + U theory may be regarded as an approximation to the **GWA**. This method is successfully used when describing the band gap of **Mott insulator**s such as transition metal and rare-earth compounds.
First described in: V. I. Anisimov, J. Zaanen, O. K. Andersen, *Band theory and Mott insulators: Hubbard U instead of Stoner I*, Phys. Rev. B **44**(3), 943–954 (1991).

LED – an acronym for a **light-emitting diode**.

LEED – an acronym for a **low-energy electron diffraction**.

Legendre polynomial has the form

$$L_n(x) = \frac{(2n)!}{n!^2 2^n} \left[x^n - \frac{n(n-1)}{2(2n-1)} x^{n-2} + \frac{n(n-1)(n-2)(n-3)}{2 \cdot 4(2n-1)(2n-3)} x^{n-4} \cdots \right],$$

where n is its degree. The first Legendre polynomials are: $L_0(x) = 1, L_1(x) = x, L_2(x) = 1/2(3x^2 - 1), L_3(x) = 1/2(5x^3 - 3x), L_4(x) = 1/8(35x^4 - 30x^2 + 3)$.
First described in: A. M. Legendre, *Recherches sur la figure des planetes* (Paris 1784).

Lennard–Jones potential describes total potential energy of two atoms at separation r (originally proposed for liquid argon): $V(r) = \epsilon[(r_0/r)^{12} - 2(r_0/r)^6]$, where ϵ is the energy at the minimum of the potential energy well and r_0 is the intermolecular separation at this energy, which is in fact the equilibrium distance between atoms. The terms with $(1/r)^{12}$ and $(1/r)^6$ represent repulsive and attractive forces, respectively.

The parameter ϵ governs the strength of the interaction and r_0 defines a length scale. The interaction repels at close range, then attracts, and is eventually cut off at some critical separation. While the strongly repulsive core arising from the nonbonded overlap between the electron clouds has rather arbitrary form (and other powers and functional forms are sometimes used), the attractive tail actually represents the **van der Waals** interaction due to electron correlations. The interactions involve individual pairs of atoms. Each pair is treated independently. In practice, both ϵ and r_0 are considered as fitting parameters for a particular atomic system.

First described in: J. E. Lennard-Jones, *Cohesion*, Proc. Phys. Soc. **43**, 461–482 (1931).

More details in: I. G. Kaplan, *Intermolecular Interactions: Physical Picture, Computational Methods and Model Potentials* (John Wiley & Sons, Chichester, 2006).

Lennard–Jones–Keesom potential = Keesom potential.

Lenz law states that when a flux of a magnetic field linking a closed electric circuit is changed, a current is induced in the circuit in such a direction as to produce a magnetic flux which opposes the changing magnetic field applied.

First described in: H. F. E. Lenz, Mem. Acad. Imp. Sci. (St. Petersbourg) **2**, 427 (1833); H. F. E. Lenz, *Über die Bestimmung der Richtung der durch elektrodynamische Vertheilung erregten Ströme*, Ann. Phys. **31**, 483–493 (1834).

Leonardo da Vinci law states that friction force between two bodies is independent of the area of their contact. See also **Amonton's law**.

LET – an acronym for a **light-emitting transistor**.

levitation – holding up an object with no visible support.

Lévy process – a random walk in space with infinite second moment of the jump size probability, or a random process with the first moment of the waiting time distribution between events being infinite.

First described in: P. Lévy, *Sur les integrales dont les elements sont des variables aléatoires indépendentes*, Ann. R. Scuola Norm. Super. Pisa, Sez. Fis. e Mat., Ser. **2**, 337–366 (1934), Ser. **4**, 217–218 (1935).

LFM – an acronym for a **lateral force microscopy**.

Lie algebra

Lie algebra – the algebra of vector fields on a manifold with additive operation given by pointwise sum and multiplication by the **Lie brackets**.

First described by S. Lie in 1888. See also in S. Lie, F. Engel, *Theorie der Transformationsgruppen* (Teubner, Leipzig, 1888–1893).

More details in: B. Hall, *Lie Groups, Lie Algebras, and Representations* (Springer, Berlin, 2004).

Lie bracket – the vector field $Z = XY − YX$ for given vectors X and Y.

Lie group – a topological group which is also a differentiable manifold in such a way that the group operations are themselves analytic functions.

lift-off process → **electron beam lithography**.

light-emitting diode (LED) – a semiconductor device that emits incoherent monochromatic light when electrically biased. Depending on the semiconductor used the color of the emitted light can be varied: yellow, orange, green, turquoise, blue-violet, and even white.

First described by: O. V. Losev for SiC in 1927 – principle possibility; N. Holonyak Jr, S. F. Bevaqua, *Coherent (visible) light emission from $Ga(As_{1−x}P_x)$ junctions*, Appl. Phys. Lett. **1**(1), 82–83 (1962) – device realization.

Recognition: Hailed as "the father of semiconductor light-emitter technology in the western world" Nick Holonyak Jr. received several awards; he is one of 13 Americans who have been awarded both the National Medal of Science (1990) and the National Medal of Technology (2002).

light-emitting silicon – Silicon, the most important elemental semiconductor, crystallizes in the **diamond structure** with a lattice constant of 0.5341 nm. Silicon has an indirect band gap of 1.12 eV in the infrared region, that is ideal for room temperature operation, and an oxide that allows the processing flexibility to place today more than 10^8 transistors on a single chip. The extreme integration levels reached by the silicon microelectronics industry have permitted high-speed performance and unprecedented interconnection levels. The present interconnection degree is sufficient to cause interconnect propagation delays, overheating and information latency between single devices. To overcome this bottleneck, **photonic integrated circuits** and photonic materials, in which light can be generated, guided, modulated, amplified and detected, need to be integrated with standard electronics circuits to combine the information processing capabilities of electronic data transfer and the speed of light. In particular, chip-to-chip or even intra-chip optical communications all require the development of efficient optical functions and their integration with state-of-art electronic functions. Silicon is the desired material, because silicon optoelectronics would open the door to faster data transfer and higher integration densities at very low cost. Silicon microphotonics has boomed these last years. Almost all the various photonic devices have been demonstrated, but the main limitation of silicon photonics remains the lack of any practical silicon-based light sources. Several attempts have been employed to engineer luminescing transitions in an otherwise indirect material. In particular, after the initial impulse given by the pioneering work on photoluminescence from **porous silicon** in 1990, nanostructured silicon has received extensive attention. This activity is mainly centered on the possibility of getting relevant optoelectronic properties from nanocrystalline Si. It is generally accepted that **quantum confinement** is essential for visible light emission in Si nanostructures. The optical gain observed in Si nanocrystals embedded in a SiO_2 matrix has given a further impulse to these studies.

More details in: S. Ossicini, L. Pavesi, F. Priolo, *Light Emitting Silicon for Microphotonics*, Springer Tracts on Modern Physics **194** (Springer-Verlag, Berlin, 2003); L. Pavesi, L. Dal Negro, C. Mazzoleni, G. Franzó, F. Priolo, Nature **408**, 440–446 (2000), L. Dal Negro, M. Cazzanelli, L. Pavesi, S. Ossicini, D. Pacifici, G. Franzó, F. Priolo, F. Iacona, *Dynamics of stimulated emission in silicon nanocrystals*, Appl. Phys. Lett. **82**, 4636–4638 (2003) – first observation of optical gain in silicon.

light-emitting transistor (LET) unlike conventional transistors, which include an electrical input port and an electrical output port, has also an infrared optical output port as illustrated in Figure L.5. This makes it possible to interconnect optical and electrical signals for display or communication purposes.

The LET can be made of indium gallium phosphide and gallium arsenide. Although it produces light essentially in the same way as a **light-emitting diode**, there is a possibility to modulate light at much higher speeds. While frequency modulation of about 1 MGz was demonstrated, higher speeds are theoretically possible.

First described in: M. Feng, N. Holonyak Jr., W. Hafez, *Light-emitting transistor: Light emission from InGaP/GaAs heterojunction bipolar transistors*, Appl. Phys. Lett. **84**(1), 151–153 (2004).

Figure L.5 Light-emitting transistor. The squiggly line in the figure represents an infrared photon emitted from the optical output port. After M. Feng, N. Holonyak Jr., W. Hafez, *Light-emitting transistor: Light emission from InGaP/GaAs heterojunction bipolar transistors*, Appl. Phys. Lett. **84**(1), 151–153 (2004).

Linde rule states that the increase in electrical resistivity of a monovalent metal produced by a substitutional impurity per atomic percent impurity is equal to

Linear combination of atomic orbitals (LCAO)

$a + b(v - 1)2$, where a and b are constants for a given solvent metal and a given row of the Periodic Table for the impurity and v is the valence of the impurity.

First described in: J. O. Linde, *Elektrische Eigenschaften verdünnter Mischkristall-legierungen. I. Goldlegierungen*, Ann. Phys. **10**, 52 (1931); J. O. Linde, *Elektrische Eigenschaften verdünnter Mischkristalllegierungen. III. Widerstand von Kupfer- und Goldlegierungen. Gesetzmässigkeiten der Widerstandserhöhungen*, Ann. Phys. **15**, 219–248 (1932).

Linear combination of atomic orbitals (LCAO) – a typical realization of the **tight-binding approach** for calculation of electronic band structures of solids. It is assumed that the atomic potential in the solid under consideration is so strong that electrons are essentially localized at their atoms and the wave functions for neighboring atomic sites have little overlap. Thus, the total wave function $\psi_k(\mathbf{r})$ and energy E for the solid are $\psi_k(\mathbf{r}) = \sum_n \exp(i\mathbf{k} \cdot \mathbf{r}_n)\phi(\mathbf{r} - \mathbf{r}_n)$ and $E = \int \psi_k^*(\mathbf{r})\mathbf{H}\psi_k(\mathbf{r})\mathrm{d}r / \int \psi_k^*(\mathbf{r})\psi_k(\mathbf{r})\mathrm{d}r$, where $\mathbf{r}_n$ is the radius vector to the n-th atom, is the single-electron wave function of the atom. The Hamiltonian for electrons of mass m: $\mathbf{H} = -\hbar^2/2m\nabla^2 + \sum_n V(\mathbf{r} - \mathbf{r}_n)$ includes the sum of single-atom potentials $V(\mathbf{r} - \mathbf{r}_n)$.

More details in: N. Peyghambarian, S. W. Koch, A. Mysyrowicz, *Introduction to Semiconductor Optics* (Prentice Hall, Englewood Cliffs, New Jersey, 1993), pp. 27–34.

linear Hanle effect → Hanle effect.

linearly polarized light → polarization of light.

linear muffin-tin orbital (LMTO) method – one of the most powerful methods for self-consistent electronic properties calculations of solids. Through the use of the **variational principle** for the **Schrödinger equation** with energy-independent basis functions, the secular equations become linear in energy, that is all eigenvalues and eigenvectors can be found simultaneously by diagonalization procedures. The trial basis functions are linear combinations of energy-independent **muffin-tin orbitals**.

The starting point is to approximate the full crystal potential with a muffin-tin potential that is spherically symmetric inside the radius r near the atomic sites and flat outside, in the interstitial region. In the atomic sphere approximation (LMTO-ASA) the radius r of the muffin-tin spheres equals the Wigner–Seitz radius. The spheres then overlap with each other and the interstitial region can be neglected. Moreover, in the ASA approximation the kinetic energy outside the spheres is chosen to be zero.

The LMTO method is particularly suited for close-packed structures. If the crystal is not close packed it is necessary to include the so-called empty spheres to reduce the overlap of muffin-tin spheres. The method combines desirable features of the **linear combination of atomic orbitals (LCAO)**, **Korringa–Kohn–Rostoker (KKR)** and cellular methods.

In the full-potential version of the LMTO (FP-LMTO) method the unit cell is divided into nonoverlapping muffin-tin spheres and the interstitial region. Inside the spheres the potential and electron densities are expanded in a way that no longer requires spherical symmetry. Differently from the **FLAPW** method, the FP-LMTO schemes use for the interstitial region localized functions derived from scattering theory (such as Hankel functions) instead of plane waves.

First described in: O. K. Andersen, *Linear methods in band theory*, Phys. Rev. B **12**(8), 3060–3083 (1975) – LMTO method; M. Methfessel, C. O. Rodriguez, O. K. Andersen, *Fast full-potential calculations with a converged basis of atom-centered linear muffin-tin orbitals: Structural and dynamic properties of silicon*, Phys. Rev. B **40**(3), 2009–2012 (1989) – FP-LMTO method.

More details in: H. L. Skriver, *The LMTO Method* (Springer-Verlag, Berlin, 1984).

linearized augmented plane wave (LAPW) method – one of the most accurate methods for calculations of electronic structure of solids. It is based on the **density functional theory** for the treatment of exchange and correlation effects.

The method is based on solving the **Kohn–Sham equations** for the ground state density, total energy, and eigenvalues of a many-electron system by introducing a basis set which is especially adapted to the problem. This adaptation is achieved by dividing the unit cell into 1. nonoverlapping atomic spheres (centered at the atomic sites) and 2. an interstitial region. In the two types of regions different basis sets are used:

1. Inside an atomic sphere a linear combination of radial functions times spherical harmonics $Y_{lm}(r)$ is used

$$\phi_{k_n} = \sum_{lm} \left[A_{lm} u_l(r, E_l) + B_{lm} \dot{u}_l(r, E_l) \right] Y_{lm}(\hat{r}),$$

where $u_l(r, E_l)$ is the regular solution of the radial Schrödinger equation for energy E_l (chosen normally at the center of the corresponding band with l-like character) and the spherical part of the potential inside sphere t; $\dot{u}_l(r, E_l)$ is the energy derivative of u_l evaluated at the same energy E_l. A linear combination of these two functions constitutes the linearization of the radial function; the coefficients A_{lm} and B_{lm} are functions of k_n determined by requiring that this basis function matches each plane wave of the corresponding basis function of the interstitial region.

2. In the interstitial region a plane wave expansion is used $\phi_{k_n} = 1/\sqrt{\omega} \exp(ik_n r)$, where $k_n = k + K_n$; K_n are the reciprocal lattice vectors and k is the wave vector inside the first Brillouin zone. Each plane wave is augmented by an atomic-like function in every atomic sphere. The solutions to the Kohn–Sham equations are expanded in this combined basis set of linearized augmented plane waves according to the linear variation method. In its general form the

LAPW method expands the potential in the following form:

$$V(\mathbf{r}) = \begin{cases} \sum_{lm} V_{lm}(\mathbf{r}) Y_{lm}(\hat{r}) & \text{inside the sphere} \\ \sum_{k} V_{k} \exp(i\mathbf{Kr}) & \text{outside the sphere} \end{cases}$$

First described in: O. K. Andersen, *Linear methods in band theory*, Phys. Rev. B **12**(8), 3060–3083 (1975); D. D. Koelling, G. O. Arbman, *Use of energy derivative of the radial solution in an augmented plane wave method: application to copper*, J. Phys. F: Met. Phys. **5**(11), 2041–2054 (1975).

linewidth enhancement factor – the ratio between the carrier-induced variations of the real and imaginary parts of susceptibility:

$$\alpha = \frac{\partial \left[\operatorname{Re}\chi(n) \right]}{\partial n} \Big/ \frac{\partial \left[\operatorname{Im}\chi(n) \right]}{\partial n},$$

where the susceptibility $\chi(n)$ is a function of the carrier density n.

Liouville's equation → **Liouville's theorem in statistical mechanics**.

Liouville's theorem in statistical mechanics states that for a classical ensemble characterized by the distribution function $f(r_1, \ldots, r_n, p_1, \ldots, p_n)$ with generalized co-ordinates $r_1, \ldots, r_n$ and generalized momenta $p_1, \ldots, p_n$ in the defined phase space: $df/dt = 0$. This is called **Liouville's equation**.

First described in: J. Liouville, *Mémoire sur la détermination des Intégrales dont la valeur est algébrique*, J. Ecole Polytechn. **14**(22), 124–193 (1833); J. Liouville, *Mèmoire sur les transcendantes elliptiques de première et de seconde espéce considérés comme fonctions de leur amplitude*, J. Ecole Polytechn. **14**(23), 37–83 (1834).

lipids – fatty acids and their derivatives. They constitute a class of relatively water-insoluble organic molecules, which are basic components of biological membranes.

Lipids consist of a polar or **hydrophilic** (attracted to water) head and one to three nonpolar or **hydrophobic** (repelled by water) tails. Since lipids have both functions, they are called amphiphilic. The hydrophobic tail consists of one or two (in triglycerides, three) fatty acids. These are unbranched chains of carbon atoms (with the correct number of H atoms), which are connected by single bonds alone (saturated fatty acids) or by both single and double bonds (unsaturated fatty acids). The chains are usually 14–24 carbon groups long.

Three groups of lipids are distinguished in biological membranes according to the composition of the hydrophilic head they have: (i) **glycolipids**, whose heads contain an oligosaccharide with 1–15 saccharide (sugar) residues; (ii) **phospholipids**, whose heads contain a positively charged group that is linked to the tail by a negatively charged phosphate group; (iii) **sterols**, whose heads contain a planar steroid ring, for example, cholesterol (only in animals).

In an aqueous environment, the heads of lipids are turned toward the environment and the tails are turned toward a hydrophobic region of another molecule. Driven by hydrophilic and hydrophobic forces, the nonpolar tails of lipids tend to cluster together, forming a lipid bilayer or a **micelle**, as schematically shown in Figure L.6. The polar heads face the aqueous environment. Micelles are spheres and can only reach a certain size, whereas bilayers have no limit to their extension. They can also form tubules.

Figure L.6 Self-organized lipid structures.

Lippmann effect – a change in surface tension that results from a potential difference across the interface between two immiscible conductors.

First described in: G. Lippmann, *Relations entre les phénomènes électriques et capillaires*, Ann. Chim. Phys. **5**, 494 (1875).

Lippmann–Schwinger equation describes scattering of nonrelativistic particles in the form $\mathbf{T} = \mathbf{V} + \mathbf{VGV}$, where $\mathbf{T}$ is the scattering operator, $\mathbf{V}$ is the scattering potential, and $\mathbf{G}$ is the full Green function of the system. It is appropriate to simulate scattering of one or two particles.

First described in: B. A. Lippmann, J. Schwinger, *Variational principles for scattering processes*, Phys. Rev. **79**(3), 469–480 (1950).

liquid crystal – a class of molecules that, under some conditions, inhabit a phase in which they exhibit isotropic, fluid they like behavior characterized by little long-range ordering. In other conditions they demonstrate one or more phases with significant anisotropic structure and long-range ordering, but still with an ability to flow. Liquid crystals find wide use in displays which rely on the change in the optical properties of certain liquid crystalline molecules in the presence or absence of an electric field. In the presence of an electric field, these molecules align with it, altering the polarization of the light in a certain way. The ordering of liquid-crystalline phases is extensive on the molecular scale, but does not extend to the macroscopic scale as might be found in classical crystalline solids. The ordering in a liquid crystal might extend along one dimension, but along another dimension might have significant disorder. Important types of liquid crystals include: the nematic phase

liquid-phase epitaxy

(most nematics are uniaxial but biaxial ones are also known), the cholesteric phase, the smectic phase (smectic A, smectic C and hexatic), and columnar phases.

First described in F. Reinitzer, *Beiträge zur Kenntnis des Cholesterins*, Monatsh. Chem. **9**, 421 (1888).

Recognition: in 1991 P.-G. De Gennes received the Nobel Prize in Physics for discovering that methods developed for studying order phenomena in simple systems can be generalized to more complex forms of matter, in particular to liquid crystals and polymers.

See also www.nobel.se/physics/laureates/1991/index.html.

liquid-phase epitaxy – an epitaxial deposition of thin solid films of a material transported from the liquid phase, typically from the melt, onto a suitable solid substrate. The melt actually contains the deposited material in its original composition or in atomic components. The epitaxial deposition is performed in a heated pencil box like the arrangement schematically shown in Figure L.7.

Graphite is an appropriate material for such a pencil box. The upper sliding part of the box contains voids filled with different melted solutes. A solute is composed so as to have a melting point far below the melting point of the substrate material. For instance, for liquid-phase epitaxy of $A^{III}B^{V}$ semiconductors, a group III metal (usually Ga or In) is used as a solvent for group V elements and dopants. When a particular melt contacts the substrate, the dissolved material condenses at the substrate surface.

The advantage of liquid-phase epitaxy is that the required equipment is very cheap, and easy to set up and operate. Meanwhile, it is difficult to achieve the level of control over the epitaxial growth conditions and impurity pollution of the deposited film.

Figure L.7 A setup for liquid-phase epitaxy.

lithography – a process of pattern formation. For more details for electronics → **electron-beam lithography, optical lithography.**

LMTO method = **linear muffin-tin orbital (LMTO) method.**

local density approximation (LDA) expresses the potential at a given location of an electron as a function of the the electron density at the same site. The function is determined by use of the knowledge of the total energy of the jellium model in which the nuclear charges are smeared out and uniformly distributed in the whole space.

The approximation is used to describe the exchange potential. This treats the inhomogeneous electron case as uniform locally, and for the exchange-correlation energy it gives $E_{xc} = \int n(r\epsilon_{xc}(n(\mathbf{r})))d\mathbf{r}$, where ϵ_{xc} is the exchange-correlation energy per electron for a uniform electron gas of density n.

First described in: W. Kohn, L. J. Sham, *Self-consistent equation including exchange and correlation effects*, Phys. Rev. **140**(4A), A1133–A1138 (1965).

localized state – a quantum mechanical state of a single electron in a strongly random potential that is localized in space so that the state amplitude is concentrated near a single point in the sample.

local spin–density approximation (LSDA) is used to describe the exchange potential accounting for the electron spin

$$E_{xc} = \int n(\mathbf{r})\epsilon_{xc}(n_\uparrow(\mathbf{r}), n_\downarrow(\mathbf{r})d\mathbf{r}),$$

where $\epsilon_{xc}(n_\uparrow(\mathbf{r}), n_\downarrow(\mathbf{r}))$ is the exchange and correlation energy per particle of a homogeneous spin-polarized electron gas with spin-up and spin-down densities.

logic gate – a device implementing a **logic operator** of **Boolean algebra**. Any logic gate realizing a particular logic operation has a specific form in drawings. Figure L.8 presents a convention for the basic logic gates. In the same way as the basic operators of the **Boolean algebra** make up more complicated expressions, the basic gates are combined to construct complex circuits. Note that additional gates that duplicate the input values on wires are frequently needed. These are called FANOUT or COPY gates. In classical computation, these are obvious gates, for they simply correspond to splitting the wire into two or more leads, which is an easy operation.

See also **quantum logic gate**.

More details in: A. Galindo, M. A. Martin-Delgado, *Information and computation: Classical and quantum aspects*, Rev. Mod. Phys. **74**(2), 347–423 (2002).

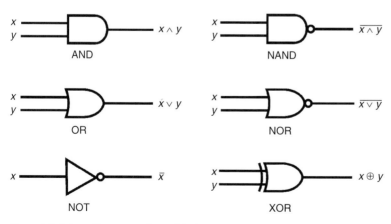

Figure L.8 Schematic presentations of logic gates.

logic operator (function)

Table L.1 Truth table for the unary logic operator **NOT**.

x	NOTx
0	1
1	0

Table L.2 Truth table for the binary logic operators **AND**, **OR**.

x	y	xANDy	xORy
0	0	0	0
0	1	0	1
1	0	0	1
1	1	1	1

Table L.3 Truth table for the binary logic operators **NAND**, **NOR** and **XOR**.

x	y	xNANDy	xNORy	xXORy
0	0	1	1	1
0	1	1	0	1
1	0	1	0	1
1	1	0	0	0

logic operator (function) supposes an operation with logic variables. A logic operator or function f is an application $f : \{0,1\}^n \longmapsto \{0,1\}^m$ which maps an input of n-bit-valued operands into an m-bit-valued output. When the target space of f is $\{0,1\}$, one usually says that f is a **Boolean** operator or function.

An operator with one operand is called a unary operator, with two operands is a binary operator. There are three basic Boolean or logic operators: (1) The unary operator **NOT**: **NOT**$x = \bar{x} = 1 - x$, also denoted by overlining the argument; (2) the binary operator **AND**: x**AND**$y = x \wedge y = xy$, also denoted by the symbol $\wedge$; and (3) the binary operator **OR**: x**OR**$y = x \vee y = x + y - xy$, denoted also by the symbol $\vee$. The action of a logic operator is represented by a truth table. A **truth table** contains as many columns as input operands and output **bits**, and $2^{\#\text{operands}}$ rows. The inputs are shown on the left, and the outputs are shown on the right. The truth tables for the basic operators are shown above.

With the basic set comprising **NOT**, **AND**, **OR** operators one can build any logic function, so that it is considered as a universal set. Moreover, the application of the **NOT** operation to **AND** and **OR** ones gives another universal set including **NAND** and **NOR** operators.

London equations determine the relation between a supercurrent I_s, which is the electrical current in a superconductor, and magnetic **H** and electrical **E** fields applied in the form:

$$\Lambda \nabla \times I_s = -\frac{1}{c} \mathbf{H}, \qquad \Lambda \frac{\partial}{\partial t} I_s = \mathbf{E}.$$

Here Λ is a constant characteristic of the superconductor but temperature dependent. The equations were shown to have local validity as they relate to the supercurrent at a given point in the superconductor.

First described in: F. London, H. London, *The electromagnetic equations of the superconductor*, Proc. R. Soc. A **149**(866), 71–88 (1935).

longitudinal acoustic mode → phonon.

longitudinal optical mode → phonon.

Lorentz force – the force acting on an electron moving in a magnetic field perpendicular to the trajectory of the electron. It is $e(\mathbf{v} \times \mathbf{B})$, where e is the electron charge, **v** is the electron velocity, **B** is the induction of the magnetic field (note that **v** and **B** are vectors).

First described in: H. A. Lorentz, *Versuch einer Theorie der electrischen und optischen Erscheinungen in bewegten Körpern*, (Leiden, 1895), H. A. Lorentz, *Simplified theory of electrical and optical phenomena in moving systems*, Konik. Akad. Wetensch. Amsterdam, **1**, 427–442 (1899).

Lorentz frame – any of the family of inertial coordinate systems with three space coordinates and one time coordinate used in the special theory of relativity. Each frame is in uniform motion with respect to all the other Lorentz frames and the interval between any two events is the same in all frames.

Lorentz gauge – the condition determining the relation between the scalar potential ϕ and vector potential **A** of an electromagnetic field. It is

$$\nabla \mathbf{A} + \frac{1}{c} \frac{\partial \phi}{\partial t} = 0.$$

Lorentzian – a frequency-dependent function presented in the form:

$$A(\Gamma/2)^2 / [(\nu - \nu_0) + (\Gamma/2)^2],$$

where A is an overall strength, G is the full width at half maximum, ν is the frequency, and ν_0 is a center frequency.

Lorentz rule defines the relationship between interspecies distances in a mixture composed of like (l_{ii}, l_{jj}) and unlike (l_{ij}) species in the form $l_{ij} = (l_{ii} + l_{jj})/2$.

Lorentz invariance

Lorentz invariance – the property possessed by the laws of physics and of certain physical quantities of being the same in any **Lorentz frame** and thus unchanged by a **Lorentz transformation**.

Lorentz transformation – a group operation that is used to transform the space and time coordinates of one inertial reference frame, say S, into those of another one, S', with S' travelling at a relative speed u to S. If an event has space-time coordinates of (x, y, z, t) in S and (x', y', z', t') in S', then these are related according to $x = \gamma(x - ut), y' = y, z' = z, t' = \gamma(t - ux/c^2)$, where $\gamma = (1 - u^2/c^2)^{-1/2}$. These equations only work if u is pointed along the x-axis of S. In other cases, it is generally easier to perform a rotation so that u does point along the x-axis. Moreover, the position of the origins must coincide at 0. What this means is that $(0,0,0,0)$ in frame S must be the same as $(0,0,0,0)$ in S'. The transformation forms the basis for the special theory of relativity. The speed of light is the same in all reference frames here.

First described in: H. A. Lorentz, *Electromagnetic phenomena in system moving with velocity less than that of light*, Konik. Akad. Wetensch. Amsterdam **12**, 986–1009 (1904).

Lorenz number → **Wiedemann–Franz–Lorenz law**.

Lorenz relation gives a relationship between the magnetic vector potential **A** and scalar potential ϕ in the form $\nabla \cdot A + \epsilon_0 \mu_0 \partial \phi / \partial t = 0$ (in MKS), $\nabla \cdot A + (1/c)\partial \phi / \partial t = 0$ (in cgs).

First described in: L. Lorenz, *On the identity of the vibrations of light with electrical currents*, Phil. Mag. **34**, 287–301 (1867).

low-energy electron diffraction (LEED) – the coherent reflection of electrons in the energy range 5–500 eV from the uppermost few atomic layers of crystalline solids. Diffraction occurs because the electron wavelength λ (in nanometers) is related to the accelerating voltage V (in Volts) via $\lambda = 1.22638/V^{1/2}$, providing that λ in the low energy range is of the order of interatomic distances in solids.

LEED measurements are schematically illustrated in Figure L.9.

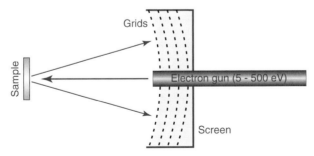

Figure L.9 Principle of the arrangement for low-energy electron diffraction.

Electrons from an electron gun diffract off the sample and head back toward the electron gun and the grids surrounding it. The two middle grids are set at an energy slightly less than the electron beam to prevent any inelastically scattered electrons from reaching the screen. The screen is biased positive between 0 and 7 keV accelerating the electrons towards it. The screen is coated in phosphor which glows green when electrons hit it. The resulting pattern is viewed from the rear end of the apparatus. It is one of experimental techniques for determination of atomic geometries of crystalline surfaces and ordered overlays thereon.

First described in: C. J. Davisson, C. H. Kunsman, *The scattering of low-speed electrons by platinum and magnesium*, Phys. Rev. **22**(1), 242–258 (1923); G. P. Thomson, A. Reid, *Diffraction of a cathode ray by a thin film*, Nature **119**, 890 (1927).

Recognition: in 1937 C. J. Davisson and G. P. Thomson received the Nobel Prize in Physics for their experimental discovery of the diffraction of electrons by crystals.

See also www.nobel.se/physics/laureates/1937/index.html.

luminescence – a process where a system in an excited state emits photons in order to relax to a lower energy state. It consists of three separate steps: (i) *excitation* of the medium providing electrons and holes to be excited by an external source of energy, (ii) *thermalization* of the excited electrons and holes resulting in their energy relaxation towards quasi thermal equilibrium, (iii) *recombination* of electrons with holes via radiative mechanisms resulting in the light emission. According to the excitation mechanism the luminescence processes are distinguished as **photoluminescence** (excited by light), **electroluminescence** (excited by electrical injection of electrons and/or holes), **thermoluminescence** (excited by heating), **cathodoluminescence** (excited by electron irradiation), **sonoluminescence** (excited by sound waves), and **chemiluminescence** (excited by chemical reactions).

luminophor – a luminescent material that converts part of the primary absorbed energy into emitted luminescent radiation.

LUMO – an acronym for a **l**owest **u**noccupied **m**olecular **o**rbital.

Luttinger liquid – a universality class of liquids characterized by the following relation between spectral properties and correlation exponents of the **Luttinger model**: $v_j v_n / v_s^2 = 1$, where v_j, v_n and v_s are three distinct parameters, all with dimension of velocity. The parameter v_s is related to the variation of the total energy of the system with the total momentum ($v_s = \delta E / \delta P$). The parameter v_j is related to the variation of the energy with the shift of all the particles in the space of momentum. It is the renormalized Fermi velocity. Finally v_n relates to changes in chemical potential to those in the Fermi vector k_F: $\delta \mu = v_n \delta k_f$. Thus, the only parameter that determines the low-energy properties of the system is the ratio $\exp(-2\phi) = v_n / v_s = v_s / v_j$.

Luttinger model

It is used for interacting quantum systems in one dimension that have many distinctive features not present in higher dimensions. In particular, gapless (metallic) one-dimensional systems of interacting fermions does not behave as a normal **Fermi liquid**. Luttinger liquids are low-energy collective excitations of the entire electron system in which electrons move in concert.

Lattice models identified as Luttinger liquids include the one-dimensional **Hubbard model** and variants such as the $t - J$ model or the extended Hubbard model. Moreover, one-dimensional electron-phonon systems, metals with impurities and edge states in the **quantum Hall effect** can be Luttinger liquids.

First described in: F. D. M. Haldane, *General relation of correlation exponents and spectral properties of one- dimensional Fermi systems: application to the anisotropic S = 1/2 Heisenberg chain,* Phys. Rev. Lett. 45(16), 1358–1362 (1980).

More details in: J. Voit, One-dimensional Fermi liquids, Rep. Prog. Phys. **58**(9), 977–1116 (1995).

Luttinger model accounts for the fact that one-dimensional spin 0 **fermion** systems show the remarkable phenomenon of the anomaly of the ground state, that is a ground state with a density of states which does not have a discontinuity at the Fermi momentum, but its graph becomes vertical (Fermi surface anomaly). A scheme was introduced by Tomonaga to illustrate this theory, via an exactly soluble model for the low-energy properties of these quantum systems. The model contains fairly realistic interaction between pairs of fermions. The **Hamiltonian** is given by:

$$H = \int_{-L/2}^{L/2} dx \sum_{\omega=\pm} \psi^+_{\omega,x} v_F (i\omega\delta_x - p_F) \psi^-_{\omega,x}$$

$$+\lambda \int_{-L/2}^{L/2} dxdy \psi^+_{+,x} \psi^-_{+,x} \psi^+_{-,y} \psi^-_{-,y} v(x - y),$$

where λ is a coupling constant, $v(x - y)$ is a smooth short-range pair potential, and v_F is the velocity at the Fermi surface. $\psi^a_{\omega,x}$ are fields where $a = \pm$ specifies the creation and annihilation operators for fermions located at a point x in the interval $[-L/2, L/2]$ and distinguished by $\omega = +$ or $\omega = -$. The calculated momentum distribution in the ground state shows no discontinuity at the Fermi surface and possesses an infinite slope at the Fermi surface.

First described in: S. Tomonaga, *Remarks on Bloch's method of sound waves applied to many fermion problems,* Prog. Theor. Phys. **5**, 544–569 (1950); J. M. Luttinger, *An exactly soluble model of a many-fermion system,* J. Math. Phys. **4**(9), 1154–1162 (1963).

Historical note: The exact solution of Luttinger was not really correct because of an error in the canonical commutation relations which, in infinitely many degrees of freedom systems, do not have a unique representation. The exact solution was obtained in: D. Mattis, E. Lieb, *Exact solution of a many-fermion system and its associated boson field,* J. Math. Phys. **6**, 304–310 (1965). The real value of the anomaly and Luttinger's result agree to the second order in λ, but differ in higher orders.

More details in: G. Gallavotti, *The Luttinger model: in the RG-theory of the one-dimensional many-body Fermi system*, J. Statist. Phys. **103** (2), 459–483 (2001).

Luttinger theorem states that the **Fermi surface** in a momentum space, which separates unoccupied **quasiparticle** states from the occupied ones, must enclose the same volume as the Fermi surface of the independent electrons.

Lyddane–Sachs–Teller relation $\rightarrow$ **phonon**.
First described in: R. H. Lyddane, R. G. Sachs, E. Teller, *On the polar vibrations of alkali halides*, Phys. Rev. **59**(8), 673–676 (1941).

Lyman series $\rightarrow$ **Rydberg formula**.
First described in: Th. Lyman, *The spectrum of hydrogen in the region of extremely short wave lengths*, Astrophys. Jour. **23**, 181–210 (1904); Th. Lyman, *An extension of the spectrum in the extreme ultra-violet*, Nature **93**, 241 (1914).

M

From Mach–Zender interferometer to Murrell–Mottram potential

Mach–Zehnder interferometer – an optical device used to determine the phase shift caused by a small sample which is placed in the path of one of two collimated light beams from a coherent light source. It is schematically shown in Figure M.1 in the calibration mode when no sample is placed in the interferometer.

Adding up all the contributions for the two undisturbed paths, one can see that they are the same. Thus light entering detector 1 via the two paths is in phase. As a result there is a constructive interference for the light entering detector 1. Doing the same analysis for the light coming to detector 2 one sees that the light goes through one additional phase change of one-half a wavelength. Therefore, there will be complete destructive interference, and no light will reach detector 2. Thus, regardless of the wavelength of the light, it all goes to the detector 1 if no sample is placed in the way of the light. But if it is, the phase shift caused by the sample alters the phase relationships between the two beams. There is no longer complete destructive interference at detector 2. Measuring the relative amount of light entering detectors 1 and 2 allows a calculation of the phase shift produced by the sample.

First described in: L. Zehnder, *Ein neuer Interferenzrefraktor,* Z. Instrumentenkunde **11**, 275–285 (1891); L. Mach, *Über einen Interferenzrefraktor,* Z. Instrumentenkunde **12**, 89–93 (1892).

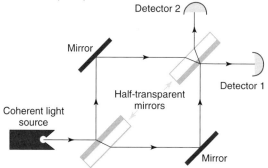

Figure M.1 A Mach–Zehnder interferometer.

macroscopic long-range quantum interference → Josephson effect.

What is What in the Nanoworld. Victor E. Borisenko and Stefano Ossicini
Copyright © 2008 WILEY-VCH Verlag GmbH & Co. KGaA, Weinheim
ISBN: 978-3-527-40783-5

Madelung constant

Madelung constant arises when one expresses the energy per mole of a crystal, for two atom crystals, due to the Coulomb interaction in the form $E = \alpha N(q^+ + q^-)/d$, where α is the Madelung constant, N is the Avogadro number, q^+ and q^- are the charges of positive and negative ions respectively, and d is the nearest-neighbor distance. The constant is dimensionless, of the order of 1, depends only upon the type of crystal structure and is independent of the lattice dimensions.

First described in: E. Madelung, *Molekulare Eigenschwingungen*, Gesell. Wiss. Göttingen, Nachr., Math.-Phys. Klasse **1**, 43–58 (1910); E. Madelung, *Das elektrische Feld in Systemen von regelmässig angeordneten Punktladungen*, Phys. Z. **19**, 524–532 (1918).

Madelung energy – electrostatic energy per ion pair

$$E = \frac{1}{2N_p} \sum_{i,j} \frac{(\pm)e^2}{r_{ij}},$$

for $i \neq j$, where N_p is the number of ion pairs.

First described in: E. Madelung, *Das elektrische Feld in Systemen von regelmässig angeordneten Punktladungen*, Phys. Z. **19**, 524–532 (1918).

Maggi–Righi–Leduc effect – a change in the thermal conductivity of a conductor when it is placed in a magnetic field.

First described in: G. A. Maggi, Arch. de Genève **14**, 132 (1859); A. Righi, Mem. Acc. Lincei **4**, 433 (1887); M. A. Leduc, J. Phys. 2e série **6**, 378 (1887).

magnetic force microscopy – a scanning probe technique for mapping magnetic field on a micro- and nanometer scale. The technique is an offspring of **atomic force microscopy (AFM)**. It is based on the use of a magnetized tip. The technique has been developed for micromagnetic stray field imaging by using a fine magnetic probe based on the magnetic balance and AFM. Its main features are (1) to detect a very small magnetic force (10^{-13} N) or its force gradient (around 10^{-4} N/m) by using the small magnetic probe, which acts strongly on the stray field from a sample surface (magnetic domain) in a near-field range, and (2) to obtain the distribution image of the stray field magnetic domain.

The images are obtained by measuring the interaction force between the magnetic tip and the magnetic sample. The field gradient is mapped as a variation of the force on the tip, as the tip is scanned in a raster pattern, at a constant height over the sample. Since magnetic interactions are long-range, they can be separated from the topography by scanning at a certain constant height (typically around 20 nm) above the surface, where the z component of the sample stray field is probed. Therefore, any magnetic force microscope is always operated in a noncontact mode. The signal from the cantilever is directly recorded while the z feedback is switched off. A static or dynamic mode can be employed. A lateral resolution below 50 nm can be routinely achieved.

First described in: Y. Martin, H. K. Wickramasinghe, *Magnetic imaging by "force microscopy" with 1000 Å resolution*, Appl. Phys. Lett. **50**(20), 1455–1457 (1987); J. J. Sáenz, N. García, P. Grütter, E. Mayer, H. Heinzelmann, R. Wiesendanger, L. Rosenthaler, H. R. Hidber, H. J. Güntherodt, *Observation of magnetic forces by the atomic force microscope*, J. Appl. Phys. **62**(10), 4293–4295 (1987).

More details in: D. Rugar, H. J. Mamin, P. Guenthner, S. E. Lambert, J. E. Stern, I. McFadyen, T. Yogi, *Magnetic force microscopy: General principles and application to longitudinal recording media*, J. Appl. Phys. **68**(3), 1169–1183 (1990); *Roadmap of Scanning Probe Microscopy*, edited by S. Morita (Springer Verlag, Berlin Heidelberg, 2007).

magnetic groups = **Shubnikov groups**.

magnetism – a set of phenomena representing a response of matter to an external magnetic field. Magnetic states of materials are primarily distinguished according to the sign, magnitude and temperature dependence of the magnetic susceptibility defined as $\chi = M/H$, where M is the induced magnetization and H is the applied magnetic field. The four basic magnetic states of matter are diamagnetism, paramagnetism, ferromagnetism, and antiferromagnetism. They are listed in the Table M.1 with their main characteristics.

The ferrimagnetic state, which is not represented in the Table M.1, is, except in detail, similar to the antiferromagnetic state. Below the **Curie point** or **Neél point**, the ferromagnetic, ferrimagnetic, and antiferromagnetic materials have an ordered arrangement of the magnetic moments associated with each atom, schematically shown in Figure M.2.

At temperatures above the Curie point or Neél point, these materials exhibit paramagnetic behavior.

Recognition: in 1970 L. Neél received the Nobel Prize in Physics for fundamental work and discoveries concerning antiferromagnetism and ferrimagnetism which have led to important applications in solid-state physics.

See also www.nobel.se/physics/laureates/1970/index.html.

Table M.1 Classification of materials according to their magnetic susceptibility.

$\chi \rightarrow$ State ↓	Sign	Magnitude	Temperature dependence
diamagnetic	−	small	$\chi \neq f(T)$
paramagnetic	+	small to medium	$\chi \neq f(T)$
		large	$\chi = const/T$
ferromagnetic	+	large	for $T < T_C^*$ – complex
			for $T > T_C$ $\chi = const/(T - T_C)$
antiferromagnetic	+	small	for $T > T_C$ $\chi = const/(T + const)$

* T_C is the **Curie point** or **Neél point** for an antiferromagnetic.

magnetooptical Kerr effect (MOKE)

Ferromagnetic   Ferrimagnetic   Antiferromagnetic

Figure M.2 Arrangement of magnetic moments in different magnetic materials below the Curie temperature.

magnetooptical Kerr effect (MOKE) → **Kerr effects**.

magnetoresistance – the electrical resistance of a material or structure measured in the presence of a magnetic field. The size of the magnetoresistance effect is expressed as a percentage, $\Delta R / R_0$, where ΔR is the change in the resistance, and R_0 is usually the resistance at zero magnetic field. Positive magnetoresistance refers to an increase in resistance when the magnetic field is applied, while negative magnetoresistance is a decrease in resistance.

First described in: W. Thomson (Lord Kelvin), *On the electro-dynamic qualities of metals: effects of magnetization on the electric conductivity of nickel and of iron*, Proc. R. Soc. Lond. **8**, 546–550 (1856).

magnetostriction effect – a change in linear dimensions of a ferromagnetic appearing when an external magnetic field is applied in a specified direction.

First described in: J. P. Joule, Sturgeon's Ann. Electricity **8**, 219 (1842).

magnetotactic bacteria – a class of bacteria that exhibit the peculiar ability to orient themselves along the magnetic field lines of Earth's magnetic field. They contain linear arrays of **antiferromagnetic** nanocrystals of Fe_3O_4, known as magnetite, or Fe_3S_4, known as greigite. The arrays are oriented parallel to the long direction of bacteria. *Magnetospirillum magnetotacticum* is an example of magnetotactic bacterium.

First described in: S. Bellini, *Su di un particolare comportamento di batteri d'acqua dolce*, Instituto di Microbiologia dell'Universitá di Pavia (1963) – first obseravtion; R. Blakemore, *Magnetotactic bacteria*, Science **190**, 377–379 (1975)–rediscovering and birth of the term magnetotactic.

magnetotaxis – specific orientation of bacteria and their navigation along a magnetic field. The microorganisms with such properties are called **magnetotactic bacteria**.

First described in: R. P. Blakemore, *Magnetotactic bacteria*, Science **190**, 377–379 (1975).

magnon – the unit of energy of a spin wave.

Mahan effect (MND effect) – a dynamical response of the Fermi sea to the core hole in a band-to-band optical absorption by **degenerate semiconductors**. **Exciton** states, arising from the electron–hole Coulomb attraction, are considered to affect the optical absorption. The calculations show that exciton states cause a logarithmic **singularity** in the absorption at the Burstein edge. This singularity is present at a moderate density of electrons or holes in the degenerate band, but it gradually disappears in the high-density limit. Lifetime broadening could make the logarithmic singularity difficult to observe at higher densities.

First described in: J. D. Mahan, *Excitons in degenerate semiconductors*, Phys. Rev. **153**(3), 882–889 (1967); P. Noziéres, C. T. De Dominicis, *Singularities in the X-ray absorption and emission of metals. III. One-body theory exact solution*, Phys. Rev. **178**(3), 1097–1107 (1969) – more complete theory.

More details in: J. D. Mahan, *Many-Particle Physics*, 2nd edition (Plenum, New York, 1990).

majorization – the mathematical procedure involving two real m-dimensional vectors $\mathbf{x} = (x_1, \ldots, x_m)$ and $\mathbf{y} = (y_1, \ldots, y_m)$. The vector $\mathbf{x}$ is majorized by $\mathbf{y}$ (equivalently $\mathbf{y}$ majorizes $\mathbf{x}$), denoting $\mathbf{x} \prec \mathbf{y}$, if for each n in the range $1, \ldots, m$: $\sum_{j=1}^{n} x_j^{\uparrow} \leq \sum_{j=1}^{n} y_j^{\uparrow}$, with equality holding when $n = m$. The superscript "$\uparrow$" indicates that elements are to be taken in descending order, so, for example, $x_1^{\uparrow}$ is the largest element in $(x_1, \ldots, x_m)$. The majorization relation is a partial order on real vectors, with $\mathbf{x} \prec \mathbf{y}$ and $\mathbf{y} \prec \mathbf{x}$ if and only if $x^{\uparrow} = y^{\uparrow}$.

More details in: R. Bhatia, *Matrix Analysis* (Springer-Verlag, New York, 1997).

Marangoni effect – the mass transfer on or in a liquid layer due to surface tension differences. Since a liquid with a high surface tension pulls more strongly on the surrounding liquid than one with a low surface tension, the presence of a gradient in surface tension will naturally cause the liquid to flow away from regions of low surface tension. The surface tension gradient can in turn be caused by a concentration gradient or by a temperature gradient.

First described in: J. Thomson, *On certain curious motions observable on the surfaces of wine and other alcoholic liquours*, Phil. Mag. **10**, 330 (1855); C. Marangoni, *Sull'espansione delle goccie di un liquido galleggiante sulla superficie di altro liquido*, (Fusi, Pavia, 1865); C. Marangoni, *Über die Ausbreitung der Tropfen einer Flüssigkeit auf der Oberfläche einer anderen*, Ann. Phys. **219**, 337–354 (1871).

Marcus formula presents the charge transfer rate in molecules between a donor and an acceptor, each with a discrete number of localized vibronic states in the form: $k_{da} = 2\pi/\hbar |V_{da}|^2 F$. Here V_{da} stands for the electronic coupling and the nuclear factor F contains the reorganization energy accounting for vibrionic effects.

First described in: R. A. Marcus, N. Sutin, *Electron transfer in chemistry and biology*, Biochim. Biophys. Acta 811(3), 265–322 (1985).

Recognition: in 1992 R. A. Marcus received the Nobel Prize in Chemistry for his contributions to the theory of electron transfer reactions in chemical systems.

Markov process

See also www.nobel.se/chemistry/laureates/1992/index.html.

Markov process – a **stochastic process** which assumes that in a series of random events the probability of an occurrence of each event depends only on the immediately preceding outcome. It is characterized by a matrix of transition probabilities. It measures the probability of moving from one state to another.

First described in: A. A. Markov, *Rasprostranenie zakona bol'shih chisel na velichiny, zavisyaschie drug ot druga*, Izvestiya Fiziko-matematicheskogo obschestva pri Kazanskom universitete, 2-ya seriya, **15**, 135–156, (1906).

Marx effect – the energy of photoelectrons emitted from an illuminated surface is decreased when the surface is simultaneously illuminated by light of lower frequency than that causing the emission.

First described in: E. Marx, *On a new photoelectric effect in alkali cells*, Phys. Rev. **35**(9), 1059–1060 (1930).

maser – an acronym for a device producing **m**icrowave **a**mplification by **s**timulated **e**mission of **r**adiation.

First described in: N. G. Basov, A. M. Prokhorov, *The theory of a molecular oscillator and a molecular power amplifier*, Disc. Faraday Soc. **19**, 96–99 (1955); A. L. Schawlow, C. H. Townes, *Infrared and optical masers*, Phys. Rev, **112**(6), 1940–1949 (1958); A. M. Prokhorov, *A molecular amplifier and generator of submillimetre waves*, Zh. Eksper. Teor. Fiz. 34(6), 1658–1659 (1958)–in Russian; N. G. Basov, B. M. Vul, Y. M. Popov, *Quantum-mechanical semiconductor generators and amplifiers of electromagnetic oscillations*, Sov. Phys. JEPT **37**(2), 416 (1960).

Recognition: in 1964 C. H. Townes, N. G. Basov and A. M. Prokhorov received the Nobel Prize in Physics for fundamental work in the field of quantum electronics, which has led to the construction of oscillators and amplifiers based on the maser–laser principle.

See also www.nobel.se/physics/laureates/1964/index.html.

mass-detection NEMS – **nano-electromechanical systems** used for the detection of ultrasmall masses. The devices consist of an oscillating very tiny cantilever; when a small particle is absorbed onto the cantilever it alters the frequency at which the cantilever oscillates. The frequency variation can be measured using laser light reflected off the cantilever, and consequently the mass of the particle can be calculated. A sensitivity in the **atto**gram (10^{-18} grams) regime has been achieved. The **zepto**gram range (10^{-21} grams) is envisioned, thus allowing the detection of viruses.

master equation – a phenomenological first-order differential equation describing the time evolution of the probability of a system to occupy each one of a discrete set of states: $dP_k/dT = \sum_l T_{kl}P_l$, where P_k is the probability for the system to be in the state k, while the matrix T_{kl} is filled with a grid of transition-rate constants.

Matthiessen rule empirically states that total resistivity of a solid is the sum of the resistivities related to different scattering mechanisms. It means that the scattering rate of electrons moving in solids is additive for different scattering mechanisms, that is $\tau^{-1} = \sum_i \tau_i^{-1}$ with τ representing the energy relaxation time. In terms of the electron mobility it has the form $\mu^{-1} = \sum_i \mu_i^{-1}$, where μ is the integral electron mobility accounting for partial electron mobilities μ_i corresponding to different scattering mechanisms. Note that it fails when the outcome of one scattering process influences the outcome of another and one or more scattering processes depend on the electron wave vector.

First described in: A. Matthiessen, C. Vogt, *Über den Einfluss der Temperatur auf die elektrische Leitungsfähigkeit der Legierungen*, Ann. Phys. **122**, 19–78 (1864).

matrix element – the notation $H_{mn} = \langle m|\mathbf{H}|n\rangle \equiv \int \psi_m^* \mathbf{H} \psi_n$, where $\mathbf{H}$ is the operator, and ψ_m and ψ_n are the wave functions corresponding to the states m and n. The notation is due to Dirac, where $|n\rangle$ is called a **ket** and denotes a state, and the **bra** $\langle m|$, is its complex conjugate. It bridges the gap between matrices and differential equations used in operators. In fact, it is a scalar product of a member of a complete orthogonal set of vectors representing states with a vector, which results from applying a specified operator to another member of this set.

Maxwell–Boltzmann distribution (statistics) gives the probability to find a particle of mass m and energy E in an ideal gas in thermal equilibrium at a temperature T

$$f(E) = \frac{1}{\sqrt{2\pi m k_B T}} \exp\left(-\frac{E}{k_B T}\right).$$

It can be derived from the more general **Gibbs distribution**. The Maxwell-Boltzmann statistics are valid only at temperatures and densities for which the probability of any given level being occupied is very small.

First described in: L. Boltzmann, *Einige allgemeine Sätze über das Wärmegleichgewicht*, Sitzungsber. Akad. Wiss. Wien: Math.-Naturwiss. Klas. **63**(II), 679 (1871).

Maxwell distribution (statistics) gives the probability to find a particle of mass m and velocity v in an ideal gas in thermal equilibrium at a temperature T. The Maxwellian distribution function is:

$$f(v_x, v_y, v_z) = \left(\frac{m}{2\pi k_B T}\right)^{3/2} \exp\left(-\frac{m(v_x^2 + v_y^2 + v_z^2)}{2k_B T}\right).$$

The Maxwellian velocity distribution is:

$$N(r,v)\mathrm{d}^3 r \mathrm{d}^3 v = \frac{N}{V} \left(\frac{m}{2\pi k_B T}\right)^{3/2} \exp\left(-\frac{mv^2}{2k_B T}\right) \mathrm{d}^3 r \mathrm{d}^3 v,$$

Maxwell equations

where N is the total number of particles, and V is the volume of the gas. The distribution of an individual velocity component is:

$$N(v_x)dv_x = \frac{N}{V}\left(\frac{m}{2\pi k_B T}\right)^{1/2}\exp\left(-\frac{mv_x^2}{2k_B T}\right)dv_x,$$

and the distribution of speeds is:

$$N(v)dv = 4\pi\frac{N}{V}\left(\frac{m}{2\pi k_B T}\right)^{3/2}v^2\exp\left(-\frac{mv^2}{2k_B T}\right)dv.$$

First described in: J. C. Maxwell, *Illustrations of the dynamical theory of gases*, Phil. Mag. **19**, 19–32 (1860), and **20**, 21–37 (1860).

Maxwell equations connect electric charges and electric and magnetic fields in any medium. In the differential form they are:

$$\nabla\mathbf{D} = \rho, \qquad \nabla\mathbf{B} = 0$$

$$\nabla\times\mathbf{E} = -\frac{\partial\mathbf{B}}{\partial t}, \qquad \nabla\times\mathbf{H} = j + \frac{\partial\mathbf{D}}{\partial t}$$

$$\mathbf{D} = \epsilon_0\mathbf{E} + \mathbf{P}, \qquad \mathbf{B} = \mu_0\mathbf{H} + \mathbf{M},$$

where ∇ is the **Nabla-operator**, $\mathbf{D}$ is the electric displacement, $\mathbf{B}$ is the magnetic induction or magnetic flux density, $\mathbf{E}$ is the electric field strength, $\mathbf{H}$ is the magnetic field strength, $\mathbf{P}$ is the polarization density of the medium (that is the electric dipole moment per unit volume), $\mathbf{M}$ is the magnetization density of the medium (that is the magnetic dipole moment per unit volume), and j is the electrical current density.

The first pair of equations show that free electric charges ρ are the sources of the electric displacement and that the magnetic induction is source free. The second pair of equations demonstrate how temporally varying magnetic and electric fields generate each other. In addition, the magnetic field $\mathbf{H}$ can be created by a macroscopic current with density j. The third pair of equations are material equations in their general form. They show that electric displacement in the material is given by the sum of electric field and polarization, while the magnetic flux density is given by the sum of magnetic field and magnetization.

First described in: J. C. Maxwell, *A Treatise on Electricity and Magnetism*, vol. II (Clarendon Press, 1873).

Maxwell's demon – the name given by W. Thomson (Lord Kelvin) to the mysterious creature separating molecules according to their velocities in the thought experiment proposed by J. C. Maxwell. The demon opens and closes a trap door between two compartments of a chamber containing gas, and pursues the subversive policy of only opening the door when fast molecules approach it from the right, or slow ones from the left. In this way he establishes a temperature difference between the two compartments without doing any visible work, in violation of the second law of thermodynamics, and consequently permitting a host of contradictions.

In fact, the second law is not violated, as it was shown in the twentieth century. It becomes clear when one considers the tricks of the demon in terms of information processing. In order to select molecules, the demon stores the random results of his observations of the molecules. The information about the molecule's location must be present in the demon's memory. The demon's memory thus gets hotter. The irreversible step is not the acquisition of information, but the loss of information if the demon later clears his memory. The information erasure operation increases the entropy in the environment, as required by the second law. This is the thermodynamic cost of elementary information manipulations.

First described in: J. C. Maxwell, *Theory of Heat* (Longmans, Green and Co., London, 1871).

Maxwell triangle – a diagram used to represent the trichromatic variables of the components in a three-color combination.

First described in: J. C. Maxwell, *On the theory of compound colours, and the relations of the colours of the spectrum*, Phil. Trans. R. Soc. London **150**, 57–84 (1860).

Maxwell–Wagner polarization – a space charge polarization in a material consisting of electrically conducting grains and insulating grain boundaries.

First described in: K. W. Wagner, *Erklärung der dielectrischen Nachwirkungsvorgänge auf Grund Maxwellscher Vorstellung*, Arch. Electrotech. **2**, 371–387 (1914); J. C. Maxwell, *A Treatise on Electricity and Magnetism*, vol. I (Clarendon Press, 1873).

Maxwell–Wagner relaxation – relaxation of a charge **polarization** characterized by a set of different relaxation times. It is typical for the relaxation in inhomogeneous dielectrics with regions of different conductivity, as it takes place in ceramics containing conducting grains of different size with insulating grain boundaries. The total relaxation current in such systems is interpreted as the sum of exponential decays according to **Debye relaxation** (with a single relaxation time), thus yielding a nonexponential time dependence of the current.

First described in: K. W. Wagner, *Erklärung der dielectrischen Nachwirkungsvorgänge auf Grund Maxwellscher Vorstellung*, Arch. Electrotech. **2**, 371–387 (1914); J. C. Maxwell, *A Treatise on Electricity and Magnetism*, vol. I (Clarendon Press, 1873).

McFarland and Tang solar cell – a **solar cell** in which the creation of electron–hole pairs does not occur in the semiconductor. The model is based on a multilayer mean free path

photovoltaic device, where photons are absorbed in a photoreceptive layer, in one case consisting of semiconductor **quantum dots** and in another case consisting of a photosensitive **dye** deposited on the surface of an ultrathin metal–semiconductor Schottky **diode**. The photoexcited electrons are transferred to the metal and travel ballistically to, and over, the **Schottky barrier**. The high ballistic mean free path for hot carriers in Au and the high dielectric constant of the TiO_2 semiconductor make the interface Au/TiO_2 a good candidate for the realization of such a photovoltaic device.

First described in: E. W. McFarland, J. Tang, *A photovoltaic device structure based on internal electron emission*, Nature **421**, 616–618 (2003).

mean free path – an average distance covered by a moving particle between two scattering events.

mechanic molecular machines are molecular assemblies that are chemically, photochemically, and/or electrochemically driven.

More detail in: V. Balzani, M. Venturi, A. Credi, *Molecular Devices and Machines: A Journey into the Nanoworld* (Wiley-VCH, Weinheim 2003).

mega- – a decimal prefix representing 10^6, abbreviated M.

Meissner–Ochsenfeld effect – when a superconducting specimen is placed in a not too large magnetic field and is then cooled through the transition temperature for superconductivity, the magnetic flux originally present is ejected from the specimen.

First described in: F. W. Meissner, R. Ochsenfeld, *Ein neuer Effekt bei Eintritt der Superaleitfähigkeit*, Naturwiss. **21**, 787–788 (1933).

MEMS – an acronym for a **m**icro-**e**lectro**m**echanical **s**ystem. It is a marriage of semiconductor processing to mechanical engineering at the micrometer scale. The principal components of MEMS are mechanical elements (for example a cantilever) and transducers. The transducers convert mechanical energy into electrical or optical signals and vice versa. Currently MEMS technology produces several objects, from digital projectors containing millions of electrically driven micromirrors to microscale motion sensors for use with airbags.

mesodesmic structure – a structure of an ionic crystal in which one of the cation-anion bonds is much stronger than the rest → **anisodesmic** and **isodesmic** structures.

mesoscopic – the term being used to indicate solid-state structures and devices that are large compared to the atomic scale but small compared to the macroscopic scale upon which normal **Boltzmann transport** theory has come to be applied.

260

Dimensions of such structures and devices are comparable to an elastic scattering length of electrons. In this regime, an electron can travel a great distance before suffering a scattering event that randomizes the phase of its wave function.

metal is distinguished from other substances by its large thermal and electrical conductivity, optical reflecting power, and opacity.

metallocene – an **organometallic compound** consisting of parallel cyclopentadienyl rings bound to a metal, forming **cyclopentadienyl complex**es. The ring molecule donates or accepts a pair of electrons to form a coordinate covalent bond with the central metal atom of coordination complexes and organometallic compounds.

There can be numerous organometallic coordination compounds distinguished by the number of cyclopentadienyl rings (with or without substituents) bonded to a central transition metal atom, the types of **transition metals** and the type of the bridge. Metallocenes possess transition metal atoms whose bonding involves an overlap of ns, (n−1) d, and np orbitals of the metal with molecular orbitals of appropriate symmetry of each aromatic ring. Metals known to form metallocene complexes are Ti, Zr, Hf, V, Cr, Mo, W, Mn, Fe, Ru, Os, Co, Rh and Ni.

Metallocenes having a catalytic site in a systematic molecular structure are used as polymerization catalysts to produce uniform polymers with unique structures and physical properties. Moreover, metallocenes are of practical interest in many electrochemical techniques, high-temperature chemistry, photolysis chemistry, structural chemistry, organic light-emitting devices, biochemistry and pharmaceutical manufacturing.

More details in: N. J. Long, *Metallocenes: An Introduction to Sandwich Complexes* (Blackwell Science, 1998).

metalloid – an element in which properties of both metallic and nonmetallic behavior can be discerned. Examples: Sb and Se. Chemical compounds of metalloids may also have intermediate properties.

metal–organic chemical vapor deposition (MOCVD) – a **chemical vapor deposition** techniques using **metal–organic compounds** as precursors for fabrication of high-quality semiconductor superlattices with abrupt interfaces and layer thicknesses down to the monolayer range. The term "metal–organic" denotes the broad group of compounds that contain metal–carbon bonds (known as organometallic compounds), those with metal–oxygen carbon bonds (the alkoxides) and the coordination compounds of metals and organic molecules. In practice, metal alkyls, preferentially with methyl (CH_3) and ethyl (C_2H_5) radical groups, are mainly used. Appropriate compounds are mentioned in the metal–organic compounds section. Most of them are liquid around room temperature. They are introduced into a reaction chamber by a carrier gas saturated with their vapor by passing through a bubbler. The deposition can be performed at atmospheric or lower pressure. The apparatus for MOCVD is sketched in Figure M.3. The classical

metal–organic chemical vapor deposition (MOCVD)

Figure M.3 Simplified schematic diagram of an MOCVD machine.

design appropriate for growth of GaAs and GaAlAs heterostructures is shown. Trimethyl gallium ($(CH_3)_3Ga$) and trimethyl aluminum ($(CH_3)_3Al$) are introduced into reactor chamber by bubbling supply metals for the compound formation. Group V components are often introduced in the form of hydrides, like AsH_3 in the example shown. Other widely used hydrides are discussed in the **hydride** section. Hydrogen is usually employed as a carrier gas in order to protect the pyrolysis of alkyl and hydride molecules by involving their common hydrogen ambient. The integrated scheme of the chemical reactions occurring at the heated surface of the substrate is as follows:

$$(CH_3)_3Ga + AsH_3 \xrightarrow{650\,^\circ C} GaAs \downarrow +3CH_4$$

Acceptor impurities, such as Zn or Cd, can be introduced into the reaction chamber within alkyls, and donor impurities such as Si, S or Se can be introduced as hydrides. For deposition of group III nitrides (AlN, GaN, InN) ammonia (NH_3) is used as a source of nitrogen.

In order to form abrupt interfaces, for changes in either the solid composition or the doping level, the composition of the gas ambient is changed by closing the lines supplying certain gaseous reactants. The volumes of the mixing system and reactor must be reasonably minimized in order to change the composition of the gas, and the resulting material grows rapidly and gives sharp interfaces. The mixture of gases must be changed without affecting the total flow through the reaction chamber. Improved sharp interfaces are also formed when rapid thermal processing regimes are employed. Heating cycles as short as 30–60 s provide appropriate conditions for material deposition, but eliminate atomic interdiffusion mixing at interfaces.

MOCVD is capable of depositing highly stoichiometric materials of virtually all $A^{III}B^V$ semiconductors and the ternaries and quaternaries based on them. It is also successfully used for the growth of other semiconductors, like $A^{II}B^{VI}$, and oxides. Continuous epitaxial films, as well as **quantum wire** and **quantum dot** structures, are fabricated by this technique via mechanisms presented in **self-organization (during epitaxial growth)**. There is a possibility of growing on a large number of wafers simultaneously, which meets the requirements of commercial mass production. Restrictions associated with MOCVD include contamination by carbon and serious practical issue of safety related to handling of the highly toxic and explosive hydrides.

First described in: H. M. Manasevit, *Single-crystal gallium arsenide on insulating substrates*, Appl. Phys. Lett. **12**(4), 156–159 (1968).

metal–organic compounds (for chemical vapor deposition) – compounds consisting of metals attached to organic molecules, that is having at least one primary metal carbon bond. Those suitable for chemical vapor deposition are listed in the Table M.2.

First described by R. Bunsen in 1839.

metal–organic molecular beam epitaxy (MOMBE) → molecular beam epitaxy.

metamaterial – an artificially structured material with electromagnetic properties governed by elements that are patterned on a macroscopic scale and that take the place of atoms or molecules in naturally occurring materials. It can therefore be engineered so that it has a much wider range of electromagnetic responses, including a negative index of refraction → **negative-index material**. As an example of a metamaterial, copper rings and wires deposited on a circuit board substrate can be mentioned.

Metropolis algorithm aims to generate a trajectory in a phase space (let it be **R**), which samples from a chosen statistical ensemble, when multidimensional integral are numerically evaluated. In order to do the evaluation, it is necessary to sample complicated probability distributions in many-dimensional spaces. The normalizations of these distributions are unknown and are so complicated that they cannot be sampled directly. The Metropolis rejection algorithm allows an arbitrarily complex distribution with a probability density $P(\mathbf{R})$ to be sampled in a straightforward way without knowledge of its normalization. It generates the sequence of sampling points $\mathbf{R}_m$ by moving a single walker according to the following steps:

Step 1. Start the walker at a random position **R**.

Step 2. Make a trial move to a new position $\mathbf{R}'$ chosen from some probability density function $T(\mathbf{R}' \leftarrow \mathbf{R})$. After the trial move the probability that the walker initially at **R** is now in the volume element $d\mathbf{R}'$ is $d\mathbf{R}' \times T(\mathbf{R}' \leftarrow \mathbf{R})$.

Step 3. Accept the trial move to $\mathbf{R}'$ with probability

$$A(\mathbf{R}' \leftarrow \mathbf{R}) = \min\left(1, \frac{T(\mathbf{R}' \leftarrow \mathbf{R})P(\mathbf{R}')}{T(\mathbf{R}' \leftarrow \mathbf{R})P(\mathbf{R})}\right).$$

If the trial move is accepted, the point $\mathbf{R}'$ becomes the next point on the walk; if the trial move is rejected, the point **R** becomes the next point on the walk. If $P(\mathbf{R})$ is high, most trial moves away from **R** will be rejected and the point **R** may occur many times in the set of points making up the random walk.

Step 4. Return to step 2 and repeat the procedure.

Table M.2 Metal-organic compounds for chemical vapor deposition of semiconductors and metals.

Element	Compound formulae	Compound name	Melting point (°C)	Boiling point (°C)	Vapor pressure equation[a] $\log P = A - B/T$ A	B
Be	(C₂H₅)₂Be	diethylberyllium	solid at RT dissociate at 85°C		14.496	5102
Mg	(C₂H₅)₂Mg	diethylmagnesium	solid at RT dissociate at 175°C		9.121	2832
	(C₅H₄CH₃)₂Mg	ditolylmagnesium			0.2T at 20°C 1.8T at 70°C	
Zn	(CH₃)₂Zn	dimethylzinc	−29.2	44.0	7.802	1560
	(C₂H₅)₂Zn	diethylzinc	−30.0	117.6	8.280	2109
Cd	(CH₃)₂Cd	dimethylcadmium	−4.5	105.5	7.764	1850
Al	(CH₃)₃Al	trimethylaluminum	15.4	126.1	8.224	2135
	(C₂H₅)₃Al	triethylaluminum	−52.5	185.6	8.999[b]	2361
	iso-(C₄H₉)₃Al	tri-isobutylaluminum	4.3		7.121[c]	1710
	(CH₃)₂NH₂CH₂Al	trimethylaminealuminum				
Ga	(CH₃)₃Ga	trimethylgallium	−15.8	55.7	8.07	1703
	(C₂H₅)₃Ga	triethylgallium	−83.2	142.8	8.224	2222
In	(CH₃)₃In	trimethylindium	88.4	135.8	10.52	3014
	(C₂H₅)₃In	triethylindium	−32	144	8.930	2815
	(C₃H₇)₃In	tripropylindium	−51	178		
Si	(CH₃)₄Si	tetramethylsilicon	−99.1	26.6		
	(C₂H₅)₄Si	tetraethylsilicon		153.7		

					A	B
Ge	$(CH_3)_4Ge$	tetramethylgermanium	-88	43.6		
	$(C_2H_5)_4Ge$	tetraethylgermanium	-90	163.5		
Sn	$(CH_3)_2Sn$	dimethytin	liquid at RT		7.445	1620
	$(C_2H_5)_2Sn$	diethytin	dissociate at 150° C		6.445	1973
P	$(CH_3)_3P$	trimethylphosphine		37.8		
	$(C_2H_5)_3P$	triethylphosphine		127		
	$(t\text{-}C_4H_9)_3P$	tertiarybutylphosphine		215	7.586	1539
As	$(CH_3)_3As$	trimethylarsine		51.9	7.405	1480
	$(C_2H_5)_3As$	triethylarsine		140		
	$(CH_3)_4As_2$	tetramethyldiarsine		170		
	$(C_4H_9)_4As$	tetrabutylarsine			7.500	1562
Sb	$(CH_3)_3Sb$	trimethylantimony		80.6	7.707	1697
	$(C_2H_5)_3Sb$	triethylantimony		160		
Te	$(CH_3)_2Te$	dimethyltellurium	-10	82	7.970	1865
	$(C_2H_5)_2Te$	diethyltellurium		137.5		
Fe	$(C_5H_5)_2Fe$	ferrocene	solid at RT sublimate at 179° C			

[a] P in Torr, T in K. [b] $\log P = A - B/(T - 73.82)$. [c] $\log P = A - B/(T - 73.82)$.

micella

The initial points generated by this algorithm depend on the starting point and should be discarded. Eventually, however, the simulation settles down and sets of points snipped out of the random walk are distributed according to $P(\mathbf{R}')$.

First described in: N. Metropolis, A. W. Rosenbluth, M. N. Rosenbluth, A. H. Teller, E. Teller, *Equation of state calculations by fast computing machines*, J. Chem. Phys. **21**(6), 1087–1092 (1953).

More details in: D. Frenkel, B. Smit, *Understanding Molecular Simulation* (Academic Press, San Diego, 1996); W. M. C. Foulkes, L. Mitas, R. J. Needs, G. Rajagopal, *Quantum Monte Carlo simulations of solids*, Rev. Mod. Phys. **73**(1), 33–83 (2001).

micella = **micelle**.

micelle (micella) – an aggregate of **surfactant** molecules dispersed in a liquid **colloid**. A typical micelle in aqueous solution forms an aggregate with the **hydrophilic** "head" regions in contact with surrounding solvent, sequestering the **hydrophobic** tail regions in the micelle centre. This type of micelle is known as a normal phase micelle (oil-in-water micelle). Inverse micelles have the head groups at the centre with the tails extending out (water-in-oil micelle). Micelles are approximately spherical in shape. Other phases, including shapes such as ellipsoids, cylinders, and bilayers are also possible. The shape and size of a micelle is a function of the molecular geometry of its surfactant molecules and solution conditions such as surfactant concentration, temperature, pH, and ionic strength.

A **micelle** is formed when a variety of molecules including soaps and detergents are added to water. The molecule may be a fatty acid, a salt of a fatty acid (soap), a **phospholipid**, or other similar molecule. The process of forming micellae is known as **micellization**. It is part of the phase behavior of many **lipids** according to their polymorphism.

Micellar systems have the unique property of being able to solubilize both **hydrophobic** and **hydrophilic** compounds.

micellization → **micelle**.

Michelson interferometer produces interference fringes by splitting a beam of coherent light into two paths, bouncing the beams back and recombining them. It is schematically shown in Figure M.4.

The different paths may be of different lengths or be composed of different materials to create alternating interference fringes on a detector.

The interference is sensitive to differences in the phase of waves, which varies over distances less than one micrometer for visible light, corresponding to very short times (about 2 fs). Therefore it is a sensitive method used to detect and measure small deformations, displacements, etc.

See also www.phy.davidson.edu/StuHome/cabell_f/ diffractionfinal/pages/Mic helson.html.

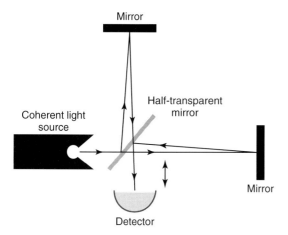

Figure M.4 A Michelson interferometer.

First described in: A. A. Michelson, *The relative motion of the earth and the luminiferous ether,* Am. J. Sci. (3)**22**, 120–29 (1881).

Recognition: in 1907 A. A. Michelson received the Nobel Prize in Physics for his optical precision instruments and the spectroscopic and metrological investigations carried out with their aid.

See also www.nobel.se/physics/laureates/1907/index.html.

micro- – a decimal prefix representing 10^{-6}, abbreviated μ.

microcavity (in a semiconductor light-emitting structure) – **quantum wells** sandwiched between two multilayer semiconductor mirrors that reflect light with an efficiency of over 99%. Any light entering the cavity containing the quantum wells is therefore trapped inside, provided that its wavelength is twice the thickness of the cavity. The light bounces back and forth between the mirrors, with a fraction less than 0.1% leaking through the mirrors at each bounce. The ability to trap photons and electrons in the cavity is used for engineering of **vertical-cavity surface-emitting lasers** and other optoelectronic devices.

micro-electromechanical system → MEMS.

Miller indices are used to define various planes and directions in a crystal. For a plane they are determined by first finding the intercepts ma, nb, pc of the plane with the three basis axes in terms of the lattice constant $(\mathbf{a}, \mathbf{b}, \mathbf{c})$ – see Figure M.5.

Then the reciprocals of these numbers are taken and the results are reduced to the smallest three integers h, k, l having the same ratio $(h : k : l = 1/m : 1/n : 1/p)$. The obtained integers are enclosed in parentheses (hkl) as the Miller indices for a single plane or set of parallel planes. If the plane intersects the axes at negative coordinates the minus sign is placed above the number to which it applies. The

Miller law

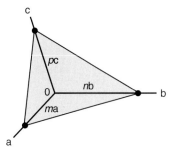

Figure M.5 The intercepts on the axes of a lattice plane with the Miller indices.

designation {*hkl*} is used for planes of equivalent symmetry (for example {100} for (100), (010), ($\bar{1}$00), (0$\bar{1}$0) and (00$\bar{1}$) in cubic symmetry). The most important low-index planes in a cubic crystal are shown in Figure M.6.

The direction is indicated as [*hkl*], with $h:k:l = \cos\alpha : \cos\beta : \cos\gamma$, where α, β, γ are the angles between this direction and the related axes in the lattice. The designation < *hjl* > is used for a full set of equivalent directions. In cubic crystals the [*hkl*] direction is normal to the (*hkl*) plane.

First described in: W. H. Miller, *A Treatise on Crystallography* (Deighton, Cambridge, 1839).

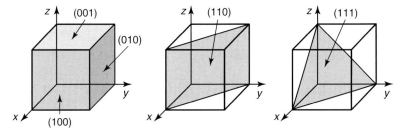

Figure M.6 Miller indices of the most important planes in a cubic crystal.

Miller law states that if the edges formed by the intersections of three faces of a crystal are taken as the three reference axes, then the three quantities formed by dividing the intercept of a fourth face with one of these axes by the intercept of a fifth face with the same axis are proportional to small whole numbers, rarely exceeding 6. It is also known as **law of rational intercepts**.

First described in: R.-J. Haüy, *Essai d'une Théorie sur la Structure des Cristaux,* (1784) and *Traité de Minéralogie* (1801), W. H. Miller, *A Treatise on Crystallography* (Deighton, Cambridge, 1839).

milli- – a decimal prefix representing 10^{-3}, abbreviated m.

Minkowski space – the four-dimensional space including time. In this space the unit of length may vary significantly with the direction in which the measurement occurs. It has the metric whose signature is $(+, -, -, -)$. It is equally often chosen to be $(-, +, +, +)$. If one denotes the coordinates of a point in Minkowski space as (t, x, y, z), then the distance l between this point and the origin is defined as $l^2 = t^2 - x^2 - y^2 - z^2$. The signs of the four terms correspond to the signature of the metric. Notice that the right side may be negative, making the distance imaginary.

It is used in the theory of special relativity.

First described in: H. Minkowski, *Die Grundgleichungen für die elektromagnetischen Vorgänge in bewegten Körpern*, Gött. Nachr. **2**, 53 (1908).

MND effect = Mahan effect.

mobility edge – a sharp boundary between states in which electrons are localized and can only move by thermally excited hopping to another localized state and the state that allows freedom of movement, albeit with a very short free path.

modality – distinct types of behavior that a system can exhibit under identical or nearly identical conditions.

modulation doped field effect transistor (MODFET) → high electron mobility transistor (HEMT).

modulation-doped structure – a semiconductor heterojunction based structure, in which the region of a heavily doped semiconductor producing charge carriers and the region where their low scattering transport is needed are separated in space. A heterostructure formed by semiconductors with different band gaps is used for this. This approach is illustrated in Figure M.7.

In the modulation-doped structure, donor impurity is introduced into the semiconductor with the larger band gap. The structure is neutral everywhere as long as the electrons sit on their donor atoms. When being released they diffuse away, cross the junction and fall into the nearest region with a lower potential energy. There they lose energy and become trapped because they cannot climb the barrier presented by ΔE_c. Thus, the electrons appear to be space separated from the positively charged donors which have produced them. The fallen electrons set up an electrostatic potential that tends to drive them back. As a result, a roughly triangular potential well for electrons is formed at the heterojunction. This well is of the order of some nanometers wide. The energy levels in it for motion along the x direction are quantized in a similar way to those in a square well. Often only the lowest level is occupied. All electrons then occupy the same state for motion in x, but remain free to move in the other two directions. A **two-dimensional electron gas** is formed.

The heterostructure fabricated as a superlattice composed of n-AlGaAs (greater band gap material) and undoped GaAs represents a classical example of modulation-doped structures. It demonstrates some order of magnitude increase in the electron

Figure M.7 Conduction band profiles around a heterojunction formed by semiconducting small gap material A and semiconducting large gap material B.

mobility. The observed low-temperature mobility of electrons as high as 2×10^7 cm^2 V^{-1} s^{-1} seems to be a record for GaAs. Meanwhile, the electron density achieved in the two-dimensional electron gas is about 5×10^{11} cm^{-2}.

The modulation doping achieves two benefits. First, it separates electrons from their donor atoms, thus reducing scattering of electrons by ionized impurities. Second, it confines electrons in the near junction region in two dimensions.

First described in: R. Dingle, H. L. Störmer, A. C. Gossard, W. Weigmann, *Electron mobilities in modulation doped semiconductor heterojunction superlattices*, Appl. Phys. Lett. **33**(7), 665–667 (1978).

MOKE – an acronym for **magnetooptical Kerr effect.**

molecular antenna – an array of molecular components capable of absorbing light and delivering the resulting electronic energy to a predetermined component of the array.

molecular battery – a molecule able to accumulate charge.

molecular beam epitaxy (MBE) – a technique of material deposition in ultrahigh vacuum. In principle, it has appeared as an advanced variant of **chemical vapor deposition** providing better purity and crystalline perfection, as well as improved composition and thickness control of the epitaxial heterostructures formed. It is performed in a chamber evacuated to ultrahigh vacuum, typically of the order of 10^{-11} Torr or better. At such low pressure, atoms and molecules injected into the chamber travel in straight lines without collisions until they impinge on the substrate or the chamber walls. A sketch of the apparatus for MBE is shown in Figure M.8.

Figure M.8 Simplified schematic diagram of an MBE machine.

Molecular beams are usually created by **effusion cells** mounted in the machine. An MBE machine is typically equipped with several effusion cells containing the elements necessary for forming the semiconductor and for doping. Their orifices are directed to the heated substrate. Growth commences when the shutters shadowing the cells are opened. The composition of the film can be varied by controlling the flux of each component, which is operated by the temperature inside the cell. The temperature of the substrate is important for composition and quality of the deposited layers. Growth-related defects will have no time to be annealed at low temperatures, while if the temperature is too high incongruent component evaporation from the surface can modify its atomic composition and unwanted interdiffusion will blur the interfaces. The substrate may be rotated during deposition in order to have better composition and thickness uniformity over the deposited layers.

An important feature of MBE is that it is performed in ultrahigh vacuum, which allows many electron beam and ion beam diagnostic techniques to be used for monitoring of the film growth process and for *in situ* analysis of the grown films. Some of the often employed electron beam probing techniques are Auger electron spectrometry (**AES**), low-energy electron diffraction (**LEED**), reflection high-energy electron diffraction (**RHEED**), and X-ray and ultraviolet photoemission spectroscopy (**XPS** and **UPS**). The ion-based probe is usually secondary ion mass spectrometry (**SIMS**).

The one most commonly used for monitoring the growth process in MBE machines is RHEED. It is performed with the electron beam directed at nearly grazing incidence to the surface of the sample. The pattern of diffracting electrons is viewed on a fluorescent screen. As the surface changes in a periodic way as a new monolayer is grown, it can be observed in the intensity and pattern of the RHEED signal. The diffraction pattern reveals the structure of the surface.

A wide range of methods has been developed from the basic MBE. The examples are **chemical beam epitaxy (CBE)** and **metal–organic molecular beam epitaxy (MOMBE)**, where the solid sources of conventional MBE are replaced by similar precursors to those used in CVD and MOCVD processes. All are aimed at improving control of the thickness and composition of the layers to improve the performance of the resulting devices.

The MBE and related techniques are usually used for fabrication of high-quality **superlattices** for **quantum wells** on the basis of $A^{III}B^V$ and Si-Ge heterostructures. Their advantages include highly abrupt junctions between different materials and precise control of the particular layer thickness. Moreover, periodically arranged self-organized **quantum wires** and **quantum dots** of these materials can also be fabricated under the conditions presented in **self-organization (during epitaxial growth)**. The main drawback of the MBE techniques in commercial applications is their slow throughput and high cost.

First described in: B. A. Unvala, *Epitaxial growth of silicon by vacuum evaporation*, Nature **194**, 966 (1962) – first attempt to grow epitaxial silicon in a vacuum with a techniques similar to MBE; J. R. Arthur Jr., *Interaction of Ga and As$_2$ molecular beams with GaAs surfaces*, J. Appl. Phys. **39**(8), 4032–4034 (1968) – demonstration of the technique that is now called molecular beam epitaxy; in the late 1960s A. Cho and J. R. Arthur Jr. at Bell Laboratories developed the MBE technique to grow multilayer heterojunction structures with atomically abrupt interfaces and precisely controlled compositional and doping profiles over distances as short as few tens of angstroms.

molecular crystal – a solid formed by inactive atoms, such as the inert gases, and saturated with molecules, such as hydrogen, methane and benzene. They are usually bound together by weak electrostatic forces, have low melting and boiling points and generally evaporate in the form of stable molecules.

molecular device – an assembly of a discrete number of molecular components (a **supramolecule**) designed to perform a specific function.

More details in: V. Balzani, M. Venturi, A. Credi, *Molecular Devices and Machines: A Journey into the Nanoworld* (Wiley-VCH, Weinheim, 2003); A. Credi, *Artificial nanomachines based on interlocked molecules*, J. Phys. Condens. Matter **18**, S1779–S1795 (2006).

molecular dynamics (simulation) – a technique to compute the equilibrium and transport properties of a classical many-body system. "Classical" means that the motion of the constituent particles obeys the laws of classical mechanics.

Molecular dynamics simulation often realizes the search of the optimal atomic configuration by means of step-by-step motion of atoms to the sites providing zero balance of the static forces acting on each atom in the structure. This configuration has the lowest energy and thus it has to be stable. The motion of the i-th atom is described by Newton's equations:

$$\frac{d\mathbf{r}_i(t)}{dt} = \mathbf{v}_i(t),$$

$$\frac{d\mathbf{v}_i(t)}{dt} = \frac{1}{M} \mathbf{F}_i \left[\mathbf{r}_1(t), \ldots, \mathbf{r}_N(t); \mathbf{v}_i(t) \right],$$

where $\mathbf{r}_i(t)$ is the coordinate, $\mathbf{v}_i(t)$ the velocity, and M the mass of the i-th atom. The number of atoms in the structure is N. Given that the classical N-body problem lacks a general analytical solution, the only path open is numerical integration.

The force $\mathbf{F}_i$ acting on every atom is calculated using many-body potential for the atomic interaction in the system under consideration. If it is not known, pairwise additive interactions between the i-th atom and all other ones (j-ths) is assumed to represent the many-body case

$$\mathbf{F}_i = \sum_j \mathbf{F}_{ij}.$$

It depends on the positions of all atoms and the velocity of the i-th one. The force $\mathbf{F}_{ij}$, representing interaction between i-th and j-th atoms, as in the molecular dynamics approach, is determined by a many-body potential or more easily by the potential of pair atomic interaction ϕ_{ij}

$$\mathbf{F}_{ij} = \frac{dU_{ij}}{dr_{ij}} \frac{\mathbf{r}_{ij}}{r_{ij}},$$

where $\mathbf{r}_{ij} = \mathbf{r}_i - \mathbf{r}_j$. The pair potential U_{ij} can be given in the form of **Lennard–Jones, Born–Mayer, screened Coulomb potential**, or in another form adequate to the peculiarities of interaction between the atoms to be simulated.

As soon as all forces acting between the atoms have been described, Newton's equations of motion can be integrated. The solution is specified completely once the law of force between interacting particles is given. To start the simulation, we should assign initial positions and velocities to all particles in the system. The atom positions should be chosen compatible with the structure that we are aiming to simulate. A particularly simple choice is to start with the atoms at the sites of a regular undisturbed lattice. Random directions and a fixed magnitude based on temperature are assigned to the initial velocities. They are also adjusted to ensure the center of mass of the system to be at rest, so any overall flow is eliminated.

The choice of a particular algorithm for the numerical integration depends on the complexity of the structure to be simulated, available computer facilities and desired accuracy of the simulation. Meanwhile, the integration procedure supposes that after each time step new positions of atoms are computed and their coordinates are updated simultaneously. Newton's equations of motion are solved until the acting static forces no longer change with time, so one can conclude that the structure has equilibrated. Actual positions of atoms in the equilibrated structure are calculated.

More details in: D. C. Rapaport, *The Art of Molecular Dynamics Simulation* (Cambridge University Press, Cambridge, 1995).

molecular electronics – a field of research and development of information processing devices with single molecules. The basic idea of molecular electronics is to use individual molecules as wires, switches, rectifiers, and memories. Another conceptual idea that is advanced by molecular electronics is the switch from a **top-down approach**, where the devices are extracted from a single large-scale building block, to a **bottom-up approach** in which the whole system is composed of small basic building blocks with **recognition**, structuring, and **self-assembly**. The great advantage of molecular electronics is the intrinsic nanoscale size of the molecular building blocks that renders this technology potentially competitive with the conventional integrated circuit technology.

First described in: A. Aviram, M. A. Ratner, *Molecular rectifiers*, Chem. Phys. Lett. **29**(2), 277–283 (1974).

More details in: A. Aviram, M. A. Ratner (eds), *Molecular electronics science and technology*, Annals of the New York Academy of Sciences, vol. 852 (The New York Academy of Sciences, New York 1998), A. Aviram, M. A. Ratner, V. Mujica (eds), *Molecular electronics II*, Annals of the New York Academy of Sciences, vol. 960 (The New York Academy of Sciences, New York 2002); *Introducing Molecular Electronics*, edited by G. Cuniberti, G. Fagas, K. Richter (Springer Verlag, Berlin Heidelberg, 2005).

molecular engineering – control of the structure of matter at the molecular level or any means of manufacturing molecules.

molecular machine – a particular type of molecular device, where the relative positions of the component parts can change as a result of some external stimulus.

More details in: V. Balzani, M. Venturi, A. Credi, *Molecular Devices and Machines: A Journey into the Nanoworld* (Wiley-VCH, Weinheim, 2003); A. Credi, *Artificial nanomachines based on interlocked molecules*, J. Phys. Condens. Matter **18**, S1779–S1795 (2006).

molecular mechanics (simulation) – a technique to compute equilibrium atomic arrangements of solids. It supposes no permanent motion of atoms during searching for their equilibrium arrangement. A static relaxation procedure is used in this approach. Its methodology is based on the analysis of the total potential energy of the structure. This energy is calculated for different atomic arrangements. The atomic configuration characterized by the lowest energy is concluded to be the equilibrium configuration.

Initial positions of atoms and the potential of their pair interaction used for calculations are chosen using the same criteria as in molecular dynamics. Coordinates of atoms can be changed simultaneously or atom-by-atom. The accuracy of the prediction of the equilibrium atomic configuration depends to a great extent on the experience and intuition of the person performing the simulation and the number of arrangements calculated. Such an approach can be used not only within classical mechanics formalism but also when quantum mechanics calculations of total energy of the structure under consideration are performed. An application of the **Monte Carlo method** makes the results obtained statistically weighted.

molecular nanoelectronics → molecular electronics.

molecular nanomagnet – a molecular nano-object with tailored magnetic characteristics. The term **single molecule magnet** is also applicable to it if the magnet is formed by a single molecule. Synthetic chemistry has developed an extraordinary capability to produce new metal–organic compounds. A molecular nanomagnet is a **cluster**, typically with size ∼1 nm, where a finite number of magnetic centers (**transition-metal, rare-earth** ions, or even organic **radicals**) are strongly coupled by organic bridges that work as paths for the **exchange interaction**. In most cases, the intra-cluster magnetic interactions are stronger than those between clusters: the bearing of the molecule is therefore unrelated to the environment. This **bottom-up approach** allows us therefore to construct magnetic materials starting from the molecular level. The molecules are typically packed in ordered crystals of macroscopic size. One useful feature of single-molecule magnet behavior is that the properties observed in bulk crystals essentially reflect those of the single molecule. Measurement techniques on bulk sample thus have access to microscopic properties. Interest in these materials has increasingly grown owing to their possible use for high-density information storage, as well as the possibility that some member of this family of materials could provide the **qubit**s needed for **quantum computation**. Mn_{12}-acetate [$Mn_{12}O_{12}(CH_3COO)_{16}(H_2O_4)$], **antiferromagnetic** spin clusters are the most studied systems because of high spin, $S = 10$, that arises by the interplay of **ferromagnetic** and antiferromagnetic interaction among the manganese atoms of the cluster, and its high anisotropy barrier. Another interesting class of magnetic molecules is constituted by antiferromagnetic rings: **ferric wheel**s, for example Fe_6, Fe_{10}, Fe_{12}, Fe_{18} (with Fe^{3+} ($s = 5/2$)), **Cr-based molecular rings** (also called **chromic wheel**s) like Cr_8 and Cr_{10} (comprising Cr^{3+} ($s = 3/2$)) and **V-based molecular rings** like V_8 and V_{10} (V^{3+} ($s = 1$)) were synthesized. [$Cr_8F_8(O_2CC(CH_3)_3)_{16}$]$0.25C_6H_{14}$, referred to as Cr_8, (see Figure M.9a) represents a paradigmatic case for this category in which **carboxylate** and fluoride bridges between the eight Cr^{3+} ions give rise to a planar and regular octahedron. The common feature of these systems is the small number n of **paramagnetic** ions and the nearest-neighbor **Heisenberg exchange interaction**. The coupling between spin centers leads to a characteristic ground state total spin $S = \sum_{i=1}^{n} s_i = 0$, due to the exact compensation between the spin centers within the ring (see Figure M.9b).

molecular orbital

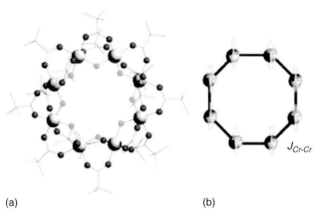

(a)   (b)

Figure M.9 (a) The structure of Cr_8 obtained by X-ray diffraction. The eight Cr^{3+} ions are bonded by two carboxylates and one fluoride bridges. H-atoms are omitted for clarity. (b) Schematic of the Heisenberg exchange interaction between the Cr^{3+} ions. Courtesy of A. Ghirri and M. Affronte.

More details in: O. Kahn, *Molecular Magnetism*, (Wiley VCH, New York 1993); D. Gatteschi, R. Sessoli, J. Villain, *Molecular Nanomagnets*, (Oxford University Press, 2006).

molecular orbital – an orbital formed by atomic orbitals when atoms are coupled in a molecule.

σ **orbital (σ state)** is a molecular orbital that has cylindrical symmetry around the internuclear axis. It resembles an s **atomic orbital** when viewed along the axis. The s orbital combinations have no angular momentum around the molecular axis.

π **orbital (π state)** is a molecular orbital which is perpendicular to the internuclear axis. The notation π is the analog of p for orbitals in atoms.

The atomic orbitals producing molecular orbitals may overlap constructively or destructively thus resulting in bonding or antibonding orbitals, respectively.

The molecular orbitals of homonuclear diatomic molecules can be labeled with a subscript g or u that specifies their **parity** indicating their behavior under inversion. The subscript g stands for gerade (German for "even") and u for ungerade (German for "odd") representing the fact that the wave function of the orbital can be even or odd when inverted through a point midway between the atoms. The bonding combination for σ orbitals is gerade. For π orbitals, the bonding combination is ungerade. A gerade π orbital is zero on the plane bisecting the bond. Examples of bonding combinations of atomic orbitals and resultant molecular bond designations are sketched in Figure M.10.

molecular orbital theory $\rightarrow$ **Hund-Mulliken theory.**

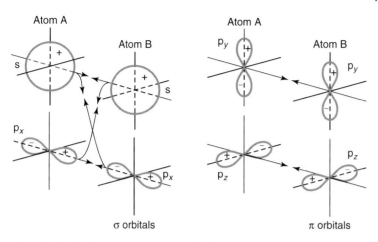

Figure M.10 Coupling of atomic orbitals in diatomic molecules.

molecular rectifier – a **diode** at molecular level→ **Aviram–Ratner model**.

First described in: A. Aviram, M. A. Ratner, *Molecular rectifiers*, Chem. Phys. Lett. **29**(2), 277–283 (1974).

More details in: A. Aviram, M. A. Ratner (eds), *Molecular electronics science and technology*, Annals of the New York Academy of Sciences, vol. 852 (The New York Academy of Sciences, New York 1998), A. Aviram, M. A. Ratner, V. Mujica (eds), *Molecular electronics II*, Annals of the New York Academy of Sciences, vol. 960 (The New York Academy of Sciences, New York 2002).

Möller scattering – scattering of electrons by electrons.

First described in: C. Möller, *Über den Stoss zweier Teilchen unter Berücksichtigung der Retardation der Kräfte*, Z. Phys. **70**(11–12), 786–795 (1931); C. Möller, *Zur Theorie des Durchgangs schneller Elektronen durch Materie*, Ann. Phys. **14**, 531 (1932).

momentum operator –

$$\mathbf{p} = \frac{\hbar}{i}\nabla = \frac{\hbar}{i}\left(\frac{\partial}{\partial x} + \frac{\partial}{\partial y} + \frac{\partial}{\partial z}\right).$$

monoclinic lattice – a **Bravais lattice** with a unit cell having a single two-fold symmetry axis or a single symmetry plane.

monochromatic – having a single color. The term is generally used to refer to electromagnetic radiation of a single wavelength.

monolayer (ML) – the thinnest layer containing a stoichiometrically complete set of atoms, that is one molecular thick.

monomer – one (mono) molecular unit.

Monte-Carlo method – a numerical simulation technique employing a random sample of the configurational state to obtain an approximate value of the average of a physical quantity. The **Metropolis algorithm** can be applied for sampling. The name "Monte Carlo" was chosen because of the extensive use of random numbers in the calculations.

The method is readily used for simulation of natural stochastic processes, like particle transport in matter, and for numerical calculations of many-dimensional integrals, sums, integral and matrix equations.

First described in: N. Metropolis, S. Ulam, *The Monte Carlo method*, J. Am. Stat. Ass. **44**(247), 335–341 (1949).

More details in: D. Frenkel, B. Smit, *Understanding Molecular Simulation* (Academic Press, San Diego, 1996).

Moore's law states that the maximum number of individual integrated circuit components which can be integrated within the most advanced single silicon chip is doubling every year (see Figure M.11).

First described in: G. E. Moore, *Cramming more components onto integrated circuits* Electronics **38**(8), 114-116 (1965).

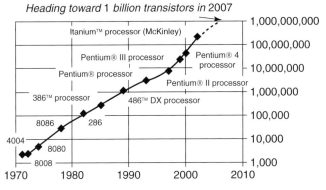

Figure M.11 Evolution of the number of transistors in a single CPU (central processing unit) versus the year. The graph is based on the Intel CPU.

Morse potential – an interatomic pair potential of the form: $V(r) = A \exp(-ar) + B \exp(br)$, where r is the interatomic distance and A, a, B, b are fitted parameters.

First described in: P. M. Morse, *Diatomic molecules according to the wave mechanics. II. Vibrational levels*, Phys. Rev. **34**(1), 57–64 (1929).

More details in: I. G. Kaplan, *Intermolecular Interactions: Physical Picture, Computational Methods and Model Potentials* (John Wiley & Sons, Chichester, 2006).

Morse rule – an empirical relationship between an equilibrium interatomic distance r in a diatomic molecule and its equilibrium vibrational frequency ω in the form $\omega r^3 = 3000 \pm 120 \text{ cm}^{-1}$.

MOS – an acronym for a **m**etal/**o**xide/**s**emiconductor structure or electron device.

MOSFET – an acronym for a **m**etal/**o**xide/**s**emiconductor **f**ield **e**ffect **t**ransistor.

Mössbauer effect – recoil-free resonance absorption and emission of gamma-rays by nuclei. It is illustrated in Figure M.12.

A gamma-ray resulting from a nuclear excitation following the beta-decay of a radioactive isotope parent is removed from a collimated beam by resonantly re-exciting another or the same species. The emitted radiation is then absorbed or scattered by the target nuclei. The transmitted fraction is recorded with the use of a detector.

When a radioactive isotope is placed in a material, it may emit gamma-radiation of relatively low energy without creating or annihilating any phonons. The gamma-ray carries the total energy of the transition and may be absorbed by the nuclei in the reverse process in a recoil-free fashion. Thus, resonance occurs between two nuclei. Every phase containing **Mössbauer isotopes** exhibits an energy absorption spectrum with characteristic hyperfine interaction parameters, which are fingerprints of the phase. It is used for phase analysis of matter. The Mössbauer isotopes are ^{57}Fe, ^{119}Sn, ^{121}Sb, ^{125}Te, ^{129}I(radioactive), ^{151}Eu, ^{161}Dy, ^{170}Yb, ^{197}Au, and ^{237}Np(radioactive).

First described in: R. L. Mössbauer, *Kernresonanzabsorption von Gammastrahlung in Ir^{191}*, Naturwiss. **45**(22), 538–539 (1958); R. L. Mössbauer, *Kernresonanzfluoreszenz von Gammastrahlung in Ir^{191}*, Z. Phys. **151**(2), 124–143 (1958).

Recognition: in 1961 R. L. Mössbauer received the Nobel Prize in Physics for his researches concerning the resonance absorption of gamma radiation and his discovery in this connection of the effect which bears his name.

See also www.nobel.se/physics/laureates/1961/index.html.

Mössbauer isotopes → Mössbauer effect.

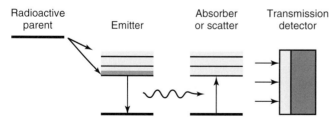

Radioactive parent   Emitter   Absorber or scatter   Transmission detector

Figure M.12 The Mössbauer effect.

Mott–Hubbard transition

Mott–Hubbard transition – a transition from metallic to insulator properties in solids due to a sufficiently strong electron-electron interaction that produces a gap for current-currying charge excitations in systems with an integral average number of electrons for a lattice site. Consider a model consisting of hydrogen atoms forming a simple cubic lattice with adjustable lattice parameter a_0. The many-body wave function is described, within a tight-binding scheme, in terms of atomic 1s orbitals centered on the nuclei R_i. The effects on the total energy as a consequence of electrons hopping onto neighboring sites can be estimated. If the electrons have more room, increasing the lattice constant, their kinetic energy will reduce. On the other hand the Coulomb repulsion energy of the electrons will increase by an energy U for each double occupancy. If a_0 exceeds a critical value a_c, the system will become an insulator. This idea, due to Mott → **Mott metal–insulator transition**, can also be understood within the one-band Hubbard model → **Hubbard Hamiltonian**. When there is one electron per lattice site on average and the interaction U is large compared to the bandwidth W, a charge gap of the order of $\Delta = U - W$ will be produced in the system. This metal–insulator transition is hence called the Mott–Hubbard transition.

First described in: N. F. Mott, *The basis of the electron theory of metals, with special reference to the transition metals*, Proc. Phys. Soc. A **62**, 416–422 (1949); J. Hubbard, *Electron correlations in narrow energy bands. I*, Proc. R. Soc. A **276**, 238–257 (1963); J. Hubbard, *Electron correlations in narrow energy bands. II*, Proc. R. Soc. A **277**, 237–259; J. Hubbard, *Electron correlations in narrow energy bands. III*, Proc. R. Soc. A **281**, 401–419 (1964); J. Hubbard, *Electron correlations in narrow energy bands. IV*, Proc. R. Soc. A **285**, 542–560 (1965).

Mott insulator – a material that in the band description should behave like a **metal**, but nonetheless is an **insulator** because of localization of electrons on individual atoms by the Coulomb interaction.

Mott law → **hopping conductivity**.

Mott metal–insulator transition – a solid loses its metallic character when the concentration of electrons in it is so large that the average electron separation $n^{-1/3}$ becomes significantly larger than four Bohr radii $n^{-1/3} \gg 4a_B$. Above a certain critical electron density the screening length becomes so small that electrons can no longer remain in a bound state. This produces metallic behavior. Below this critical electron concentration the potential well of the screened field extends far enough for a bound state to be possible. There is a critical value of the lattice constant (lattice of hydrogen atoms at absolute zero) that separates the metallic state (smaller values) from the insulating state (larger values) $a_c = 4.5a_B$. The metallic state appears when a conduction electron starts seeing a screened Coulomb interaction from each proton.

First described in: N. F. Mott, *On the transition to metallic conduction in semiconductors*, Can. J. Phys. **34**(12A), 1356–1368 (1956).

Recognition: in 1977 P. W. Anderson, N. F. Mott and J. H. van Vleck received the Nobel Prize in Physics for their fundamental theoretical investigations of the electronic structure of magnetic and disordered systems.

See also www.nobel.se/physics/laureates/1977/index.html.

Mott scattering – scattering of identical particles due to a **Coulomb force**. It is typical for transition metals whose d band is only partially occupied and, therefore, the **Fermi level** in these metals intersects not only the conduction but also the d bands. Moreover, since the atomic wave functions of d levels are more localized than those of the outer levels, they overlap much less, which means that the d band is narrow and the corresponding density of states is high. This opens up a new very effective channel for scattering of conduction electrons into the d band. This scattering mechanism explains why all transition metals are poor conductors compared with noble metals.

First described in: N. F. Mott, *Electrons in transition metals*, Adv. Phys. **13**, 325–422 (1964).

Mott–Wannier exciton = Wannier–Mott exciton.

mRNA – messenger RNA→ **RNA**.

muffin-tin approach – an approximation of atomic potentials in solids by the spherically symmetrical function of the radius r_{mt}, which is the muffin-tin radius, and by a constant in the region between spheres. This potential function is called the **muffin-tin potential**. The potentials of two neighboring atoms do not overlap.

First described in: J. C. Slater, *Wave functions in a periodical potential*, Phys. Rev. **51**(10), 846-851 (1937).

muffin-tin orbital – a piecewise function consisting of the solution of the **Schrödinger equation** for an isolated atom within a sphere of given radius and a zero kinetic energy tail function outside this region.

muffin-tin potential → **muffin-tin approach.**

Mulliken transfer integral – an energy involved in moving an electron from one site in a crystal to a nearest-neighbor site.

First described in: R. S. Mulliken, *Overlap integrals and chemical binding*, J. Am. Chem. Soc. **72**(10), 4493–4503 (1950).

multi-asperity contact model → **Greenwood-Williamson theory.**

multiferroic material – a material having both **ferromagnetic** and **ferroelectric** properties in the same phase. It has spontaneous magnetization which can be switched by an applied magnetic field, spontaneous polarization which can be switched by an applied electric field, and often some coupling between the two.

In general, the transition metal d electrons, which are essential for magnetism, reduce the tendency for off-center ferroelectric distortion. Consequently, an additional electronic or structural driving force must be present for ferromagnetism and ferroelectricity to occur simultaneously.

The examples of multiferroic materials include $Ni_3B_7O_{13}I$, $Pb_2(CoW)O_6$, Pb_2 $(FeTa)O_6$, $(1 - x)Pb(Fe_{2/3}W_{1/3})O_3 - xPb(Mg_{1/2}W_{1/2})O_3$.

First described in: G. A. Smolensky, A. I. Agranovskaya, V. A. Isupov, *New complex ferroelectrics of the $A_2^{2+}(B_I^3+B_{II}^{5+})O_6$ type I*, Fiz. Tverdogo Tela 1(1) 170–171 (1959) (in Russian); Sov. Phys. Sol. Stat. 1(1), 149 (1959) (in English) – predicted; E. Ascher, H. Rieder, H. Schmid, H. Stössel, *Some properties of ferromagnetoelectric nickel-iodine boracite, $Ni_3B_7O_{13}I$*, J. Appl. Phys. 37(3), 1404–1405 (1966) – experimentally observed.

More details in: R. Ramesh, N. A. Spaldin, *Multiferroics: a magnetic twist for ferro-electricity*, Nature Materials 6, 13–20 (2007); R. Ramesh, N. A. Spaldin, *Multiferroics: progress and prospects in thin films*, Nature Materials 6, 21–29 (2007).

multijunction solar cell a device where multiple **solar cell**s are stacked on each other in decreasing energy band gap. The first material absorbs the high energy photons, but transmits photons with lower energies, which then get absorbed in the cells below. The total voltage is the sum of the voltages but the current is determined by the worst of the two sub-cells.

multimodal nanoparticle – a nanoparticle that can be simultaneously detected by magnetic resonance imaging (that is **nuclear magnetic resonance**) and optical imaging (that is **fluorescence microscopy**). These nanoparticles integrate the advantages of high sensitivity, due to the optical imaging, with the potential of three-dimensional imaging, due to the magnetic resonance detection. Thus they can play an important role in the understanding of biological processes at molecular level and in the progress of diagnostic tools. Examples are **dye**-doped silica nanoparticles, gold nanoparticles, and **quantum dots**.

More details in: P. Sarma, S. Brown, G. Walter, S. Santra, B. Moudgil, *Nanoparticles for bioimaging*, Adv. Coll. Int. Sci. **123–126**, 471–485 (2006).

multiplet – a collection of closely spaced energy levels which appear from of a single energy level as a result of its splitting due to a relatively weak interaction.

multiquantum well → **quantum well**

Murrell–Mottram potential – the many-body expansion of the potential energy function truncated at the three-body level. For N interacting atoms it has the form

$$V = \sum_{i}^{N-1}\sum_{j>i}^{N} V_{ij}^{(2)} + \sum_{i}^{N-2}\sum_{j>i}^{N-1}\sum_{k>j} V_{ijk}^{(3)},$$

where $V_{ij}^{(2)}$ and $V_{ij}^{(3)}$ are the two- and three-body terms, respectively.

The two-body term being a Rydberg energy function is $V_{ij}^{(2)} = -D(1 + a\rho_{ij})exp$ $(-a\rho_{ij})$ with $\rho_{ij} = (r_{ij} - r_e)/r_e$, where D is the energy scaling parameter which determines the well-depth, a is the parameter determining the range of the interaction, r_{ij} is the interatomic distance, and r_e is the equilibrium pair separation. The three-body term represents a repulsive interaction by a polynomial in terms of the separations between the three atoms: r_{ij}, r_{ik}, r_{jk}. This term must be symmetrical with respect to the interchange of like atoms and therefore the polynomial is not a simple function of the separations but is an expansion in terms of symmetry coordinates Q_i defined as:

$$\begin{pmatrix} Q_1 \\ Q_2 \\ Q_3 \end{pmatrix} = \begin{pmatrix} \sqrt{1/3} & \sqrt{1/3} & \sqrt{1/3} \\ 0 & \sqrt{1/2} & -\sqrt{1/2} \\ \sqrt{2/3} & -\sqrt{1/6} & -\sqrt{1/6} \end{pmatrix} \begin{pmatrix} \rho_{ij} \\ \rho_{jk} \\ \rho_{ki} \end{pmatrix},$$

The three-body term then becomes $V_{ijk}^{(3)} = DP(Q_1, Q_2, Q_3)F(b, Q_1)$. Here $P(Q_1, Q_2, Q_3)$ is the polynomial in the symmetry coordinates and $F(b, Q_1)$ is the damping function (with a damping parameter b) which constrains the three-body term to approach zero as Q_1 becomes large.

All totally symmetric polynomials can be written as sums of products of the following three functions: $Q_1 Q_2^2 + Q_3^2 Q_3 - 3Q_3 Q_2^2$.

The polynomial term in $V_{ijk}^{(3)}$ can be cubic or quartic. The cubic one is given by $P(Q_1, Q_2, Q_3) = c_0 + c_1 Q_1 + c_2 Q_1^2 + c_3(Q_2^2 + Q_3^2) + c_4 Q_1^3 + c_5 Q_1(Q_2^2 + Q_3^2) + c_6$ $(Q_3^3 - 3Q_3 Q_2^2)$ and the quartic polynomial has the additional terms $c_7 Q_1^4$ $+ c_8 Q_1^2(Q_2^2 + Q_3^2) + c_9(Q_2^2 + Q_3^2)^2 + c_10(Q_3^3 - 3Q_3 Q_2^2)$.

The damping function can be any of the following $F(b, Q_1) = \exp(-bQ_1)$, $F(b, Q_1) = \exp(-bQ_1^2)$, $F(b, Q_1) = 0.5 - 0.5\tanh(0.5bQ_1)$, or $F(b, Q_1) = \text{sech}(bQ_1)$, which all decay exponentially at long range though they have different short range behavior.

As the potential is empirical the parameters are to be fitted to experimental data.

The potential is successfully applied to represent interactions in a wide range of atomic systems including metals from groups II and III of the Periodic Table, group IV semiconductors and transition metals.

First described in: J. N. Murrell, R. E. Mottram, *Potential energy functions for atomic solids*, Mol. Phys. **69**(3), 571–585 (1990).

More details in: G. A. Mansoori, *Principles of Nanotechnology: Molecular-based Study of Condensed Matter in Small Systems* (World Scientific, Singapore, 2005).

N

From NAA (neutron activation analysis) to Nyquist–Shannon sampling theorem

NAA – an acronym for **neutron activation analysis**.

Nabla-operator – the differential operator, which in Cartesian coordinates is $\nabla = (\partial/\partial x + \partial/\partial y + \partial/\partial z)$. Its applications to scalar $f(\mathbf{r})$ or vector $\mathbf{A}(\mathbf{r})$ fields are $\nabla f(\mathbf{r}) = \mathrm{grad} f$, $\nabla \cdot \mathbf{A}(\mathbf{r}) = \mathrm{div}\mathbf{A}$, $\nabla \times \mathbf{A}(\mathbf{r}) = \mathrm{curl}\mathbf{A}$.

NAND operator → **logic operator**.

nano- – a decimal prefix representing 10^{-9}, abbreviated n.

nanobiotechnology – a specialized field of **nanotechnology**, focusing on the improved and novel physical, chemical, and biological properties of biological objects at the nanoscale.
  More details in: Nanobiotechnology: Concepts, Applications and Perspectives, edited by C. M. Niemeyer, C. A. Mirkin (Wiley-VCH, Weinheim, 2004).

nanocomposite – a material containing several compact components with their spatial organization at the nanoscale or molecular level.

nanocrystallite – a crystal of nanometer (1 nm $= 10^{-9}$ m) dimensions.

nano-electromechanical system → **NEMS**.

nanoelectronics – the field of science and engineering dealing with fabrication, study and application of nanosize electronic devices for information technologies. Quantum effects generally determine operational principles of these devices.
  More details in: R. Companó (ed.) *Technology Roadmap for Nanoelectronics*; Office for Official Publications of the European Communities: Luxembourg 2000.

What is What in the Nanoworld. Victor E. Borisenko and Stefano Ossicini
Copyright © 2008 WILEY-VCH Verlag GmbH & Co. KGaA, Weinheim
ISBN: 978-3-527-40783-5

nanoethics

nanoethics – the field related to the social, ethical and broader human implications of nanotechnology.

More details in: J. Moor, J. Weckert, *Nanoethics: assessing the nanoscale from an ethical point of view* in *Discovering the Nanoscale*, edited by D. Baird, A. Nordmann, J. Schummer (IOS Press, Amsterdam 2004).

nano-flash memory device – a flash memory device similar in structure to a MOS field effect transistor → **MOSFET** except that it is a three-terminal device with two gate electrodes, one on top of the other. The top electrode forms the control gate, below which a floating gate is capacitively coupled to the control gate. The memory cell operation involves putting charge on the floating gate or removing it, corresponding to two logic levels. Nanoflash devices utilize single or multiple nanoparticles as the charge storage element.

More details in: B. G. Streetman, S. Banerjee, *Solid State Electronic Devices* (Prentice Hall, New Jersey, 2000); R. Companó (ed.) *Technology Roadmap for Nanoelectronics*; Office for Official Publications of the European Communities: Luxembourg 2000.

nanoimprinting → **imprinting**.

nanolithography – **lithography** at the nanometer scale.

nanomagnetism – a subfield of nanoscience dealing with magnetic structures having dimensions in the nanometer range. It is already central to data storage, sensor and device technologies but is increasingly being used in the life sciences and medicine.

More details in: D. L. Mills, J. A. C. Bland, *Nanomagnetism* (Elsevier, Amsterdam 2006); S. D. Bader, *Colloquium: Opportunities in nanomagnetism*, Rev. Mod. Phys. **78**(1), 1–15 (2006).

nanomedicine – includes the medical diagnosis, monitoring and treatment of the human body at the level of single molecules or molecular assemblies that provide structure, control, signaling, homeostasis, and motility at the nanoscale. More generally nanomedicine addresses the important challenges and advances in medical nanoscale-structured materials and devices, biotechnology devices and **molecular machine** systems and nanorobotics.

More details in: R. A. Freitas Jr, *What is nanomedicine?*, Nanomedicine: Nanotech. Biol. Med. **1**(1), 2–9 (2005).

nanoparticle-based solar cell – a **photovoltaic** device that relies on the properties of nanostructures → **Grätzel solar cell, intermediate band solar cell, McFarland and Tang solar cell, plasmonic effects based solar cell, up- and down-conversion solar cell**.

nanopeapod – a nanostructure containing an interior space filled by foreign atom(s) or molecule(s). Typically these are **carbon nanotubes** and **fullerenes** with dopants inside.

First described in: P. M. Ajayan, S. Iijima, *Capillarity-induced filling of carbon nanotubes*, Nature **361**, 333–334 (1993).

nanophotonics – the field of science and engineering dealing with fabrication, study and application of optical phenomena at the nanoscale. It deals with structures and processes which are spatially localized in domains smaller than the wavelength of visible radiation. The field includes nanoscale confinement of radiation → **NSOM** and **photonic crystals**, nanoscale confinement of matter, nanoscale physical or photochemical transformations.

nanopiezotronics utilizes the coupled **piezoelectric** and semiconducting properties of nanowires and nanobelts for designing and fabricating electronic devices and components, such as field-effect **transistors** and **diodes**.

More details in: Z. L. Wang, *Nanopiezotronics*, Adv. Matter. **19**, 889–892 (2007).

nanopolymers – a new class of materials that represents the nanoscale equivalent of **polymers**. They were obtained by the introduction of defects onto two opposing areas on the surface of spherical-shaped metallic nanoparticles. These divalent nanoparticles were then chained together making freestanding films.

First described in: G. A. DeVries, M. Brunnbauer, Y. Hu, A. M. Jackson, B. Long, B. T. Neltner, O. Uzun, B. H. Wunsch, F. Stellacci, *Divalent metal nanoparticles*, Science **315**, 358–361 (2007).

nanostructure – an ensemble of bonded atoms that have at least one dimension in the range of one to some hundreds of nanometers (1 nm = 10 Å = 10^{-9} m).

nanotechnology – a set of methods and techniques providing the fabrication of structures consisting of individual atoms, molecules or macromolecular blocks in the length scale of approximately 1–100 nm. It is applied to physical, chemical and biological systems in order to explore their novel and differentiating properties and functions arising at a critical length scale of matter typically under 100 nm.

nanotoxicology – the scientific field related to the potential for human and environmental exposure, hazard and risk associated with the use and development of nanostructured materials.

More details in: V. Stone, K. Donaldson, *Nanotoxicology: sign of stress*, Nature Nanotechnology **1**(1), 23–24 (2006).

nanotribology represents the study of friction, wear and lubrication at atomic length and time scales.

nanotube radio

More details in: J. Krim, *Friction at the nanoscale*, Physics World **18**, 31–34 (2005).

nanotube radio – a fully functional, fully integrated radio receiver, orders-of-magnitude smaller than any previous radio, from a single **carbon nanotube**. The single nanotube serves, simultaneously, as all major components of a radio: antenna, tuner, amplifier, and demodulator. Moreover, the antenna and tuner are implemented in a radically different manner than traditional radios, receiving signals via high frequency mechanical vibrations of the nanotube rather than through traditional electrical means.

First described in: K. Jensen, J. Weldon, H. Garcia, A. Zettl, *Nanotube radio*, Nano Lett. **7**(11), 3508–3511 (2007).

nascent state – the condition of a chemical element at the instant of its liberation as a result of a chemical reaction. In some cases, the element will remain in the atomic state for an appreciable time before assuming its normal molecular condition. During this interval further chemical reactions are facilitated.

natural motors – play the role of a motor inside the cells. There are linear motors like dynein, kinesin and myosin, and rotory motors, like ATP syntase.

More details in: V. Balzani, M. Venturi, A. Credi, *Molecular Devices and Machines: A Journey into the Nanoworld* (Wiley-VCH, Weinheim, 2003); E. L. Wolf, *Nanophysics and Nanotechnology – Second Edition* (Wiley VCH, Weinheim, 2006).

near-field region – an area closest to an aperture or source of electromagnetic waves where the diffraction pattern differs substantially from that observed at an infinite distance.

near-field scanning optical microscopy (NSOM) – a type of scanning probe microscopy that maps the near-field electrical and optical properties of semiconductor structures on a nanometer scale. NSOM (also often termed **SNOM – scanning near-field microscopy**) is a technique in which a tapered optical fiber is held atop a sample within a fraction of wavelength and scanned across its surface, thus circumventing the far-field wave physics and its limitations → **Abbe's principle**. It is illustrated in Figure N.1.

At each point in the scan, the optical radiation emitted by the sample is collected from the near field through the small aperture created by the tip of the tapered fiber. It is an aperture-type NSOM. The spatial resolution of the scan is determined by the tip–sample distance and by the tip size. With the most widely used aperture probe, resolutions well below 100 nm have been demonstrated.

In addition, different apertureless techniques were also developed, some of which resulted in resolutions down to a few nanometers. Scattering type NSOM is one of them. It uses a sharpened homogeneous metal tip as a probe. When the tip is illuminated with the polarization parallel to the tip axis, the electric field at the

(a)   Farfield imaging   (b) Nearfield imaging

Figure N.1 Comparison of far-field and near-field imaging of a sample. (a) Far-field imaging. Light is diffracted so that the area illuminated is larger than the aperture. Sub-λ details of the image are lost. (b) Near-field imaging. The area illuminated corresponds well with the aperture of the optics. From http://www.thermomicro.com/products/aurora.html.

apex is induced and enhanced owing to the large surface charge accumulation at the apex. Spatial resolution defined by the apex diameter is expected. The field enhancement effect can also contribute to the detection of extremely weak signals, such as Raman scattering from single nanoobjects. However, because the illumination area is limited by diffraction generating unwanted background, elimination of signal from the strong background is critical for imaging with sufficient contrast.

Photon-emission **scanning tunneling microscopy (STM)** and light-illumination STM can be categorized into NSOM from the viewpoint of essential contributions of light to them. They allow the atomic resolution associated with from STM operation. In photon-emission STM, one detects photons created by inelastic scattering of electrons in the tunneling gap applying a bias voltage. Photon-emission spectra provide electronic state information of a nanoscale region, such as a local density of states. Light-illuminating STM measures the tunneling current of photoexcited carriers. Precise spectroscopy of local electronic states can be carried out by means of highly controlled light. By using polarized photons one can couple to a specific spin state. Ultrafast optical pulses enables one to observe electron dynamics with a high spatiotemporal resolution.

First described in: E. H. Singe, *A suggested method for extending microscopic resolution into the ultra microscopic region*, Phil. Mag. J. Sci. **6**, 356–362 (1928) – the suggestion; E. A. Ash, G. Nicholls, *Super-resolution aperture scanning microscope*, Nature **237**, 510–512 (1972) – demonstration for micro-wave radiation; D. W. Pohl, W. Fischer, M. Lanz, *Optical stethoscopy: image recording with resolution λ/20*, Appl. Phys. Lett. **44**(7), 651–653 (1984) and E. Betzig, A. Lewis, A. Q. Harootunian, M. Isaacson, E. Kratschmer, *Near-field scanning optical microscopy (NSOM)*, Biophys. J. **49**, 269–279 (1986) – practical development for the optical range.

More details in: R. C. Dunn, *Near-field scanning optical microscopy*, Chem. Rev. **99**(10), 2891–2927 (1999); *Roadmap of Scanning Probe Microscopy*, edited by S. Morita (Springer Verlag, Berlin Heidelberg, 2007).

Néel point (temperature)

Néel point (temperature) – the temperature above which particular **antiferromagnetic** properties of a substance disappear.

First described in: L. Néel, *Influence of the fluctuations of the molecular field on the magnetic properties of bodies*, Ann. Phys. (Paris) **18**, 1–105 (1932); *Propriétés magnétiques de l'état métallique et énergie d'interaction entre atomes magnétiques*, Ann. Phys. (Paris) **5**, 232–279 (1936).

Recognition: in 1970 L. Néel received the Nobel Prize in Physics for fundamental work and discoveries concerning antiferromagnetism and ferrimagnetism which have led to important applications in solid state physics.

See also www.nobel.se/physics/laureates/1970/index.html.

Néel wall – a boundary between two magnetic domains in a thin film in which the magnetization vector remains parallel to the faces of the film in passing through the wall.

First described in: L. Néel, *Energies des parois de Bloch dans les couches minces*, Compt. Rend. **241**, 533–536 (1955).

negative differential resistance – a property of some electric circuit components and networks which enables them to behave like a power source and supply energy to systems connected their terminals. It is typically a dynamic effect represented by current voltage characteristics shown in Figure N.2.

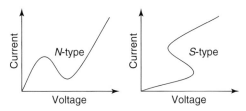

Figure N.2 Current voltage characteristics representing two types of negative differential resistance of an electronic component or circuit.

negative index material – a material with a negative index of **refraction**. It is also called left-handed material. Contrary to the usual "positive" refraction when a wave emerges on the opposite side of the line that runs perpendicular to the refracting interface (see Figure N.3), a wave coming to the interface with a negative index material emerges on the same side of the surface normal. The material composed of copper rings and wires deposited on a circuit board substrate can be an example of materials with a **negative refractive index**.

A slab of negative index material can be used to focus the rays coming from a nearby source. This is illustrated in Figure N.4.

If the source is placed on one side of a slab with a negative refractive index let it be $(n = -1)$, the waves are refracted in such a way as to produce a focus inside the slab and then another just outside the slab. Focusing should be much

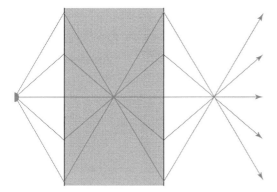

Figure N.3 Refraction of waves coming to the interface between a vacuum and a positive refractive index medium (a) and between a vacuum and a negative refractive index medium (b).

Figure N.4 Focusing of waves passing through a slab of a negative-index material.

better than positive index optics restricted by diffraction. It could lead to a perfect lens.

*First described in:*V. G. Veselago, *The electrodynamics of substances with simultaneously negative values of e and* μ, Uspekhi Fizicheskih Nauk **92**(3), 517–526 (1967) – in Russian, Sov. Phys. Uspekhi **10**(4), 509–514 (1968) – in English, theoretical prediction; R. A. Shelby, D. R. Smith, S. Schultz, *Experimental verification of a negative index of refraction*, Science **292**, 77–79 (2001) – experimental observation.

nematic phase – a form of **liquid crystal** with an appearance of moving, threadlike structures, particularly visible when observed in thick specimens with polarized light. During this phase, the molecules of the crystal are parallel and able to travel past each other following the direction of their longitudinal axes. This form has one optical axis that occupies the direction of an applied magnetic field. The transition from the isotropic liquid state to the nematic phase is a first-order phase transition.

NEMS – an acronym for a nano-electromechanical system. Like **MEMS** (microelectromechanical system), it is a marriage of semiconductor processing to mechanical engineering, but at the nanometer scale. NEMS open the way to very-high-frequency mechanical devices, ultra-low power signal processing, and high sensitivity to applied forces.

More details in: K. L. Ekinci, M. L. Roukes, *Nanoelectromechanical systems*, Rev. Sci. Inst. **76**, 061–101 (2005).

Nernst effect – development of a transverse electric field in an electric conductor with zero current passing through it when there is a temperature gradient in the conductor and a transverse magnetic field is applied to it. This field is defined as $E = Q \nabla T \times \mathbf{H}$, where Q is the Nernst coefficient. It is also known as the **Ettingshausen–Nernst effect**.

First described in: A. von Ettingshausen, W. Nernst, *Über das Auftreten electromotorischer Kräfte in Metallplatten, welche von einem Wärmestrome durchflossen werden und sich im magnetischen Felde befinden*, Ann. Phys. **265**(10), 343 (1886).

Nernst–Thomson rule states that in a solvent having a high dielectric constant the attraction between anions and cations is small so that dissociation is favored, while the reverse is true in solvents with a low dielectric constant.

First described in: W. Nernst, *Dielektrizitätskonstante und chemisches Gleichgewicht*, Z. Phys. Chem. **13**, 531 (1894); J. J. Thomson, *On the effect of electrification and chemical action on a steam-jet and of water-vapor on the discharge of electricity through gases*, Phil. Mag. **36**, 320 (1893).

Neumann–Kopp's rule states that the heat capacity of 1 mole of a solid substance is approximately equal to the sum over the elements forming the substance of the heat capacity of a gram atom of the element times the number of atoms of the element in a molecule of the substance.

First described in: F. E. Neumann, *Untersuchung über die specifische Wärme der Mineralien*, Ann. Phys. **23**, 1–60 (1831); H. Kopp, *Geschichte der Chemie, Dritter Theil*, (1845) p. 138.

Neumann principle states that the symmetry elements of the point group of a crystal are included among the symmetry elements of any property of the crystal.

First described in: F. E. Neumann, *Vorlesungen über die Theorie der Elastizität der festen Körper und des Lichtäthers* (Teubner-Verlag, Leipzig, 1885).

neural network – a computing paradigm that attempts to process information in a manner similar to that of the brain. It differs from artificial intelligence in that it relies not on preprogramming but on the acquisition and evolution of interconnections between nodes.

neutron – a subatomic particle with no net electric charge, spin $\hbar/2$ and a mass of $1.6749543 \times 10^{-27}$ kg (slightly greater that of a **proton**). A neutron is classified as a **baryon** and consists of two down quarks and one up quark.

First described in: J. Chadwick, *Possible existence of a neutron*, Nature **129**, 312 (1932).

Recognition: in 1935 J. Chadwick received the Nobel Prize in Physics for his discovery of the neutron.

See also www.nobel.se/physics/laureates/1932/index.html.

neutron activation analysis (NAA) is based on bombarding a sample with neutrons, resulting in the production of a radioactive isotope of the element of interest. Thermal neutrons (energy about 0.025 eV) are usually used for this. Gamma-rays emitted by the radioactive isotope are then analyzed. The energy of the radiation is characteristic of the radioactive isotope. It is used for the identification of chemical elements, that is for qualitative analysis. The number of gamma rays emitted correlates to the number of particular atoms in the sample, which is appropriate for quantitative analysis.

Newton's laws of motion – (1) a particle not subjected to external forces remains at rest or moves with constant speed in a straight line; (2) the acceleration of a particle is directly proportional to the resultant external force acting on the particle and is inversely proportional to the mass of the particle; (3) if two particles interact, the force exerted by the first particle on the second particle (called the action force) is equal in magnitude and opposite in direction to the force exerted by the second particle on the first particle (called the reaction force). They form the basis of classical, or Newtonian, mechanics, and have proved valid for all mechanical problems not involving speeds comparable with the speed of light.

First described in: I. Newton, *Philosophiae Naturalis Principia Mathematica* (The Mathematical Principles of Natural Philosophy), 1687.

Newton–Raphson iteration – a technique to find out the solution satisfying the equation $f(E) = 0$. In this technique, if E_n is a first guess at the solution, then a better estimate is $E_{n+1} = E_n - f(E_n)/f'(E_n)$. The new estimate E_{n+2} is then used to generate a second approximation E_{n+3}, and so on, until the successive estimates converge to a required accuracy.

First described in: I. Newton, *Methodus fluxionum et serierum infinitarum* (1664–1671); J. Raphson, *Analysis aequationum universalis* (London, 1690).

Nielsen theorem states that a state $|\psi\rangle$ transforms to $|\phi\rangle$ using local operations and classical communication if and only if λ_ψ is majorized by $\lambda_\phi \rightarrow$ **majorization**. More succinctly, $|\psi\rangle \rightarrow |\phi\rangle$ if $\lambda_\psi \prec \lambda_\phi$, where λ_ψ denotes the vector of eigenvalues of $\phi_\psi \equiv \mathrm{tr}(|\psi\rangle\langle\psi|)$.

First described in: M. F. Nielsen, *Conditions for a class of entanglement transformations*, Phys. Rev. Lett. **83**(2), 436–439 (1999).

nitrate – a salt of nitric acid (HNO_3). For inorganic salts the general formula of nitrates can be expressed as $Me(NO_3)_n$, where Me is a metal or NH_4^+ group and n represents the valency of the metal ($n = 1$ for Me $= NH_4^+$).

nitride – a compound formed by combination of an element with nitrogen.

nitrite – a salt of nitrous acid (HNO_2).

NMR – an acronym for **nuclear magnetic resonance.**

noble gases = **inert gases.**

noble metals – metals with resistance to attack by general chemical reagents: Hg, Ag, Au, Pt, Pd, Ru, Rh, Ir, Os.

Noether theorem states that every time a physical system is left invariant under a continuous transformation, a conserved quantity can be derived.
First described in: E. Noether, *Invariante Variationsprobleme*, Gött. Nachr., Math. Phys. Klasse, 235–252 (1918).

noiseless coding theorem → **Shannon's theorems.**

noisy channel coding theorem → **Shannon's theorems.**

noncloning theorem states that the linearity of quantum mechanics forbids exact replication of quantum states. This means that quantum information cannot be duplicated or cloned (accurately copied).
First described in: W. K. Wootters, W. H. Zurek, *A single quantum cannot be cloned*, Nature **299**, 802–803 (1982).

noncrossing rule states that potential energy curves of diatomic molecules which are of the same electronic species never cross.
First described in: J. von Neumann, E. Wigner, *Über merkwürdige diskrete Eigenwerte*, Phys. Z. **30**, 465–467 (1929).

nonlinear Hanle effect → **Hanle effect.**

NOR operator → **logic operator.**

Nordheim's rule states that the residual resistivity of a binary alloy containing mole fraction x of one element and $1 - x$ of another is proportional to $x(1 - x)$.
First described in: L. Nordheim, *Zur Elektronentheorie der Metalle. II*, Ann. Phys. **9**, 641–678 (1931).

normal distribution = **Gaussian distribution.**

NOT operator → **logic operator.**

Notre Dame architecture → **quantum cellular automata**.

NRA – an acronym for **nuclear reaction analysis**.

NSOM – an acronym for **near-field scanning optical microscopy**.

nuclear magnetic resonance (NMR) – some nuclei placed in a static magnetic field resonantly absorb energy from a radio frequency field at certain characteristic frequencies. The magnetic field forces magnetic moments, *i.e.* spins, of all the nuclei in the sample to line up and precess around the direction of the field. The spins all precess at the same frequency but with random phases. The radio frequency radiation directed at the sample disturbs the alignment of the spins and produces a state in which the phases are coherent. Thus, the frequency of the resonance coincides with the **Larmor frequency** of precession of the nuclei in the magnetic field and is proportional to the strength of the field.

Nuclear magnetic resonance spectroscopy is used to study molecular structures of matter.

First described in: E. M. Purcell, H. C. Torrey, R. V. Pound, *Resonance absorption by nuclear magnetic moments in a solid*, Phys. Rev. **69**(1), 37–38 (1946); F. Bloch, *Nuclear induction*, Phys. Rev. **70**(7/8), 460–474 (1946).

More details in: C. P. Slichter, *Principles of Magnetic Resonance* (Springer Verlag, Berlin, 1990).

Recognition: in 1952 F. Bloch and E. M. Purcell received the Nobel Prize in Physics for their development of new methods for nuclear magnetic precision measurements and discoveries in connection therewith; in 2003 P. C. Lauterbur and P. Mansfield received the Nobel Prize in Physiology or Medicine for their discoveries concerning magnetic resonance imaging.

See also www.nobel.se/physics/laureates/1952/index.html.

See also www.nobel.se/medicine/laureates/2003/index.html.

nuclear reaction analysis (NRA) involves probing a sample with light ions accelerated to approximately one tenth of the speed of light and measuring the energies and yields of resulting nuclear reaction products. It is often used instead of **Rutherford backscattering spectroscopy** when the matrix contains elements heavier than the isotope of interest. The list (see Table N.1) of usually used particle–particle nuclear reactions include:

For many elements, some being oxygen, nitrogen, carbon, aluminum, magnesium and sulfur, the use of a deuterium probing beam (rather than protons or helium) can give enhanced sensitivity and accuracy owing to larger nuclear reaction cross sections, better defined spectral features and less background interference.

As the technique is isotope specific, it is independent of chemical or matrix effects, providing depth profiling non-destructively.

nucleon

Table N.1 Usually used particle–particle nuclear reactions.

With protons	With deuterons	With helium-3 ions
^{6}Li(p,^{4}He)^{3}He	D(d,p)T	D(^{3}He,p)^{4}He
^{7}Li(p,^{4}He)^{4}He	D(d,n)^{3}He	^{12}C(^{3}He,d)^{14}N
^{11}B(p,^{4}He)^{8}Be	^{3}He(d,p)^{4}He	^{12}C(^{3}He,^{4}He)^{11}C
^{15}N(p,^{4}He)^{12}C	^{6}Li(d,^{4}He)^{4}He	^{13}C(^{3}He,p)^{15}N
^{18}O(p,^{4}He)^{15}N	^{11}B(d,^{4}He)^{9}Be	^{13}C(^{3}He,d)^{14}N
	^{12}C(d,p)^{13}C	^{13}C(^{3}He,^{4}He)^{12}C
	^{12}C(d,^{4}He)^{10}B	^{14}N(^{3}He,p)^{16}O
	^{14}N(d,p)^{15}N	^{14}N(^{3}He,^{4}He)^{13}N
	^{14}N(d,^{4}He)^{12}C	^{16}O(^{3}He,^{4}He)^{15}O
	^{16}O(d,p)^{17}O	^{16}O(^{3}He,p)^{18}F
	^{16}O(d,^{4}He)^{14}N	
	^{19}F(d,^{4}He)^{17}O	
	^{28}Si(d,p)^{29}Si	
	^{31}P(d,p)^{32}P	
	^{32}S(d,p)^{33}S	

nucleon – a collective name for a **proton** or a **neutron**, which are main constituents of atomic nuclei.

nucleotide – a building unit of nucleic acids consisting of a heterocyclic base, a sugar (ribose or deoxyribose), and phosphoric acid → **DNA**.

nutation – nodding of the axis of a top about the cone of **precession** caused by impulses or rearrangement of the mass of the top.

Nyquist criterion – in image acquisition, it postulates that the pickup sampling frequency must be a minimum of twice the rate of brightness change of any detail to be resolved.
First described in: H. Nyquist, *Thermal agitation of electric charge in conductors,* Phys. Rev. **32**(1), 110–114 (1928).

Nyquist formula describes thermal generation of noise, also called **Johnson–Nyquist (or Nyquist) noise**, in an electrical resistor R related to the charge fluctuation in it at temperature T: $\overline{\Delta V^2} = 4k_B TR\Delta f$. The fluctuation is regarded as being produced by a zero impedance voltage generator ΔV in series with the resistor producing frequency components in a narrow bandwidth Δf.
First described in: H. Nyquist, *Thermal agitation of electric charge in conductors,* Phys. Rev. **32**(1), 110–114 (1928).

Nyquist frequency – a frequency equal to twice the frequency of the highest frequency significant spectral component of a signal → **Nyquist–Shannon sampling**

theorem. Signals sampled at rates higher than the Nyquist frequency can theoretically be accurately reconstructed from the sampled data.

Nyquist interval → Nyquist–Shannon sampling theorem.

Nyquist noise → Johnson–Nyquist noise.

Nyquist–Shannon interpolation formula states that if a function $s(x)$ has a Fourier transform $F[s(x)] = S(f) = 0$ for $|f| > W$, then $s(x)$ can be recovered from its samples s_n as

$$s(x) = \sum_{n=-\infty}^{\infty} s_n \frac{\sin[\pi(2Wx - n)]}{\pi(2Wx - n)}.$$

Nyquist–Shannon sampling theorem states that, when an analog signal is converted into the digital one (or otherwise sampling a signal at discrete intervals), the sampling frequency must be originated perfectly from the sampled version. If the sampling frequency is less than this limit, then frequencies in the original signal that are above half the sampling rate will be "aliased" and will appear in the resulting signal as lower frequencies. Therefore, an analog low pass filter (called an anti-aliased filter) is typically applied before sampling to ensure that no components with frequencies greater than half the sample frequency remain. The theorem also applies when reducing the sampling frequency of an existing digital signal.

If a function $s(x)$ has a **Fourier transform** $F[s(x)] = S(f) = 0$ for $|f| > W$, then it is completely determined by giving the value of the function at a series of points spaced $1/(2W)$ apart. The values $s_n = s(n/(2W))$ are called the samples of $s(x)$. The function $s(x)$ can be recovered from its samples by the **Nyquist–Shannon interpolation formula**.

The minimum sample frequency that allows reconstruction of the original signal, that is $2W$ samples per unit distance, is known as the **Nyquist frequency**, (or Nyquist rate). The time between samples is called the **Nyquist interval**.

It has to be noted that even if the concept of "twice the highest frequency" is the more commonly used idea, it is not absolute. In fact the theorem holds for "twice the bandwidth", which is totally different. Bandwidth is related to the range between the first frequency and the last frequency that represent the signal. Bandwidth and highest frequency are identical only in baseband signals, that is, those that go very nearly down to direct current. This concept leads to what is called **undersampling**, often used in software-defined radio systems, which are radio communication systems employing software for modulation and demodulation of radio signals.

The theorem is a fundamental tenet in the field of information theory.

First described in: H. Nyquist, *Certain topics in telegraph transmission theory*, Trans. AIEE **47**(4), 617–644 (1928) – first formulated; C. E. Shannon, *Communication in the presence of noise*, Proc. Inst. Radio Eng. **37**(1), 10–21 (1949) – formally proved.

From octet rule to oxide

octet rule states that atoms combine to form molecules in such a way as to give each atom an outer shell of eight electrons, that is an octet.

First described by I. Langmuir in 1919.

Recognition: in 1932 I. Langmuir received the Nobel Prize in Chemistry for his discoveries and investigations in surface chemistry.

See also www.nobel.se/chemistry/laureates/1932/index.html.

Ohm's law states that there is a linear relation between electric current density **J** in a matter and electric field **E** applied to it in the form $\mathbf{J} = \sigma\mathbf{E}$, where σ is the matter conductivity.

First described in: G. S. Ohm, J. Chem. Phys. (Schweigger's J.) **46**, 137 (1826); *Die galvanische Kette mathematisch gearbeitet* (Berlin, 1827).

OLED – an acronym for an **organic light emitting diode**.

oligomer – a molecule composed of a finite number of molecular subunits (**monomer**s) linked together, in contrast to a **polymer** consisting of an unbounded number of subunits. The actual figure is a matter of debate, but often a value between 10 and 100 is referred to.

oligonucleotide – a **polymer** formed by a few **nucleotide**s.

one-atom laser – a **laser** operating with a single atom. An atom (for example cesium) is cooled and trapped in an optical resonant cavity. Then a laser is used to excite the atom, which in its decay to an intermediate state emits a photon into the cavity. Another laser is used to excite the atom into another **excited state** from which the atom decays to the **ground state**, and the process starts again. In contrast to a, conventional laser there is no threshold for lasing. The output flux from the cavity mode exceeds that from atomic **fluorescence** by more than tenfold. The operation mode is sketched in Figure O.1.

What is What in the Nanoworld. Victor E. Borisenko and Stefano Ossicini
Copyright © 2008 WILEY-VCH Verlag GmbH & Co. KGaA, Weinheim
ISBN: 978-3-527-40783-5

opal glass

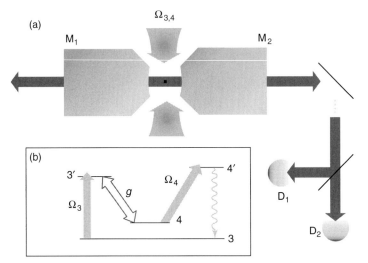

Figure O.1 A simplified schematic of operation of a one-atom laser. **(a)** A cesium atom (black dot) is trapped inside a high finesse optical cavity formed by the curved, reflective surfaces of mirrors M_1 and M_2. Light generated by the atom's interaction with the resonant cavity mode propagates as a gaussian beam to single-photon detectors D_1 and D_2. **(b)** The relevant transitions involve the $6S_{1/2}$, $F = 3$, 4; $6P_{3/2}$, $F' = 3',4'$ levels of the D_2 line at 852.4 nm in atomic cesium. Strong coupling at rate g is achieved for the lasing transition $F' = 3' \rightarrow F = 4$ near a cavity resonance. Pumping of the upper level $F' = 3'$ is provided by the field Ω_3, while recycling of the lower level $F = 4$ is achieved by way of the field Ω_4 and spontaneous decay back to $F = 3$. After J. McKeever, A. Boca, A. D. Boozer, J. R. Buck, H. J. Kimble, *Experimental realization of one-atom laser in the regime of strong coupling*, Nature **425**, 268–271 (2003).

First described in: J. McKeever, A. Boca, A. D. Boozer, J. R. Buck, H. J. Kimble, *Experimental realization of one-atom laser in the regime of strong coupling*, Nature **425**, 268–271 (2003).

opal glass – a glass which contains a disperse phase of small particles of different refractive index from that of the matrix has. They determine reflecting and scattering of light in the glass.

operator (mathematical) – a term used to mean something that operates on a function.

optical confinement factor – a fraction of the electromagnetic energy of the guided mode that exists within an active region of a semiconductor **laser**. It is calculated as a ratio of the emission profile within the active region to the emission profile within the cavity.

Figure O.2 Optical microcavity.

optical lithography – a process for creating chemical patterns on a surface. Actually optical lithography is the dominant lithography technique. Using projection optics, the resolution depends on the source wavelength; decreasing the wavelength decreases the critical dimensions of the patterning. Using optical proximity corrections and phase-shifting masks, structures smaller than the critical dimensions can be created. Actually optical lithography reaches the 100 nm region. Extreme ultraviolet lithography is a natural continuation of conventional optical lithography toward smaller wavelengths. Masks are used at wavelengths in the range 10 to 15 nm and reflection optics. Using wavelengths of the order of 1 nm, X-ray proximity lithography represents the last step in the decrease of the photon wavelength for **nanolithography** technology.

optical microcavity – a structure confining electromagnetic radiation of a particular wavelength from the optical range. It is used to select and amplify a particular wavelength in light-emitting devices. The basic principle is that an optically active material or light-emitting structure is placed in an optical cavity where only a single photon mode has a spectral overlap with the emission band. Radiative recombination in the material can only take place in that photon mode. The emitted light has the spectral characteristics of the photon mode, that is it can be extremely narrow as determined by the photon lifetime in the cavity. The fate of all electron–hole pairs not matching the energy and the momentum of the mode photon is to either recombine nonradiatively or to reach the photon-matched states by momentum and energy relaxation before a nonradiative recombination takes place.

Semiconductor microcavity structures have simple one-dimensional implementation: one uses planar **Fabry–Pérot resonator**s where the mirrors are multi quarter wave stacks distributed **Bragg reflector**s as shown in Figure O.2.

optical phonon

The usual multi-pass **interference** description of the **Fabry–Pérot resonator** shows that the resonant photon mode for which all reflected waves in the cavity are in phase has an intensity increased by a factor $4/(1 - R)$, where R is the mirror reflectivity. Conversely, those nonresonant photon modes are suppressed by a factor of about $(1 - R)$. The cavity acts as if it concentrates the electric field from a nonresonant mode into the resonant mode. The net effect is therefore that the spontaneous lifetime has hardly changed, as long as efficient relaxation toward the photon-coupled electron hole state prevents a too strong **hole-burning** phenomenon.

optical phonon – a quantum of excitation related to an optical mode of atomic vibrations in solids → **phonon**.

optical tweezers – an instrument using light pressure, within a highly focused laser beam, to trap tiny objects. It can be used to manipulate micrometer-sized dielectric particles and individual molecules with nanometer precision and measure sub-piconewton forces, making it possible to explore areas of research previously inaccessible to experimental investigation. Optical tweezing was also used by S. Chu for cooling and trapping atoms. Currently, optical tweezers are emerging as one of the most powerful tools to study physical properties of biomolecules.
First described in: A. Ashkin, *Acceleration and trapping of particles by radiation pressure*, Phys. Rev. Lett. **24**, 156–159 (1970).
More details in: A. Ashkin, *Optical trapping and manipulation of neutral particles using lasers*, PNAS **94**, 4853–4860 (1997).
Recognition: in 1997 S. Chu shared with C. Cohen-Tannoudij and W. D. Phillips the Nobel Prize in Physics for development of methods to cool and trap atoms with laser light.
See also www.nobel.se/physics/laureates/1997/index.html.

optics – the science of the behavior of light.

OPW – an acronym for an **orthogonalized plane wave**.

orbital → **atomic orbital**.

organelle – a discrete part of a biological cell having specialized functions.

organic light-emitting diode (OLED) – a solid-state light-emitting device in which an organic material is used to generate light by passing an electric current through it. It has much in common with a semiconductor **light-emitting diode**, but where the active semiconductor region is replaced by an organic thin film or a stacked layer of such films. The organic molecules emit light when they are electrically excited.
First described by C. W. Tang, US Patent 4356429 (1980).

organic radical battery – a new type of battery notable for its flexibility, rapid charging, and small size. Organic radical batteries are so named because, unlike traditional metal batteries, they use an organic radical **polymer** to produce energy. Developed by NEC Corporation in 2005.

organic solar cell = **polymer solar cell**.

organic spin valve – a **spin valve** that consists of a thin layer of an organic semiconductor sandwiched between two **ferromagnetic** electrodes. It has been demonstrated that the **spin relaxation time** in an organic **nanowire** spin valve is at least 1000 times longer than that reported in any other systems. These materials could be ideal for use in **spintronics**.

First described in: S. Pramanik, C.-G. Stefanita, S. Patibandla, S. Bandyopadhyay, K. Garre, N. Harth, M. Cahay, *Observation of extremely long spin relaxation times in organic nanowire spin valve*, Nature Nanotechnology **2**, 216–219 (2007).

organometallic compounds → **metal–organic compounds**.

orientation polarization → **polarization of matter**.

OR operator → **logic operator**.

orthogonal – perpendicular, or some concept analogous to it.

orthogonalized plane wave – a **plane wave** function made orthogonal to the atomic core functions.

orthogonalized-plane-wave method supposes solving the **Schrödinger equation** for electrons in an atom by building up the wave function out of only a few traveling plane waves, each constructed so as to be intrinsically orthogonal to the core functions. An expansion in terms of such orthogonalized plane waves does indeed exhibit vastly improved convergence properties over simple plane waves.

First described in: A. Sommerfeld, H. Bethe, *Handbuch der Physik*, Vol. 24 (Springer, Berlin, 1933) – the idea.

orthogonal transformation supposes that a linear transformation

$$y_1 = p_{11}x_1 + p_{12}x_2 + \ldots + p_{1n}x_n$$

$$y_2 = p_{21}x_1 + p_{22}x_2 + \ldots + p_{2n}x_n$$

$$\ldots$$

$$y_n = p_{n1}x_1 + p_{n2}x_2 + \ldots + p_{nn}x_n$$

orthogonal wave functions (orbitals)

for every choice of the argument x satisfies the relation

$$y_1^2 + y_2^2 + \ldots + y_n^2 = x_1^2 + x_2^2 + \ldots + x_n^2.$$

orthogonal wave functions (orbitals) – the **wave functions (orbitals)** for which the **overlap integral** is zero or more generally when

$$\int_a^b \psi_m(x)\psi_n(x)\mathrm{d}x = 0,$$

if $m \neq n$. Two different atomic orbitals on the same atom are always orthogonal.

orthohydrogen – the form of molecular hydrogen in which the nuclei of the atoms spin in the same direction. Its content in molecular hydrogen is about 75% at normal conditions. The rest is **parahydrogen.**

orthorombic lattice – a **Bravais lattice** in which the three axes of a unit cell are mutually perpendicular and two have not the same length. It is also known as a **rhombic lattice.**

oscillator strength – the dimensionless term introduced to characterize electron transitions from one energy state i to another energy state j. It is essentially the "number" of oscillators with frequency ω_{ji}:

$$f_{ij} = 2\frac{|P_{ij}|^2}{m\hbar\omega_{ji}} = 2h^2\frac{\left|\langle\psi_i|\partial/\partial x|\psi_j\rangle\right|^2}{m(E_i - E_j)}.$$

P_{ij} is the matrix element of the dipole, and m is the mass of the electron.

In fact, it is a dimensionless measure of the intensity of a spectral band: a classical concept (giving the effective number of electrons taking part in a certain transition) adapted to wave mechanics. It is defined as the ratio of the strength of atomic or molecular transition to the theoretical transition strength of a single electron using a harmonic-oscillator model.

oscillatory exchange coupling appears in a multilayer structure in which magnetic layers are separated by a nonmagnetic layer of a certain thickness. As a result, the orientation of the magnetic moments of the magnetic layers oscillates between parallel **(ferromagnetic)** and antiparallel **(antiferromagnetic)** as a function of the nonmagnetic layer thickness.

First described in: S. S. P. Parkin, N. More, K. P. Roche, *Oscillations in exchange coupling magnetoresistance in metallic surface structures: Co/Ru, Co/Cr and Fe/Cr,* Phys. Rev. Lett. **64**(19), 2304–2307 (1990).

osmosis – a spontaneous flow of a chemical species through a semi-permeable membrane from a region of high concentration to one of lower concentration. Semi-permeable membranes differ from ordinary membranes in that they are selective: they permit one or more types of molecular species to pass through but are impermeable to others.

Ostwald ripening – a growth of big islands in a thin film at the expense of the small ones.

First described in: W. Ostwald, *Lehrbuch der Allgemeinen Chemie*, vol. **2**, part 1 (Leipzig, 1896).

Overhauser effect – when a radio frequency field is applied to a substance in an external magnetic field, whose nuclei have spin 1/2 and which has unpaired electrons, at the electron spin resonance frequency, the resulting polarization of the nuclei is as great as if the nuclei had a much larger electron magnetic moment.

First described in: A. W. Overhauser, *Paramagnetic relaxation*, Phys. Rev. **89**(4), 689–700 (1953) – theoretical prediction; T. R. Carver, C. P. Slichter, *Polarization of nuclear spins in metals*, Phys. Rev. **92**(1), 212–213 (1953) – experimental confirmation.

overlap integral has the form $\int \psi(A)^* \psi(B) d\tau$. It is used as a measure of the extent to which two wave functions (orbitals) belonging to different atoms A and B overlap.

Ovshinsky effect – the characteristic of a particular thin-film solid-state switch that responds identically to both positive and negative polarities of the external voltage applied, so that the current can be made to flow in both directions equally.

First described in: S. R. Ovshinsky, *Reversible electrical switching phenomena in disordered structures*, Phys. Rev. Lett. **21**(20), 1450–1453 (1968).

oxide – a compound formed by an element with oxygen.

P

From PALM (photoactivable localization microscopy) to pyrrole

PALM – an acronym for **photoactivable localization microscopy**.

paraffins – saturated compounds of carbon and hydrogen of general formula C_nH_{2n+2}. The simplest examples are methane CH_4 and ethane C_2H_6.

parahydrogen – the form of molecular hydrogen in which the nuclei of the two atoms have spin in opposite directions. Its content in molecular hydrogen is about 25% at normal conditions. The rest is **orthohydrogen**.
First described in: K. F. Bonhoeffer, P. Harteteck, *Experimente über Para- und Orthowasserstoff*, Naturwiss. **17**, 182 (1929).

paramagnetic – a substance which in an external magnetic field demonstrates magnetization in the same direction as the field. The magnetization is proportional to the applied field → **magnetism**.
First described in: M. Faraday, *Experimental researches in electricity.–Twentieth Series*. Phil. Trans. **136**, 31–40 (1846), M. Faraday, *Experimental researches in electricity.–Twenty-first Series*. Phil. Trans. **136**, 41–62 (1846).

parity – a symmetry property of a wave function. The parity is 1 (or even) if the wave function is unchanged by an inversion (reflection with respect to the origin) of the coordinate system. It is −1 (or odd) if the wave function is only changed in sign. All degenerate states have a definite parity → **molecular orbital**.

parity conservation law states that if the wave function describing the initial state of a system has even (odd) parity, the wave function describing the final state has even (odd) parity. It is violated by the weak interactions.

PAS – an acronym for **positron annihilation spectroscopy**.

Paschen–Back effect – a change in the **Zeeman effect** arising as the strength of the external magnetic field reaches a certain critical value. A new complicated line

What is What in the Nanoworld. Victor E. Borisenko and Stefano Ossini
Copyright © 2008 WILEY-VCH Verlag GmbH & Co. KGaA, Weinheim
ISBN: 978-3-527-40783-5

pattern appears as soon as the magnetic separation reaches a value comparable with the natural separation of doublet or triplet lines. Under these conditions the coupling between orbital spin angular momentum of the electrons is reduced, the new lines being predominantly due to transitions between energy levels of electron orbits.

First described in: F. Paschen, E. Back, *Normale und anomale Zeemaneffekte*, Ann. Phys. **39**(5), 897–932 (1912).

Paschen series → **Rydberg formula**.

First described in: F. Paschen, *Zur Kenntnis ultraroter Linienspektra. I. (Normalwellenlängen bis 27000 Å.-E.)*, Ann. Phys. **332**(13), 537–570 (1908).

Pauli exclusion principle states that there can never exist two or more equivalent electrons in the atom for which the values of all quantum numbers coincide.

First described in: W. Pauli, *Über den Zusammenhang des Abschlusses der Elektronengruppen im Atom mit der Komplexstruktur der Spektren*, Z. Phys. **31**(10), 765–783 (1925).

Recognition: in 1945 W. Pauli received the Nobel Prize in Physics for the discovery of the Exclusion Principle, also called the Pauli Principle.

See also www.nobel.se/physics/laureates/1945/index.html.

Pauling rule states that the number of ions of opposite charge in the neighborhood of a given ion in an **ionic crystal** is in accordance with the requirement of local electrical neutrality of the structure.

First described in: L. Pauling, *The principles determining the structure of complex ionic crystals*, J. Am. Chem. Soc. **51**(4), 1010–1026 (1929).

More details in: L. Pauling, *The Nature of the Chemical Bond and the Structure of Molecules and Crystals; An Introduction to Modern Structural Chemistry* (1960).

Recognition: in 1954 L. Pauling received the Nobel Prize in Chemistry for his research into the nature of the chemical bond and its application to the elucidation of the structure of complex substances.

See also www.nobel.se/chemistry/laureates/1954/index.html.

Pauli principle → **Pauli exclusion principle**.

Pauli spin matrices arise with the introduction of electron spin into nonrelativistic wave mechanics. The wave function ψ of a single-electron system is then a function not only of time t and position $\mathbf{r}$, but also of an additional variable σ. This spin variable is not permitted to vary continuously like the others, but is restricted to the two values, ± 1. The principal attitude of spin is of course angular momentum. It is necessary to have operators $\hbar s_x, \hbar s_y, \hbar s_z$ of spin angular momentum corresponding to the orbital angular momentum operators like

$$\hbar l_x = -i\hbar \left(y\frac{\partial}{\partial z} - z\frac{\partial}{\partial y} \right).$$

It is found that the required operators can be defined by the equations:

$$\sigma_x \psi(\sigma, \mathbf{r}, t) = \psi(-\sigma, \mathbf{r}, t)$$
$$\sigma_y \psi(\sigma, \mathbf{r}, t) = -i\sigma \psi(-\sigma, \mathbf{r}, t)$$
$$\sigma_z \psi(\sigma, \mathbf{r}, t) = \sigma \psi(\sigma, \mathbf{r}, t)$$

Now instead of regarding ψ as an ordinary function with an additional argument σ, one may equally regard it as a two-component function of the old variables:

$$\psi(\sigma, \mathbf{r}, t) \equiv \left(\begin{array}{c} \psi_1(\mathbf{r}, t) \\ \psi_2(\mathbf{r}, t) \end{array} \right), \quad \text{where} \quad \psi_1(+1, \mathbf{r}, t), \psi_2(-1, \mathbf{r}, t).$$

When arranged as in the last equation, the two components form a column vector, and also (in the present connection) a **2-spinor**. The column vector resulting when is operated on by σ_x, σ_y or σ_z, will have components, which are linear combinations of the originals, and may therefore be obtained by multiplying the original vector by a matrix. The matrix representations are:

$$\sigma_x \equiv \left(\begin{array}{cc} 0 & 1 \\ 1 & 0 \end{array} \right), \quad \sigma_y \equiv \left(\begin{array}{cc} 0 & -i \\ i & 0 \end{array} \right), \quad \sigma_z \equiv \left(\begin{array}{cc} 1 & 0 \\ 0 & -1 \end{array} \right).$$

These are the Pauli matrices. Among their properties is that of forming, together with the unit matrix, a complete set as a linear combination of which any other 2×2 matrix can be expressed. They also, either as matrices or operators, satisfy the important algebraic relations:

$$\sigma_x \sigma_y = -\sigma_y \sigma_x = i\sigma_z$$
$$\sigma_y \sigma_z = -\sigma_z \sigma_y = i\sigma_x$$
$$\sigma_z \sigma_x = -\sigma_x \sigma_z = i\sigma_y$$

First described in: W. Pauli, *Zur Quantenmechanik des magnetischen Elektrons,* Z. Phys. **43**(9/10), 601–623 (1927).

Paul trap – a device for the dynamic stabilization of free ions. It was originally named *Ionenkäfig,* German for ion trap. The trap allows observations of isolated ions, even a single one, over a long period of time, thus enabling measurements of their properties with extremely high accuracy. The potential configuration of the trap is generated by a structure consisting of two parts, as schematically shown in Figure P.1.

The first part is given by a hyperbolically shaped ring, whose symmetry is about the $x - y$ plane at $z = 0$. The inner surface of the ring electrode is a time-dependent electrical equipotential surface. Then one has two hyperbolic rotationally symmetric end caps. The distance form the origin to the ring focus of the hyperboloid is d, whereas the distance from the origin to the focuses of the two hyperbolic caps

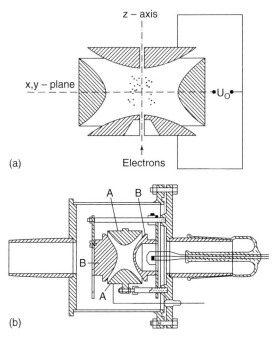

Figure P.1 Schematic view of the Paul trap (a) and a cross section of the first Paul trap (1955) (b). After W. Paul, *Electromagnetic traps for charged and neutral particles*, Rev. Mod. Phys. **62**(3), 531–540 (1990).

is usually the square root of *d*. The two end cap surfaces are time-dependent equipotential surfaces with a sign opposite to that of the ring. The electrical field in the trap is thus a quadrupole field. Through the application of oscillatory potential, one can obtain the dynamic stabilization of a charged particle.

The dynamic stabilization of ions in the Paul trap can be easily understood through a mechanical analog, due to Paul himself. The equipotential lines in the trap form a saddle surface. A mechanical ball is put at the center of the surface. With no motion of the surface, the ball will fall off. However, rotating the saddle with an appropriate frequency about the axis normal to the surface renders the ball stable. The ball makes small oscillations and can be kept stable over a long time.

First described in: W. Paul, H. Steinwedel, Z. Naturforschung Teil A **8**, 448 (1953); W. Paul, H. Steinwedel. German Patent No. 944 900; US Patent 2939958 (1953).

More details in: W. Paul, *Electromagnetic traps for charged and neutral particles*, Rev. Mod. Phys. **62**(3), 531–540 (1990).

Recognition: in 1989 W. Paul shared the Nobel Prize in Physics with H. Dehmelt for the development of the ion trap technique.

See also www.nobel.se/physics/laureates/1989/index.html.

PCR – an acronym for a **polymerase chain reaction.**

peak-to-valley ratio → resonant tunneling.

Peierls gap → Peierls transition.

Peierls–Nabarro force – the force required to displace a dislocation along its slip plane.
First described in: R. Peierls, *The size of a dislocation*, Proc. R. Phys. Soc. **52**, 34–37 (1940); F. R. N. Nabarro, *Dislocations in a simple cubic lattice*, Proc. R. Phys. Soc. **59**, 256–272 (1947).

Peierls temperature → Peierls transition.

Peierls transition – a structural distortion or phase transition in solids driven by nonuniformities in the electron distribution resulting from electron–phonon interactions. It is characterized by the **Peierls temperature** T_P.

Below this temperature, either a dimerization into two sets of unequal inter-particle distances d_1 and d_2, such that the result is $d = d_1 + d_2$, or some other structural distortions must take place. The electronic energy of the atomic chain is lowered by the formation of charge density waves. Their X-ray signature is diffuse reflection similar to thermal diffuse scattering streaks in a reciprocal lattice. This phase transition opens up an energy gap, called the **Peierls gap**, in the dispersion relation for the energy. Below T_P, the locked (static) charge density wave of the conduction electrons at $2k_F$ or $4k_F$, where k_F is the **Fermi wave vector**, couples with the other atomic or molecular electrons in the lattice. The resulting slight lattice distortions cause extra X-ray reflection peaks.

The Peierls distortion disrupts the periodicity in the lattice and can transform a metal (above T_P) into a semiconductor or insulator (below T_P).
First described in: R. Peierls, *On Ising's model of ferromagnetism*, Proc. Camb. Phil. Soc. **32**, 477–481 (1936).
More details in: R. E. Peierls, *Quantum Theory of Solids* (Clarendon, Oxford, 1955).

Pekar mechanism – the appearance of an effective A.C. electric field and related additional coupling of acoustic **phonons** with **electrons** in crystalline solids in the presence of an external electric field. It is associated with the dependence of dielectric permittivity of the materials on the strain. This mechanism of the electron–phonon interaction is related directly to the **electrostriction** effect. The induced A.C. electric field has a piezo-like dependence on the phonon wavevector and can roughly be characterized by an effective piezo-constant $\beta_{eff} \sim \epsilon^2 E p$, where ϵ is the dielectric permittivity of the unstrained crystal, E is the external electric field, and p is the photoelasticity constant.

The mechanism is essential only for materials having very large ϵ. In bulk semiconductors it appears only in a current-carrying state. In nanostructures, however, much stronger perpendicular electric fields are often encountered, which

Peltier coefficient

do not cause current and, on the other hand, make the Pekar interaction essential even in materials having ordinary dielectric parameters.

First described in: S. I. Pekar, *Electron–phonon interaction proportional to the external applied field and sound amplification in semiconductors*, Zh. Eksp. Teor. Fiz. **49**(2), 621–629 (1965) – in Russian, Sov. Phys. JETP **22**(2), 431–438 (1966); A. A. Demidenko, S. I. Pekar, V. N. Piskovoi, B. E. Tsekvava, *Voltage current curve of a semiconductor with electron–phonon coupling proportional to the applied field*, Zh. Eksp. Teor. Fiz. **50**(1), 124–130 (1966) – in Russian, Sov. Phys. JETP **23**(1), 84–89 (1966).

Peltier coefficient → **Peltier effect.**

Peltier effect – heat liberation or absorption when electric current flows across a junction composed of two different materials, as shown in Figure P.2.

When two different materials, such as conductors or semiconductors, are joined at two points to make a closed circuit and an electric current is passed round this circuit, heat is evolved at one of the junctions and absorbed at the other. The heat produced or absorbed per a unit area of the junction is given by the product of the current density and the **Peltier e.m.f.** (e.m.f. = electromotive force), also called the **Peltier coefficient**, which is a characteristic of the materials in question.

First described in: J. C. Peltier, *Nouvelles experiences sur la caloricité des courants electriques*, Ann. Chim. Phys. (2nd Ser.) **56**, 371–386 (1834).

Figure P.2 Schematic view of the Peltier effect.

Peltier e.m.f – Peltier electromotive force → **Peltier effect.**

Penn gap – an average energy gap used in calculations of the wave-number-dependent **dielectric function**. It arises in the isotropic version of the nearly free electron model in which both **Bragg reflections** and **Umklapp processes** were included. The dielectric function depends only on this parameter, which can be determined from optical data. Note that for small wave numbers Umklapp processes give the major contribution to the dielectric function, whereas for large wave numbers normal processes dominate.

First described in: D. R. Penn, *Wave number dependent dielectric function of semiconductors*, Phys. Rev. **128**(5), 2093–2097 (1962).

Penning ionization – the ionization of gas atoms or molecules in collisions with metastable atoms.

First described in: F. M. Penning, *Über Ionisation durch metastabile Atome*, Naturwiss. **15**, 818 (1927).

Penning trap – a quadrupole trap for charged particles schematically shown in Figure P.3. If one applies to the trap only a D.C. voltage such that ions perform stable oscillations in the z direction, the ions are unstable in the $x-y$ plane. Through the application of a magnetic field in the axial direction, the z motion remains unchanged but the ions perform a cyclotron motion in the $x-y$ plane because of the **Lorentz force** directed towards the center. This force is partially compensated by the radial electric force. As long as the magnetic force is much larger than the electric one, there is stability also in the $x-y$ plane.

The main difference between the Penning trap and the **Paul trap** is the use of static fields in order to confine charged particles. This helps in overcoming the problem called "*rf* heating" in the Paul traps, thus allowing laser cooling. In 1973, H. Dehmelt succeeded in observing for the first time a single electron in such a trap.

First described in: F. M. Penning, *Introduction of an axial magnetic field in the discharge between two coaxial cylinders*, Physica **3**, 873–894 (1936).

More details in: H. Dehmelt, *Experiments with an isolated subatomic particle at rest*, Rev. Mod. Phys. **62**(3), 525–530 (1990).

Recognition: in 1989 H. Dehmelt shared with W. Paul the Nobel Prize in Physics for the development of the ion trap technique.

See also www.nobel.se/physics/laureates/1989/index.html.

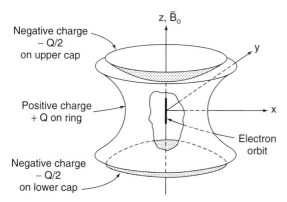

Figure P.3 Penning trap. The simplest motion of an electron in the trap is along its symmetry axis, along a magnetic field line. Each time it comes too close to one of the negatively charged caps it turns around, originating harmonic oscillations. After H. Dehmelt, *Experiments with an isolated subatomic particle at rest*, Rev. Mod. Phys. **62**(3), 525–530 (1990).

peptide – a natural or synthetic compound containing two or more **amino acids** linked by the **carboxyl** group of one amino acid to the amino group of another.

peptide bond

peptide bond – the C–N chemical bond formed between two molecules containing a **carboxyl** group and an **amino** group as shown in Figure P.4.

A molecule of water is released in the reaction. This is a dehydration synthesis reaction usually occurring between **amino acids**. The resulting molecule is called an **amide**.

The bond has a partial double bond character (with the nitrogen atom attaining a partial positive charge and the oxygen atom a partial negative charge) and the molecule can normally not rotate around this bond. The whole arrangement of the four C, O, N, H atoms as well as the two attached carbons in a peptide bond is planar.

$$
\underset{\text{OH}\ \ \text{H}}{R-\overset{\displaystyle O}{\overset{\displaystyle \|}{C}}} + \underset{}{\overset{\text{H}}{N}-R'} \longrightarrow R-\overset{\displaystyle O}{\overset{\displaystyle \|}{C}}-\underset{\text{H}}{N}-R' + H_2O
$$

Figure P.4 Formation of the peptide bond in the dehydration synthesis reaction.

peritectic – a crystalline structure of certain binary alloys in which a secondary crystal formed during cooling envelops the primary crystal and prevents its contact with the melt. As a result, the structure is not in equilibrium and does not obey the phase rule.

permanent spectral hole-burning → spectral hole-burning.

permanganate – a salt of the general formula $XMnO_4$, where X is monovalent.

permeability (absolute permeability of medium) – the ratio of the magnetic flux density B to the magnetizing force H applied to the medium: $\mu = B/H$.

permittivity (absolute permittivity of a medium) = dielectric constant of a medium – the ratio of the electric displacement D in the medium to the strength of the electric field E applied: $\epsilon = D/E$. It is a measure of the ability of the medium to be polarized → polarization.

perovskite structure – a structure of cubic symmetry possessed by crystals with chemical composition ABO_3. Typical examples are $CaTiO_3$ (perovskite) and $BaTiO_3$ (above $120\,^\circ C$).

peroxide – a compound containing an oxygen–oxygen single bond per molecule thus forming O_2^{2-} anion. Among inorganic compounds hydrogen peroxide (H_2O_2) and sodium peroxide (Na_2O_2) are typical examples. In organic chemistry peroxides contain a specific functional group or a molecule with an oxygen–oxygen single bond (R–O–O–R′).

Peroxides are powerful oxidizers, and usually fairly unstable. Ionic peroxides react with water and diluted acids to form hydrogen peroxide. Organic compounds are oxidized to carbonates, even at normal temperatures.

persistent reversible hole-burning → spectral hole-burning.

Persson–Tosatti model studies the influence of surface roughness on the **adhesion** of elastic solids. Most real surfaces have roughness on many different length scales, and this fact is taken into account in the analysis. Considering in detail the case when the surface roughness can be described as a self-affine **fractal**, one can show that when the fractal dimension $D_f > 2.5$, the adhesion force may vanish, or be at least strongly reduced.

First described in: B. N. J. Persson and E. Tosatti, *The effect of surface roughness on the adhesion of elastic solids*, J. Chem. Phys. **115**(12), 5597–5610 (2001).

perturbation method – an approximate solution of equations for a complicated system by first solving the equations for another similar system, chosen so that its solution is relatively easy, and then considering the effect of small changes or perturbations on this solution.

Pettifor map – a heuristic scheme that assigns a numerical chemical scale value to each element allowing a binary alloy $A_x B_y$ to be mapped on Cartesian axes of the chemical scales for element A and element B. Alloys with similar stable structure will cluster together in the map, so the unknown stable structure for a new alloy system can be predicted by examining the stable structures of its nearby neighbours.

First described in: D. G. Pettifor, *Structure maps for pseudobinaries and ternary phases*, Mater. Sci. Technol. **4**(8), 675–691 (1988).

More details in: http://www.tothcanada.com/PettiforMaps/.

phage → virus.

phase coherence length – a distance over which the electron maintains its phase memory.

phase velocity → wave packet.

phenols – a class of organic compounds that have a hydroxy group (–OH) attached to an **aromatic ring** (benzene ring).

Phillips model describes electron space distribution in covalent crystals with point charges localized at midpoints of covalent bonds. It is used for simulation of an interatomic interaction.

phonon

First described in: J. C. Phillips, *Covalent bond in crystals*, Phys. Rev. **166**(3), 832–838, 917–921 (1968).

phonon – a quantum of lattice vibrations. Phonons have a wave-like nature that allows them to be described in terms of wave mechanics. Four different modes of the vibrations are distinguished, as illustrated in Figure P.5.

Two modes characterized by the neighboring atoms to vibrate in phase are called acoustic, and consequently the phonons are acoustic phonons. In the **longitudinal acoustic mode**, abbreviated as LA, the atomic displacements are in the same direction as the direction of the energy transfer (**k**-vector direction). In the **transverse acoustic mode**, abbreviated as TA, the atomic displacements are perpendicular to this direction.

Two other modes with the displacements of neighboring atoms in opposite phase are called optical, with optical phonons representing them. They can be **longitudinal optical** (LO) and **transverse optical** (TO). In polar crystals a long-wavelength longitudinal optical phonon involves uniform displacement of the charged atoms within the primitive cell. Such relative displacement of oppositely charged atoms generates a macroscopic electric field. This electric field can then interact with electrons which is known as the **Fröhlich interaction**. The ratio of LO and TO phonon frequencies (ω_L and ω_T) is governed by the **Lyddane–Sachs–Teller** relation $\omega_L/\omega_T = (\epsilon_0/\epsilon_\infty)^{1/2}$, where ϵ_0 and ϵ_∞ are respectively low- and high-frequency dielectric constants of the material (high frequency supposes that it is much higher than the vibrational frequencies but below the frequency corresponding to electronic excitation energy). Optical lattice vibrations coinciding with an electric dipole moment are called **polar optical phonons**. As quantum

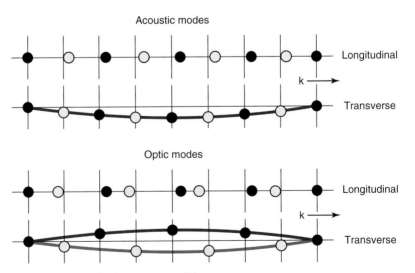

Figure P.5 Atomic displacements in solids.

particles, phonons belong to bosons, so their energy distribution follows the **Bose–Einstein statistics**.

First described in: J. Frenkel, *Wave Mechanics* (Clarendon, Oxford, 1932), p. 267.

phosphide – a compound of a metal with phosphorus in which phosphorous exerts a valency of three.

phosphite – a salt of dibasic phosphorous acid (H_3PO_3 or $HPO(OH)_2$).

phospholipids → **lipids**.

phosphor = **luminophor**.

phosphorescence – classically the continuous emission of a characteristic radiation by a substance after (usually after 10^{-8} s) an excitation. In today words, it is the light emission caused by a spin-forbidden radiative deactivation of **excited state**s.

photoactivable localization microscopy (PALM) → **fluorescence nanoscopy**.

First described in: E. Betzig, G. H. Patterson, R. Sougrat, O. W. Lindwasser, S. Olenych, J. S. Bonifacino, M. W. Davidson, J. Lippincott-Schwartz, H. F. Hess, *Imaging intracellular fluorescent proteins at nanometer resolution*, Science **313**, 1642–1645 (2006); S. T. Hess, T. P. K. Girirajan, M. D. Mason, *Ultra-high resolution imaging by fluorescence photoactivation localization microscopy*, Biophys. J. **91**(11), 4258–4272 (2006).

photochemistry – the study of chemical reactions induced by light.

photoconduction – electrical conduction produced in a solid by the influence of light (from ultraviolet to infrared).

photodissociation – dissociation of chemical compounds into simpler molecules or free atoms by the action of light (from ultraviolet to infrared).

photoelectric effect – a change of electronic properties of a substance subjected to irradiation with light. It can be external, when an ejection of electrons from a solid surface being irradiated takes place, or internal, when the substance being irradiated changes its internal electronic properties. The external photoelectric effect is also called photoemission.

First described in: H. Hertz, *Über einen Einfluss des ultravioletten Lichtes auf die electrische Entladung*, Ann. Phys. (Leipzig) **31**, 293 (1887). First described quantum mechanically in: A. Einstein, *Zur allgemeinen molekularen Theorie der Wärme*, Ann. Phys. **14**, 354–362 (1904), A. Einstein, *Über einen die Erzeugung und Verwandlung des Lichtes betreffenden heuristischen Gesichtspunkt*, Ann. Phys. **17**, 132–148 (1905).

photoemission

Recognition: in 1921 A. Einstein received the Nobel Prize in Physics for his services to Theoretical Physics, and especially for his discovery of the law of the photoelectric effect.

See also www.nobel.se/physics/laureates/1921/index.html.

photoemission – a process in which a photon is absorbed by a solid and an electron is emitted. The photoemitted electron escapes the solid with a maximum energy $E = h\nu - E_I$, where ν is the frequency of the photon and E_I is the photothreshold (or ionization) energy.

photoluminescence – light emission from semiconductors and dielectrics excited by photons of energies higher than their band gaps. The emitted photons have energies lower than the excitation photons, which is the **Stokes law of photoluminescence**.

photon – a particle representing one quantum of light. It has the energy $E = h\nu$, where ν is the frequency of the radiated energy. The rest mass of a photon is zero.

First described in: A. Einstein, *Über einen die Erzeugung und Verwandlung des Lichtes betreffenden heuristischen Gesichtspunkt*, Ann. Phys. **17**, 132–148 (1905); A. Einstein, *Strahlungs-Emission and Absorption nach der Quantentheorie*, Verh. Dtsch. Phys. Ges. **18**, 318–323 (1916); A. Einstein, *Zur Quantentheorie der Strahlung*, Phys. Z. **18**, 121–128 (1917).

photonic crystal – a structure with a periodic space modulation of dielectric function with a period of the photon wavelength. Formation of allowed bands of particle states and appearance of the forbidden band gap in between, which is well known for conventional solids, is not due to the charge carried by an electron but is basically the common property of all waves in a periodic potential. In the case of optical waves, space modulation of the refractive index is an analog of electrostatic potential in the case of electrons. Simple examples of photonic crystals of different dimensionality are shown in Figure P.6.

The photon **density of states** is significantly modified when alternating layers, needles or balls are arranged in a one-, two-, or three-dimensional lattice, respectively. The terms "one-, two- or three-dimensional" refer to the fact that the structure has a periodic change in the refractive index in the indicated number of directions—one, two or three.

One-dimensional   Two-dimensional   Three-dimensional

Figure P.6 Photonic crystals.

A one-dimensional photonic crystal is nothing else but a well-known dielectric mirror with alternating $\lambda/2$ layers. Because of the Bragg condition $\rightarrow$ **Bragg equation**, certain wave numbers are relevant to standing waves. These waves have to be reflected backward because they cannot propagate throughout the medium. A photonic band gap corresponding to the forbidden wavelength is formed, as illustrated in Figure P.7.

If the structure consists of alternating layers of two materials with different refraction indices n_1 and n_2 with the same thickness $a/2$, the band gap appears at wave numbers $k_m = m\pi/a$, where m is an integer. The larger ratio n_1/n_2 corresponds to a wider photonic band gap. Because of the periodicity of the structure, wave numbers differing by $2m\pi/a$ are equivalent. Thus, the photonic band structure can be presented in the reduced form, as in the right-hand plot, in the wave number interval from $-\pi/2$ to $+\pi/2$, which is the first Brillouin zone for photons. Photonic band gaps in two- and three-dimensional photonic crystals are formed under the same regularities.

A photonic crystal is an optical analogy of a periodic crystal potential for electrons, in which the role of the potential for light is played by the lattice of macroscopic dielectric media instead of atoms. If the dielectric constants of the materials forming the lattice are different enough, and the absorption of light by the material is minimal, then scattering at the interface can produce many of the same phenomena for photons as the atomic potential does for electrons.

First described in: E. Yablonovitch, *Inhibited spontaneous emission in solid-state physics and electronics*, Phys. Rev. Lett. **58**(20), 2059–2062 (1987); S. John, *Strong localization of photons in certain disordered dielectric superlattices*, Phys. Rev. Lett. **58**(23), 2486–2489 (1987).

More details in: J. D. Joannopoulos, R. D. Meade, J. N. Winn, *Photonic Crystals* (Princeton University Press, Princeton, 1995).

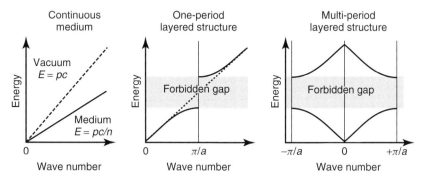

Figure P.7 Energy wave number relations demonstrating formation of the photonic band gap in a one-dimensional photonic crystal. S. V. Gaponenko, *Optical Properties of Semiconductor Nanocrystals* (Cambridge University Press, Cambridge, 1998).

photonic stopband $\rightarrow$ **distributed Bragg reflector**.

photorefractive effect – a change in the refractive index of medium as a function of the light intensity. Materials possessing such a property are called **photorefractive materials**.

First described in: S. J. Wawilov, W. L. Lewschin, *Die Beziehungen zwischen Fluoreszenz und Phosphoreszenz in festen und flüssigen Medien*, Z. Phys. **35**, 920–936 (1926).

photorefractive materials → **photorefractive effect**.

photovoltaic cell = **solar cell**.

photovoltaic effect – generation of electric current in materials or structures exposed to light.

First described in: E. Becquerel, *Mémoire sur les effets électriques produits sous les influences des rayons solaires*, C.R. **9**, 561–567 (1839).

phthalocyanine – a macrocyclic compound having an alternating nitrogen atom –carbon atom ring structure shown in Figure P.8.

The molecule is able to coordinate hydrogen and metal cations in its center by coordinate bonds with the four isoindole nitrogen atoms. The central atoms can carry additional ligands. Most of the elements have been found to be able to coordinate to the phthalocyanine macrocycle. Therefore, a variety of phthalocyanine complexes exist. Phthalocyanines are highly colored compounds which have found widespread commercial application.

More details in: A. L. Thomas, *Phthalocyanine Research and Applications* (CRC, 1990).

Figure P.8 A phthalocyanine molecule. M = Me (Ni, Co, Cu, Fe, Zn), 2H.

physisorption – an **adsorption** process characterized by weak **van der Waals forces** between adsorbed atoms and substrate atoms at the adsorbing surface.

pi or π theorem → **Buckingham's theorem**.

pico- – a decimal prefix representing 10^{-12}, abbreviated p.

piezoelectric – a material (usually dielectric) in which an electric charge is induced when it is subjected to pressure or tension in certain directions. A converse effect is also observed, that is if an external electric field is applied to a sample made of such material, the mechanical strain is induced in it leading to the change of its size.

First described in: J. Curie, P. Curie, *Sur l'électricité polaire dans les cristaux hémièdres à faces inclinées*, C. R. **91**, 294–295 (1880).

pink noise = **flicker noise** = $1/f$ **noise**.

pixel – contraction of "picture element." A small element of a scene, often the smallest resolvable area, in which an average brightness value is determined and used to represent that portion of the scene. Pixels are arranged in a rectangular array to form a complete image.

Planck law states that energy associated with electromagnetic radiation is emitted or absorbed in a discrete amount, which is proportional to the frequency of radiation.

First described in: M. Planck, *Zur Theorie des Gesetzes der Energieverteilung im Normalspektrum*, Verh. Dtsch. Phys. Ges. **2**, 237–245 (1900).

Recognition: in 1918 M. Planck received the Nobel Prize in Physics in recognition of the services he rendered to the advancement of Physics by his discovery of energy quanta.

See also www.nobel.se/physics/laureates/1918/index.html.

Planck length → **Planck scale**.

Planck mass → **Planck scale**.

Planck's radiation formula (for energy distribution) determines the energy density irradiated by a black body at the temperature T in the range from λ to $\lambda + d\lambda$ as:

$$dU = \frac{8\pi hc}{\lambda^5}\left[\exp\left(\frac{hc}{\lambda k_B T}\right) - 1\right]^{-1} d\lambda.$$

First described in: M. Planck, *Zur Theorie des Gesetzes der Energieverteilung im Normalspektrum*, Verh. Dtsch. Phys. Ges. **2**, 237–245 (1900).

Planck scale – the scale at which the gravitational interaction between individual particles becomes as strong as other interactions. At this scale, the description of sub-atomic particle interactions in terms of quantum field theory breaks down because of the nonrenormalizability of gravity.

The scale is represented by the energy scale of about 1.22×10^{19} GeV (corresponding to the **Planck mass** $m_P = (hc/2\pi G)^{1/2} = 2.2 \times 10^{-5}$ g), **Planck length** of 1.616×10^{-35} m calculated as $L_P = (hG/2\pi c^3)^{1/2}$ and **Planck time** of 5.4×10^{-44} s calculated as $t_P = (hG/2\pi c^5)^{1/2}$. Here G is the constant of gravitation.

Planck time

At the Planck scale, the strength of gravity is expected to become comparable to the other forces, and it is theorized that all the fundamental forces are unified at that scale, but the exact mechanism of this unification remains unknown.

Planck time → **Planck scale.**

plane wave – a wave which is constant in any plane perpendicular to the direction of the wave propagation and periodic along the lines parallel to the direction of the wave propagation.

plasmon – a quantum of plasma oscillation arising as a result of collective excitation in a plasma-like system composed of positive ions and virtually free electrons.

plasmonic effects based solar cell – **solar cell**, in which metal nanoparticles are deposited onto its surface such that the surface **plasmon** can be excited in the nanoparticles layer. Such surface plasmon excitation results in large electromagnetic field enhancement near the metal surface where the absorption of photons increases. Nanoparticles can be varied in shape and size to enhance the absorption of a photon.
   First described in: O. Stenzel, A. Stendal, K. Voigtsberger, C. von Borczyskowski, *Enhancement of the photovoltaic conversion efficiency of copper phthalocyanine thin film devices by incorporation of metal clusters*, Solar Energy Materials and Solar Cells **37**, 337–348 (1995).

Pockels effect – a change in the refractive properties of certain crystals in an applied electric field, which is proportional to the strength of the field.
   First described in: F. Pockels, *Über den Einfluss des electro-statischen Feldes auf das optische Verhalten piezoelektrischer Kristalle*, Abh. Göttingen Ges. Wiss., Math. Physik. Kl. **39**, 1–204 (1893).

Poincaré sphere – the model construction used for the representation of light polarization. The Stokes coordinate system is used for this. The **Stokes parameters** S, Q, U, and V are real numbers, directly related to the intensity of light:

$$S = I_x + I_y = I$$
$$Q = I_x - I_y$$
$$U = I_{45°} - I_{-45°}$$
$$V = I_{LCP} - L_{RCP}.$$

Here S is the total intensity of light, Q is the difference in intensity between horizontal and vertical linearly polarized components, U is the difference in intensity between linearly polarized components oriented at $+45°$ and $-45°$, and V is the difference between the intensity of left and right circularly polarized light.

The Stokes parameters are measurable quantities. For completely polarized light one has:

$$Q^2 + U^2 + V^2 = S^2.$$

For entirely unpolarized light:

$$Q^2 + U^2 + V^2 = 0.$$

Thus, it is possible to identify the parameters Q, U and V as the coordinates of a point of a sphere with radius corresponding to the intensity of the polarized part as illustrated in Figure P.9.

On the Poincaré sphere, each point represents a specific polarization state. Points on the equator define linearly polarized states, south and north poles represent left and right circularly polarized states. Points on the lower and upper hemisphere denote the degree of left and right elliptically polarized states. Any pair of antipodal points on the sphere defines states with orthogonal polarization.

First described in: G. G. Stokes, *On the composition and resolution of streams of polarized light from different sources*, Trans. Camb. Phil. Soc. **9**, 399–423 (1852) – Stokes parameters were introduced; H. Poincaré, *Theorie Mathématique de la Lumiére*, Georges Carrié Ed. Paris 1892, Vol. 2, Chap. 12, pp. 275–285 – the sphere was described.

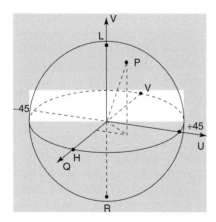

Figure P.9 The Poincaré sphere and the Stokes coordinate system. The point *P* lies on the surface of the sphere and represents a given polarization state.

point group – a collection of symmetry operations which, when applied about a point in the crystal, will leave the crystal structure invariant. It does not contain any translation operations.

More details in: G. F. Koster, J. O. Dimmock, R. G. Wheeler, H. Statz, *Properties of the Thirty-two Point-Groups* (MIT Press, Cambridge, MA, 1963).

Poisson distribution

Poisson distribution – a probability distribution whose mean and variance have a common value k and whose frequency is $f(n) = k^n \exp(-k)/n!$ for $n = 0, 1, 2, \ldots$

First described in: S. D. Poisson, *Recherches sur la Probabilité des Jugements en Matière Criminelle et en Matière Civile*, Paris (1837).

Poisson equation $\nabla^2 \phi = f(x, y, z)$. It occurs in electrostatics, where ϕ is the potential due to a charge distribution of volume density ρ and $f(x, y, z) = -4\pi\rho$. Poisson's equation reduces to the **Laplace equation** when $\rho = 0$.

First described in: S. D. Poisson, *Remarques sur une équation qui se présente dans la théorie des attractions des sphéroïdes*, Nouveau Bull. Soc. Philomathique de Paris **3**, 388–392 (1813).

polariton – a quantum **quasiparticle** formed by **exciton–photon coupling**. In fact, it is an oscillation of the elastically vibrating crystal lattice coupled to the electromagnetic field. Being part light, part matter, polaritons possess new properties that are not seen in either excitons or photons. An exciton combined with a photon can produce a high-energy polariton, also called the upper polariton, or a low-energy one called the lower polariton.

polarization of light is determined by the direction of the electric field vector, disregarding the magnetic field since it is almost always perpendicular to the electric field. The light can be linearly, circularly or elliptically polarized.

The light is said to be **linearly polarized** if the direction of oscillation of the electric field E is fixed. There are two possible linear polarization states, with their E fields orthogonal to one another. Any other angle of linear polarization can be constructed as a superposition of these two states. The direction of polarization is arbitrary with respect to the light itself. It is usual to label the two linear polarization states in accordance with some other external reference. For example, the terms horizontally and vertically polarized are generally used when light is propagating in free space. If the light is interacting with a surface, the terms s- and p-polarized are used in order to identify sagittal plane and tangential plane polarization, respectively.

The light is said to be **circularly polarized** if the direction of the electric field E is not fixed, but rotates as the light propagates. Two possible independent circular polarization states exist, termed left-hand or right-hand circularly polarized depending on whether the electric field is rotating in a counter-clockwise or clockwise sense, respectively, when looking in the direction of the light propagation.

Elliptical polarization is a combination of circular and linear polarization.

polarization of matter – a separation of positive and negative charge barycentres of bound charges forming electric dipoles. A macroscopic polarization of matter is defined by the vector sum of the individual dipole moments of atoms, molecules, ions and bigger charged atomic particles constituting the matter. An individual dipole moment is $\mathbf{p} = \alpha E_{loc}$, where α is the polarizability and E_{loc} is the local electric field at the site of the polarizing particle.

Four main mechanisms of polarization are distinguished in solids:

1. **Electronic polarization** is related to the displacement of the negatively charged electron shell of an atom against the positively charged nucleus. It is typical for dielectrics. The electronic polarizability is approximately proportional to the volume of the electron shell. It is practically temperature independent.

2. **Ionic polarization** takes place in materials with ionic bonds when positive and negative ion sublattices are displaced with respect each other by an electric field. The ionic polarizability is usually weakly positive because of the thermal expansion of the lattice.

3. **Orientation polarization** appears due to an alignment of permanent dipoles in matter when an electric field is applied. The average degree of orientation is a function of the applied field and temperature. The mean polarizability in this case is given by $< \alpha >= p^2/3k_b T$, where p is the value of the permanent dipole moment. The strong temperature dependence of this mechanism is its fingerprint.

4. **Space charge polarization** is a polarization effect in a dielectric material with spatial inhomogeneities of charge carrier densities. It also occurs in ceramics with electrically conducting grains and insulating grain boundaries, which is called **Maxwell–Wagner polarization**, and in composite materials in which metallic particles are isolated in polymer or glass matrices.

polarizer – an optical device that transmits the component of incident light having an electric field vector parallel to the transmission axis of the device.

polar molecule – a molecule composed of negatively charged electrons and positively charged nuclei in a state of vibration about certain fixed points in space.

polaron – an equilibrium assembly of an electron or a hole and the polarization field it induces in a solid.
First described in: L. D. Landau, *Über die Bewegung der Elektronen im Kristallgitter*, Phys. Z. Sowjetunion **3**, 664–666 (1933).

polar optical phonon → phonon.

polyelectrolyte – a polymer with repeating units bearing electrolyte groups. These groups dissociate in aqueous solutions, making the polymer charged. Polyelectrolyte properties are thus similar to both electrolytes and polymers. Because of this they are sometimes called polysalts. Like salts, their solutions are electrically conductive. Like polymers, their solutions are often viscous. Many biological molecules are polyelectrolytes.

polymer – a molecule composed of an unbounded number of molecular subunits (**monomers**) linked together, in contrast to an **oligomer** consisting of a finite number of subunits.

polymer–clay nanocomposite → clay–polymer nanocomposite.

polymer nanocontainer – a hollow polymer particle. Such nanocontainers are useful for biomedical applications as confined reaction vessels, protective shells for **enzymes**, traps for the selective recovery of biotransformation or polymerase chain reaction products, or drug delivery carriers.
  More details in: A. Graff, S. M. Benito, C. Verbert, W. Meier, *Polymer nanocontainer,* in: *Nanobiotechnology. Concepts, Applications and Perspectives,* edited by C. M. Niemeyer and C. A. Mirkin (Wiley-VCH, Weinheim, 2004).

polymer solar cell – a **solar cell** built from thin films of organic semiconductors such as polymers and small-molecule compounds.
  More details in: J. J. M. Halls, R. H. Friend, *Organic photovoltaic devices* in: *Clean Electricity from Photovoltaics, Series on Photoconversion of Solar Energy,* Vol. 1, edited by M. D. Archer, R. Hill, (Imperial College Press, 2001) pp. 377–432.

polymerase chain reaction (PCR) – the method for the generation of unlimited copies of any fragment of **DNA** or **RNA**. PCR exploits the natural function of the enzymes called polymerases. Polymerases are present in all living things and their action is to copy genetic material. The PCR works in three steps (see Figure P.10).
  First, the target genetic material must be denatured, that is the strands of its helix must be unwound and separated by heating; second, two primers (short chains of the four different chemical components that construct any strand of genetic material) bind to their complementary bases on the single-stranded DNA by hybridization or annealing; finally, DNA synthesis is achieved by a polymerase which, starting from the primer, can read the template strand and match it with complementary nucleotides.
  First described in: R. K. Saiki, S. J. Scharfs, F. A. Faloona, K. B. Mullis, G. T. Horn, H. A. Ehrlich, N. Arnheim, *Enzymatic amplification of β globin genomics sequences and restriction site analysis for diagnosis of sickle-cell anaemia,* Science **230**, 1350–1354 (1985); K. B. Mullis, F. A. Faloona, *Specific synthesis of DNA in vitro via a polymerase catalyzed chain reaction,* Meth. Enzymol. **155**, 335–350 (1987).
  More details in: H. A. Erlich, *PCR Technology - Principles and Applications for DNA Amplification* (Oxford University Press, London, 1999).
  Recognition: in 1993 K. B. Mullis shared the Nobel Prize in Chemistry with M. Smith for contributions to the developments of methods within DNA-based chemistry, in particular for his invention of the polymerase chain reaction (PCR) method.
  See also www.nobel.se/chemistry/laureates/1993/index.html.

Figure P.10 Polymerase chain reaction amplification cycles.

polymorphism – different crystal structures of the same chemical compound. For single element solids, the term **allotropy** is used.

polyrotaxane → rotaxane.

polysalt = **polyelectrolyte**.

p orbital

p orbital → atomic orbital.

porosity – the ratio of the net volume of pores to the total volume of the sample containing these pores.

porous silicon – a porous material traditionally produced by anodic electrochemical etching of monocrystalline silicon in hydrofluoric acid (HF) solutions. It can be also fabricated by stain etching of silicon in HF–HNO$_3$ mixtures and by ion implantation of hydrogen or **noble gases**. This material has unique physical and chemical properties determined by a dense network of nanosize pores in the crystalline matrix and specific surface structure of the pore walls. Depending on the pore size, macroporous (>50 nm), mesoporous (2–50 nm) and nanoporous (≤2 nm) structures are distinguished. **Quantum confinement** and surface effects in nanostructures, that is **quantum dots** and **quantum wires**, formed in porous silicon, force this material to behave like a direct-gap semiconductor demonstrating rather effective photo- and electroluminescence. It is used in light-emitting devices integrated into monocrystalline silicon, for biological and medical applications, for **micro-electromechanicals** and **nano-electromechanical systems**, and in fuel cells.

The simplest cell for the anodization process consists of a chemically inert bath filled with an aqueous or ethanolic solution of HF, in which a silicon wafer and a platinum electrode are placed. In order to initiate electrochemical etching, the wafer is anodically polarized with respect to the platinum electrode. The process is usually performed with monitoring of the anodic current as it allows an appropriate control of both the porosity and thickness and good reproducibility from run to run. If the wafer is simply dipped into the solution the porous layer is formed on both wafer surfaces and the cleaved edges contacting with the solution. Since the electric current flows through the bulk of the wafer, there is a potential drop between the top and bottom parts of the wafer resulting in the decrease in the current density from the top to the bottom, which induces porosity and thickness gradients.

Better uniformity and formation of the porous layer on one side only are achieved in a cell with a planar electric contact to the silicon wafer being anodized. Such a cell, with a solid contact to the back of the wafer, is schematically shown in Figure P.11.

Only the front of the sample is exposed to the electrolyte. A metal or graphite external contact is gently pressed on to the back of the wafer. For low-resistivity silicon (typically lower than a few mΩcm), rather good uniformity is obtained without additional metal films or specially doped layers on the back surface, which indeed become important for providing a uniform current density through high-resistivity wafers. The porous layer homogeneity is also improved by agitating the electrolyte in order to remove bubbles and dissolved products of reactions. The reactions responsible for local electrochemical dissolution of silicon in HF solutions assume a hole(h^+) electron(e^-) exchange to be involved according to the

Figure P.11 Simplified schematic diagram of a conventional single-tank anodization cell.

following general scheme

$$Si + 2HF + lh^+ \rightarrow SiF_2 + 2H^+ + (2-l)e^-$$
$$SiF_2 + 2HF \rightarrow SiF_4 + H_2 \uparrow$$
$$SiF_4 + 2HF \rightarrow SiH_2F_6,$$

where l is the number of charges exchanged during the elementary step. The silicon dissolution requires HF and holes. Light illumination of the anodized surface is often performed in order to generate enough holes and electrons in silicon. It is in particular important for n-type and lightly doped (below $10^{18} cm^{-3}$) p-type materials. Gaseous hydrogen and soluble H_2SiF_6 are main products of the reactions. When purely aqueous HF solutions are used, the hydrogen bubbles stick to the surface and induce lateral and in-depth inhomogeneity of the porous layer. One of the most efficient ways to eliminate the bubbles is to add a surfactant to the solution. Absolute ethanol at a concentration not less than 15% is good for this. Acetic acid is another efficient surfactant, which allows better control of the solution pH. Only a few percent of this acid is required for bubble removal.

All the properties of a porous layer, such as its thickness, porosity (fraction of voids within the layer), diameter of pores and their structure are strongly dependent on the silicon wafer characteristics and anodization conditions. The most influencing factors are the type of conductivity, resistivity and crystallographic orientation of silicon, as well as HF concentration, pH of the solution and its chemical composition, temperature, current density, duration of anodization, stirring conditions, and illumination during anodization. Optimal control of the fabrication process and its reproducibility are only possible by accounting for these factors.

There can be three principal types of structure of the porous layer, and these are illustrated in Figure P.12. They are: (i) cylinder-like structure with ordered nonconnected pore channels, (ii) tree-like structure with quasi-ordered branching pore channels, and (iii) sponge-like structure with a net of randomly running pores. The cylinder-like structure is usually formed in low-doped n- and p-type

porous silicon

Figure P.12 Typical morphology of porous silicon layers: cylinder-like (a), tree-like (b), sponge like (c) structures. The images reproduced by courtesy of Doctors V. V. Starkov and S. K. Lazarouk from their private collections.

monocrystalline silicon anodically etched in the dark (without illumination). It is characterized by a small branching density corresponding to porosity below 10%. The tree-like structure consisting of long voids of about 10 nm in diameter running perpendicular to the surface are formed on heavily doped silicon (resistivity lower than 0.05 Ω cm) either n- or p-type. Layers with porosity up to 60% can be produced. For lightly doped silicon, the situation is somewhat different. The porous layer fabricated on lightly doped p-type wafers and on n-type ones under illumination consists of apparently random arrays of very small holes of about 2–4 nm. Achievable porosity is higher.

The crystallographic orientation affects the morphology of the porous layer only in the case of n-type monocrystalline silicon determining directions of main pore channels. In all other cases, the porous material formed from amorphous, polycrystalline and monocrystalline silicon with identical type and concentration of major charge carriers have quite similar properties, while their behavior during anodization is different. Porosity is increased with an increase in the anodization current typically ranging from 10 to 200 mA cm^{-2}. An increase in the HF concentration in an electrolyte reduces porosity. Thickness of the porous layer is linearly increased with the anodization time. Porous layers as thick as tens of nanometers up to hundreds of micrometers can be formed. Patterned Si$_3$N$_4$ thin films or others resistant to HF masks are used for local formation of porous regions in silicon wafers and films.

Interpore regions in the monocrystalline silicon matrix preserve its crystalline structure. Thus, pores themselves and their branching produce a variety of nanocrystallites in the form of crystalline clusters and wires. They are randomly distributed within the porous layer with the density and size distribution determined by the matrix material and anodization conditions. Individual nanostructures are impossible to select within any porous layer. That is why practical applications of porous silicon are limited by optoelectronic devices based on statistical behavior of quantum size particle ensembles.

First described in: A. Ulhir, Jr., *Electrolytic shaping of germanium and silicon,* Bell Syst. Tech. J. **35**(2), 333–347 (1956) and D. R. Turner, *Electropolishing silicon in hydrofluoric acid solutions,* J. Electrochem. Soc. **105**(7), 402–408 (1958) – first formation of porous silicon; C. Pickering, M. I. J. Beale, D. J. Robbinson, P. J. Pearson,

R. Greef, *Optical studies of porous silicon films formed in p-type degenerate and non-degenerate silicon*, J. Phys. C **17**(35), 6535–6552 (1984) – first observation of visible photoluminescence from porous silicon at 4.2 K; L. T. Canham, *Silicon quantum wire array fabrication by electrochemical and chemical dissolution of wafers*, Appl. Phys. Lett. **57**(10), 1046–1048 (1990) – first observation of visible photoluminescence from porous silicon at room temperature; A. Richter, P. Steiner, F. Kozlowski, W. Lang, *Current-induced light emission from a porous silicon device*, IEEE Electron Device Lett. **12**(12), 691–692 (1991) and A. Halimaoui, C. Oules, G. Bomchil, A. Bsiesy, F. Gaspard, R. Herino, M. Ligeon, F. Muller, *Electroluminescence in the visible range during anodic oxidation of porous silicon films*, Appl. Phys. Lett. **59**(3), 304–306 (1991) – first light-emitting devices employed electroluminescence from porous silicon with solid and liquid contacts, respectively; S. Lazarouk, P. Jaguiro, V. Borisenko, *Integrated optoelectronic unit based on porous silicon*, Phys. Stat. Sol. (a) 165(1), 87–90 (1998) – first integrated in silicon optoelectronic unit including porous silicon light emitter and porous silicon light detector connected by alumina waveguide.

More details in: *Properties of Porous Silicon*, edited by L. Canham (INSPEC, The Institution of Electrical Engineers, London, 1997); O. Bisi, S. Ossicini, L. Pavesi, *Porous Silicon, a quantum sponge structure for silicon based optoelectronics*, Surf. Sci. Rep. 38(1–3), 1–126 (2000).

porphyrin – a macrocyclic compound having an alternating nitrogen atom–carbon atom ring structure shown in Figure P.13. A porphyrin in which no metal is inserted in its cavity is called a free base.

Porphyrins exhibit strong light absorption, and thus play an important role in the light harvesting process of green plants, in the photoinduced charge separation at the photosynthetic reaction center as oxygen carriers, and as **redox** mediators. They can be used as a component molecule for molecular- based conductors, magnets, light emitting devices, and molecular memories.

Figure P.13 A porphyrin molecule. M=Me (Ni, Co, Cu, Fe, Zn,), 2H.

Pöschl–Teller potential describes potential relief for charge carriers of an effective mass m in barriers and hole-like wells produced by impurity diffusion

$$V(z) = -\frac{\hbar^2}{2m}\alpha^2 \frac{\lambda(\lambda-1)}{\cosh^2 \alpha z},$$

positron

where α is the width parameter, and λ is the depth parameter. Values of $0 < \lambda < 1$ correspond to potential barriers, $\lambda = 1$ gives a flat band, and values of $\lambda > 1$ correspond to potential wells. The eigenvalues of the resulting **Schrödinger equation** for a well of width a are:

$$E_n = -\frac{\hbar^2 \alpha^2}{2m}(\lambda - 1 - n)^2 a.$$

First described in: G. Pöschl, E. Teller, *Bemerkungen zur Quantenmechanik des anharmonischen Oszillators*, Z. Phys. **83**(3–4), 143–151 (1933).

More details in: I. G. Kaplan, *Intermolecular Interactions: Physical Picture, Computational Methods and Model Potentials* (John Wiley & Sons, Chichester, 2006).

positron – the anti-particle of the **electron** having mass and spin that are identical to those of an electron but a charge and magnetic moment that are opposite.

positron annihilation spectroscopy (PAS) is based on registration of gamma photons emitted during annihilation of **positrons** with **electrons** in condensed matter. These gamma photons carry information on the electronic environment at which the positron annihilates, in particular on defects in a crystal lattice. The most frequently used source of positrons is the radioisotope ^{22}Na, which has a 2.6 year half-life. It emits 0.545 MeV end-point energy positrons simultaneously with a 1.28 MeV gamma photons.

The penetration depth of the source emitted positrons is sufficient to probe the bulk of a solid sample. The positrons slow down and thermalize in the solid within a few picoseconds. After diffusing in the lattice, a thermalized positron annihilates with an electron creating two gamma quanta with energies of 511 keV, which is equivalent to the rest mass of the particles. These two quanta are emitted in approximately opposite directions. The deviation from $180°$ is caused by the momentum of the electron, since the momentum of the positron is negligible due to thermalization. In the case of core electrons, the momentum is large while for the valence electrons it is small.

The approaches employed to characterize defects from the positron annihilation include positron annihilation lifetime spectroscopy, two-dimensional angular correlation radiation (2D-ACAR) measurements and Doppler broadening measurements.

The lifetime experiments are capable of distinguishing different kinds of defects but provide no direct information on chemical variations. The number of exponential decay terms in the lifetime spectrum is the number of defect states plus one. The lifetimes deduced from the spectra are then checked against fingerprint values of known defects.

2D-ACAR data allow the characterization of defects through the analysis of the momentum distribution of conduction- and valence-band electrons in the perturbing fields of defects.

Doppler broadening of annihilation radiation provides a sensitive method of defect characterization by measuring the momentum distribution of the electrons.

Unlike 2D-ACAR, which is mainly sensitive to low-momentum electrons, it allows the examination of high-momentum core electrons. The core electrons of the atoms near a defect site retain the properties of the free atoms, including the momentum distribution. The principle of the method lies in the analysis of the positron annihilation line shape, which directly corresponds to the distribution of momentum of electron–positron pairs. The momentum itself is measured from the amount of the Doppler shift of the emitted photons.

More details in: A. Harrich, S. Jagsch, S. Riedler, W. Rosinger, *Positron annihilation coincidence Doppler broadening spectroscopy*, American Journal of Undergraduate Research **2**(3), 13–18 (2003).

power efficiency → **quantum efficiency**.

Poynting's vector – the vector **S** giving direction and magnitude of energy flow in an electromagnetic field:

$$\mathbf{S} = \frac{c}{4\pi}|\mathbf{E}| \times |\mathbf{H}|,$$

where **E** and **H** are the mutually orthogonal vectors of electric and magnetic field, respectively. If energy dissipation takes place during the propagation, this vector becomes complex with a real part equal to the average energy flow.

First described in: J. H. Poynting, *On the transfer of energy in the electromagnetic field*; Phil. Trans. **175**, 343 (1884).

precautionary principle – the statement that we should not go ahead with a new technology, or persist with an old one, unless we are convinced it is safe. It covers cases where scientific evidence is insufficient, inconclusive or uncertain. Emerging in European environmental policies in the late 1970s, the principle has becomes enshrined in numerous international treaties and declarations. The giant advances in communications technology have fostered public sensitivity to the emergence of new risks before scientific research has been able to fully illuminate the problems. The principle will play a more and more important role for future emerging technologies.

More details in: http://europa.eu.int/comm/off/com/healthconsumer.

precession – an effect shown by a rotating body when torque is applied to it in such a way as to tend to change the direction of its axis of rotation. If the speed of rotation and the magnitude of the applied torque are constant, the axis, in general, describes a cone, its motion at any instant being at right angles to the direction of the torque.

precipitation – formation of a new condensed phase in liquids and solids.

precision chemical engineering – a combination of **top-down** lithographic techniques with **bottom-up** assembly methodologies. Its principles are illustrated in Figure P.14.

In the beginning, a **self-assembled monolayer** (SAM) is fabricated on a substrate and then lithographically processed. Nanolithographic radiative techniques employing ultraviolet light, X rays or electron beams are used for this. Molecules in the SAM can undergo either removal (process A), crosslinking, damage, partial removal without exposing any new surface chemical functionality (process B) or modification of an internal and/or terminal functional group (process C) upon the irradiation. In a second stage, selective self assembling of nanocomponents onto the templated SAM can be achieved on the unirradiated (processes D, F and G) or irradiated (processes E and H) regions.

A suitable choice of the self-assembling material and lithographic technique provides fabrication of the structures with a lateral resolution down to 5 nm.

More details in: P. M. Mendes, J. A. Preece, *Precision chemical engineering: integrating top-down and bottom-up methodologies,* Cur. Opin. Colloid Interfacial. Sci. **9**, 236–248 (2004); N. P. Mahalik, *Micromanufacturing and Nanotechnology* (Springer, Berlin Heidelberg, 2005).

Figure P.14 Precision chemical engineering (From P. M. Mendes, J. A. Preece, Curr. Opin. Colloid Interfacial Sci. **9**, 236–248 (2004)).

pretzelane → catenane.

primer – a short segment of **DNA** or **RNA** that anneals to a single strand of DNA in order to initiate template-directed synthesis and extend a new DNA strand by the enzymatic action of DNA polymerase to produce a duplex-stranded molecule.

primitive cell – a minimum-volume parallelepiped defined by primitive axes, which being translated will reproduce a crystal structure.

primitive translations are used to construct theoretically a crystal lattice. A perfect crystal is characterized by a set of lattice translations T that, if applied to the crystal, take every ion (except those near the surface) to a position previously occupied by an equivalent ion. The three shortest such translations that are not coplanar are called primitive translations τ_1, τ_2, τ_3. The position of every ion in the crystal can be given by some lattice translation $T = n_1\tau_1 + n_2\tau_2 + n_3\tau_3$, where n_1, n_2 and n_3 are integers.

principal function – the stationary value of the integral $P = \int_{t_1}^{t_2} \mathbf{L}(r_i, v_i, t) dt$, where $\mathbf{L}(r_i, v_i, t)$ is the Lagrangian and, r_i and v_i are generalized coordinates and velocities that describe the state of a conservative system. The stationary value of the principal function establishes the path traversed by the system between two fixed points $r_i(t1)$ and $r_i(t2)$.

projection operator – an **operator** that projects (maps) vectors in **Hilbert space** onto a subspace.

prokaryote – a single-celled organism without a nucleus, like bacteria and archaea. The name came from the Greek root *karyon*, meaning "nut", combined with the prefix *pro-*, meaning "before".

propagator – a function used in quantum field theory to characterize propagation of a relativistic field or a quantum of that field from one point to another.

prosthetic group – a non**protein** component covalently bound to a catalytic center of an **enzyme** and essentially involved in the catalytic mechanism.

protein – a high-molecular weight biological molecule composed of a **polymer** of **amino acids** linked via **peptide** bonds. The name originated from the Greek *proteios*, meaning "precedence", "holding the first place", which reflects their great importance in all living organisms. Proteins are formed by linear polymerization of α-amino-acid $RCHNH_2CO_2H$, a molecule of water being eliminated between neighboring amino- (NH) and carboxyl- (CO) groups to form the peptide link: $-CHR_1ONHCHR_2-$.

The side chains R are of some twenty different kinds including aliphatic and aromatic polar and nonpolar types. All amino acids share the same backbone, differing in terms of their side chains. Seven amino acids have an aliphatic or aromatic side chain making them strongly **hydrophobic**: valine (V), isoleucine (I), leucine (L), phenylalanine (F), methionine (M), tyrosine (Y), and trypto-phane (W). Six amino acids have a strongly **hydrophilic** side chain: aspartic acid (D), glumatic acid (E), asparagines (N), glutamine (Q), lysine (K), and arginine (R). The other seven have intermediate properties: alanine (A), cysteine (C), threonine (T), glycine (G), proline (P), serine (S), and histidine (H). Such a distribution of hydrophobicity/hydrophilicity offers an extremely useful range of blocks with which to build macromolecules exhibiting remarkable properties.

There are two cases when the side chain is bonded back to the main chain nitrogen atom to form an amino acid. The peptide link is generally planar, with N–H and C=O trans to the short C–N bond.

Under normal conditions, any fairly long polypeptide (from a few dozen to a few hundred amino acids) folds spontaneously in the presence of water into globular domains with a stable three-dimensional architecture. Some of them can also fold specifically, often in helical form, within lipid membranes. It is the dichotomy between hydrophobicity and hydrophilicity that acts as the driving force for these processes.

The succession of different types of amino acids along the polymer that is individual for each protein is called the primary structure or sequence. This information is sufficient for the polypeptide chain to adopt a stable and unique three-dimensional structure in a suitable medium, mainly water, with occasional exceptions.

Two classical groups of proteins are distinguished: **fibrous proteins**, which have the characteristic proteins of polymers, and **globular proteins**, which are molecular. The natural fibrous proteins have simpler analogs in synthetic polypeptide fibers containing only one kind of side chain. In a globular protein the chain is folded to make an approximately spherical shape. Larger molecules may be formed by association of a few such subunits, either similar or different.

First described by G. J. Mulder in an article published in the Bulletin des Sciences Physiques et Naturelles en Neerlande on July 30, 1838. Mulder appears to have taken the term from S. Berzelius, who, on July 10, sent Mulder a letter in which he suggested the name "protein".

More details in: T. Simonson, *Electrostatics and dynamics of proteins*, Rep. Prog. Phys. **66**(5) 737–787 (2003); H. Bisswanger, *Proteins and enzymes*, in: *Encyclopedia of Applied Physics*, Vol. 15, edited by G. L. Trigg (Wiley-VCH, Weinheim, 1996).

protein engineering – a highly visible expression of genetic engineering as well as a quintessential development of classical molecular biology. Its main applications are in medicine, namely in disease prevention and therapy, as well as in industry, notably in catalysis.

The field was pioneered by M. Smith in the late 1970s and early 1980s. The two main strategies currently underpinning the discipline are **structure-based engineering** (also called **rational design**) and **directed molecular evolution**.

More details in: J. A. Brannigan, A. J. Wilkinson, *Protein engineering 20 years on*, Nat. Rev. Molec. Cell Biol. **3**, 964–970 (2002).

Recognition: in 1993 M. Smith received the Nobel Prize in Chemistry for his fundamental contributions to the establishment of oligonucleotide-based, site-directed mutagenesis and its development for protein studies.

See also www.nobel.se/chemistry/laureates/1993/index.html.

protein folding – spontaneous rearrangements of **protein** chains related to their functional three-dimensional structure.

protein unfolding – stimulated rearrangements of **protein** chains related to their functional three-dimensional structure.

proteome – the term used to refer to the collection of **protein**s in certain biological systems. For example, all of the proteins in a **virus** can be called a viral proteome. A complete proteome for an organism can be conceptualized as the complete set of proteins from all of the various cellular proteomes.

A proteome is constantly changing through its biochemical interactions with the genome. One biological system will have radically different protein expression in its different parts and in different stages of its life cycle.

proteomics – the field of study of **protein**s, particularly their structure and functions.

proton – a subatomic particle with a single positive electric charge, spin 1/2 and a mass of $1.6726231 \times 10^{-27}$ kg. It is a **baryon**. The name originates from Greek *protos* first.

protonated water cluster – an association of a **proton** (H^+) with a finite number of water molecules described as $H_3O^+(H_2O)_n$. The basic component of the cluster is the **hydronium ion** H_3O^+, which is formed by a strong covalent bond between proton and oxygen. The ion has C_{3v} symmetry and an O–H bond dissociation energy of about 260 kcal mol^{-1}. Despite this relatively large bond strength, the hydronium ion is the key species in the transfer of protons between molecules in aqueous acid–base chemistry.

pseudomorphic superlattices $\rightarrow$ **superlattice**.

pseudopotential – a potential in the form $V_p = V(r) + V_R$, where $V(r)$ is the periodic crystal potential and V_R has the character of a repulsive potential, which cancels in part the large attractive **Coulomb potential**, such that in the region of the core the higher Fourier coefficients of the pseudopotential V_p are small enough to be neglected in the first-order approximation. In fact, the true potential is replaced by an effective potential producing the correct energies of the valence electrons and giving the right behavior of the corresponding wave functions except in the vicinity of the ion cores.

pseudopotential method

pseudopotential method supposes separation of the crystal **eigenfunctions** into core states and valence states and the solution of the **Schrödinger equation** with the resulting **Hamiltonian** for energies and wave functions within the valence subspace. It is used in the numerical simulation of electron energy band structure of solids.

pseudorotaxane → rotaxane.

pseudo-spin valve → giant magnetoresistance effect.

pull-off force → Johnson–Kendall–Roberts theory.

Purcell effect – a spontaneous emission of radiation from excited atoms enhanced or inhibited by placing the atoms in a specially designed cavity or between mirrors. It has the following explanation. A two-level system decays spontaneously by interaction with a vacuum continuum at a rate proportional to the spectral density of modes per volume evaluated at the transition frequency. Within a cavity, the density of modes is modified and huge variations in its amplitude can occur. In fact the maximal density of cavity modes occurs at quasi-mode resonant frequencies and is enhanced with respect to the corresponding free-space density. It was observed that a single mode occupies within a cavity of volume V a spectral linewidth ν/Q, where Q is the quality factor of the cavity. Normalizing a resulting cavity-enhanced mode density per unit volume to the mode density of the free space reveals that an atom whose transition falls within the mode linewidth will experience an enhancement of its spontaneous decay rate given by the factor $P = (3\lambda^3/4\pi^2)Q/V$, termed the **Purcell spontaneous emission enhancement factor**.

First described in: E. M. Purcell, *Spontaneous emission probabilities at radio frequencies*, Phys. Rev. **69**(1/2), 681 (1946); E. M. Purcell, H. C. Torrey, R. V. Pound, *Resonance absorption by nuclear magnetic moments in a solid*, Phys. Rev. **69**(1/2), 37–38 (1946).

Recognition: in 1952 E. M. Purcell shared with F. Bloch the Nobel Prize in Physics for their development of new methods for nuclear magnetic precision measurements and discoveries in connection therewith.

See also www.nobel.se/physics/laureates/1952/index.html.

Purcell spontaneous emission enhancement factor → Purcell effect.

pyridine – a simple heterocyclic aromatic organic compound C_5H_5N. Structurally it is a six-membered aromatic ring with one CH group replaced by a nitrogen atom as shown in Figure P.15.

Figure P.15 Pyridinium cation.

Pyridine has an equatorial lone pair of electrons at the nitrogen atom that does not participate in the aromatic π-system. That makes it a good solvent and a starting material for synthesis of other compounds.

pyroelectric – a material in which a spontaneous electric moment is developed when it is heated. The spontaneous polarization in such materials is temperature dependent. Therefore, any temperature change ΔT results in the change of the polarization $\Delta P = \alpha \Delta T$, where α denotes the pyroelectric coefficient.
First described by D. Brewster in 1824.

pyrolysis – decomposition of compounds by heat.

pyrrole – a heterocyclic organic compound of five-membered diunsaturated ring structure composed of four carbon atoms and one nitrogen atom (C_4H_5N). It is shown in Figure P.16.
The pyrrole ring system is involved in colored natural products (green pigment in chlorophyll, red in hemoglobin). Pyrrole and its derivatives are widely used as intermediates in the synthesis of pharmaceuticals, medicines, agrochemicals, **dyes**, photographic chemicals, perfumes and other organic compounds. They are also used as catalysts for polymerization processes, corrosion inhibitors, preservatives, and as solvents for resins and terpenes.

Figure P.16 Pyrrole molecule.

From Q-control to qubit

Q-control → Q-factor.

Q-factor – the quality factor describing in **dynamic force microscopy** the number of cantilever oscillation cycles after which the damped oscillation amplitude decays to $\exp(-1)$ of the initial amplitude with no external excitation. It is $Q = m\omega_r/\alpha$, where m is the effective mass of the cantilever, ω_r is its resonant frequency, and α is the damping coefficient. The minimum detectable force gradient is inversely proportional to the square root of Q.

An additional feedback circuit in a dynamic force microscope allows the control of the quality factor that is referred to as **Q-control** feedback. It makes possible high-resolution, high-speed, or low-force scanning during atomic imaging of a sample surface.

More details in: A. Schirmeisen, B. Anczykowski, H. Fuchs, *Dynamic force microscopy*, in: *Handbook of Nanotechnology*, edited by B. Bhushan (Springer, Berlin, 2004), pp. 449–473.

Q-switching – a modification of the resonance characteristics of the laser cavity. The name comes from the Q-factor used as a measure of the quality of a resonance cavity in microwave engineering.

quantum cascade laser is based on a fundamentally different principle to normal semiconductor **lasers**. It uses only one type of charge carrier, namely electrons, and is thus called a unipolar laser. It operates like an electronic waterfall that is initiated in a **multiquantum well** structure. Electrons cascade down a series of identical energy steps built into the wells, emitting a photon each step. The energy levels involved in the transitions are created by **quantum confinement** in the well, this being the active region of the laser.

So far, the quantum cascade laser has been demonstrated using two different approaches. The original design was based on a three-quantum well active region separated from the injection/relaxation region by a tunneling barrier. The second design has the active region consisting of a superlattice, which for the best performance can be chirped to compensate for the applied electric field (see Figure Q.1).

What is What in the Nanoworld. Victor E. Borisenko and Stefano Ossini
Copyright © 2008 WILEY-VCH Verlag GmbH & Co. KGaA, Weinheim
ISBN: 978-3-527-40783-5

quantum cascade disk-lasers

GaAs/Al$_{0.33}$Ga$_{0.67}$As

F = 48 kV/cm

Figure Q.1 Schematic conduction-band diagram of a portion of the laser heterostructures at threshold bias. The thicknesses of the layers are indicated in nm, and the underlined numbers denote the four layers which are *n*-type doped. The wavy arrow indicates the transition 3→2 responsible for the laser action. The solid curves represent the moduli squared of the relevant wave functions. After C. Sirtori, P. Kruck, S. Barbieri, P. Collot, J. Nagle, *GaAs/Al$_x$Ga$_{1-x}$As quantum cascade lasers*, Appl. Phys. Lett. **73**, 3486–3488 (1998).

In both cases, the necessity to confine the upper state of the active region requires the laser to be separated into an active region followed by an injection/relaxation region. Quantum cascade lasers have been obtained using AlInAs/GaInAs/InP and AlGaAs/GaAs superlattices. They have demonstrated pulsed operation at room temperature in the midinfrared wavelength range (3–19 µm).

Quantum cascade structures are also employed in **quantum cascade distributed-feedback lasers** and microcavity **quantum disk lasers**.

First described in: R. F. Kazarinov, R. A. Suris, *Possibility of the amplification of eletromagnetic waves in a semiconductor with a superlattice*, Fiz. Tekh, Poluprovodn. 5(4), 797–800 (1971) – idea (in Russian); J. Faist, F. Capasso, D.L. Sivco, C. Sirtori, A.L. Hutchinson, and A.Y. Cho, *Quantum Cascade Lasers*, Science **264**, 553–556 (1994) – realization.

More Details in: J. Faist, F. Capasso, C. Sirtori, A. Cho, in *Intersubband Transitions in Quantum Wells: Physics and Device Applications II*, edited by H. Liu and F. Capasso, (Academic, New York, 2000) vol. 66, pp. 1–83.

quantum cascade disk-lasers – a quantum cascade laser arranged as a freestanding-in-air disk cut off a thin quantum cascade structure → **quantum cascade laser**. The disk geometry allows small devices and low threshold currents. Moreover, laser lasing in this case occurs in the so-called "whispering gallery" modes that correspond to light travelling close to the perimeter of the disk impinging onto its edge at angles large than the critical angle of refraction. This results in a high-quality resonator, with only low losses due to tunneling of light, scattering from surface roughness and intrinsic waveguide loss.

First described in: J. Faist, C. Gmachl, M. Striccoli, C. Sirtori, F. Capasso, D. L. Sivco, C. Sirtori, and A. Y. Cho, *Quantum cascade disk lasers*, Appl. Phys. Lett. **69**(17), 2456–2458 (1996).

quantum cascade distributed-feedback laser is formed by incorporation of a grating in a **quantum cascade laser**. This allows tunable operation of the laser at and above room temperature with a linewidth consistent with the desired applications, such as remote chemical sensing and pollution monitoring.

First described in: J. Faist, C. Gmachl, M. Striccoli, F. Capasso, C. Sirtori, D. L. Sivco, J. N. Baillargeon, and A. Y. Cho, *Distributed feedback quantum cascade lasers*, Appl. Phys. Lett. **70**(20), 2670–2672 (1997).

quantum cellular automata – binary logical devices and computers designed as an array of **quantum-dot** cells that are connected locally by the Coulomb interaction of the electrons confined within the dots. It is also called the **Notre Dame architecture**. A basic cell consists of four (or five) quantum dots in a square array, as shown in Figure Q.2, coupled by tunnel barriers (in the case of five dots the fifth one is placed in the center of the square).

Electrons are able to tunnel between the dots, but cannot leave the cell. When two excess electrons are injected into the cell, Coulomb repulsion separates them and places them in dots on opposite corners. There are thus two energetically equivalent ground state polarizations, which can be labeled logic "0" and logic "1". If two cells are brought close together, the Coulomb interaction between the electrons forces the cells to get the same polarization. When the polarization of one of the cells is gradually changed from one state to another, the second cell exhibits a highly bistable switching of its polarization. It allows digital processing of information.

Some examples of logic elements based on these cells are shown in Figure Q.3. Since the cells are capacitively coupled to their neighbors, any change in the state of the input cell results in a change of the output. In a line of cells no metastable states, where only a few cells flip, are possible. Transmission of binary information, signal inversion, majority voting, splitting and other logic functions can be realized.

One meets two main problems when implementing quantum cellular automata devices: individual adjustment of each cell and the limit on the operating temperature. The individual adjustment is needed because of fabrication tolerance, the presence of stray changes, and the need for exactly $4N + 2$ excess electrons in each

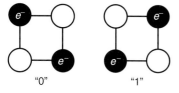

"0"   "1"

Figure Q.2 Basic cell of quantum cellular automaton and its two ground polarization states.

quantum computations

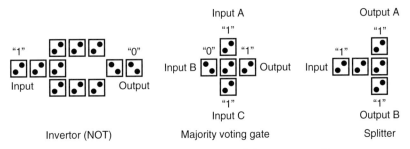

Figure Q.3 Logic elements composed of quantum cellular automata cells.

cell. For a cell on the 100 nm scale, layout errors of less than a tenth of a nanometer would be sufficient to completely disrupt cell operation. The limitation on the operation temperature is determined by the weakness of the dipole interaction between the cells, which must be significantly larger than $k_B T$. Along quantum dot structures fabricated within a semiconductor-based nanotechnology, single bistable molecules have also prospects for implementation quantum cellular automata. In the latter case, the limitations related to fabrication tolerance are easily overcome, as molecules are intrinsically precise, and bound electrons are sufficiently localized to act as excess electrons without the need for external leads.

First described in: P. D. Tougaw, C. S. Lent, W. Porod, *Bistable saturation in coupled quantum dot cells*, J. Appl. Phys. **74**(5), 3558–3566 (1993).

quantum computations – a procedure of massively parallel computations exploiting the specific properties of quantum states, such as quantum superposition, **coherence** and **entanglement**. Quantum parallelism supposes a **quantum computer** to operate on quantum superposition states of all inputs simultaneously. The same mathematical operation on different input numbers is performed within a single quantum computational step. The result is a superposition of all the corresponding outputs. Extraction of information in a quantum computer is nontrivial, because simply measuring a quantum state with an exponential number of nonzero amplitudes will yield a highly random result. In certain algorithms, however, appropriate quantum logic gates cause the amplitudes to interfere so that only a few amplitudes survive in the end. Following a measurement (or a limited number of repeated measurements on identical runs), the result (or distribution of results) can depend on a global property of all inputs.

Parallel quantum information processing is evidently more efficient than any classical computational procedure.

First described in: Yu. I. Manin, *Computed and Non-computed* (Sovetskoe Radio, Moscow, 1980) – in Russian; R. P. Feynman, *Simulating physics with computers*, Int. J. Theor. Phys. **21**(6/7), 467–488 (1982).

More details in: M. A. Nielsen, I. L. Chang, *Quantum Computation and Quantum Information* (Cambridge University Press, Cambridge, 2000).

quantum computer – a machine realizing **quantum computation**.

quantum-confined Stark effect → Stark effect.

quantum confinement (effect) – the phenomenon of the nonzero lowest energy and quantization of the allowed energy levels in low-dimensional structures, arising from the confinement of electrons within a limited space. It is illustrated as follows.

A free electron moving in three dimensions has a kinetic energy corresponding to the space components of its impulse p_x, p_y, p_z:

$$E = \frac{1}{2m^*}(p_x^2 + p_y^2 + p_z^2) \quad \text{or in wave terms} \quad E = \frac{\hbar^2}{2m}(k_x^2 + k_y^2 + k_z^2),$$

where m^* is the effective mass of the electron, which in solids is usually appreciably less than m_0, the electron rest mass, and k_x, k_y, k_z are the space components of the wave vector. Free motion of electrons in a low-dimensional structure is restricted at least in one direction. In this direction, let it be the x direction, the forces which confine the electrons can conveniently be described by an infinitely deep potential well, as illustrated by Figure Q.4.

Noting that a is the width of the well in the x direction, the electrons have zero potential energy in the region $0 < x < a$. Infinitely high impenetrable potential barriers prevent them from straying beyond this region. Thus, the electron wave function drops to zero at $x = 0$ and a. Only a restricted set of **wave functions** meet these requirements. They can be only standing waves whose wavelength λ takes one of the discrete values given by $\lambda_n = 2a/n$ (where $n = 1, 2, \ldots$). The allowed wave vectors of the waves are $k_n = 2\pi\lambda_n = n\pi/a$. They result in the set of discrete energy states of electrons defined by:

$$E_n = \frac{\hbar^2 k_n^2}{2m} = \frac{\hbar^2 \pi^2 n^2}{2ma^2}.$$

The integer n is a **quantum number** that labels the state. It means that electrons confined to a region of space can occupy only discrete energy levels. The lowest

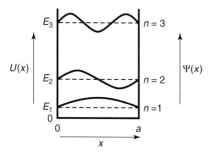

Figure Q.4 Potential well and wave functions of electrons confined in it.

quantum confinement (effect)

state has the energy

$$E_1 = \frac{\hbar^2 \pi^2}{2ma^2},$$

which is always above zero. This contrasts with classical mechanics, where the lowest state of a particle sitting on the bottom of the well is equal to zero. Such behavior would violate the uncertainty principle in quantum mechanics. In order to satisfy the uncertainty relation $\Delta p \Delta x \geq \hbar/2$ (in our case $\Delta x = a$), the electron must have the uncertainty in its momentum $\Delta p \geq \hbar/2a$. The latter corresponds to a minimum amount of energy $\Delta E = (\Delta p)^2/2m = \hbar^2/8ma^2$, which resembles E_1 to an accuracy of $\pi^2/4$.

In addition to increasing the minimum energy of the electron, confinement also causes its excited state energies to become quantized proportionaly to n^2.

The number of directions remaining free of the confinement is used for classification of elementary low-dimensional structures within three groups. These are quantum films, quantum wires and quantum dots schematically shown in Figure Q.5.

Quantum films are two-dimensional (2D) structures in which quantum confinement acts only in one direction, which is the z direction in Figure Q.5 corresponding to the film thickness. Charge carriers are free to move in the xy plane of the film giving their total energy to be a sum of the quantum confinement induced and kinetic components:

$$E = \frac{\hbar^2 \pi^2 n^2}{2ma^2} + \frac{\hbar^2 k_x^2}{2m} + \frac{\hbar^2 k_x^2}{2m}.$$

In k-space it is represented by a set of parabolic bands called **sub-bands**. It can be seen that the sub-bands overlap in energy. The minimum energy that an electron can have in the n-th sub-band is $E_n = (\hbar^2 \pi^2 n^2)/(2ma^2)$ when there is no motion in the plane.

Density of states in a quantum film has a characteristic staircase shape, replacing the parabolic shape typical for free electrons in three-dimensional (3D) structures. Electrons in quantum films are usually considered as a **two-dimensional electron gas (2DEG)**.

Quantum wires are one-dimensional (1D) structures. As compared to quantum films, one more dimension of the structure appears to be so small as to provide quantum confinement. Charge carriers are free to move only along the wire. Thus, only one kinetic component along with the confined energy contributes to the total energy. As a consequence, the density of states has $E^{1/2}$ dependence for each of the discrete pairs of states in the confined directions.

Quantum dots are zero-dimensional (0D) structures in which carriers are confined in all three directions. Their energy states are quantized in all directions and the density of states is represented by series of discrete, sharp picks resembling that of an atom. This comparison is the origin of labeling of quantum dots as **artificial atoms**. Quantum dots are usually formed by a definite number of atoms. They are typically represented by atomic clusters or **nanocrystallites**. Considering coupling of quantum dots with external electric circuits, they can be classified as open dots and almost-isolated or closed quantum dots. In an open dot, the coupling is strong and the movement of electrons across dot-lead junctions is classically allowed. When point contacts, connecting a dot with an external circuit, are pinched off, effective barriers are formed and conductance occurs only by tunneling. These are almost-isolated or closed quantum dots.

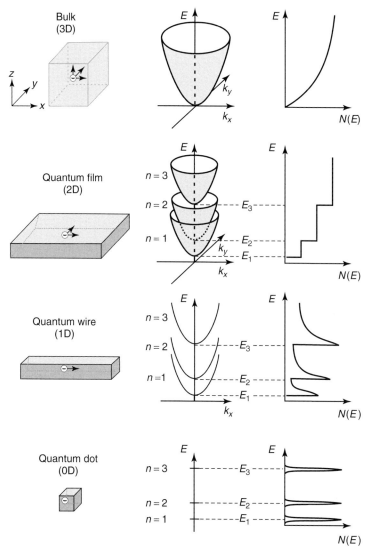

Figure Q.5 Elementary low-dimensional structures, their electron energy diagrams $E(k)$ and electron density of states $N(E)$ in comparison with those in three dimensions.

quantum critical point → **quantum phase transition**.

quantum cryptography – the art of rendering a message unintelligible to any unauthorized party by the use of **Heisenberg's uncertainty principle, quantum entanglement** and the fact that it is not possible to take a measurement of a quantum system without perturbing the system. Hence, an eavesdropper cannot get any information about the communication without introducing perturbations that would reveal his presence.

In general, security of a cryptotext depends entirely on the secrecy of the key, which is a set of specific parameters defining encrypting/decrypting procedures. The key is supplied together with the plain text as an input to the encrypting algorithm, and together with the cryptogram as an input to the decrypting algorithm. The main quantum cryptosystems developed for the key distribution use: (i) encoding based on two non-commuting observables → **BB84 protocol**, (ii) encoding built upon **quantum entanglement** and **Bell's theorem** → **E91 protocol**, and (iii) encoding based on two nonorthogonal state vectors → **B92 protocol**.

Quantum cryptography provides means for two parties to exchange an enciphering key over a private channel with complete security of communication. While classical cryptography employs various mathematical techniques to restrict eavesdroppers from learning the contents of encrypted messages, in quantum cryptography, the information is protected by the laws of quantum physics.

First described in: S. Wiesner, *Conjugate coding*, SIGACT News **15**(1), 78–88 (1983). The original manuscript was written circa 1970, but rejected by the journal to which it was submitted.

More details in: N. Gisin, G. Ribordy, W. Tittel, H. Zbinden, *Quantum cryptography*, Rev. Mod. Phys. **74** (1) 145–195 (2002).

quantum decoherence → decoherence (quantum).

quantum dot → quantum confinement.

quantum dot laser operates on the basis of the fact that population inversion necessary for lasing in semiconductors occurs more efficiently when the active layer material scales down from three-dimensional (bulk) to zero-dimensional (**quantum dot**) structures. The structure of the active layer of the laser is schematically shown in Figure Q.6 where the GaAs/InGaAs system is used as a prominent example.

In fact, a quantum dot laser combines the length scales defining the confinement of photons (100 nm) and the length scales defining the confinement of electrons and holes (10 nm). Moreover, quantum dot lasers exhibit reduced temperature

Figure Q.6 Composition of the active layer of a quantum dot laser.

sensitivity. For the fabrication, it is necessary to form high quality, uniform quantum dots in the active layer. This is provided by the self-assembly of nanoscale three-dimensional clusters through the **Stranski–Krastanov growth mode**. The number of layers with the vertically aligned dots can be 3–10 or even more.

First described in: Y. Arakawa, H. Sakaki, *Multidimensional quantum well laser and temperature dependence of its threshold current*, Appl. Phys. Lett. **40**(11), 939–441 (1982) – idea; N. Kirstaedter, N. N. Ledentsov, M. Grundmann, D. Bimberg, V. M. Ustinov, S. S. Ruvimov, M. V. Maximov, P. S. Kopev, Z. I. Alferov, U. Richter, P. Werner, U. Gosele, J. Heydenrieich, *Low threshold, large T_0 injection laser emission from (InGa)As quantum dots*, Electron. Lett. **30**(17), 1416–1417 (1994) – realization.

More details in: N. N. Ledentsov, V. M. Ustinov, V. A. Schukin, P. S. Kopev, Z. I. Alferov, D. Bimberg, *Heterostructures with quantum dots: fabrication, properties, lasers (Review)*, Semiconductors **32**(4), 343–365 (1998).

Recognition: in 2000 Z. I. Alferov received the Nobel Prize in Physics for basic work on information and communication technology: for developing semiconductor heterostuctures used in high-speed and optoelectronics.

See also www.nobel.se/physics/laureates/2000/index.html.

quantum dot solid – an ensemble of **quantum dots** spatially arranged with a period (the distance between the nearest dots) comparable or shorter than the electron **de Broglie wavelength**. The basic properties of such dense quantum dot ensembles are expected to reproduce features inherent in solids, namely formation of energy bands in a perfect lattice and coexistence of localized and delocalized electron states in disordered quantum dot structures. Electronic and optical properties of quantum dot solids are determined by electron confinement in each dot and collective effects arising from periodic special organization of the dots. The quantum dot solid can be considered as a particular form of **colloidal crystals**.

First described in: C. B. Murray, C. R. Kagan, M. G. Bawendi, *Self organization of CdTe nanocrystallites into three-dimensional quantum dot superlattices*, Science **270**, 1335–1338 (1995).

quantum efficiency – a capability of electrons and holes to generate photons by radiative recombination in a **light-emitting device (LED)**. Internal and external quantum efficiency are distinguished.

The **internal quantum efficiency** is defined as the ratio of the number of photons generated to the number of minority carriers injected into the region where recombination mostly occurs. In fact, it is the number of excited carriers that recombine radiatively expressed as a fraction of the total recombination rate $v_{in} = R_r/R = \tau_{nr}/(\tau_{nr} + \tau_r)$, where R_r and R are the radiative and total recombination rates, and τ_{nr} and τ_r are the nonradiative and radiative relaxation times of carriers. It is used to characterize the internal generation of photons. It does not take into account (a) how many charge carriers injected into the LED do not excite the material to the photon emitting states, or (b) how many generated photons get lost before exiting the device (for example by reabsorption processes). In fact, for those photons that are generated, there can still be loss through absorption within the LED material, reflection loss when light passes from a semiconductor to air due to differences in refractive index and total internal reflection of light at angles greater than the critical angle defined by the **Snell's law**.

The **external quantum efficiency** is the ratio of the number of photons actually emitted by an LED and the number of electrical carriers entering the LED per unit time. By using the external quantum efficiency ν_{ext} one can derive the **power efficiency**, which is the ratio of the radiant flux (the power carried by the light beam) emitted by the LED to the input electrical power supplied to it: $\nu_{pow} = \nu_{ext}\hbar\omega/2\pi eV$, where V is the applied voltage. The power efficiency of an LED should be at least 1% for display applications, whereas for optical interconnects it should be at least 10%, because of the requirement for a small thermal budget on the chip. Operating voltage can be high for display applications, while it should be low (1–5 V) for interconnects.

The power efficiency is sometimes reported in lumens per watt. It is also called **wall-in-plug efficiency** because it is intended that the power should be directly taken by the plug-in. It should be noticed that this general definition can be also used for nonelectrically pumped luminescence mechanisms.

quantum ensemble – a number of identically prepared particles.

quantum entanglement – the term introduced by E. Schrödinger for the description of nonseparable quantum states. It is an ability of quantum systems to exhibit correlations between states within a superposition. If there are two **qubits** each in a superposition of 0 and 1 state, the qubits are said to be entangled if the measurement of one qubit is always correlated with the result of the measurement of the other qubit. It means that two spatially separated and noninteracting quantum systems that have interacted in the past may still have some locally inaccessible information in common–information which cannot be accessed in any experiment performed on either of them alone.

Entanglement is a feature of quantum systems that allows much stronger correlation between quantum particles than classical physics permits. A measurement on one particle in an entangled system instantly reveals the properties of the other particle even if the two particles are widely separated. The quantum entanglement is used for quantum processing of information.

First described in: E. Schrödinger, *Die gegenwärtige Situation in der Quantenmechanik*, Naturwiss. **23**(48), 807–812 (1935).

quantum film → **quantum confinement**.

quantum Fourier transform decomposes a quantum state of the form $|\phi\rangle = \sum_x a_x|x\rangle$ into

$$QFT|\phi\rangle = \sum_x a_x \sum_k \frac{1}{2^{N/2}} e^{i2\pi kx}|k\rangle = \sum_K \left(\sum_x \frac{1}{2^{N/2}} e^{i2\pi kx} a_x \right) |k\rangle$$

Here the term in the brackets is just the discrete Fourier transform of a_x, the representation $|k\rangle$ being the bit reversed of the representation of x. One could do a bit reversal operation to get $|k\rangle$ into the same bit order as $|x\rangle$.

More details in: M. A. Nielsen, I. L. Chuang, *Quantum Computation and Quantum Information* (Cambridge University Press, Cambridge, 2000).

See also en.wikipedia.org/wiki/Quantum_Fourier_transform.

quantum (quantized) Hall effect → **Hall effect**.

quantum leap – a leap-like transition of a quantum particle from one state to another under the influence of an interaction typical for this type of particle. It is often used to indicate an abrupt change, sudden increase, or dramatic advance.

First described in: N. Bohr, *On the constitution of atoms and molecules. I*, Phil. Mag. **26**, 1–25 (1913).

quantum logic gate – a device that performs as a quantum **unitary operator** acting on the states of a certain set of **qubits**. If the number of such qubits is n, the evolution operator for this quantum gate is represented by a $2^n \times 2^n$ matrix in the same unitary group. It is thus a reversible gate: one can reverse the action, thereby recovering the initial quantum state from the final one. Generically a quantum logic gate can have any final number of input qubits. Schematically a quantum logic gate is usually represented by a box with a number of input and output lines. The performing operation is symbolically written inside the box.

Some representative gates in increasing order of complexity are shown below. One-qubit gates are the simplest possible gates because they take one input qubit and transform it into one output qubit. Their schematic representations are shown in Figure Q.7.

The quantum NOT gate is a one-qubit gate. Its unitary evolution operator is $U_{NOT} = \textbf{PM} = \begin{pmatrix} 0 & 1 \\ 1 & 0 \end{pmatrix}$ where **PM** is the **Pauli spin matrice**. This quantum gate coincides with its classical counterpart → **logic gate**. However, there is a basic underlying difference: the quantum gate acts on qubits while the classical gate operates on bits.

This difference allows us to introduce a truly quantum one-qubit gate: the $\sqrt{\textbf{NOT}}$ gate. Its matrix representation is

$$U_{\sqrt{NOT}} = \frac{e^{i\pi/4}}{\sqrt{2}}(1 - i\textbf{PM}) = \begin{pmatrix} \frac{1+i}{2} & \frac{1-i}{2} \\ \frac{1-i}{2} & \frac{1+i}{2} \end{pmatrix}.$$

Figure Q.7 Quantum one-qubit logic gates.

quantum logic gate

This gate, when applied twice, gives $\mathbf{NOT} = U_{\sqrt{NOT}}U_{\sqrt{NOT}} = \begin{pmatrix} 0 & 1 \\ 1 & 0 \end{pmatrix}$
$= U_{NOT}$. This gate has no counterpart in classical computers since it implements nontrivial superpositions of basis states.

Another one-qubit gate without analog in classical circuitry and heavily used in quantum computations is the so-called **Hadamard gate**. It is defined by the operator $U_H = \frac{1}{\sqrt{2}}(\mathbf{PM} + \mathbf{PM}) = \frac{1}{\sqrt{2}}\begin{pmatrix} 1 & 1 \\ 1 & -1 \end{pmatrix}$ where **PM** is the corresponding **Pauli spin matrice**.

Examples of quantum binary gates are presented in Figure Q.8.

CNOT   CPHASE   SWAP

Figure Q.8 Quantum binary logic gates.

The two qubits CNOT (controlled-NOT) gate is equivalent to the XOR (exclusive-OR) one. It is instructive to give the unitary action of this gate in several forms. Its action on the two-qubit basis states is $U_{CNOT}|00\rangle = |00\rangle$, $U_{CNOT}|01\rangle = |01\rangle$, $U_{CNOT}|10\rangle = |11\rangle$, $U_{CNOT}|11\rangle = |10\rangle$. From this definition one can see that the name of this gate is quite apparent, for it means that it executes a NOT operation on the second qubit conditioned to have the first qubit in the state. Its matrix representation is

$$U_{CNOT} = U_{XOR} = \begin{pmatrix} 1 & 0 & 0 & 0 \\ 0 & 1 & 0 & 0 \\ 0 & 0 & 0 & 1 \\ 0 & 0 & 1 & 0 \end{pmatrix}.$$

The CNOT gate plays a paramount role in both the theory and experimental realization of quantum computers. It allows the implementation of conditional logic at a quantum level.

Unlike the CNOT gate, there are two-qubit gates with no classical analog. One example is the controlled-phase gate or CPHASE:

$$U_{CPHASE} = \begin{pmatrix} 1 & 0 & 0 & 0 \\ 0 & 1 & 0 & 0 \\ 0 & 0 & 1 & 0 \\ 0 & 0 & 0 & e^{i\phi} \end{pmatrix}.$$

It implements a conditional phase shift $e^{i\phi}$ on the second qubit. An important result is that it is possible to reproduce the CNOT gate with a controlled-phase gate of $\phi = \pi$ and two **Hadamard transformations** on the target qubits. This is a simple consequence of the relation $U_H\mathbf{PM}U_H = \mathbf{PM}$. Other interesting two-qubit gates are the **SWAP** gate, which interchanges the states of the two qubits, and the $\sqrt{\text{SWAP}}$ gate, whose matrix representations are

$$
U_{\text{SWAP}} = \begin{pmatrix} 1 & 0 & 0 & 0 \\ 0 & 0 & 1 & 0 \\ 0 & 1 & 0 & 0 \\ 0 & 0 & 0 & 1 \end{pmatrix}, \quad U_{\sqrt{\text{SWAP}}} = \begin{pmatrix} 1 & 0 & 0 & 0 \\ 0 & \frac{1+i}{2} & \frac{1-i}{2} & 0 \\ 0 & \frac{1-i}{2} & \frac{1+i}{2} & 0 \\ 0 & 0 & 0 & 1 \end{pmatrix}.
$$

Among three qubit gates the most representative are depicted in Figure Q.9.

Figure Q.9 Quantum three-qubit logic gates.

An immediate extension of the CNOT construction to three qubits yields the CCNOT gate (or C^2NOT), which is also called the **Toffoli gate**[1]. Its matrix representation is a one-qubit extension of the CNOT gate:

$$
U_{\text{CCNOT}} = U_T = \begin{pmatrix} 1 & 0 & 0 & 0 & 0 & 0 & 0 & 0 \\ 0 & 1 & 0 & 0 & 0 & 0 & 0 & 0 \\ 0 & 0 & 1 & 0 & 0 & 0 & 0 & 0 \\ 0 & 0 & 0 & 1 & 0 & 0 & 0 & 0 \\ 0 & 0 & 0 & 0 & 1 & 0 & 0 & 0 \\ 0 & 0 & 0 & 0 & 0 & 1 & 0 & 0 \\ 0 & 0 & 0 & 0 & 0 & 0 & 0 & 1 \\ 0 & 0 & 0 & 0 & 0 & 0 & 1 & 0 \end{pmatrix}.
$$

The associated **truth table** is shown below.

Table Q.1 Truth table for the Toffoli gate.

x	y	z	x'	y'	z'
0	0	0	0	0	0
0	0	1	0	0	1
0	1	0	0	1	0
0	1	1	0	1	1
1	0	0	1	0	0
1	0	1	1	0	1
1	1	0	1	1	1
1	1	1	1	1	0

quantum mirage

The first two input qubits x, y are copied to the first two output qubits x', y', while the third output qubit z' is the **XOR** of the third input z and the **AND** of the first two inputs x, y. The **Deutsch gate**[2] is also an important three-qubit gate. It realizes a controlled-controlled-S or $C_2 S$ operation, where $U_S = i \cos(\theta/2) + \mathbf{PM} \sin(\theta/2)$ is a unitary operation that rotates a qubit about the x axis by an angle θ and then multiplies it by the factor i. The angle θ is required to be incommensurate to π, that is, not a rational multiple of π.

Another useful instance of a three-qubit gate is the **Fredkin gate**[3]. It performs a controlled **SWAP** operation represented by the matrix

$$
U_F = \begin{pmatrix}
1 & 0 & 0 & 0 & 0 & 0 & 0 & 0 \\
0 & 1 & 0 & 0 & 0 & 0 & 0 & 0 \\
0 & 0 & 1 & 0 & 0 & 0 & 0 & 0 \\
0 & 0 & 0 & 1 & 0 & 0 & 0 & 0 \\
0 & 0 & 0 & 0 & 1 & 0 & 0 & 0 \\
0 & 0 & 0 & 0 & 0 & 0 & 1 & 0 \\
0 & 0 & 0 & 0 & 0 & 1 & 0 & 0 \\
0 & 0 & 0 & 0 & 0 & 0 & 0 & 1
\end{pmatrix}.
$$

The above unitary linear gates act not only on the individual basis states, but also on any linear combination of them. The simple gates can be assembled into a network-like arrangement that enables one to perform more complicated quantum operations than those initially carried out by those gates. This is the basic idea of a quantum circuit[4].

[1] *First described in:* T. Toffoli, *Bicontinuous extension of reversible combinatorial functions*, Math. Syst. Theory **14**(1), 13–23 (1981).

[2] D. Deutsch, *Quantum computational networks*, Proc. R. Soc. London A **425**, 73–90 (1989).

[3] E. Fredkin, T. Toffoli, *Conservative logic*, Int. J. Theor. Phys. **21**, 219–253 (1982).

[4] *More details in:* A. Galindo, M. A. Martín-Delgado, *Information and computation: Classical and quantum aspects*, Rev. Mod. Phys. **74**(2), 347–423 (2002); M. Nielsen, I. Chuang, *Quantum Computation and Quantum Information* (Cambridge University Press, Cambridge, 2000).

quantum mirage – a phenomenon of coherent projection of electronic structure in atomic scale arrangements. It is illustrated in Figure Q.10 for the case in which it was first observed.

A **scanning tunneling microscope** (STM) was used to position 36 cobalt atoms in an elliptical ring. Then another cobalt atom at one of the two focal points of the ring was placed. The **Kondo effect** was measured by the STM at both focal points. It can be seen to take place not only at the focal point with the lone cobalt atom, but also at the vacant focal point, where it is much weaker. It looks like a mirage, which gives gave the name to the phenomenon.

The phenomenon promises to make possible data transfer without conventional electrical wiring.

First described in: H. Manoharan, C. Lutz. D. Eigler, *Quantum mirages formed by coherent projection of electronic structure*, Nature **403**(5), 512–515 (2000).

Figure Q.10 Visualization of the quantum mirage.

quantum number – an integer (in the case of spin a half-integer) that labels the state of a quantum system.

quantum phase transition – a phase transition taking place at absolute zero temperature and driven solely by quantum fluctuations associated with **Heisenberg's uncertainty principle**. The fluctuations, being quantum mechanical in nature, exhibit scale invariance in both space and time. The location of the transition is indicated as a quantum critical point. At the critical point the critical fluctuations extend over the entire system. Unlike classical critical points, where the critical fluctuations are limited to a narrow region around the phase transition, the influence of a **quantum critical point** is felt over a wide range of temperatures above the quantum critical point, so the effect of quantum criticality is felt without ever reaching absolute zero.

Quantum phase transitions can lead to non-Fermi behavior of electrons and unusual states of matter.

More details in: S. Sachdev, *Quantum Phase Transitions* (Cambridge University Press, Cambridge, 2001).

quantum point contact – a narrow constriction of a two-dimensional conducting channel. An example of such a structure and its related conductance curve are illustrated in Figure Q.11. A short constriction in the burred **two-dimensional electron gas** is defined by the electrically connected surface gates shaped like opposed knives. The gates are negatively biased in order to deplete the two-dimensional electron gas under them and form a narrow conducting channel. For

quantum teleportation

Figure Q.11 Schematic layout and conductance of the quantum point contact induced in burred two-dimensional electron gas by surface gates. After B. J. van Wees, H. van Houten, C. W. J. Beenakker, J. G. Williamson, L. P. Kouwenhoven, D. Van der Marel, C. T. Foxon, *Quantized conductance of point contacts in a two-dimensional electron gas*, Phys. Rev. Lett. **60**(9), 848–850 (1988).

the channel with N transmitting modes its conductance is $G = N(2e^2/h)$ –it is quantized in units of $2e^2/h$.

The externally applied gate voltage V_g controls the channel width. By an increase of the negative bias, the width of the point contact can be gradually reduced, until it is fully pinched off. As the conducting channel is widened, the number of transverse eigenstates below the Fermi level increases. Conductance steps corresponding to increasing values of N appear. Note that if one is dealing with a nonideal channel in which scattering events reducing total current take place, it is important to account for the probability of transmission.

quantum teleportation – the "disembodied transport" of a quantum state (particle) from one place to another. It is an alternative to the hardly realized direct move of such states through space. Direct transmission would have to be along a channel that maintains quantum coherence, which requires the complete isolation of the transmitted quantum object. An approach based on the measurement of all observables of the quantum system, their subsequent transmission in the form of conventional digital codes and restoring the original quantum system by a receiver on the basis of measured observables, meets fundamental limitations. It cannot be applied to transmission of an unknown quantum system because any measurement made on a quantum system uncontrollably perturbs it, destroying the original. Accurate copying (cloning) of an unknown quantum state is also not possible (**non-cloning theorem**) because it would be equivalent to a simultaneous sharp measurement of all observables of the system, including non-commuting ones, which is forbidden by the uncertainty principle. Quantum teleportation avoids these restrictions.

A system for quantum teleportation should have a source of **Einstein–Podolsky –Rosen (EPR) pairs**, quantum decoder and quantum synthesizer connected with a sender (conventionally named Alice) and a receiver (conventionally named Bob) by communication channels, as schematically shown in Figure Q.12. In order to teleport an unknown quantum state Ψ represented by a quantum particle from Alice to Bob, they are each given one particle of the entangled EPR pair. Then Alice brings together her particle and the particle in an unknown state and performs jointly on those two particles a special measurement using the quantum decoder.

This measurement has four possible outcomes – it is, in fact, the same measurement that is performed at the end of the two-bit communication process. Alice then communicates the result of the measurement to Bob by any classical channel. According to these data, Bob, who has the other member of the EPR pair, performs one of four operations on his particle using the quantum synthesizer. As a result, Bob gets the particle in exactly the same state that Alice's particle was originally in. Thus, if Alice and Bob share an Einstein–Podolsky–Rosen pair Bob can reproduce Alice's unknown quantum state with the assistance of only two classical bits of information sent by Alice to Bob through a classical channel. Quantum teleportation supposes sending an unknown quantum state from one place, where it vanishes, to another place, where it shows up, without really traversing the intermediate space.

First described in: C. H. Bennett, G. Brassard, C. Crépeau, R. Jozsa, A. Peres, W. K. Wootters, *Teleporting an unknown quantum state via dual classical and Einstein–Podolsky–Rosen channels*, Phys. Rev. Lett. **70**(13), 1895–1899 (1993) – theory; D. Bouwmeester, J. W. Pan, K. Mattle, M. Eibl, H. Weinfurter, A. Zeilinger, *Experimental quantum teleportation*, Nature **390**, 575–579 (1997) – experiment.

Figure Q.12 Principle of quantum teleportation.

quantum well – a finite region of space restricted by potential barriers. In semiconductor electronics this is typically a structure composed of semiconductors with different band gaps or a semiconductor and a dielectric, in which the material with a small gap in a low-dimensional form is placed in between the regions of the large gap materials. It acts as a well for charge carriers. The material with the small gap creates a well, while the large gap material forms potential barriers for this well. Such an arrangement can be translated in space resulting in **multiquantum wells**.

Superlattices fabricated by deposition of semiconductors with dissimilar electronic properties are considered to be a classical example of solid-state quantum wells. Nevertheless, quantum wells are also formed when semiconductor low-dimensional structures are imbedded into a dielectric matrix even without lattice

quantum well

matching of the materials. Silicon nanoclusters imbedded into silicon dioxide can be an example of these.

According to the confinement energy schemes for electrons and holes, quantum wells are usually classified in two main groups, which are labeled **type I** and **type II**. Energy relationships in these groups are illustrated by Figure Q.13, where the material A is expected to have smaller band gap than the material B.

In the **type I** or straddling alignment, the bottom of the conduction band in the well (material A) lies lower in energy than in the barrier (material B), but for the top of the valence band it is quite the opposite: in the well it is higher than in the barrier. Electrons and holes are both localized and confined in the same region, which is the well. This type of the well can be called a "spatially direct gap semiconductor" having energy diagrams very similar to the band structure of direct gap semiconductors.

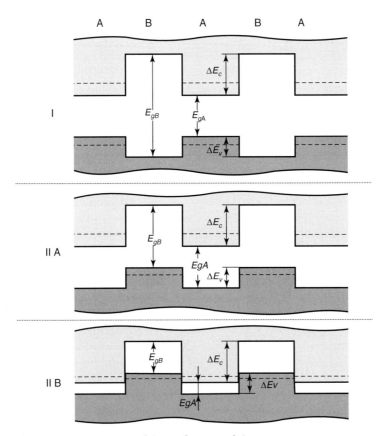

Figure Q.13 Arrangements of the confinement of electrons and holes in quantum wells formed by the small gap material A and large gap material B with the band gaps E_{gA} and E_{gB}, respectively.

In **type II**, both the bottom of the conduction band and the top of the valence band in the well are located lower in energy than those in the barrier. As a result, electrons appear to be localized and confined in the well, while holes come to the barrier. Meanwhile, the difference in energy between the bottom of the conduction band in the barrier and the top of the valence band in the well can be positive or negative in this case. If it is positive, the energy gap extends through the whole structure, as shown in the figure for type IIA (staggered alignment). Such a structure can be referred to as "spatially indirect gap semiconductor". If the energy difference is negative (or zero), as shown for type IIB (misaligned) structures, these can be considered as "spatially gapless semiconductors".

Quantum wells are one of the most important elements of many of nanoelectronic and optoelectronic devices.

quantum well laser – a semiconductor laser that has an active region composed of ultrathin layers (~20 nm or less) forming narrow potential wells for injected carriers. Quantization of electronic states in the wells, among other benefits, brings reduction of the threshold current as compared to semiconductor lasers with a uniform active region.

First described in: J. P. van der Ziel, R. Dingle, R. C. Miller, W. Wiegmann, W. A. Nordland, Jr, *Laser oscillations from quantum states in very thin GaAs-Al$_{0.2}$Ga$_{0.8}$As multilayer structures*, Appl. Phys. Lett. **26**(8), 463–465 (1975) – first experimental demonstration.

quantum well solar cell – a multiple band gap **photovoltaic** device with intermediate properties between a **heterojunction solar cell** and a **tandem solar cell**.

First described in: K. W. J. Barnham, G. Duggan, *A new approach to high efficiency multi-band-gap solar cells*, J. Appl. Phys. **67**, 3490–3493 (1990).

quantum wire → **quantum confinement**.

quantum yield – the number of times that a defined radiation-induced event occurs per photon absorbed by the system. It is a measure of the efficiency with which the absorbed light produces some effect. In most processes quantum yield is less than 1 since not all photons are absorbed productively. Quantum yields greater than 1 are possible for radiation-induced avalanche reactions, in which a single photon may trigger a long chain of transformations.

quantum Zeno paradox – a continuously observed quantum system cannot make transitions. It is a consequence of the **Zeno paradox** that motion is impossible because before reaching a goal one has to get halfway there. Before that, one had to get halfway to the midpoint, and so on. The wave function of a quantum system evolves continuously according to the Schrödinger equation. When a measurement is made, there is a discontinuous reduction of the wave packet. Thus, if one could make a continuous set of measurements on a quantum system, this system would never evolve. Another name for the paradox is the **watched pot effect**, after the proverb "A watched pot never boils".

quasicrystal – a structure which cannot be indexed to any **Bravais lattice** and which has a symmetry intermediate between a crystal and a liquid. It has a well-defined, discrete point group symmetry, like a crystal, but one which is explicitly incompatible with periodic translational order (for example exhibiting five-, eight-, or twelve-fold symmetry axes). Instead, quasicrystals possess a particular kind of translational order known as quasiperiodicity.

First described in: D. Shechtman, I. Blech, D. Gratias, J. W. Cahn, *Metallic phase with long range orientational order and no translation symmetry*, Phys. Rev. Lett. **53**(20), 1951–1953 (1984).

First described in: R. McGrath, U. Grimm, R. D. Diehl, *The forbidden beauty of quasicrystals*, Physics World **17**(12), 23–27 (2004).

quasiparticle – a single particle excitation.

quasi-steady-state hole-burning → **spectral hole-burning**.

qubit – an acronym for a **qu**antum **bit**. It is a quantum system that, like an ordinary computer bit, has two accessible states. But unlike a classical system, a qubit can exist in a superposition of these two states at the same time. Qubits are just quantum two-level systems such as the spin of an electron, the nuclear spin or the polarization of a photon and can be prepared in a coherent superposition state corresponding to "0" or "1". Important times characterizing performance of different two-level quantum systems are tabulated below (Table Q.2).

The time t_{sw} is the minimum time required to execute one quantum gate. It is estimated as $h/\Delta E$, where ΔE is the typical energy splitting in the two-level system.

Table Q.2 Switching (t_{sw}) and phase coherence (t_{ph}) times of two-level quantum systems (D. P. DiVincenzo, *Quantum computation*, Science **270**, 255–261 (1995)).

Quantum system	$t_{sw}(s)$	$t_{ph}(s)$	t_{ph}/t_{sw}
Mössbauer nucleus	10^{-19}	10^{-10}	10^{9}
Electrons: GaAs	10^{-13}	10^{-10}	10^{3}
Electrons: Au	10^{-14}	10^{-8}	10^{6}
Trapped ions: In	10^{-14}	10^{-1}	10^{13}
Optical cavity	10^{-14}	10^{-5}	10^{9}
Electron spin	10^{-7}	10^{-3}	10^{4}
Electron quantum dot	10^{-6}	10^{-3}	10^{3}
Nuclear spin	10^{-3}	10^{4}	10^{7}

The duration of a π tipping pulse cannot be shorter than this uncertainty time for each system. The phase coherence time t_{ph} is the upper bound on the time scale over which a complete quantum computation can be executed accurately. The ratio of these two times gives the largest number of steps permitted in a quantum computation using these quantum bits.

First described in: B. Schumacher, *Quantum coding*, Phys. Rev. A **51**(4), 2738–2747 (1995).

R

From Rabi flopping to Rydberg gas

Rabi flopping – the sinusoidal population transfer in two-level systems → **Rabi frequency**.

Rabi frequency arises in an analysis of a system with two energy levels. The two-level system is a basic model approximating accurately many physical systems, in particular those with resonance phenomena. The basic two-level system is a spin 1/2 particle (spin-up and spin-down). Consider the motion of the spin 1/2 particle in the presence of a uniform but time-dependent magnetic field. Suppose the particle has a magnetic moment $\mu = \gamma \mathbf{S}$, where γ is the gyromagnetic ratio and $\mathbf{S}$ is the spin operator. In a static magnetic field $\mathbf{B}_0$ lying along the z axis the motion of the magnetic moment can be described through

$$\mu_x = \mu \sin\theta \cos\omega_0 t; \quad \mu_y = \mu \sin\theta \sin\omega_0 t; \quad \mu_z = \mu \cos\theta.$$

Here θ is the angle between μ and $\mathbf{B}_0$ and $\omega_0 = -\gamma B_0$ is the **Larmor frequency**. Now we introduce a magnetic field $\mathbf{B}_1$ and this field rotates in the $x - y$ plane at the Larmor frequency. The total magnetic field is $\mathbf{B} = (B_1 \cos\omega_0 t, -B_1 \sin\omega_0 t, B_0)$ (called the **Rabi pulse**) and we will find the motion of μ in this field. The solution is simple in a rotating coordinate system where one has that the moment μ precesses about the field at the rate:

$$\omega_R = \gamma B.$$

These frequency is called the **Rabi frequency**. If the moment initially lies along the z axis, then its tip traces a circle in the $y - z$ plane. At time t it has precessed through an angle $\phi = \omega_r t$ and the component z of the moment is given by $\mu_z(t) = \mu \cos\omega_R t$. At time $T = \pi/\omega_R$, the moment points along the negative z axis: it has turned over. This phenomenon is called **Rabi flopping**.

If the magnetic field rotates at a generic frequency different from the resonance frequency ω_0, it is straightforward to demonstrate that the moment precesses about the field at a rate (called the **effective Rabi frequency**)

$$\omega_R^{eff} = \sqrt{\omega_R^2 + (\omega_0 - \omega)^2}.$$

What is What in the Nanoworld. Victor E. Borisenko and Stefano Ossini
Copyright © 2008 WILEY-VCH Verlag GmbH & Co. KGaA, Weinheim
ISBN: 978-3-527-40783-5

Rabi oscillations

The z-component of the moment oscillates but, unless $\omega = \omega_0$, never completely inverts. The rate of oscillation (**Rabi oscillation**) depends on the magnitude of the rotating filed. Its amplitude depends on the frequency difference $\delta = \omega - \omega_0$ with respect to ω_R, called detuning. The moment in the field oscillates between two states at the effective Rabi frequency, where the contrast of the oscillation is determined by detuning.

The Rabi resonance method can be used to measure the energy splitting between two states. It plays an important role in quantum optics. It is fast becoming popular in the analysis and design of spin **qubit** systems.

First described in: I. I. Rabi, *On the process of space quantization*, Phys. Rev. **49**(4), 324–328 (1936); I. I. Rabi, *Space quantization in a gyrating magnetic field*, Phys. Rev. **51**(8), 652–654 (1937).

Recognition: in 1944 I. I. Rabi received the Nobel Prize in Physics for his resonance method for recording the magnetic properties of atomic nuclei.

See also www.nobel.se/physics/laureates/1944/index.html.

Rabi oscillations – oscillations in a two-level system $\rightarrow$ **Rabi frequency**.

Rabi pulse – if the oscillating field entering the determination of the **Rabi frequency** is resonant, that is it satisfies $\omega = \omega_0$, then it is called a Rabi pulse $\rightarrow$ **Rabi frequency**.

racemization – a process that converts an **enantiomer** into a racemic mixture of enantiomers.

Rademacher functions – the system of orthogonal functions defined as $f_n(x) = f_0(2^n x)$ with $n = 1, 2, 3 \ldots$; $f_0(x) = 1$ for $0 \leq x < 1/2$; $f_0(x) = -1$ for $1/2 \leq x < 1$; $f_0(x+1) = f_0(x)$ for $-\infty < x < \infty$. This system is not a complete one. Its completion is achieved within the system of **Walsh functions**.

First described in: H. A. Rademacher, *Einige Sätze über Reihen von allgemeinen Orthogonalenfunktionen*, Math. Ann. **87**, 112–138 (1922).

radical (chemical) – a group of atoms, which occurs in different compounds and remains unchanged during chemical reactions.

Rafii–Tabar and Sutton potential – the potential energy of an atom i in a solid accounting for its long range interaction with neighbors and having the form: $V_i = \epsilon/2 \sum_j (a/r_{ij})^n - c\epsilon[\sum_j (a/r_{ij})^m]^2$, where a is chosen to be the lattice constant, and r_{ij} is the distance between the i-th and j-th atoms. The constant c is chosen so that the energy of an atom in the bulk has its minimum value when the r_{ij} assume the values they would have in a **face-centered cubic** lattice with a lattice constant of a. The parameters ϵ, n and m are obtained by fitting the cohesive energy and the bulk modulus, while n and m are required to be integers with m greater than or equal to 6. The potential is suitable for alloys.

First described in: H. Rafii-Tabar, A. P. Sutton, *Long-range Finnis-Sinclair potentials for fcc metallic alloys*, Phil. Mag. Lett. **63**, 217–224 (1991).

More details in: G. A. Mansoori, *Principles of Nanotechnology: Molecular-based Study of Condensed Matter in Small Systems* (World Scientific, Singapore, 2005).

Raman effect – the light scattered from solids contains components which are shifted in frequency with respect to the incident monochromatic light, as illustrated in Figure R.1.

The difference between the frequency of the incident **photon** and the frequency of the scattered photon is called **Raman frequency**. The shifts are caused by the gain or loss of characteristic amounts of energy when atoms or molecules change their rotational or vibrational motion. Light scattering involves two photons–one in (ω), one out (ω'). A photon is scattered inelastically by a crystal, with creation or annihilation of a **phonon** (Ω) or **magnon**. The photon at $\omega' = \omega - \Omega$ is called the **Stokes line** and the process of the phonon emission is a **Stokes process**. When a phonon is absorbed during scattering, such process is called an **anti-Stokes process** and corresponding photons with $\omega' = \omega + \Omega$ form the **anti-Stokes line**.

The process is called **Brillouin scattering** when an **acoustic phonon** is involved and **polariton** scattering when an optical phonon is involved. Similar processes occur with **magnons**.

The scattered radiation vanishes for certain choices of the light polarization and scattering geometries. These so-called **Raman selection rules** are a result of the symmetries of the medium and of the vibrational modes involved in the scattering.

First described in: C. V. Raman, *A new radiation*, Indian J. Phys. **2**, 387–398 (1928) and G. Landsberg, L. Mandelstam, *Über die Lichtzerstreuung in Kristallen*, Z. Phys. **50**(11/12), 769–780 (1928).

Recognition: in 1930 only C. V. Raman received the Nobel Prize in Physics for his work on the scattering of light and for the discovery of the effect named after him.

See also www.nobel.se/physics/laureates/1930/index.html.

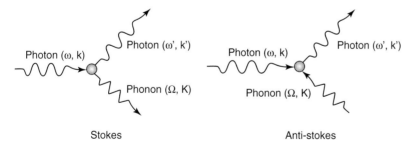

Figure R.1 Raman scattering of a photon with emission or absorption of a phonon.

Raman frequency $\rightarrow$ **Raman effect**.

Raman selection rules $\rightarrow$ **Raman effect**.

Raman spectroscopy

Raman spectroscopy detects vibrational frequencies in molecules or molecule-like structures. The idea is that when an intense beam of light enters a sample its scattering due to the **Raman effect** occurs at frequencies higher and lower than the frequency of the incident light. The difference between the frequencies of the incident and scattered light is equal to the vibrational frequency of the molecular bond responsible for the Raman scattering. Thus, by analyzing the spectral dependence of Raman scattering, different molecular bonds and various functional groups in the molecule can be recognized.

rapid single flux quantum device – a two-terminal device formed by a Josephson junction → **Josephson effects** where a flux quantum is used as a **bit** of information. Logical 1 and 0 are represented by the existence or absence of a single flux quantum. Such devices have unusual dynamics due to the macroscopic quantum nature of **Cooper pairs** (that have an integer spin) employed as charge carriers. They exhibit a clock frequency beyond hundreds of gigahertz and extremely low power consumption. The main limit for a practical use of the devices is their low-temperature operation mode (4–5 K).
First described in: D. Zinoviev, K. Likharev, *Feasibility study of RSFQ-based selfrouting nonblocking digital switches,* IEEE Trans. Appl. Superconductivity **7**(2), 3155–3163 (1997).

rare earth elements – 17 metallic elements from Group IIIA of the Periodic Table ranging in atomic number from 21 (scandium) to 71 (lutetium). The elements from atomic number 57 (lanthanum) to 71 (lutetium) are also called **lanthanides**.
First described in: S. A. Arrhenius, Sitz. Ber. Wiener Akad., II Abteilung **96**, 831 (1887).

Rashba coupling = Rashba effect.

Rashba effect – spin splitting of electron energy levels in solids without external magnetic field as a result of enhanced spin-orbit coupling in the material. In bulk solids it is connected with a lack of inversion symmetry of the crystal potential. In low-dimensional structures, in particular in **quantum wells**, this effect can be superimposed by an inversion asymmetry of the confinement potential. Thus, two components, that is material-related spin–orbit coupling and quantum confinement, may contribute to the total spin splitting.

An asymmetric quantum well has an interface electric field, which is described along the normal of the well plane and lifts the spin degeneracy of the two-dimensional electron energy bands by coupling the electron spin and orbital motion. This spin–orbital coupling is described by a **Hamiltonian** $H_{so} = \alpha_s (\sigma \times \mathbf{k}) \cdot \mathbf{z}$, where σ is the **Pauli spin matrix**, $\mathbf{z}$ is the unit vector along the surface field direction, and $\mathbf{k}$ is the electron wave vector along the plane. This Hamiltonian is usually referred to as a **Rashba term**. The spin–orbit coupling constant is implicitly proportional to the strength of the built-in surface electric field. The effective mass Hamiltonian with the spin–orbit term has the form:

$$H = \frac{\hbar^2 k^2}{2m} + \alpha_s(\boldsymbol{\sigma} \times \mathbf{k}) \cdot \mathbf{z}.$$

It yields the electron energy dispersion related to the motion along the plane:

$$E(k) = \frac{\hbar^2 k^2}{2m} \pm \alpha_s k.$$

The important characteristic of this dispersion relation is that the spins are degenerate at $k = 0$ and the spin splitting linearly increases with k, where α_s is the spin–orbit coupling constant, being a measure of the strength of this coupling.

First described in: E. I. Rashba, *Properties of semiconductors with a loop of extrema*, Fiz. Tverd. Tela (Lenigrad) **2**(6), 1224–1238 (1960) – in Russian – introduction of the spin–orbit term; Y. A. Bychkov, E. I. Rashba, *Oscillatory effects and the magnetic susceptibility of carriers in the inversion layers*, J. Phys. C **17**(33), 6039–6045 (1984) – first demonstration of the role of spin coupling in two-dimensional structures.

Rashba term → **Rashba effect**.

rational design → **structure-based engineering**.

Rayleigh–Jeans law expresses the energy density of black body radiation of wavelength λ corresponding to a temperature of the black body T as $P(\lambda) = 8\pi k_B T/\lambda^4$.

First described in: J. W. S. Rayleigh, *Remarks upon the law of complete radiation*, Phil. Mag. **49**, 539–540 (1900); J. H. Jeans, *On the partition of energy between matter and ether*, Phil. Mag. **10**, 91–98 (1905), and independently in: A. Einstein, *Über einen die Erzeugung und Verwandlung des Lichtes betreffenden heuristischen Gesichtspunkt*, Ann. Phys. **17**, 132–148 (1905).

Rayleigh law of scattering → **Rayleigh scattering (of light)**.

Rayleigh limit – the restriction of wavefront error to within a quarter of a wavelength of a true spherical surface to assure essentially perfect image quality.

Rayleigh line – the element of a spectrum line in scattered radiation having a frequency equal to that of the corresponding incident radiation, due to ordinary or **Rayleigh scattering**.

Rayleigh range – the axial distance from the point of minimum beam waist W to the point where the beam diameter has increased to $(2W)^{1/2}$ in the region of a Gaussian beam focus by a diffraction-limited lens.

First described in: Lord Rayleigh (J. W. Strutt), *Images formed without reflection or refraction*, Phil. Mag. **11**, 214–218 (1881).

Rayleigh scattering (of light)

Rayleigh scattering (of light) – the term employed to describe scattering of light waves by a particle much smaller than their wavelength. The scattered wave amplitude is directly proportional to the volume of the particle and inversely proportional to the square of the wavelength. The scattering intensity is thus proportional to the inverse fourth power of the wavelength, which is the **Rayleigh law of scattering**. It predicts that short-wavelength light will be scattered by a medium containing small particles more efficiently than long-wavelength radiation. This peculiarity of light scattering is responsible for the blue color of the sky on a sunny day and a red sunset. It results from air pollutants or more generally from entropy fluctuations. Blue light, having half of the wavelength of red, is scattered sixteen times as well. The red sunset is due to the fact that such selective scattering removes the blue rays from the direct beam much more effectively than the red. The great thickness of the atmosphere traversed is sufficient to produce the visible effect of giving the transmitted light its intense red color.

First described in: J. W. S. Rayleigh, *On the light from the sky, its polarization and colour*, Phil. Mag. **41** 107–120 and 274–279 (1871).

Rayleigh's equation (for group velocity of waves) expresses the group velocity as a function of the wave velocity v and wavelength λ in the form $v_{gr} = v - \lambda(dv/d\lambda)$.

First described in: J. W. S. Rayleigh, *On waves propagating along the plane surface of an elastic solid*, Proc. Lond. Math. Soc. **17**, 4–11 (1887).

RBS – an acronym for **Rutherford backscattering spectroscopy**.

reactance → impedance.

real contact area → Bowden–Tabor law.

reciprocal lattice – an invariant mathematical construction defined by the direct lattice of a crystal. If $\mathbf{a}_I (I = 1, 2, 3)$ are the primitive lattice vectors of the crystal in the direct lattice space, then its reciprocal lattice vectors $\mathbf{a}_j^* (j = 1, 2, 3)$ are given by the relation $\mathbf{a}_I \cdot \mathbf{a}_j^* = \delta_{ij}$, where $\delta_{ij} = 1$ if $i = j$, and $\delta_{ij} = 0$ if $i \neq j$. The two vectors, one in the direct lattice space (the $\mathbf{a}_I$) and the other in the reciprocal space (the $\mathbf{a}_j^*$) are orthogonal to each other. Thus, the vectors of a reciprocal lattice are $\mathbf{a}_1^* = (\mathbf{a}_2 \times \mathbf{a}_3)/V, \mathbf{a}_2^* = (\mathbf{a}_3 \times \mathbf{a}_1)/V, \mathbf{a}_3^* = (\mathbf{a}_1 \times \mathbf{a}_2)/V$, where $V = \mathbf{a}_1(\mathbf{a}_2 \times \mathbf{a}_3)$ is the volume of the unit cell in direct lattice space. Each set of lattice planes in a crystal is represented by a point in the reciprocal lattice. All symmetry properties possessed by the direct lattice are inherent in the reciprocal lattice.

reciprocity failure → Bunsen–Roscoe law.

recognition – molecular recognition describes the capability of a molecule to form selective bonds with other molecules or with substrates, based on the information stored in the structural features of the interacting partners. Molecular recognition processes may play a key role in molecular devices (in particular **molecular electronics**) by driving the fabrication of devices and integrated circuits from elementary building blocks, incorporating them into supramolecular arrays, allowing for selective operations on given species acting as dopants, and controlling the response to external perturbations represented by interacting partners or applied field.

reconstruction → **surface reconstruction**.

red shift – a systematic shift of an optical spectrum toward the red end of the optical range.

redox – the term originated from the two concepts of reduction and oxidation. These two terms go together, because in a chemical reaction, one cannot occur without the other. Electrons lost by one compound must be gained by another. Reduction can also be considered to be the reducing of an atom's positive charge, and oxidation its opposite (gaining positive charge). Reduction describes the uptake of an electron by a molecule, atom or ion. Oxidation describes a loss of an electron by the same species.

Reed–Muller algebra – operations within a universal basis including constant 1, **XOR**, **AND**, and **NOT** operations over Boolean → **Boolean algebra** variables. Expressions in this algebra are classified as fixed and mixed polarity expressions.
   First described in: I. Reed, *A class of multiple-error-correcting codes and a decoding scheme*, IEEE Trans. Inform. Theory **4**, 38–49 (1954); D. Muller, *Applications of Boolean algebra to switching circuit design and to error correction*, IRE Trans. Elec. Comp. **3**, 6–12 (1954).

Reetz method provides electrochemical synthesis of metal nanocrystals via an electrorefining process. It consists of six elementary steps illustrated in Figure R.2.
   These are: (1) oxidative dissolution of metal anode, (2) migration of metal ions to the cathode, (3) reduction of the ions to a zero-valent state, (4) formation of particles by nucleation and growth, (5) arrest of the growth by capping agents, (6) precipitaton of the particles. The capping agents are typically quaternary ammonium salts containing long-chain alkanes such as tetraoctyl-ammonium bromide.
   The size of the nanocrystals can be tuned by altering the current density, the distance between the electrodes, the reaction time, the temperature, and the polarity of the solvent. Usually, low current densities yield larger particles, while high current densities yield smaller particles.
   The method has been successfully employed to synthesize both Ni, Co, Fe, Ti, Ag, Au nanoparticles and bimetallic colloids like Pd-Ni, Fe-Co, Fe-Ni and others

reflection high-energy electron diffraction (RHEED)

Figure R.2 Schematic illustration of the electrochemical synthesis of metal nanocrystals by the Reetz method.

using two anodes consisting of either metals. Moreover, mono and bimetallic nanoparticles can be fabricated by reduction of their salts dissolved in the electrolyte.

First described in: M. T. Reetz, W. Helbig, *Size-selective synthesis of nanostructured transition metal clusters,* J. Am. Chem. Soc. **116**(16), 7401–7402 (1994).

reflection high-energy electron diffraction (RHEED) – diffraction of electrons reflected from a solid surface when the incident electron beam (5–100 keV) strikes the surface at a grazing angle. The diffraction pattern is used to distinguish monocrystalline, polycrystalline and amorphous structures of solid surfaces. It is widely employed in monitoring monoatomic layer growth from changes in surface atomic roughness. As the film thickness increases during the growth, the intensity of the specular and diffracted electron beams oscillates with a period equal to the atomic or molecular layer thickness. High scattered intensities are associated with flat surfaces, which occur when an atomic layer is complete. Low intensities correspond to atomically rough surfaces with a high density of monolayer thick islands. The technique is indeed useful for *in situ* monitoring of the growth of **superlattice** films.

More details in: P. K. Larsen and P. J. Dobson (eds) *Reflection High-Energy Electron Diffraction and Reflection Electron Imaging of Surfaces* (Plenum Press, New York, 1988).

reflectivity – the ratio of the intensity of the total reflected radiation to that of the total incident radiation.

refraction – a change in the direction of the wave propagation when a wave passes an interface between two media with different **refractive indexes**. At the interface the wave changes direction, its wavelength increases or decreases, but its frequency remains constant.

The phenomenon is schematically illustrated in Figure R.3 where a ruler is placed in water. The light waves travel more slowly in the water than in the air. As a result, the wavelength of the outcoming light decreases and the waves bend at the interface. This causes the end of the ruler to appear shallower than it really is.

Refraction is also responsible for rainbows and for splitting up of white light into a rainbow spectrum as it passes through a transparent prism. Different frequencies of light travel at different speeds, causing them to be refracted at different angles at the prism surface. The different frequencies correspond to different colors observed.

The amount that the light bends during refraction is calculated using **Snell's law**.

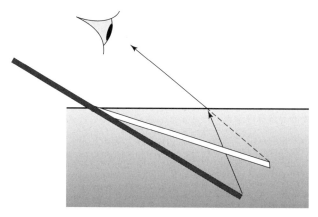

Figure R.3 Refraction of light waves coming out of water.

refractive index in the complex form provides a complete description of the linear optical response of the material. Its real part is the real index of refraction, namely the ratio of the speed of light in vacuum to the speed of light in the material. The imaginary part is the **extinction coefficient** → **dielectric function**.

relativistic particle – a particle moving at a speed comparable with the speed of light.

remanent magnetism – a magnetization left in a specimen after the removal of a magnetic field.

Renner–Teller effect – splitting in the vibrational levels of molecular entities due to even terms in the vibronic perturbation expansion. This is generally a minor effect for non-linear molecular entities compared to the **Jahn–Teller effect**, which is due to the odd terms. For linear molecular entities it is the only possible vibronic effect characteristic of degenerate electronic states.

First described in: R. Renner, *Zur Theorie der Wechselwirkung zwischen Elektronen- und Kernbewegung bei dreiatomigen, stabförmigen Molekülen*, Z. Phys. **92**, 172–193 (1934).

research misconduct – fabrication, falsification, or plagiarism in proposing, performing or reviewing research, or in reporting research results. A finding requires that: there be a significant departure from accepted practices of the relevant research community; and the misconduct be committed intentionally or knowingly, or recklessy; and the allegation be proven by a preponderance of evidence. This new definition of scientific misconduct arose in recent years after the discovery of several cases of misconduct also in nanoscience, as in the case of the nanophysicist Jan Hendrik Schön regarding the discovery of new nanodevices at Bell Laboratories in 1999–2002 and the south-Korean cloning researcher Woo Suk Wang regarding the creation of embryonic stem cells in 2005–2006.

resistance (electrical) – the property of an electric circuit element to obstruct current flow by causing electrical energy to be dissipated in the form of heat.

resistivity – the ratio of the electric field in a material to the density of the electric current passing through it. Phenomenologically, substances are broadly classified into three categories according to the value and temperature dependence of their resistivities: (i) **conductors** (10^{-6}–10^{-4} Ω cm, positive temperature coefficient of resistivity), (ii) **semiconductors** (10^{-4}–10^9 Ω cm, negative temperature coefficient of resistivity), and (iii) **insulators** ($>10^{10}$ Ω cm, in general negative temperature coefficient of resistivity).

RESOLFT – an acronym for a **reversible saturable optically linear fluorescence transitions microscopy**.

resonance – a phenomenon exhibited by a system acted upon by an external periodic driving force in which the resulting amplitude of oscillation of the system becomes large when the frequency of the driving force approaches a certain value, which is usually a natural free oscillation frequency of the system.

resonant scattering – scattering of a photon by a quantum mechanical system (usually an atom or nucleus) in which the system first absorbs the photon by undergoing a transition from one of its energy state to one of higher energy and subsequently reemits the photon by the exact inverse transition.

resonant tunneling – a charge carrier tunneling process in which the electron transmission coefficient through a structure is sharply peaked about certain energies, analogous to the sharp transmission peaks, as a function of wavelength evident through optical filters such as a **Fabry–Pérot resonator**. It usually arises when charge carriers pass through a quantum well in which a quantized energy level is lined up with the energy level in the injected electrode when the external voltage is increased.

The simplest device demonstrating the phenomenon, which is a **resonant tunneling diode**, consists of a **quantum well** formed in a double tunnel barrier structure with emitter and collector contacts. Its energy band diagrams and current voltage characteristic are shown in Figure R.4.

The barriers are thin enough so that under certain circumstances electrons can tunnel from the emitter through the barrier into the well. At the same time the well is thin enough for the density of states to break up into a number of two-dimensional sub-bands. In the example considered here, only one sub-band is localized in the well at an energy $E_1 = \hbar^2 \pi^2 / (2m^* a^2)$ above the conduction band edge, where m^* is the electron effective mass and a is the well width. The important point is that at zero external bias E_1 is higher than the **Fermi energy** E_F in the contact regions on either side of barriers. Therefore, when a vanishingly small voltage is initially applied there are no states in the quantum well for the electrons to tunnel into. As a consequence, the only current flowing at low bias is that due to thermal emission over the barriers, which can be made negligibly small for high barriers. The applied voltage is dropped mostly across the resistive regions, namely the barriers. As the bias is increased the resulting electric field gives rise to appreciable band bending. The emitter region is warped upwards, and the collector region is warped downwards.

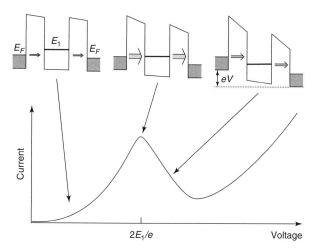

Figure R.4 Conduction band diagrams and current–voltage characteristics of a resonant tunneling diode.

Eventually at a certain voltage, the Fermi energy on the emitter contact is pulled through the resonant level E_1. As the Fermi level lines up with the resonant state in the well, a large current flows because of the increased transmission from emitter to collector. Electrons pass the well with energy and momentum conservation, which provides their highest transmission probability. At the same time, the back flow of electrons is suppressed as they meet a large potential barrier in the collector region. These conditions give an increase in the current up to the peak value.

When the applied voltage is further increased, the energy of the confined state drops below the Fermi energy in the emitter. At this point there are no electrons which can tunnel into the quantum well and still conserve energy and momentum, and the current through the device falls. The gradient of the current–voltage curve is now negative giving rise to a region of **negative differential resistance**, which makes the device useful as a high-frequency amplifier. At still higher bias the current begins to grow again because those electrons which are thermally activated over the barrier experience a large electric field and can contribute a significant current.

For an ideal double-barrier structure with identical barrier and contact parameters the current–voltage curve is symmetrical about the origin. The peak of the current corresponds to the voltage $2E_1/e$. Asymmetry is more typical for real structures arising from several technology-related sources, such as asymmetry in impurity and defect distribution through the structure or differences in roughness of interfaces.

For many applications, devices with a negative differential resistance should have a large peak current and a small valley current, where the latter is the minimum current following the peak current as the magnitude of the voltage increases. Therefore an important figure of merit for such devices is the **peak-to-valley ratio**. For resonant tunneling diodes it is in the range from $4:1$ to $60:1$ at room temperature depending on the geometry and well/barrier materials. It can be much higher at lower temperatures. Note that with a temperature increase the peak-to-valley ratio decreases until negative differential resistance completely vanishes. This reduction is simply related to the increase in off-resonant current over the barrier due to thermionic emission as well as the spreading of the distribution function around the resonant energy, which decreases the peak current. The valley current is also increased at higher temperatures as a result of phonon-assisted tunneling. Peak current density in excess of 10^5 A cm^{-2} may be obtained at room temperature. The highest peak current density and the peak-to-valley ratio are not achievable in the same structure because there is usually a trade-off between these two parameters in terms of device design.

First described in: D. Bohm, *Quantum Theory* (Prentice Hall, Englewood Cliffs, NJ, 1951), p. 283 – the phenomenon; L. V. Iogansen, *On the possibility of resonance passage of electrons in crystals through a system of barriers*, Zh. Eksp. Teor. Fiz. **45**(2), 207–213 (1963) – in Russian – theoretical description; L. Esaki, R. Tsu, *Superlattice and negative differential conductivity in semiconductors*, IBM J. Res. Dev. **14**(1), 61–65 (1970) – experimental resonant tunneling diodes.

Recognition: in 1973 L. Esaki received the Nobel Prize in Physics for his experimental discoveries regarding tunneling phenomena in semiconductors.

See also www.nobel.se/physics/laureates/1973/index.html.

rest mass – the mass of a particle in an inertial coordinate system with three space coordinates and one time coordinate (**Lorentz frame**) in which it is at rest.

Retger law states that properties of crystalline mixtures of isomorphous substances are continuous functions of the percentage composition.

First described by Retger in 1889.

---

reversible logic – logic devices organized in such a way as to permit arbitrary patterns of logically reversible information to be processed in 3D space. They entail no minimum entropy per operation. Reversible or information-lossless circuits are useful for digital signal processing, communication, computer graphics, cryptography, and quantum computation.

More details in: A. N. Al-Rabadi, *Reversible Logic Synthesis* (Springer, Berlin, 2003).

reversible saturable optically linear fluorescence transitions (RESOLFT) microscopy → fluorescence nanoscopy.

First described in: M. Hofmann, C. Eggeling, S. Jakobs, S. W. Hell, *Breaking the diffraction barrier in fluorescence microscopy at low light intensities by using reversibly photoswitchable proteins*, Proc. Natl. Acad. Sci. (PNAS) **102**(49), 17565–17659 (2005).

RHEED – an acronym for a **reflection high-energy electron diffraction**.

rhombic lattice → orthorombic lattice.

rhombohedral lattice → trigonal lattice.

ribosome – a cellular particle consisting of a **protein** and nucleic acid. It is responsible for the transition of the genetic information encoded in nucleic acids into the primary sequence of proteins.

Richardson constant → Richardson equation.

Richardson–Dushman equation → Richardson equation.

Richardson effect → Einstein–de Haas effect.

Richardson equation represents the saturation current density of electrons emitted from a hot metal at a temperature T in the form $j = AT^{1/2} \exp(-\phi/k_B T)$, where A is a constant and ϕ is the thermionic work function. It was later modified by Dushman using the quantum theory concept to give $j = BT^2 \exp(-\phi/k_B T)$, where B is the material-dependent parameter usually called the **Richardson constant**. It got the name **Richardson–Dushman equation**. Both equations have good agreement with experiment because the relationships are dominated by the exponential term. The plot $\log(j/T^2)$ versus $1/T$ is called a **Richardson plot** and helps to determine ϕ.

First described in: O. W. Richardson, *On the negative radiation from hot platinum*, Proc. Camb. Phil. Soc. **11**, 286–295 (1901); S. Dushman, *Electron emission from metals as a function of temperature*, Phys. Rev. **21**(6) 623–636 (1923).

More details in: S. Dushman, *Thermionic emission*, Rev. Mod. Phys. **2**(4), 381–476 (1930).

Recognition: in 1928 O. W. Richardson received the Nobel Prize in Physics for his work on the thermionic phenomenon and especially for the law named after him.

Richardson plot

See also www.nobel.se/physics/laureates/1928/index.html.

Richardson plot → **Richardson equation.**

Righi–Leduc coefficient → **Righi–Leduc effect.**

Righi–Leduc effect – appearance of a transverse temperature gradient in an electronic conductor when it is placed in a magnetic field **H** and a temperature difference is maintained along it. It is described by the equation $\nabla_{tr} T = S\mathbf{H} \times \nabla_{long} T$, where S is the **Righi–Leduc coefficient.**

First described in: A. Righi, Mem. Acc. Lincei **4**, 433 (1887); M. A. Leduc, J. Phys. 2e série **6**, 378 (1887).

Ritz procedure → **variation principle.**

Rivest–Shamir–Adleman algorithm forms a base of an encryption and decryption. It has the property that publicly revealing an encryption key does not thereby reveal the corresponding decryption key. In the algorithm two large prime numbers are first generated, then multiplied, and through additional operations a set of two numbers that forms the encryption key and another set that constitutes the decryption key are obtained. Once the two keys have been established, the original prime numbers can be discarded. For encryption/decryption both keys are needed, but only the owner of the decryption key needs to know it. The security of the algorithm depends on the difficulty of factoring large integers.

First described in: R. L. Rivest, A. Shamir, L. Adleman, *A method for obtaining digital signatures and public-key cryptosystems*, Commun. Assoc. Comput. Mach. **21**(2), 120–126 (1978).

R-matrix theory → **random matrix theory.**

RNA – an acronym for **r**ibo**n**ucleic **a**cid, which is a complex compound of high molecular weight that functions in cellular **protein** synthesis and replaces **DNA** as a carrier of genetic codes in some viruses. It consists of ribose nucleotides in strands of varying lengths. The structure varies from helical to uncoiled strands. The nitrogenous bases in RNA are adenine, guanine, cytosine, and uracil.

There are three main types of RNA: messenger RNA (**mRNA**), transfer RNA (**tRNA**) and ribosomal RNA (**rRNA**). In protein formation, mRNA carries codes from the DNA in the nucleus to the sites of protein synthesis in the ribosomes. Ribosomes are composed of rRNA and protein; they can "read" the code carried by the mRNA. A sequence of three nitrogenous bases in mRNA specifies incorporation of an amino acid. The compound tRNA brings the amino acids to the ribosomes where they are linked into proteins. It is sometimes called soluble or activator, and contains fewer than 100 nucleotide units. Other types of RNA contain thousands of units.

First described in: R. W. Holley, J. Apgar, G. A. Everett, J. T. Madison, M. Marquisee, S. H. Merrill, J. R. Penswick, A. Zamir, *Structure of ribonucleic acid*, Science **147**, 1462–1465 (1965) – the structure of RNA.

Recognition: in 1968 R. W. Holley, H. G. Khorana, and M. W. Nirenberg received the Nobel Prize in Physiology or Medicine for their interpretation of the genetic code and its function in protein synthesis.

See also www.nobel.se/medicine/laureates/1968/index.html.

roadmap – an extended look at the future of a chosen field of research composed from the collective knowledge of the researchers involved in that field.

Roosbroek–Schockley relation links an emission rate R_{cv} for transitions from the conduction band to the valence band and the absorption rate P_{vc} for inverse transitions from the valence to conduction band in semiconductors and dielectrics at thermal equilibrium as $P_{vc}\rho(v) = R_{cv}(v)$, where $\rho(v)$ is the photon density (photon energy density divided by the photon energy).

First described in: W. van Roosbroek, W. Schockley, *Photon radiative recombination of electrons and holes in germanium*, Phys. Rev. **94**(6), 1558–1560 (1954).

rotamers → **isomers**.

rotaxane (rotor and axle) – a **supramolecular** assembly in which one or many cycle units are threaded onto a linear segment. Different types of rotaxanes are illustrated in Figure R.5.

Stopper groups preventing the macrocyclic rings from dethreading are their typical parts. Those having no stopper groups are referred to as **pseudorotaxanes**.

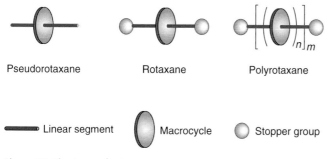

Pseudorotaxane   Rotaxane   Polyrotaxane

━━━▷ Linear segment   Macrocycle   Stopper group

Figure R.5 The types of rotaxanes.

rRNA – ribosomal RNA → **RNA**.

Ruderman–Kittel oscillations = **Friedel oscillations**.

First described in: M. A. Ruderman, C. Kittel, *Indirect exchange coupling of nuclear magnetic moments by conduction*, Phys. Rev. **99**(1), 96–102 (1954).

Russell–Saunders coupling – a process for building many-electron single-particle eigenfunctions of orbital angular momentum and spin. The orbital functions are combined to make an eigenfunction of the total spin angular momentum and then the results are combined into eigenfunctions of the total angular momentum of the system.

First described in: H. N. Russell, F. A. Saunders, *New regularities in the spectra of the alkaline earths*, Astrophys. J. **61**, 38–69 (1925).

Rutherford backscattering spectroscopy (RBS) – a technique of nondestructive elemental analysis of condensed matter. Monoenergetic light ions, usually H^+, He^+, C^+, N^+, or O^+ with energies in the range 0.5–4.0 MeV, are used to probe a sample. When these ions strikes atoms in the target sample, the energy of backscattered ions appears to be characteristic of the scattering atoms and their depth location. The energy distribution and the intensity of the scattered ion beam recorded with a suitable detector is a function of atomic number, concentration and depth distribution of the atoms scattering the ions bombarding the sample. The thickness of thin films and depth profiles of foreign atoms can be measured within a few hundred nanometers of the surface when the foreign atoms are of greater atomic mass than the host material.

Rutherford scattering – elastic scattering of charged atomic particles by the Coulomb field of an atomic nucleus.

Rutherford scattering cross section gives effective cross section for scattering of two non-relativistic particles interacting via Coulomb forces within a solid angle $d\Omega$ in the form:

$$\frac{d\sigma}{d\Omega} = \left(\frac{q_1 q_2}{2mv^2}\right)^2 \sin^{-4}\left(\frac{\theta}{2}\right),$$

where q_1 and q_2 are charges of the particles, $m = m_1 m_2 / (m_1 + m_2)$, m_1 and m_2 are masses of the particles, v is the velocity of one particle with respect to the other, and θ is the scattering angle.

First described in: E. Rutherford, *On the scattering of α and β particles by matter and the structure of the atom*, Phil. Mag. **21**, 669–688 (1911).

Rydberg atom – a cooled excited atom in which an outer electron has a radius larger than that in the ground state. An ensemble of such atoms is called a **Rydberg gas**. The more highly excited the atoms are, the more susceptible they are to environmental conditions, and the stronger their interactions with each other. Rydberg atoms do not move or collide because they are cooled, but the electron orbits of adjacent atoms can overlap. This gives rise to a distinction in properties between the Rydberg gas and a collection of the same atoms in the ground state.

Rydberg constant → **Rydberg formula**.

Rydberg formula determines the full spectrum of light emission from hydrogen as

$$\frac{1}{\lambda_{\text{vac}}} = R_H \left(\frac{1}{n_1^2} - \frac{1}{n_2^2} \right),$$

where λ_{vac} is the wavelength of the light emitted a in vacuum, R_H is the Rydberg constant for hydrogen $(2\pi^2 m_0 e^4 / (ch^3) = 109737.31 \text{ cm}^{-1})$, and n_1 and n_2 are integers such that $n_1 < n_2$. By setting n_1 equal to 1 and letting n_2 run from 2 to infinity, the spectral lines known as the **Lyman series** converging to 91 nm are obtained. In the same manner the **Balmer series** ($n_1 = 2, n_2$ runs from 3 to infinity, converging to 365 nm) and the **Paschen series** ($n_1 = 2, n_2$ runs from 4 to infinity, converging to 821 nm) are obtained. The formula can be extended for use with any chemical element:

$$\frac{1}{\lambda_{\text{vac}}} = RZ^2 \left(\frac{1}{n_1^2} - \frac{1}{n_2^2} \right).$$

Here R is the Rydberg constant for this element, and Z is the atomic number that is the number of protons in the atomic nucleus of the element.

First described in: J. R. Rydberg, *On the structure of the line spectra of the chemical elements*, Phil. Mag. **29**, 331–337 (1890).

Rydberg gas → Rydberg atom.

From Saha equation to synergetics

Saha equation → **exciton**.
First described in: M. N. Saha, *On a physical theory of stellar spectra*, Proc. R. Soc. Lond., Series A, **99**(697), 135–153 (1921); M. N. Saha, *Versuch einer Theorie der physikalischen Erscheinungen bei hohen Temperaturen mit Anwendungen auf die Astrophysik*, Z. Phys. **6**(1) 40–55 (1921).

Sakata–Taketani equation – a relativistic wave equation for a particle with spin 1 whose form is similar to that of the nonrelativistic **Schrödinger equation**.
First described in: M. Taketani, S. Sakata, *On the wave equation of meson*, Proc. Phys. Math. Soc. Jap. **22**, 757–770 (1940), (Reprinted in Suppl. Progr. Theor. Phys. **22**, 84–97 (1955)).

SAM – an acronym for a **self-assembled monolayer**.

saturated hydrocarbons → **hydrocarbons**.

saturated pattern excitation microscopy (SPEM) → **fluorescence nanoscopy**.
First described in: R. Heintzmann, T. M. Jovin, C. Cremer, *Saturated patterned excitation microscopy – a concept for optical resolution improvement*, J. Opt. Soc. Am. A **19**, 1599 (2002); M. G. L. Gustafsson, *Nonlinear structured-illumination microscopy: Wide-field fluorescence imaging with theoretically unlimited resolution*, Proc. Natl. Acad. Sci. (PNAS) **102**(37), 13081–13086 (2005).

saturated structured illumination microscopy (SSIM) → **fluorescence nanoscopy**.
First described in: R. Heintzmann, T. M. Jovin, C. Cremer, *Saturated patterned excitation microscopy – a concept for optical resolution improvement*, J. Opt. Soc. Am. A **19**, 1599 (2002); M. G. L. Gustafsson, *Nonlinear structured-illumination microscopy: Wide-field fluorescence imaging with theoretically unlimited resolution*, Proc. Natl. Acad. Sci. (PNAS) **102**(37), 13081–13086 (2005).

SAXS – an acronym for a **small angle X-ray scattering**.

What is What in the Nanoworld. Victor E. Borisenko and Stefano Ossicini
Copyright © 2008 WILEY-VCH Verlag GmbH & Co. KGaA, Weinheim
ISBN: 978-3-527-40783-5

SCALPEL

SCALPEL – an acronym for a **scattering with angular limitation projection electron-beam lithography.**

scanning capacitance microscopy (SCM) uses information obtained from the change in electrostatic capacitance between a surface of a sample and a probe scanning just above the surface. It is usually realized with an **atomic force microscope** (AFM) equipped with an ultrahigh frequency resonant capacitance sensor connected to a conducting probe as shown in Figure S.1. The technique is used to extract two-dimensional charge carrier profiles in semiconductor samples.

The fundamental principle of SCM is based on the metal-oxide-semiconductor concept. The tip of the conducting probe acts as the metal contacting the layer of insulating oxide grown on top of the semiconductor sample. The AFM employed is operated in the contact mode.

The tip–sample capacitance is probed by modulation of the carrier concentration in the semiconductor layer under the tip with a bias-containing alternating current (*ac*) and direct current (*dc*) components. As a practical matter, the capacitance variation dC/dV is measured with high accuracy because the capacitance itself is too small to be measured reliably. The mapped in scanning area dC/dV is a function of the lateral carrier concentration in the subsurface layer of the semiconductor sample.

More details in: Roadmap of Scanning Probe Microscopy, edited by S. Morita (Springer Verlag, Berlin Heidelberg, 2007).

Figure S.1 Schematic of scanning capacitance microscopy.

scanning electron microscopy (SEM) – a microscopy technique in which a beam of electrons, a few tens of nanometers in diameter or thinner, systematically sweeps over a specimen and the intensity of secondary electrons generated at the point of impact of the beam on the specimen is measured.

scanning Hall probe microscopy (SHPM) is a technique which incorporates the accurate sample approach and positioning of **STM** with a semiconductor Hall sensor. It exploits the high sensitivity of the Hall probe (down to 10^3 μ_B), which allows the pick-up of tiny magnetic signals in a wide range of temperature and magnetic fields. The main advantage of SHPM is to be directly sensitive to stray field **B** by means of the **Hall effect**, so the local field is measured by the sensor without interacting with the surface. In the first application, the corner of the SHPM sensor was gold coated and used as an STM tip for the positioning of the device on the surface, even allowing the simultaneous acquisition of SHPM and STM images. More recently other techniques have been developed for the approach to nonconducting surfaces. The lateral resolution of SHPM is related to the dimensions of the electrical channels of the Hall probe.

First described in: A. M. Chang, H. D. Hallen, L. Harriott, H. F. Hess, H. L. Kao, J. Kwo, R. E. Miller, R. Wolfe, J. van der Ziel, *Scanning Hall probe microscopy*, Appl. Phys. Lett. **61**(16), 1974–1976 (1992).

More details in: A. Oral, *Scanning Hall probe microscopy: quantitative & non-invasive imaging and magnetometry of magnetic materials at 50 nm scale*, in *Magnetic Nanostructures*, edited by B. Aktas, F. Mikailov, L. Tagirov, Springer Series in Materials Science **94** (Springer, Berlin 2007) pp. 7–14.

scanning probe lithography (SPL) – a lithographic technique employing a probe of a **scanning tunneling microscope** (STM) or **atomic force microscope** (STM) to fabricate surface structures with a size from hundreds of nanometers down to the atomic scale. The most practical approaches are based on the use of high electric field and current density between tip and substrate for **atomic engineering**, modification of **Langmuir-Blodgett films** and **self-assembled monolayers**, selective anodic oxidation of substrates and surface deposition of materials. Electrons injected from the tip apex are employed for **electron-beam lithography**. Mechanical modification of the material by the tip can be used for direct writing of grooves.

scanning SQUID microscopy (SSM) – a powerful tool for imaging magnetic fields above sample surfaces. It combines low-temperature positioning with a single-chip miniature **SQUID** magnetometer to make a sensitive and reliable magnetic field imaging instrument. There are various schemes for scanning the sample and for coupling the sample field into the SQUID sensor. One can distinguish two classes. In the first, both sample and SQUID are held at low temperatures either in a common vacuum space or in a cryogenic fluid. This allows for minimum spacing between SQUID and sample and therefore the best spatial resolution. In the second class, the SQUID is cold but the sample is warm and can be exposed to the atmosphere. In this case SQUID and sample are separated by some form of thermal insulation. This has the advantage of not requiring sample cooling, but entails some sacrifice in spatial resolution.

First described in: A. Mathai, D. Song, Y. Gim, F. C. Wellstood, *One-dimensional magnetic flux microscope based on the dc superconducting quantum interference device*, Appl. Phys. Lett. **61**(5), 598–600 (1992); J. R. Kirtley, M. B. Ketchen, K. G. Stawiasz, J. Z. Sun, W. J. Gallagher, S. H. Blanton, S. J. Wind, *High-resolution scanning SQUID microscope*, Appl. Phys. Lett. **66**(9), 1138–1140 (1995).

scanning tunneling microscopy (STM)

More details in: J. R. Kirtley, J. P. Wikswo Jr., *Scanning SQUID microscopy*, Ann. Rev. Mater. Sci. **29**, 117–148 (1999).

scanning tunneling microscopy (STM) – a technique based on the ability to position with extremely high precision an atomically sharp probe in close proximity to the sample surface. The physics behind the technique is schematically illustrated in Figure S.2.

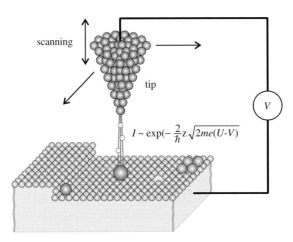

Figure S.2 Tip–sample geometry in scanning tunneling microscopy.

A sharp metal probe tip, typically made of tungsten, is positioned over the surface of a conducting or semiconducting sample at a distance of about one nanometer or less. When a bias voltage V is applied between the sample and the tip, electrons tunnel through the potential barrier separating them. The magnitude of the tunneling current depends primarily on three factors: (i) the probability that an electron with energy will tunnel through the potential energy barrier between the sample and the tip, which varies exponentially with the tip–sample separation z; (ii) the electronic density of states at the sample surface; (iii) the electronic density of states at the tip. The last two factors are represented by the barrier potential U in the current dependence in the figure. The extreme sensitivity of the tunneling current to the separation between the tip and the sample surface is used in an electronic feedback loop that permanently adjusts the tip–sample separation, maintaining a constant current when the tip is scanned over the surface. Piezoelectric scanners are used for this. Thus, it yields a profile of the local electronic density of states resembling the arrangement of atoms at the surface and their chemical nature. STM has vertical resolution of 0.01–0.05 nm and lateral resolution less than 0.3 nm. It can image areas as large as several hundreds of micrometers.

First described in: G. Binning, H. Rohrer, *Scanning tunneling microscopy*, Helv. Phys. Acta **55**(6), 726–735 (1982); G. Binning, C. Gerber, H. Rohrer, E. Weibel, *Tunneling through controllable vacuum gap*, Appl. Phys. Lett. **40**(2), 178–180 (1982);

scattering with angular limitation projection electron-beam lithography (SCALPEL)

G. Binning, H. Rohrer, Ch. Gerber, E. Weibel, *Surface studies by scanning tunneling microscopy*, Phys. Rev. Lett. **49**(1), 57–61 (1982).

More details in: Handbook of Nanotechnology, edited by B. Bhushan (Springer, Berlin, 2004).

Recognition: in 1986 G. Binning and H. Rohrer received the Nobel Prize in Physics for their design of the scanning tunneling microscope.

See also www.nobel.se/physics/laureates/1986/index.html.

scattering – a change in the direction of a propagating particle or wave caused by inhomogeneity of the transmitting medium.

scattering matrix → *S*-**matrix**.

scattering with angular limitation projection electron-beam lithography (SCALPEL) – the approach combining high resolution and wide process latitude of **electron-beam lithography** with the throughput of a parallel projection system. It is schematically shown in Figure S.3. The mask consisting of a low atomic number membrane and a high number pattern layer is uniformly irradiated with high energy electrons. They are chosen to be essentially transparent to the electron beam, so very little of the beam energy is deposited on it. The portion of the beam which passes through the high atomic number pattern layer is scattered through angles of few milliradians. An aperture in the back focal plane of the electron projection imaging lenses stops the scattered electrons and produces a high-contrast image in the plane of the semiconductor wafer. One obtains lithography results down to less than 100 nm.

First described in: S. D. Berger, J. M. Gibson, *New approach to projection-electron lithography with demonstrated 0.1 µm linewidth*, Appl. Phys. Lett. **57**(2), 153–155 (1990).

Figure S.3 Basic principle of the SCALPEL technique. After L. R. Harriot, *Scattering with angular limitation projection electron beam lithography for sub optical lithography*, J. Vac. Sci. Technol. B **15**(6), 2130–2135 (1997).

Schoenflies symbols (notations)

More details in: L. R. Harriot, *Scattering with angular limitation projection electron beam lithography for sub optical lithography,* J. Vac. Sci. Technol. B **15**(6), 2130–2135 (1997).

Schoenflies symbols (notations) – symbols representing rotation axes, mirror rotation axes, order of rotations and other elements of symmetry operations used to describe symmetry of molecules. They are widely used in spectroscopy.
First described in: A. Schoenflies, *Krystallsysteme und Krystallstruktur* (Teubner, Leipzig 1891) (Reprinted by Springer, Berlin, 1984).

Schottky barrier – a barrier to a semiconductor, usually formed by metals, characterized by the formation of a charge carrier depleted layer in the semiconductor. Under an external voltage V applied, an electric current is passing through the barrier mainly due to the thermionic emission. Within this mechanism the current density is $j = B^* T^2 \exp(-\phi^*/k_B T) \exp(eV/nk_B T)[1 - \exp(eV/k_B T)]$, where B^* is the effective **Richardson constant** for the semiconductor, ϕ^* is the effective zero bias barrier height, and n is the **ideality factor**.
First described in: W. Schottky, *Halbleitertheorie der Sperrschicht,* Naturwiss. **26**, 843 (1938).

Schottky defect – lattice vacancy.
First described in: W. Schottky, C. Wagner, *Theorie der geordneten Mischphasen,* Z. Phys. Chem. B **11**, 163 (1930).

Schottky effect – image-force-induced lowering of the potential energy for charge carrier emission from a metallic surface when an electric field is applied.
First described in: W. Schottky, *Vereinfachte und erweiterte Theorie der Randschichtgleichrichter,* Z. Phys. **118**(9–10), 539–592 (1942).

Schottky emission (of electrons) – electric field induced emission of electrons from solids.
First described in: W. Schottky, *Über kalte und warme Elektronenentladungen,* Z. Phys. **14**(1), 63–106 (1923).

Schrödinger equation – the fundamental wave equation of nonrelativistic quantum mechanics. For a single particle of mass m in a potential $V(\mathbf{r})$ its **time-dependent** form is:

$$i\hbar \frac{\partial \Psi(\mathbf{r}, t)}{\partial t} = -\frac{\hbar^2}{2m} \nabla^2 \Psi(\mathbf{r}, t) + V(\mathbf{r})\Psi(\mathbf{r}, t), \quad \text{where} \quad \nabla^2 = \frac{\partial^2}{\partial x^2} + \frac{\partial^2}{\partial y^2} + \frac{\partial^2}{\partial z^2}.$$

$\Psi(\mathbf{r}, t)$ is the wave function (each electron is represented by a wave function). It describes time evolution of electron waves. The equation is often written as $i\hbar(\partial \Psi(\mathbf{r}, t)/\partial t) = \mathbf{H}\Psi(\mathbf{r}, t)$, where $\mathbf{H} = -(\hbar^2/2m)\nabla^2 + V(\mathbf{r})$ is the **Hamiltonian**

operator. The equation has various extensions, which may be used to describe the whole of nonrelativistic dynamics, including systems that are free $(V(\mathbf{r}) = 0)$ as well as bound.

The square of the absolute value of a given wave function $|\Psi(\mathbf{r}, t)|^2 d\tau$ is interpreted as a probability of finding a particle represented by the wave function $\Psi(\mathbf{r}, t)$ in the volume element $d\tau$ at a position r at time t.

Besides, the equation provides calculations of the energy of the stationary states of a quantum system. The **time-independent Schrödinger equation** in the form

$$-\frac{\hbar^2}{2m}\nabla^2\Psi_n(\mathbf{r}) + V(\mathbf{r})\Psi_n(\mathbf{r}) = E_n\Psi_n(\mathbf{r}),$$

or equivalently $\mathbf{H}\Psi_n(\mathbf{r}) = E_n\Psi_n(\mathbf{r})$, is used for this. Here $\Psi_n(\mathbf{r})$ is the wave function of the quantum particle in an energy eigenstate of energy E_n, and n is the integer called the quantum number. No arbitrary postulates concerning quantum numbers are required; instead, integers enter automatically in to the process of finding satisfactory solutions of the wave equation.

In the realm of quantum physics the **Schrödinger equation** plays a role of similar importance to that which Newton's equation of motion does for classical physics.

First described in: E. Schrödinger, *Quantisierung als Eigenwertproblem (erste Mitteilung)*, Ann. Phys. **79**(4), 361–376 (1926).

Recognition: in 1933 E. Schrödinger shared with P. A. M. Dirac the Nobel Prize in Physics for the discovery of new productive forms of atomic theory.

See also www.nobel.se/physics/laureates/1933/index.html.

Schrödinger's cat – the cat proposed by E. Schrödinger to be considered as an object in a thought experiment showing the absurdity of applying quantum mechanics to macroscopic objects. He invited one to imagine placing a cat in a box, along with a vial of poison that is connected to a radioactive atom. When the atom decays, it releases the poison, which kills the cat. If the box is closed, we do not know whether the atom has decayed or not. It means that the atom can be in a superposition of two states – decayed and nondecayed, at the same time. Therefore, the cat can be both dead and alive at the same time. Quantum mechanically his state can be described as $|\Psi\rangle_{cat} = |\text{alive}\rangle + |\text{dead}\rangle$, which is now known as the **Schrödinger cat state**. This clearly shows that describing a macroscopic object, such as a cat, in quantum terms is ridiculous. E. Schrödinger concluded that there was obviously a boundary between the microscopic world of atoms and quantum mechanics and the macroscopic world of cats and classical mechanics. The cat is a macroscopic object, the states of which (alive or dead) could be distinguished by a macroscopic observation as distinct from each other whether observed or not. He calls this "the principle of state distinction" for macroscopic objects, which is in fact the postulate that the directly measurable system (consisting of the cat) must be classical.

First described in: E. Schrödinger, *Die gegenwärtige Situation in der Quantenmechanik*, Naturwiss. **23**(48), 807–812 (1935).

Schrödinger's cat state

Schrödinger's cat state → **Schrödinger's cat.**

Schwinger variational principle – the technique used to calculate an approximate value or a linear of a quadratic functional, such as scattering amplitude or reflection coefficient, when the function for which the functional is evaluated is the solution of an integral equation.
First described in: J. Schwinger, *On gauge invariance and vacuum polarization,* Phys. Rev. **82**(5), 664–679 (1951).

Schwoebel effect = Ehrlich–Schwoebel effect.
First described in: R. L. Schwoebel, E. J. Shipsey, *Step motion on crystal surfaces,* J. Appl. Phys. **37**(10), 3682–3686 (1966).

scintillation – a localized emission of light of short duration produced in a suitable solid or liquid medium by the absorption of a single ionized particle or photon. Most of the energy of such particles is lost in their interaction with electrons in the medium that leads to excitation and ionization of some molecules. The subsequent deexcitation and recombination result in the emission of light quanta with a characteristic spectral distribution.

SCM – an acronym for **scanning capacitance microscopy.**

screened Coulomb potential – the interatomic pair screening potential in the form

$$V(r) = \frac{Z_1 Z_2 e^2}{r} \exp\left(-\frac{r}{a}\right),$$

where r is the interatomic distance, Z_1 and Z_2 are the atomic numbers of the interacting atoms, $a = a_B k/(Z_1^{2/3} + Z_2^{2/3})^{1/2}$, $k = 0.8$–3.0 is the empirical coefficient used to fit experimental data.
More details in: I. G. Kaplan, *Intermolecular Interactions: Physical Picture, Computational Methods and Model Potentials* (John Wiley & Sons, Chichester, 2006).

screening wave number – inverse screening length defined as

$$k = \left(4\pi e^2 \frac{\partial n}{\partial \mu}\right)^{1/2},$$

where n is the density of electrons and μ is the chemical potential → **Debye–Hückel screening length** and **Thomas–Fermi screening length.**

secondary ion mass spectrometry (SIMS) – a technique of destructive elemental analysis of condensed matter. An analyzed sample is bombarded with energetic ions, such as He^+, Ar^+, Xe^+ or metal ions. They in turn cause ejection (by sputtering) of secondary ions from the sample surface. These secondary ions are then detected and analyzed by a quadrupole mass spectrometer. Thus, the technique provides information about elemental composition of the sample surface layer. The primary generated signal results from a region typically of about 1–3 nm. When it is combined with layer-by-layer etching of the surface, near surface concentration profiles of impurities can be recorded. The diameter of the ion probe varies from a few μm for **noble gas** ions to less than 50 nm for metal ions (Cs^+, Ga^+).

Seebeck coefficient → **Seebeck effect**.

Seebeck effect – a development of a voltage in a loop composed of two different materials, say A and B, (metals, semiconductors) when their junctions are held at different temperatures, as shown in Figure S.4. The developed open-circuit voltage V_{AB} is proportional to the temperature difference ΔT. The coefficient of the proportionality is the **Seebeck coefficient** defined as

$$\alpha_{AB} = \lim_{\Delta T \to 0} \frac{V_{AB}}{\Delta T}.$$

First described in: T. J. Seebeck, *Magnetische Polarisation der Metalle und Erze durch Temperatur-Differenz*, Abh. Kön. Akad. Wiss. Berlin, 265–373 (1822).

Figure S.4 Schematic of Seebeck effect.

Seebeck effect in an organic molecule was measured for the first time in 2007. Two gold electrodes were coated with molecules of benzenedithiol, dibenzenethiodol or tribenzenedithiol, and one side was then heated to create a difference in temperature. For each degree Celsius of difference, a voltage difference of ~10 μV was measured. This could open the door to more cost-effective thermoelectric converters.

First described in: P. Reddy, S.-Y. Jang, R. Segalman, A. Majumdar, *Thermoelectricity in molecular junctions*, Science **315**, 1568–1571 (2007).

selection rule – the statement specifying which electron transitions in atoms and molecules are allowed when a photon is emitted or absorbed. It was derived from the conservation of **angular momentum** of the electron, characterized by quantum numbers l and m_l. So, a transition is possible when $\Delta l = \pm 1$ and $\Delta m_l = 0, \pm 1$. The principal quantum number n can change by any amount consistent with the Δl for the transition because it does not relate directly to the angular momentum.

self-assembled monolayer (SAM) – an ordered molecular assembly formed by **adsorption** of an active surfactant on a solid surface. It is produced within a **self-assembling process**.

First described in: W. C. Bigelow, D. L. Pickett, W. A. Zisman, *Oleophobic monolayers. I. Films adsorbed from solution in nonpolar liquids*, J. Colloid Interface Sci. **1**, 513–538 (1946).

self-assembly – a process of construction of a structure from components acting under forces/motives internal or local to the components themselves, and arising through their interaction with the environment. In nanotechnology it supposes the adsorption and particular arrangement of molecules on a solid surface. It can be considered as a coordinated action of independent entities under local control of driving forces used to produce larger, ordered structures or to achieve a desired group effect. It is usually driven by chemisorption, being pronounced for highly energetic reactions between adsorbate and substrate molecules. In contrast to strong interaction with the substrate surface, the adsorbed molecules weakly interact with each other. Many examples of self-assembly can be found in the organic and inorganic world. Self-assembled monolayer films have an extremely low defect density, and are quite environmentally stable and mechanically robust. They serve as imaging layers in lithographic processes. Imaging layers are used for recording a pattern, either by masking, as with photon-assisted processes, or by direct writing, as with conventional electron beam writers and scanning probes. Nanometer scale high-resolution lithography with imaging layers fabricated via self-assembly is usually performed with the use of a **scanning tunneling microscope (STM)** or **atomic force microscope (AFM)** probes.

A monolayer film appropriate for high-resolution patterning via self-assembly should consist of three main functional parts. These are a substrate surface binding part, a spacer part and a surface functional part. While these parts are not fully interchangeable from one molecule to the next, different considerations have to be made for the selection of each. Positioning of functional groups and their recognition properties are better developed for organic materials than for inorganic ones, whereas electronic properties of the latter are much better understood. It is, therefore, envisaged that especially the combination of the different materials will permanently lead to new and fruitful approaches.

For the substrate surface binding part, a silane $RSiX_3$ ($R = CH_3$, C_2H_5, ...) is usually used to bind to hydroxyl (OH) groups, which typically terminate silicon and other technologically important surfaces. Hydrogen substituents on the silane, namely the "X" component, are most commonly methoxy groups (CH_2O), Cl or a mixture of the two. The composition of the surface binding part strongly affects film ordering, the tendency of the film to grow multilayers and, to a lesser extent, the packing density. A thiol (RSH) forms highly ordered layers on surfaces of GaAs and gold.

The spacer part has an effect on the interaction of the film with the writing tool. By moving the functional part away from the surface and closer to the high field of the tip, long spacer groups (multiple CH_2 groups, for example) can lower the dose or voltage threshold for exposure. The conjugated bonds found in phenyl groups are also somewhat conductive, and hence more amenable to STM operation.

The surface functional part actually defines properties of the "new" surface. For example, **amine** (NH_2) groups can be used as ligation sites for certain molecules. While not particularly active themselves, **halides** (Cl, I, etc.) have an extremely high cross section for electron capture and subsequent desorption of the halide fragment. Subsequent processing can be based on either replacing the halide atom with a more reactive functional group or acting upon the fragment remaining after halide desorption. Alkyl-terminated surfaces are extremely inert and **hydrophobic**, being chemically identical to paraffin. As such, they are implemented for pattern transfer as masks against wet etching and, to a limited extent, dry etching. They are also used to enhance STM-induced etching of gold.

Processing steps illustrating the use of a self-assembling monolayer film for nanopatterning on a silicon substrate are shown in Figure S.5.

Before the film deposition the silicon substrate is cleaned and hydrogen passivated in dilute HF. Then it is immersed briefly in a solution of an organosilane monomer, removed and dried in order to fabricate a self-assembling monolayer film on the substrate.

Figure S.5 Replication of a high-resolution pattern by the use of a STM-exposed self-assembling monolayer film: (a) deposition of the self-assembling monolayer film, (b) pattern generation with an STM tip, (c) ligation of Pd catalyst, (d) electroless deposition of Ni. After E. S. Snow, P. M. Campbell, F. K. Perkins, *Nanofabrication with proximal probes*, Proceedings of the IEEE, **85**(4), 601–611(1997).

The list of appropriate organosilanes includes: octadecyltrichlorosilane, phenethyltrimethoxysilane, chloromethylphenyltrimethoxysilane, chloromethyl-phenethyltrimethoxysilane, monochlorodimethoxysilane versions of phenethyl-trimethoxysilane and phenethyltrimethoxysilane.

The monolayer film with a typical thickness of about 1 nm is exposed to low-energy electrons from an STM or conducting AFM tip according to a desired pattern. Exposure threshold voltage is in the range of 2–10 V, depending on the film composition, mainly by the terminal group, and substrate surface passivation. After patterning, the sample is immersed in the colloidal palladium, which selectively attaches to the unexposed surface of the self-assembled film. The sample is then rinsed and again immersed in a nickel borate electroless plating bath, where nucleation of metallic nickel is catalyzed by the surface attached palladium islands. The subsequent growth of the depositing nickel occurs anisotropically, and rough edges reproducing boundaries of the islands become smoothed down. The patterned metallic film is used as a hard etch mask. The achievable resolution is in the range of 15–20 nm.

Most aspects of assembly and self-assembly, such as positioning of functional groups and their recognition properties, are better developed for organic materials than for inorganic ones, whereas the electronic properties of the latter are much better understood. It is, therefore, envisaged that especially the combination of the different material types will lead to fruitful new approaches. It may be the case that assembly will take place via recognition of organic parts followed by derivatization with inorganic materials, or that small inorganic units modified with organic anchoring points are assembled into circuits in a single process.

Within the self-assembling approach, the ultimate dimensions of nanoelectronic elements of integrated circuits are defined by the dimensions of atoms and molecules. At ambient temperature it is probably easier to confine a molecule through its functional groups than a single atom. The development of fast methods for performing chemistry on single molecules is therefore important. Various probe techniques, like STM and AFM, play a key role in this field.

self-organization (in bulk solids) supposes particular arranging of strongly interacting atoms in solid state structures. Spontaneous self-organization processes occurring in the bulk of solids and at their surfaces provide an effective way of nanostructure fabrication, namely of **quantum wires** and **quantum dots**. The major driving force of these processes is the tendency of the solid to attain a stable atomic configuration corresponding to the minimum free energy of the system. In solids it usually results in spontaneous crystallization. A phase transformation leading to nucleation and growth of a new crystalline phase occurs only in thermodynamically nonequilibrium conditions typically present in supersaturated or strained systems.

Formation of a crystalline nucleus lowers the energy of the system by $\Delta g = g_{am} - g_{cr}$, because the crystalline phase (characterized by the energy g_{cr}) is always more stable than an amorphous phase (characterized by the energy g_{am}). This energy decrease is opposed by the increase of the surface energy of the growing nucleus. With Δg calculated per unit volume of the new phase, an appearance of a nucleus of radius r with a specific surface energy σ^* gives the total change of free energy in the system $\triangle G = 4\pi r^2 \sigma^* - 4/3\pi r^3 \triangle g$.

The free energy change varies non-monotonically as a function of the nucleus radius. It is illustrated in Figure S.6. For the formation of the nucleus surface, work must be carried out on the system, while in the formation of the nucleus bulk work is gained from the system. The free energy change has its maximum value for an atomic cluster of critical radius r_{cr}. It corresponds to $r_{cr} = 2\sigma^*/\triangle g$.

Formation of a cluster smaller than the critical nucleus needs a positive free energy change, and therefore the system is unstable. The population of such nuclei exists in equilibrium distribution. Bigger nuclei are energetically favored to be formed. The rate of nucleation v_n is proportional to the concentration of critical nuclei and to the rate at which these nuclei originate. It can be described as $v_n \sim \exp(-\triangle G_{cr}/k_B T)\exp(-E_a/k_B T)$, where G_{cr} is the free energy of the critical nuclei formation, T the absolute temperature. The term $\exp(-E_a/k_B T)$ represents atomic diffusion contributing to the nucleation and subsequent growth of the nuclei. It is characterized by the activation energy E_a. Since G_{cr} is inversely proportional to T^2, the rate of nucleation will vary as $\exp(-1/T^3)$. Thus, it is evident that nucleation of any particular phase takes place within narrow temperature limits, below which nothing occurs and above which the reaction is extremely fast.

The spontaneous crystallization is widely implemented for fabrication of quantum dot structures without sophisticated **nanolithography**. It is realized within nanocrystal growth initiated in inorganic and organic matrices.

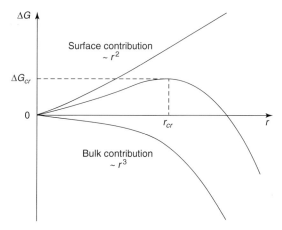

Figure S.6 Variation of the free energy change of nuclei as a function of their radii.

self-organization (during epitaxial growth) becomes apparent in an epitaxial growth on **vicinal surfaces** and when **Stranski–Krastanov growth mode** of thin films is realized. These are illustrated as follows.

A particular growth mode of an epitaxial film for a given couple of materials depends on the lattice mismatch and on the relation between surface and interface energies. It is important to note here that all energetic considerations are valid

self-organization (during epitaxial growth)

for an equilibrium state of the system. The formation of thin epitaxial films, which is often an out-of-equilibrium process, is hardly described within the pure energetic approach. Kinetic effects controlled by the deposition rate and substrate temperature can dramatically modify the growth mode. Nevertheless, the energetic consideration presented below is very useful for many practical cases, at least for predicting behavior of a system in equilibrium or quasi-equilibrium conditions.

In a lattice-matched system, the growth mode is governed by the interface and surface energies only. If the sum of the surface energy of the epitaxial film and the interface energy is lower than the energy of the substrate surface (the deposited material wets the substrate), growth occurs in the **Frank–van der Merwe growth mode**. Uniform coherent pseudomorphic and strained superlattices are fabricated in this case. It is also appropriate for fabrication of self-organized quantum wires. Monocrystalline substrates with a surface oriented slightly off one of the low-index planes (usually (001) or (311)), a vicinal surface, are used for this. Figure S.7 shows the main stages of the fabrication procedure. An as-prepared vicinal surface consists of planar equal-width terraces with low **Miller indices**. The neighboring terraces are separated by equidistant monoatomic or monomolecular steps. Such step bunching is typically homogeneous throughout a vicinal surface. The deposition starts with the wire material.

The substrate temperature is chosen so as to allow the deposited atoms to be mobile enough for diffusion around the surface. It is energetically favorable for them to stick at the steps rather than the middle of the terraces. Only a portion of the wire material corresponding to its thickness of less than one monolayer is

Figure S.7 Fabrication of quantum wires by self-organization during epitaxial growth on the (001) vicinal surface: (a) as-prepared surface, (b) deposition of the wire material, (c) half-monolayer of the wire material has been deposited, (d) complementing half-monolayer of the substrate material has been deposited, (e) burred quantum wire produced be repetition of steps c and d.

deposited in order to leave space for the subsequently deposited substrate material, which restores the steps to their original position in the plane, but a monolayer higher. By repeating the deposition of wire and then substrate materials one can fabricate in-built quantum wires. One of the main problems in practical application of the above approach is nonlinearity of the terrace edges, which usually have a wave-like shape. This naturally results in the formation of wave-like quantum wires. In order to fabricate ordered arrays of linear-type quantum wires one should facet the surfaces with artificial grooves. In this case wires nucleate and grow in the grooves.

Strained homogenous epitaxial films can start to grow layer-by-layer even when there is a certain lattice mismatch between deposited and substrate materials. Accumulation of the strain energy with an increase of the film thickness inevitably pushes the system to lower its energy by island formation, which occurs in the **Stranski–Krastanov growth mode**. In this case the self-organized growth of nanocrystalline islands on a monocrystalline substrate occurs via strain-induced transition from a two- to a three-dimensional structure (2D→3D) when the growing material has a noticeably bigger lattice constant than the substrate. Reduction of the total energy of the strained substrate/epilayer system is a driving force of the monocrystalline island formation. In contrast to a pseudomorphic two-dimensional layer, which can relax only in the direction perpendicular to the surface, the formation of three-dimensional islands leads to strain relief within and around the islands, resulting in more efficient relaxation.

Consider qualitatively the changes in the total energy of a mismatched system versus deposition time at a sufficiently low deposition rate eliminating kinetic effects. Figure S.8 shows schematically the changes in the total energy of a mismatched system versus time. The material (under compression) is deposited at a constant deposition rate until time point X is reached. Three main periods, indicated A, B and C, are distinguished.

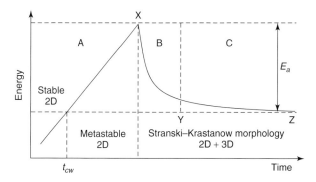

Figure S.8 Time variation of the total energy of the epitaxial structure being grown in Stranski–Krastanov mode. After W. Seifert, N. Carlsson, J. Johansson, M.-E. Pistol, L. Samuelson, *In situ growth of nano-structures by metal-organic vapour phase epitaxy*, J. Cryst. Growth **170**(1–4) 39–46 (1997).

self-organization (during epitaxial growth)

In the initial period A the epitaxial two-dimensional structure is growing layer-by-layer. The substrate is perfectly wetted. The elastic strain energy increases linearly with the volume of the deposited material. At the time point t_{cw} the critical wetting layer thickness is reached and further layer-by-layer growth becomes metastable. A supercritically thick wetting layer builds up, which means that the still continuous epitaxial layer is potentially ready to be broken under Stranski–Krastanov transition to a three-dimensional island structure. The extent of the metastability range depends primarily on the height of the transition barrier E_A.

The period B, representing 2D→3D transition, starts when the stored elastic energy is sufficient to overcome the transition barrier at the time point X. It is presumed that once the transition starts it can continue without further material supply, simply by consuming the excess material accumulated in the supercritically thick wetting layer. There are two stages, namely island nucleation and its subsequent growth, within this period. One can assume that thickness or strain fluctuations lead to the formation of the critical nuclei on a uniform surface. The thickness of the wetting layer at which spontaneous nucleation of three-dimensional islands starts mainly depends on the lattice mismatch of the substrate, deposited material and surface anisotropy. For example, in epitaxy of germanium onto monocrystalline silicon it happens when the thickness of germanium film exceeds a few monolayers. For a controlled nucleation "geometrical" or "strain" pre-patterning of the surface is needed.

The nucleation stage defines the surface density of the islands. As soon as the first overcritical nuclei have been formed, the islands themselves act as catalysts to decompose the wetting layer. An expected strain energy density profile along the interface between the substrate and the island is shown in Figure S.9. The surface

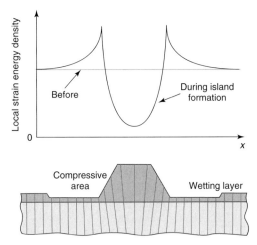

Figure S.9 Change in the local strain energy density at the substrate/epilayer interface related to the island formation and schematic of deformation of lattice planes inside and around a coherent island. After W. Seifert, N. Carlsson, J. Johansson, M.-E. Pistol, L. Samuelson, *In situ growth of nano-structures by metal-organic vapour phase epitaxy*, J. Cryst. Growth **170**(1–4) 39–46 (1997).

of the island represents the minimum in this potential, whereas the maximum is at the edge of the island. It is induced by the strain propagation along the substrate, which increases the inherent misfit between the substrate and wetting layer material around the island. The minimum is caused by the partial strain relaxation in the island, and this is a driving force for the material to crystallize at its surface. As a result, three-dimensional coherently strained dislocation-free monocrystalline islands are formed.

The subsequent growth of islands is enhanced by high supersaturation at the beginning of this period. It is characterized by rates tens times higher than conventional growth rates. Intrinsically rough surfaces grow very fast and disappear, whereas facets of low indexed planes $\{11n\}$ ($n = 0, 1, \sim3$) become rate limiting. As a result, the islands have a pyramidal shape typically with $\{113\}$ or $\{110\}$ facets or a shape of elongated truncated pyramids. Although the growing islands are usually coherent to the substrate, non-coherent islands and related misfit dislocations starting at the interface can also be produced in the case of too high a material supply rate.

Further growth of the islands proceeds within the period C, but via a ripening mechanism. The system has lost most of the excess energy accumulated by strains. A difference in the Gibbs free energy between small and large islands maintains slow growth of larger ones at the expense of small islands. The process is mediated by surface diffusion.

The epitaxial deposition in the Stranski–Krastanov mode is successfully employed for the formation of 2–40 nm quantum dots made of $A^{III}B^{V}$ and $A^{II}B^{VI}$ semiconductors, germanium, SiGe alloys with remarkably uniform size distribution and high surface density. It permits one to exceed the critical thickness limitation of pseudomorphic growth without introducing dislocations. Several methods were developed in order to provide nanoscale positioning and site selectivity of the quantum dot formation. It is achieved by selective epitaxial growth in the windows of a nanopatterned mask fabricated by conventional electron-beam or scanning probe lithography.

SEM – an acronym for a **scanning electron microscopy**.

semiconductor – a substance whose electrical resistivity (10^{-4}–10^{9} Ωcm, negative temperature coefficient of resistivity) is intermediate between that of **conductors** and **insulators**. Semiconductors have an energy gap between valence and conduction band not higher than 5 eV.

Pure semiconductor materials (say 99.999%) are called intrinsic semiconductors as their properties are intrinsic to the material itself. By adding some impurities to a semiconductor its electrical properties are changed. Doped semiconductors are called extrinsic semiconductors. Impurities that cause electron type of conductivity (n-type) are called **donors** and those producing hole type of conductivity (p-type) are called **acceptors**.

Donor impurities are elements with valence higher than that of the component they substitute, and acceptor impurities are elements with valence lower than that of semiconductor equation

the component they substitute. In $A^{III}B^V$ compound semiconductors, a tetravalent element, such as Si and Ge, can be both a donor impurity if it substitutes a trivalent component or an acceptor impurity if it substitutes a pentavalent component. Such impurities are called amphoteric.

The main properties of studied monoatomic and binary compound semiconductors are presented in Appendix.

semiconductor equation – in a semiconductor at thermal equilibrium $np = n_i^2$, where n and p are the concentrations of electrons and holes, respectively; n_i is the intrinsic carrier concentration in this material. It does not depend on doping or the position of the **Fermi level**.

semiconductor injection laser – a semiconductor device emitting coherent light. In general, it consists of a $p - n$ junction in a direct gap semiconductor placed in an optical resonator that is formed by the polished facets perpendicular to the junction plane. The light emission is produced under forward bias of the $p - n$ junction as a result of radiative recombination of electrons and holes injected into the depletion layer associated with the junction. The wavelength of the emitted radiation is basically determined by the relation $\lambda \sim hc/E_g$, where E_g is the energy band gap of the semiconductor. Optical feedback is provided by the polished facets. Lasing starts at a certain current density, called the threshold current density, that in case of a device made of a single semiconductor is usually above 10^4 A cm^{-2}.

First described in: N. G. Basov, O. N. Crokhin, Yu. M. Popov, *The possibility of use of indirect transitions to obtain negative temperature in semiconductors*, Z. Eksper. Teor. Fiz. **39**(5), 1486–1487 (1960) – idea (in Russian); R. N. Hall, G. E. Fenner, J. D. Kingsley, T. J. Soltys, R. O. Carlson, *Coherent light emission from GaAs junctions*, Phys. Rev. Lett. **9**(9), 366–368 (1962) and M. I. Nathan, W. P. Dumke, G. Burns, F. H. Dill, Jr., G. Lasher, Appl. Phys. Lett. **1**(1), 62–64 (1962) – experiment.

semimetal – a metal with a small overlap between partially filled **valence** and **conduction bands**.

Shannon's entropy $\rightarrow$ **Shannon's theorems**.

Shannon's theorems – two theorems forming the basis of classical information theory.

The **noiseless coding theorem** states that the optimal code compresses each symbol to $H(X)$ bits asymptotically. Here $H(X)$ is the **Shannon's entropy**, which is an entropy of an ensemble X consisting of strings of symbols defined as $H(X) = -\sum_i p(i) \log p(i)$, where $p(i)$ is the probability of the i-th symbol. Thus, information can be measured, and these measurements are of practical importance in choosing methods for transmission of messages.

Sent and received symbols may differ from one another because of a noise in the transmitting channel. The conditional probability to receive j-th symbol from ensemble Y while sending i-th symbol from ensemble X is denoted $p(j|i)$. The Shannon's entropy for the ensemble Y is $H(Y) = -\sum_{ij} p(i|j)p(i) \log[\sum_k p(j|k)p(k)]$ and the **joint Shannon's entropy for two ensembles** is defined as $H(X,Y) = -\sum_{ij} p(j|i)p(i) \log[p(j|i)p(i)]$.

The **noisy channel coding theorem** states that for the optimal coding one symbol transmits C bits of information at maximum and the probability of error goes to zero as the number of symbols in the message goes to infinity. The channel capacity C is defined as the maximal value of mutual information $H(X) + H(Y) - H(X,Y)$ over all possible probability distributions $p(i)$ for fixed $p(j|i)$. Information can be transmitted faithfully, even if errors are generated during the communication. The most famous example of this is given for the bandwidth-limited and power-constrained transmitting channel in the presence of Gaussian noise, usually expressed in the form $C = W \log_2(1 + R)$, where C is measured in bits/s, W is the bandwidth (in Hertz), and R is the signal-to-noise ratio.

First described in: C. E. Shannon, *A mathematical theory of communication*, Bell Syst. Tech. J. **27**(3), 379–423; C. E. Shannon, *A mathematical theory of communication*, Bell Syst. Tech. J. **27**(4), 623–656 (1948); C. E. Shannon, *Communication theory of secrecy systems*, Bell Syst. Tech. J. **28**(4), 656–715 (1949).

More details in: A. Galindo, M. A. Martín Delgado, *Information and computation: classical and quantum aspects*, Rev. Mod. Phys. **74**(2), 347–423 (2002).

Sharvin resistance – the resistance of a point electric contact in the classical ballistic regime. It is entirely due to elastic backscattering at the geometrical narrowing of the ballistic point contact. Dissipative processes, which bring the electron system into thermodynamic equilibrium, take place far away from the point contact. Therefore, the cause of the resistance in this system is spatially separated from its corresponding Joule heating. When the width of the contact is comparable to the Fermi wavelength one comes to the case of quantum point contact for which the resistance is $h/(2e^2 N)$, where N is the number of the occupied electron subbands (transmitting modes).

First described in: Yu. V. Sharvin, *A possible method for studying Fermi surfaces*, Zh. Eksp. Teor. Fiz. **48**(3), 984–985 (1965) – in Russian, Sov. Phys. JETP **21**(3), 655–656 (1965) – in English.

shear modulus → **Bowden–Tabor law**.

Shenstone effect – an increase in photoelectric emission of certain metals following passage of an electric current.

First described in: A. G. Shenstone, *Effect of an electric current on the photoelectric effect*, Phil. Mag. **41**, 916 (1921).

Shklovskii–Efros law states that temperature dependence of hopping conductivity in a disordered system has the form $\ln \sigma \sim T^{-1/2}$ at low temperatures when $\triangle E_C > k_B T$, where E_C is the gap in the density of states near the **Fermi level** of the system due to electron-electron Coulomb interaction. See more details in **hopping conductivity**.

First described in: A. L. Efros, B. I. Shklovskii, *Coulomb gap and law temperature conductivity of disordered systems*, J. Phys. C: Solid State Phys. **8**(4), L49–L51 (1975).

Shockley partial dislocation – a partial dislocation whose **Burger's vector** lies in the fault plane, so that it is able to glide, in contrast to **Frank partial dislocation**.

First described in: W. Shockley, W. T. Read, *Quantitative predictions from dislocation models of crystal grain boundaries*, Phys. Rev. **75**(4), 692–692 (1949).

Shockley–Queisser limit – the maximum efficiency one can reach with the conventional **solar cell** structure, also called **single material limit**. Using a detailed balance approach and assuming that each photon above the band gap gives rise to just one electron-hole pair, while all photons with energy below the band gap is lost, one comes to a theoretical maximum energy conversion efficiency of 30%. In particular Shockley and Queisser derived this uppermost conversion efficiency for an optimum band gap of 1.26 eV by modeling the sunlight as unconcentrated black body radiation at a temperature of 6000 K. This efficiency is relatively low, mainly because of the thermalization and transmission losses, which account for about 56% loss of photon energy. These losses are the result of the fact that a single energy band gap is used to convert a wide range of solar energy photons. While two-band systems are bounded by the Shockley–Queisser efficiency limit, higher efficiencies are possible for systems with more than two different Fermi-distributions. Examples are **multijunction solar cells**, **impurity photovoltaic effect solar cells**, **intermediate band solar cells**, **quantum well solar cells**, **up- and down-conversion solar cells**, or **hot-carrier solar cells**.

First described in: W. Shockley, H. J. Queisser, *Detailed balance limit of efficiency of p-n junction solar cells*, J. Appl. Phys. **32**(3), 510–519 (1961).

More details in: C. S. Solanki, G. Beaucarne, *Advanced solar cell concepts*, Energy for Sustainable Development – The Journal of the International Energy Initiative **11**, 17–23 (2007).

shock wave – a fully developed compression wave of large amplitude across which density, pressure and velocity change drastically. Imagine you have a projectile that flies faster than the speed of sound; it will push on the sound waves in front of it. But sound waves cannot travel faster than the speed of sound. Thus, the waves pile up against each other as they are created. These piled up waves are shock waves.

Shock waves can be produced by hurling a projectile or shooting a high-intensity **laser** at a crystal. Recently computer simulations have shown that sending shock waves through a **photonic crystal** could lead to faster and cheaper telecommunication devices, more efficient **solar cells**, and advances in **quantum computation**.

More details in: Ia. B. Zeldovich, I. P. Raizer, *Elements of Gas Dynamics and the Classical Theory of Shock Waves* (Academic Press, New York, 1968); M. Rasmussen, *Hypersonic Flow* (Wiley, New York, 1994).

Shor algorithm – the factoring algorithm, where a quantum computer would be able to factor N numbers exponentially faster than any known classical computer algorithm. It determines the period of the function $a^n \bmod N$, where a is a randomly chosen small number with no factors in common with N, and n is an integer. The value of $a^n \bmod N$ is the remainder that is left over when a^n is divided by N. Once the period of this function is known, N can be rapidly factored using classical number theory. Quantum factoring works by using quantum parallelism to try out all possible values of n simultaneously and then applying a quantum Fourier transform to find the period. In the special case of $N = 15$, the period is always either 2 or 4, making it particularly easy to implement the algorithm and determine the period.

Fast quantum number factoring has profound implications in cryptanalysis, of many popular encryption algorithms.

First described in: P. W. Shor, *Polynomial time algorithms for prime factorisation and discrete logarithms on quantum computer*, SIAM J. Comp. **26**(5), 1484–1509 (1997).

shot noise in electrical circuits is a consequence of the discrete nature of the electric charge passing through a conductor. It results from fluctuations in the number of charge carriers. The mean square fluctuation of electrical current I corresponding to individual current pulses due to arrival of individual electrons in a frequency interval $\triangle f$ is $\overline{\triangle I^2} = 2eI\triangle f$.

First described in: W. Schottky, *Über spontane Stromschwankungen in verschiedenen Elektrizitätsleitern*, Ann. Phys. **57**, 541–567 (1918).

SHPM – an acronym for a **scanning Hall probe microscopy**.

Shubnikov–de Haas effect – periodic variation of electric resistivity of **degenerate** solids (metals, semiconductors) with $1/H$, where H is the strength of a static magnetic field. The period of variation depends on the magnetic field orientation with respect to the crystal axes directions in the solid. It is $2\pi\hbar e/cS$, where S is the maximum cross-sectional area of the **Fermi surface** regarding projection of the electron momenta onto the magnetic field. The effect is observed when $\hbar e/m^* c > k_B T$ and $eH/m^* c > \tau$, where m^* is the effective mass of electrons and τ is the electron relaxation time.

The phenomenon is caused by the changing occupation of the **Landau levels** in the vicinity of the **Fermi energy** in the plane perpendicular to the direction of the magnetic field applied.

First described in: L. V. Shubnikov, W. J. de Haas, *Resistance changes of bismuth crystals in a magnetic field at the temperature of liquid hydrogen*, Proc. K. Akad. Amsterdam **33**(4), 363–378 (1930).

Shubnikov groups – points group and space groups of crystals if one adds another element of symmetry, for example two-color symmetry that switches black to white or vice versa. They are also called **black-and-white groups** or **magnetic groups**. Thus, silicide

Shubnikov groups are useful when properties such as spin are to be considered over the normal crystallographic symmetry. It is a powerful tool also for describing **quasicrystals**, crystals with imperfections and incommensurate modulated crystal phases.

First described in: A. V. Shubnikov, *On works of Pierre Curie in the field of symmetry*, Usp. Fiz. Nauk. **59**, 591–602 (1956) – in Russian, A. V. Shubnikov, *Time reversal as an operation of antisymmetry*, Kristallografiya (USSR) **5**(2), 328–333 (1960).

silicide – a compound formed by a metal with silicon. Elements of the Periodic Table form more than 100 such compounds. Most of them have electronic properties typical of metals and demonstrate electrical resistivity in the range of 15–100 $\mu\Omega$cm ($TiSi_2$, $CoSi_2$, $TaSi_2$, tetragonal $MoSi_2$, tetragonal WSi_2, $PtSi$, Pd_2Si) depending on the purity and structure of the materials. There are about a dozen silicides with semiconductor behavior. Their room temperature energy gaps range from 0.07 eV (hexagonal $MoSi_2$ and WSi_2) and 0.12 eV ($ReSi_{1.75}$) to 1.8 eV ($OsSi_2$) and 2.3 eV (Os_2Si_3).

More details in: Properties of Metal Silicides, edited by K. Maex, M. van Rossum (INSPEC, IEE, London, 1995); *Semiconducting Silicides*, edited by V. E. Borisenko (Springer, Berlin, 2000).

Silsbee's effect – the ability of an electric current to destroy a superconducting state by means of the magnetic field that it generates without raising the temperature.

First described in: F. B. Silsbee, Bull. Natl. Bur. Std. **14**, 301 (1918).

SIM – an acronym for a **structured illumination microscopy**.

SIMS – an acronym for a **secondary ion mass spectrometry**.

Sinai process – a random walk on a one-dimensional lattice with fixed probabilities to move to the right and left at each lattice site. Although these probabilities are fixed, their values are chosen from a probability distribution such that there is a strong bias at each site for the walker to go in a preferred direction. The walker makes many jumps going between two sites: a leftmost site that when approached from the right sends the walker back toward the right with high probability, and similarly a rightmost site that sends the walker back to the left. There can be many such pairs of sites, especially in a nested configuration. In the process, the mean value of $r^2(t)$ was shown to be proportional to $\ln^4(t)$.

First described in: Ya. G. Sinai, *The limiting behavior of one dimensional random walk in a random medium*, Theory Prob. Appl. **27**(2), 256–268 (1982).

sine-Gordon equation – the nonlinear differential equation with **soliton** solutions:

$$\frac{\partial^2 \psi}{\partial t^2} - a^2 \frac{\partial^2 \psi}{\partial x^2} + \omega_p^2 \sin \psi = 0,$$

where ω_p and a are parameters. If ψ is so small that the sine term can be replaced by the argument, the solution is a harmonic wave with the frequency wave vector relation given by $\omega^2 = \alpha\omega_p^2 + a^2q^2$.

In the linear approximation, a superposition of the waves can make up a wave packet localized in space. However, the components are dispersive so that after a time the width of the packet grows and the wave packet begins to unravel. The nonlinear term in the equation acts to counter this tendency. A suitably localized wave packet can maintain its shape forever or for as long as damping, which has not been included in the equation, does not come into play. Such a soliton solution is $\psi = 4\arctan\{\exp[x - x_0 - vt]/\xi\}$. The solution changes from 0 to 2π in the neighborhood of $x_0 + vt$ within a width of the order

$$\xi = \frac{a}{\omega_p}\sqrt{1 - \frac{v^2}{a^2}}.$$

It is a rounded step propagating with speed $v < a$, not otherwise dependent on the equation. Although the equation is nonlinear, a characteristic of the solitons is that two solitons can pass through each other with no changes in shape.

single-electron box – a single-electron device which consists of just one small island separated from a larger electrode, serving as an electron source, by a tunnel barrier. It is schematically shown in Figure S.10. The island is made of a conducting or semiconducting material. It has a size typical of a **quantum dot**. An external electric field is applied to the island using a gate separated from the island by a thicker insulator which does not allow noticeable tunneling. Increasing gate voltage attracts more and more electrons to the island. In an electric circuit the tunnel barrier is represented by the barrier capacitance C_t and tunneling resistance R_t. The capacitance C_g belongs to the gate.

By applying a positive bias to the gate electrode one can attract an electron to the island. When the increasing gate bias exceeds the **Coulomb blockade** limit the electron enters the island. Further increase in the gate voltage makes it possible to add one more electron to the island and so on. The discreteness of the electron transfer through the tunnel barrier necessarily makes such an increase step-like, thus producing a **Coulomb staircase** in terms of the charge accumulated in the island as a function of the voltage applied to the gate. The conditions for an experimental observation of the staircase characteristic suppose the charging energy to exceed the thermal energy $e^2/2C >> k_BT$, where $C = C_t + C_g$, and $R_t >> h/e^2$. At finite temperatures the steps and sawtooth dependence are eliminated.

The single-electron box is a basic building unit for other single-electron devices.

More details in: K. K. Likharev, *Single-electron devices and their applications*, Proc. IEEE **87**, 606–632 (1999).

single-electron device

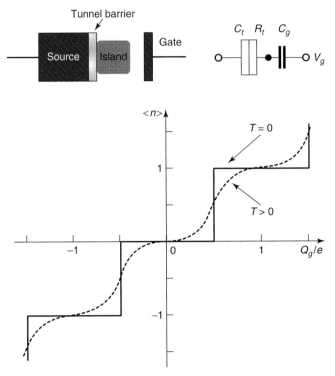

Figure S.10 Single-electron box; its representation in electric circuits (top) and (bottom) the average number of electron charges $\langle n \rangle$ in the box as a function of the gate charge $(Q_g = C_g V_g)$.

single-electron device – an electron device that can control the motion of even a single electron for information storage and processing. It typically consists of one or-few **quantum dot**s electronically interconnected and connected to other elements by tunnel junctions. Such devices retain their scalability even on an atomic scale.

More details in: K. K. Likharev, *Single-electron devices and their applications*, Proc. IEEE **87**, 606–632 (1999).

single-electron pump – a **single-electron device** which consists of a series of small gated islands separated by tunnel barriers and placed between electrodes serving as an electron source and drain. It is schematically shown in Figure S.11.

Its operation has much in common with a **single-electron trap**. Alternating voltage waveforms applied to each gate electrode are phase shifted, forming a potential wave gliding along the island array. This wave picks up an electron from the source and carries it down to the drain – a process very similar to the action of charge-coupled devices. Notice that this device does not need a DC source drain voltage: the direction of the transferred electron is determined by that of the running wave of electric potential. The pump transfers exactly one electron from the source to the drain during one cycle of the alternating gate biases. Of all single-electron devices, it dissipates the lowest energy and has the highest transfer accuracy.

First described in: H. Pothier, P. Lafarge, P. F. Orfila, C. Urbina, D. Esteve, M. H. Devoret, *Single-electron pump fabricated with ultrasmall normal tunnel junctions,* **169**(1–4), 573–574 (1991).

More details in: K. K. Likharev, *Single-electron devices and their applications,* Proc. IEEE **87**, 606–632 (1999).

Figure S.11 Single-electron pump and its representation in electric circuits.

single-electron transistor – a **single-electron device** which consists of a small conducting island with a pair of tunnel junctions separating it from the electrodes serving as an electron source and drain. It is charge operated by the third electrode acting as a gate. The transistor is schematically shown in Figure S.12 with its direct current characteristics.

Note that the characteristics shown are typical for a symmetric structure supposing identity of the source and drain tunnel junctions. The conditions to maintain n uncompensated electrons in the island are:

$$\frac{1}{C_d}\left(ne - \frac{e}{2} - C_g V_g\right) \langle V_{sd} \langle \frac{1}{C_d}\left(ne + \frac{e}{2} - C_g V_g\right)$$

$$\frac{1}{C_s + C_g}\left(-ne + \frac{e}{2} + C_g V_g\right)\rangle V_{sd}\rangle \frac{1}{C_s + C_g}\left(-ne - \frac{e}{2} + C_g V_g\right),$$

where $V_{sd} = V_d - V_s$.

The grey areas shown in the $V_{sd} - V_g$ relationship corresponds to an effective **Coulomb blockade**. The number of electrons in the island takes a fixed value

single-electron transistor

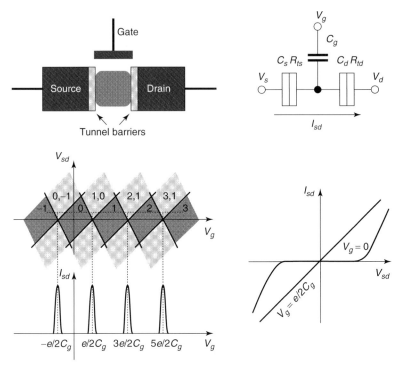

Figure S.12 Single-electron transistor, its representation in electric circuits and static electric characteristics.

indicated in the dark areas. In the neighboring light areas the island can take two electron numbers, namely one of those indicated. In this case the bigger number corresponds to preferred tunneling through the source junction, while the smaller one supposes electron tunneling through the drain junction.

Initially there are no uncompensated electrons in the island. When a finite positive source-to-drain voltage V_{sd} (indicated by the dashed line) is applied and the gate voltage is $e/2C_g + ne/C_g$, electron transport through the island occurs. For the source tunneling junction, an increase in the electron number in the island is preferable, so that an electron tunnels from the source to the island. However, for the drain tunneling junction, a decrease in the electron number in the island is preferable, thus providing an electron to tunnel from the island to the drain. As a result, an electron passes from the source to the drain, and source-to-drain current is observable at these bias conditions. The oscillating $I_{sd} - V_g$ characteristics are often called **Coulomb oscillations**.

The current–voltage (I_{sd} versus V_{sd}) characteristics are dependent on both gate voltage and temperature. When $0 < V_g < e/2C_g$ there is no current at a small source–drain voltage. This suppression of the current is controlled by the **Coulomb blockade**. At a certain threshold voltage the Coulomb blockade is overcome, and at much higher voltages the current exhibits quasi-periodic increase similar to the

Coulomb staircase in the **single-electron box**. The threshold voltage, as well as the source–drain current in its vicinity, is a periodic function of the gate voltage.

Appropriate temperature conditions for a distinct operation of the transistor needs $e^2/2C >> k_BT$, where $C = C_g + C_s + C_d$.

First described in: D. V. Averin, K. K. Likharev, *Coulomb blockade of tunneling and coherent oscillations in small tunnel junctions*, J. Low. Temp. Phys. **62**(2), 345–372 (1986) – theory; T. A. Fulton, G. J. Dolan, *Observation of single-electron charging effects in small tunneling junctions*, Phys. Rev. Lett. **59**(1), 109–112 (1987) – experimental demonstration.

More details in: K. Uchida, *Single electron devices for logic applications*, in: *Nanoelectronics and Information Technology*, edited by R. Waser (Wiley-VCH, Weinheim, 2003).

single-electron trap – a **single-electron device** which consists of several islands between the source and **single-electron box** as schematically shown in Figure S.13

The main new feature of this system is its internal memory (bi- or multistability). Certain ranges of applied gate voltage cause the system to be in one of two (or more) charged states of its edge island. This is due to electric polarization effects: an electron located in one of the islands of the array extends its field to a certain distance outside.

By applying sufficiently high voltage $V^\uparrow$ to the gate one can force an electron to drive into the edge island. If the array is not too long, other electrons feel its repulsion and do not follow. When the gate voltage is subsequently decreased to the initial level, the electron is trapped in the edge island behind the energy barrier. In order to remove the electron from the trap, the voltage has to be reduced further, to $V^\downarrow < V^\uparrow$. As a result, the dependence of the number of electrons in the trap on the the gate voltage exhibits regions of bi- or multi-stability, in which the charge state of the trap depends on its prehistory.

The lifetime of a certain state within the multi-stability region is fundamentally limited by the thermal activation over the energy barrier (E_c) and **co-tunneling**. The first effect is exponentially low in E_c/k_BT, while the second effect falls exponentially with the number of the islands in the array. As a result, electron retention time may be very long.

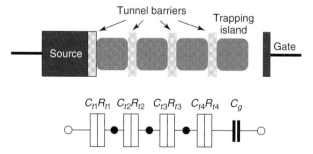

Figure S.13 Single-electron trap and its representation in electric circuits.

single-electron tunneling

First described in: T. A. Fulton, P. L. Gammel, L. N. Dunkleberger, *Determination of Coulomb-blockade resistances and observation of the tunneling of single electrons in small-tunnel-junction circuit*, Phys. Rev. Lett. **67**(22), 3148–3151 (1991).

More details in: K. K. Likharev, *Single-electron devices and their applications*, Proc. IEEE **87**, 606–632 (1999).

single-electron tunneling – electron-by-electron tunneling governed by the **Coulomb blockade** effect. It is illustrated in Figure S.14 for the system composed of two conducting electrodes separated by a tunnel junction formed by a very thin **dielectric**. Electronic processes in the structure are, in parallel, represented by the mechanical model showing formation of water droplets at a faucet on a pipe. Initially the charge at both surfaces of the dielectric is zero. When external constant bias is applied to the structure, the electric charge starts to accumulate at the surface of the dielectric. It is transferred by both continuous and discrete processes.

Electric current in solids is known to be a result of the motion of electrons through the lattice of atomic nuclei. While each electron carries the lowest fixed amount of the negative charge, denoted as $-e$, the smallest transferred charge can be lower than this value. This is due to the fact that the transferred charge is proportional to the sum of shifts of all electrons with respect to the lattice nuclei. As the shifts can be extremely small, their sum changes continuously. Thus, it can have practically any value, even a fraction of the charge of single electron.

The transferred charge flowing continuously through the conductor is accumulated on the surface of the electrode against the dielectric layer of the junction. The adjacent electrode will have equal but opposite surface charge. One can imagine this surface charge Q as a slight continuous shift of the electrons near the surface from their equilibrium positions. The charge collects in the junction just as water forms a droplet at the faucet. On the other hand, only tunneling can change this charge in a discrete way: when an electron tunnels through the dielectric barrier, the surface charge will be changed exactly by either $+e$ or $-e$, depending on the direction of tunneling.

A tunneling event becomes energetically favorable only when Q just exceeds $e/2$. At this condition the tunneling electron will change the charge abruptly from $+e/2$ to $-e/2$ and the energy of the junction remains the same. If Q is less than $+e/2$ or greater than $-e/2$, tunneling in any direction would increase the energy of the system. Thus, if the accumulated charge is within the range from $-e/2$ to $+e/2$, tunneling is forbidden by the Coulomb blockade.

When the charge reaches $e/2$, tunneling becomes possible. One electron crosses the junction, just like a water droplet falls as soon as it reaches a critical size. The accumulated charge drops to zero and the system appears at the Coulomb blockade range again. The repetition of the charge/discharge processes produces single-electron tunneling oscillations with a frequency equal to the current I divided by the charge of one electron, e.

First described in: I. O. Kulik, R. I. Shekhter, *Kinetic phenomena and charge discreteness effects in granulated media*, Zh. Eksp. Teor. Fiz. **68**(2), 623–640 (1975) – in Russian, Sov. Phys. JETP **41**(2), 308–314 (1975) – in English, theoretical prediction;

Conductor Dielectric Conductor

Tunneling

Electron

Figure S.14 Single-electron tunneling and its mechanical analogy with water falling from a pipe (from K. K. Likharev, T. Claeson, *Single electronics*, Scientific American (**6**), 50–55 (1992)).

single-electron turnstile

E. Ben-Jacob, Y. Gefen, *New quantum oscillations in current driven small junctions,* Phys. Lett. **108**A(5/6), 289–292 (1985); K. K. Likharev, A. B. Zorin, *Theory of Bloch wave oscillations in small Josephson junctions,* J. Low Temp. Phys. **59**(3/4), 347–382 (1985); D. V. Averin, K. K. Likharev, *Coulomb blockade of tunneling and coherent oscillations in small tunnel junctions,* J. Low Temp. Phys. **62**(2), 345–372 (1986) – theory; T. A. Fulton, G. J. Dolan, *Observation of single-electron charging effects in small tunneling junctions,* Phys. Rev. Lett. **59**(1), 109–112 (1987) – experiment.

single-electron turnstile – a **single-electron device** which consists of an odd number of small islands separated by tunnel barriers and placed between electrodes serving as an electron source and drain. It is schematically shown in Figure S.15.

At $V = 0$, the device operates exactly like a **single-electron trap**: one electron is pulled into the central island (at random, from either source or drain) increasing the gate voltage U beyond a certain threshold. Then it may be pushed out by decreasing U. The application of a modest drain–source bias $V \neq 0$ breaks the source–drain symmetry. In this case an electron is always picked up from the source when the gate voltage goes up and delivered to the drain when it goes back down. If the gate voltage is cycled periodically, one electron is transferred from the source to the drain in each period.

First described in: L. J. Geerligs, V. G. Anderegg, P. A. M. Holweg, J. E. Mooij, H. Pothier, D. Esteve, C. Urbina, M. H. Devoret, *Frequency-locked turnstile device for single electrons,* Phys. Rev. Lett. **64**(22), 2691–2694 (1990).

More details in: K. K. Likharev, *Single-electron devices and their applications,* Proc. IEEE **87**, 606–632 (1999).

Figure S.15 Single-electron turnstile and its representation in electric circuits.

single material limit = Shockley–Quesser limit.

single molecule magnet → molecular nanomagnet.

singlet – a state of two electrons with paired (antiparallel) spins. It has zero net spin.

singularity or singular point of a function of one or more variables is a point about which it cannot be expanded in a **Taylor series**; in particular, one at which the function or any of its partial derivatives become infinite. For example, $x = a$ is a singular point of both the functions $(x - a)^{1/2}$ and $\log(x - a)$.

skin effect – a tendency for an alternating electric current in a conductor to concentrate in the outer part, or "skin", rather than be distributed uniformly over the body of the conductor. The effect arises from the increase of internal self-inductance of the conductor with the depth below the surface of the conductor. In consequence, the magnitude of the effect increases with the frequency of the current, the diameter of the conductor and the magnetic permeability of the conductor material. Defining the penetration depth d as the distance measured from the surface of the conductor at which the current has fallen to $1/l$ of its surface value, it can be given by $d = (\pi\mu\mu_0\nu\sigma)^{-1/2}$, where μ is the relative permeability of the material, ν is the frequency of the alternating current, and σ is the conductivity of the material.

skyrmion – a **quasiparticle** corresponding to topological twists or kinks in a **spin** space. It can also be interpreted as a **soliton** with spin and statistics different from those of the underlying fields in a nonlinear field theory. Skyrmions are a characteristic of nonlinear continuum models ranging from microscopic to cosmological scales.

First described in: T. H. Skyrme, *A non-linear field theory*, Proc. R. Soc., Ser. A **260**, 127–138 (1961).

Slater determinant – a quantum mechanical wave function for N electrons, or in general for N fermions, which is an $N \times N$ determinant whose entries are N different one particle wave functions $\psi_N(r_N)$ depending on the coordinates of each of the particles in the system:

$$\Psi = \frac{1}{(N!)^{1/2}} \begin{vmatrix} \psi_1(r_1) & \psi_2(r_1) & \cdots & \psi_N(r_1) \\ \psi_1(r_2) & \psi_2(r_2) & \cdots & \psi_N(r_2) \\ \cdots & \cdots & \cdots & \cdots \\ \psi_1(r_N) & \psi_2(r_N) & \cdots & \psi_N(r_N) \end{vmatrix}.$$

First described in: J. C. Slater, *The theory of complex spectra*, Phys. Rev. **34**(10), 1293–1322 (1929).

Slater–Jastrow wave function – a product of **Slater determinants** and the Jastrow correlation factor:

$$\psi_T(\{\mathbf{r}_i\}) = \det(A_{up})\det(A_{down}) \exp\left(\sum_{i<j} U_{ij}\right).$$

Slater orbital

Here, the A_{up} and A_{down} are defined as the Slater matrices of the single particle up and down orbitals, respectively. That is

$$A = \begin{vmatrix} \phi_1(\mathbf{r}_1) & \phi_1(\mathbf{r}_2) & \phi_1(\mathbf{r}_3) & \cdots \\ \phi_2(\mathbf{r}_1) & \phi_2(\mathbf{r}_2) & \phi_2(\mathbf{r}_3) & \cdots \\ \phi_3(\mathbf{r}_1) & \phi_3(\mathbf{r}_2) & \phi_3(\mathbf{r}_3) & \cdots \\ \vdots & \vdots & \vdots & \ddots \end{vmatrix},$$

where ϕ_k are molecular orbitals centered at $\mathbf{c}_k$:

$$\phi_k(\mathbf{r}) = \exp\left(\frac{-(\mathbf{r} - \mathbf{c}_k)^2}{\omega_k^2 + v_k|\mathbf{r} - \mathbf{c}_k|}\right).$$

The **Jastrow correlation factor** U_{ij} terms are defined as

$$U_{ij} = \frac{a_{ij}r_{ij}}{1 + b_{ij}r_{ij}},$$

where $r_{ij} \equiv |r_i - r_j|$ and

$$a_{ij} = \begin{cases} e^2/8D & \text{if } ij \text{ are like spins} \\ e^2/4D & \text{if } ij \text{ are unlike spins} \\ e^2/2D & \text{if } ij \text{ are electron-nuclear pairs} \end{cases}$$

The Slater–Jastrow wave has a number of desirable properties for many-body quantum Monte Carlo calculations of electrons in the presence of ions.

First described in: J. C. Slater, *The self consistent field and the structure of atoms,* Phys. Rev. **32**(3), 339–348 (1928); R. Jastrow, *Many-body problem with strong forces,* Phys. Rev. **98**(5), 1479–1484 (1955).

Slater orbital – an approximate solution to the eigenvalue equation: $Nr^{n-1}e^{\zeta r}Y_l^m(\theta\phi)$, where N is the normalization constant, n is the principal quantum number of the orbital, $\zeta = (Z - s)/n$ is the orbital exponent, $Y_l^m(\theta\phi)$ is the angular part of the orbital, Z is the atomic number and s is a screening constant.

First described in: J. C. Slater, *Atomic shielding constants,* Phys. Rev. **36**(1), 57–64 (1930).

small angle X-ray scattering (SAXS) – X-ray diffraction technique with scattering angles less than $1°$. The diffraction information about structures with large periods lies in the region. Therefore the technique is commonly used for probing large length scale structures, such as high-molecular-weight polymers, biological macromolecules (proteins, nucleic acids, and so on) and self-assembled superstructures (such as surfactant-templated mesoporous materials). The measurements are technically challenging because of the small angular separation of the direct beam and the scattered beam. Large specimen-to-detector distances (0.5–10 m) and high quality collimating optics are used to achieve a good signal-to-noise ratio.

S-matrix = **scattering matrix**. It expresses what comes out of a system as a function what goes into it in a matrix form:

$$\begin{pmatrix} C \\ D \\ \cdots \end{pmatrix} = \mathbf{S} \begin{pmatrix} A \\ B \\ \cdots \end{pmatrix},$$

where $A, B, \ldots$ represent the incoming influence on the system and $C, D, \ldots$ represent the system reaction.

First described in: J. A. Wheeler, *On the mathematical description of light nuclei by the method of resonating group structure*, Phys. Rev. **52**(11), 1107–1122 (1937).

More details in: D. Iagolnitzer, *The S Matrix* (North Holland, Amsterdam, 1978).

Smith–Helmholtz law states that for a single refracting surface of sufficiently small aperture the product of the index of refraction, distance from the optical axis and the angle which a light ray makes with the optical axis at the object point is equal to the corresponding product at the image point. This formula is also associated with the names of Lagrange and Clausius. In a more restricted form it was also known to Huygens.

More details in: J. W. S. Rayleigh, *Notes, chiefly historical, on some fundamental propositions in optics*, Phil. Mag. **21**(5), 466 (1886); T. Smith, *The general form of the Smith-Helmholtz equation*, Trans. Opt. Soc. **31** 241–248 (1930); M. Born, E. Wolf, *Principles of Optics, Seventh Edition* (Cambridge University Press, Cambridge 1999).

Snell's law gives the relationship between the angle of incidence and the angle of refraction for a wave impinging on an interface between two media with different refractive indexes (Figure S.16): $n_1 \sin \theta_1 = n_2 \sin \theta_2$.

First described by: W. Snell in 1621 and published in: R. Descartes, *Dioptrique* (Leyden 1637).

Figure S.16 Refraction of light.

SNOM – an acronym for a scanning near-field optical microscopy. For more details → **near-field scanning optical microscopy**.

soft nanotechnology – the nanotechnology focused on soft materials.

Sohncke law states that the stress per unit area normal to a crystallographic plane needed to produce a fracture in a crystal is a constant characteristic of a crystalline substance.

sol

First described in: L. Sohncke, *Entwicklung einer Theorie der Krystallstruktur* (Leipzig, 1879).

sol → **sol-gel technology**.

solar cell – a semiconductor device converting light energy into electrical energy. The **photovoltaic effect** is used for this. The first solar cell was fabricated by Fritts in 1883 using Se as the absorbing layer with a rectifying metal contact.

Usually three generations of solar cells are distinguished. The first generation consists of a large-area, single-crystal, single-layer $p - n$ junction **diode**, capable of generating electrical energy from light sources with the wavelengths of sunlight. The first semiconductor $p - n$ junction solar cells were described in 1941 by Ohl. Most photovoltaic solar cells produced to date have been based on silicon $p - n$ junctions, although now relying on junctions formed more controllably by diffusing one polarity dopant into a wafer substrate of opposite polarity. To reduce reflection losses crystallographic texturing on the top surface has been widely used. Silicon solar cells, made from single-crystalline and multi-crystalline silicon, contribute ~94% of the world market. The maximum efficiency one can reach with the conventional solar cell structure is given by the so-called **single material** or **Schockley-Queisser limit**. The second generation of solar cell relies on the use of thin epitaxial deposits of semiconductors on lattice-matched wafers. An example is the use of **heterojunction**s formed from different $A^{III}B^V$ compounds or their alloys to produce the rectifying junction in **heterojunction solar cell**s. In particular **multijunction solar cells** are the most sophisticated solar cells to date. The dominance of the photovoltaic field by inorganic solid-state junction devices is now being challenged by the emergence of a third generation of cells, based, for example, on **polymer solar cells**, **nanoparticle-based solar cells**, and **dye-sensitized solar cell**s.

The main parameters characterizing a solar cell are open circuit voltage, short circuit current, energy conversion efficiency, fill factor, and quantum efficiency. All of them are referred to measurements under standard test conditions (STC). The most often used STC specifies a temperature of $25°C$ and an irradiance of 1000 W m^{-2} with an air mass 1.5 (AM1.5) spectrum. These correspond to the irradiance and spectrum of sunlight incident on a clear day upon a sun-facing $37°$-tilted surface with the sun at an angle of $41.81°$ above the horizon.

First described in: C. E. Fritts, *New form of selenium cell, with some remarkable electrical discoveries made by its use*, Proc. Am. Assoc. Adv. Sci. **33**, 97 (1883); C. E. Fritts, *A new form of selenium cell*, Am. J. Sci. **26**, 465 (1883), R. S. Ohl, *Light-sensitive electric device*, US Patent No. 2 402 622, 27th May 1941; R. S. Ohl, *Light-sensitive device including silicon*, US Patent No. 2 443 542, 27th May, 1941.

More details in: M. A. Green, *Photovoltaic principles*, Physica E **14**, 11–17(2002); C. S. Solanki, G. Beaucarne, *Advanced solar cell concepts*, Energy for Sustainable Development–The Journal of the International Energy Initiative **11**, 17–23 (2007); P. Würfel, *Physics of Solar Cells: From Principles to New Concepts* (Wiley-VCH, Weinheim, 2005); T. Markvart, L. Castaner, *Solar Cells: Materials, Manufacture and Operation* (Elsevier, Amsterdam, 2005).

sol–gel technology – a method of material synthesis based on the sol–gel transition. A **sol** is a solution of colloidal particles, which are solid particles with a size of 1–100 nm, in a liquid. A **gel** represents a network of closely connected polymer chains derived from the colloidal particles of sol. Porous glasses and glasses with embedded nanocrystals of semiconductors can be formed in this way in the form of thin films and bulk samples.

An example is the sol–gel transition in water solutions of silicon-organic compounds with a composition $Si(OR)_4$, where R is one of the alkyl groups (CH_3, C_2H_5, C_3H_7, . . .). Substitution of the alkyls by OH-groups is the first step of the sol–gel transformation

$$
\begin{array}{ccc}
\text{OR} & & \text{OH} \\
| & & | \\
\text{RO}-\text{Si}-\text{OR} + 4H_2O \rightarrow & & \text{HO}-\text{Si}-\text{OH} + 4ROH \\
| & & | \\
\text{OR} & & \text{OH}
\end{array}
$$

The produced molecules of silicon hydroxide ($Si(OH)_4$) then form chains and a threedimensional network by means of Si–O–Si bonds

$$
\begin{array}{cccc}
\text{OH} & \text{OH} & \text{OH} & \text{OH} \\
| & | & | & | \\
\text{HO}-\text{Si}-\text{OH} + \text{HO}-\text{Si}-\text{OH} \rightarrow & & \text{HO}-\text{Si}-\text{O}-\text{Si}-\text{OH} + H_2O \\
| & | & | & | \\
\text{OH} & \text{OH} & \text{OH} & \text{OH}
\end{array}
$$

These transformations occur at relatively low temperatures – below $400°C$. Porous glasses developed in this way contain nanometer-size voids. They can be easily impregnated with **dye** molecules or atomic clusters. Nanocrystals can be directly derived therefrom. As compared to conventional glasses, porous glasses can be saturated with semiconductor materials up to concentrations reaching 10%, which are high enough for the study and use of electron interactions between nanocrystals. Additionally, crystallites in these matrixes are expected to have a smaller amount of defects because of the lower precipitation temperature. $A^{II}B^{VI}$ and $A^{III}B^{V}$ semiconductor nanocrystals can be successfully produced by this technique. Their size distribution is rather wide.

More details in: J. D. Wright, N. A. J. M. Sommerdijk, *Sol-Gel Materials: Chemistry and Applications* (Taylor & Francis, New York, 2001).

solid phase epitaxy – solid state structural rearrangements in a thin amorphous or polycrystalline layer on a monocrystalline substrate resulting in the layer structure becoming epitaxial to the substrate. The epitaxial rearrangements starts at the interface and proceeds through the layer with a rate proportional to $\exp(-E_a/k_B T)$, where E_a is the activation energy of the solid-phase epitaxial growth. It is widely used to restore the crystalline structure in ion-implanted layers of semiconductors and to fabricate epitaxial **silicides** on monocrystalline silicon formed during their solid-phase synthesis.

solitary wave – a localized disturbance in a continuous medium that can propagate over a long distance without any change to its shape or amplitude.

soliton – a hump developed by waves in a shallow channel. It propagates as a coherent entity, continuing its course along the channel apparently without change of form. Solitons on canals can have various widths, but the smaller the width the larger the height must be and the faster the soliton travels. Thus, if a high, narrow soliton is formed behind a low, broad one, it will catch up with the low one. It turns out that, when the high soliton does so, it passes through the low one and emerges with its shape unchanged.

From the mathematical point of view, a soliton is a localized nonlinear wave that regains asymptotically (as $t \to \infty$) its original (as $t \to -\infty$) shape and velocity after interacting with any other localized disturbance. The only long-term effect on the soliton from the interaction is a phase shift. In other words, a soliton is a spatially localized object, that is a solitary wave, which is dynamically and structurally stable. Soliton waves can be obtained as solutions of the **sine-Gordon equation**.

First described by S. Russell in 1834.

More details in: S. E. Trullinger, V. E. Zakharov, V. L. Pokrovsky, *Solitons* (North Holland, Amsterdam, 1986).

Sommerfeld fine-structure constant – the nondimensional coefficient $\alpha = 2\pi e^2/(hc)$ $= 7.297 \times 10^{-3}$. With the help of this constant, the grouping of closely spaced emission lines in the spectrum of one electron atoms can be related to the **precession** of elliptic electron orbitals.

First described in: A. Sommerfeld, *Zur Quantentheorie der Spektrallinie*, Ann. Phys. **51**, 1–94 (1916).

Sommerfeld's factor $\to$ **Bir–Aronov–Pikus mechanism**.

Sommerfeld wave – a surface wave guided by the plane interface between a dielectric and a good conductor, the energy decaying exponentially normal to the surface of the dielectric. The velocity of propagation depends on the dielectric constant of the medium in which the wave flows.

First described in: A. Sommerfeld, *Über die Fortpflanzung elektrodynamischer Wellen längs eines Drahtes*, Ann. Phys. **67**, 233–290 (1899).

sonoluminescence – light emission from a substance excited by high frequency sound waves or phonons.

First described in: W. T. Richards, A. L. Loomis, *The chemical effects of high frequency sound waves I. A preliminary survey*, J. Am. Chem. Soc. **49**(12), 3086–3100 (1927); F. O. Schmitt, C. H. Johnson, A. R. Olson, *Oxidations promoted by ultrasonic radiation*, J. Am. Chem. Soc. **51**(2) 370–375 (1929); H. Frenzel, H. Schultes, *Luminescenz im ultraschallbeschickten Wasser*, Z. Phys. Chem. Abt. B **27**, 421–424 (1934).

s orbital → atomic orbital.

space charge polarization → polarization of matter.

space group – a full symmetry group of a crystalline solid containing operations of a point group and lattice translation. In fact, space groups consist of rotations, reflections, screw rotations, glide reflections and translations, which transform the crystal as a whole into itself. There are 230 such groups. These are each described in the **Hermann–Mauguin notation**, which shows essential symmetry elements, for example $P6_3/mmc$. A space group symbol consists of two parts: a letter depicting the lattice type and some essential symmetry elements.

spectral hole-burning – the optical phenomenon observed in **quantum dot** ensembles, in particular for nanocrystals (dots) of semiconductors embedded in dielectric matrices. It is photo-induced modification of an emission spectrum just under irradiation, which is registered like a hole in the spectrum.

The phenomenon can have a transient, persistent or even irreversible character. In terms of the relaxation time of the burnt spectral hole the following groups are distinguished.

1. **Transient spectral hole-burning** due to selective population of the higher electron–hole states within the same crystallite with consequent relaxation to the lower levels. The typical relaxation times are in the subnanosecond range. In order to observe this effect the following condition should be satisfied $\tau_{imp} < \tau_{rel}$, where τ_{imp} is duration of the pump and probe pulses and τ_{rel} is the energy relaxation time of excitons from the upper to the lower states. In the specific case of large energy level separation when the "phonon bottleneck" effect is pronounced, a condition $\tau_{rec} < \tau_{rel}$, where τ_{rec} is the recombination lifetime, makes it possible to observe hole-burning of this type in a quasi-steady-state regime, that is independently of the pump pulse duration. The relevant effect in molecular spectroscopy is often considered in terms of the "hot luminescence". In bulk semiconductors a similar phenomenon is possible on the femtosecond time scale when **Fermi–Dirac distribution** of electron–hole gas is not achieved. This effect is readily observable in nanocrystals of $A^{II}B^{VI}$ compounds in glass and in porous silicon.

2. **Quasi-steady-state hole-burning** due to absorption saturation of resonantly excited crystallites in an inhomogeneously broadened ensemble. The recovery process is controlled by electron–hole recombination with a characteristic time in the subnanosecond to microsecond range. This effect has been reported by a number of groups for $A^{II}B^{VI}$ and $A^{I}B^{VII}$ quantum dots in dielectric matrices. If the relaxation time is comparable to or longer than the recombination time, then a selective steady-state population of the higher states is possible similarly to case 1. Spectral hole-burning due to creation

of the first electron-hole pair in a crystallite is accompanied by an induced absorption due to the formation of a biexciton state when the pump and the probe quanta are absorbed by the same crystallite. In the case of larger dots exciton–exciton interactions results in more complicated deformation of the absorption spectrum.

3. **Persistent reversible hole-burning** due to photoionization of a crystallite and/or surface localization of the electron occurs under certain conditions. In this case optical absorption experiences modifications due to the local electric field effect (**quantum-confined Stark effect**) with a differential absorption spectrum typical of exciton absorption in the external electric field. In the case of small quantum dots, surface localization of one of the carriers leads to a problem of a biexciton and the differential absorption spectrum may be similar to case 2. The effect has a recovery time from microseconds to hours depending on the temperature and/or dot-matrix interface.

4. **Permanent spectral hole-burning** is possible due to an irreversible sequence of photochemical reactions in resonantly excited crystallites stimulated by a manifold photoionization. The phenomenon was observed in semiconductor-doped polymer and may be interpreted in terms of size-selective photolysis of nanocrystallites. At the early stage, photoionization results in the local electric field effect with a differential absorption spectrum similar to case 3.

Unlike case 1, cases 2–4 are possible only under conditions of inhomogeneous broadening. Therefore they have no analog among the intrinsic bulk semiconductors. The analogs can be found, however, in a number of systems possessing inhomogeneous broadening, such as doped semiconductors, atoms in gases, ions in dielectrics and molecules in a gas phase, in solution and in solid matrices.

In terms of the underlying microscopic mechanisms, spectral hole-burning in nanocrystals can be a result of population-induced effects (selective absorption saturation), local electric field induced effects (selective photoionization), or photochemical effects (selective photolysis). Some mechanisms of spectral hole-burning may coexist. Observation of the specific hole-burning effect does not eliminate a manifestation of the other possible hole-burning phenomena.

More details in: S. V. Gaponenko, *Optical Properties of Semiconductor Nanocrystals* (Cambridge University Press, Cambridge, 1998).

SPEM – an acronym for **saturated pattern excitation microscopy**.

spherical coordinates – a system of curvilinear coordinates in which the position of a point in space is designated by its distance r from the origin or pole, called the radius vector, the angle ϕ between the radius vector and a vertically directed polar axis, called the cone angle or colatitude, and the angle θ between the plane of ϕ and fixed meridian plane through the polar axis, called the polar angle or longitude.

spin – an intrinsic angular momentum of an elementary particle or nucleus. It exists even when the particle is at rest, as distinguished from orbital angular momentum.

First described in: A. H. Compton, *Magnetic electron*, J. Frankl. Inst. **192**, 145–155 (1921) – idea of a quantized spinning of the electron; G. Uhlenbeck, S. A. Goudsmit, *Ersetzung der Hypothese vom unmechanischen Zwang durch eine Forderung bezüglich des inneren Verhaltens jedes einzelnen Elektrons*, Naturwiss. **13**, 953–954 (1925) and G. Uhlenbeck, S. A. Goudsmit, *Spinning electrons and the structure of spectra*, Nature **117**, 264–265 (1926) – spin concept.

spin–Coulomb drag – a change in the electron transport due to Coulomb interaction between spin-up and spin-down electrons. This interaction can transfer momentum between spin-up and spin-down electrons thereby effectively introducing friction for the relative motion of the two spin components. If, for example, one of the two spin components is set in motion relative to the other, it will tend to drag the latter in the same direction. Or, if a finite spin current is set up through the application of an external field, then the Coulomb interaction will tend to equalize the net momenta of the two spin components. The most dramatic manifestation of the phenomenon is the appearance of a finite trans-resistivity defined as the ratio of the gradient of the spin-down electrochemical potential to the spin-up current density when the spin-down current is zero.

First described in: I. D'Amico, G. Vignale, *Theory of spin Coulomb drag in spin-polarized transport*, Phys. Rev. B **62**(8), 4853–4857 (2000).

spin drag = spin–Coulomb drag.

spinel – the mineral $MgAl_2O_4$.

spin glass – a magnetic alloy in which the concentration of magnetic atoms is such that below a certain temperature their magnetic moments are no longer able to fluctuate thermally in time but are still directed at random, in loose analogy to the atoms of ordinary glass. If such an alloy is cooled in an external magnetic field from above to below this transition temperature and the external field is then removed, the magnetization decays slowly to zero, typically in a matter of hours.

spin Hall effect – a deflection in opposite directions of electrons with different **spins** perpendicular to the main current flow when they are scattered at impurities, defects or other imperfections in solids as illustrated in Figure S.17.

This spin current results in an accumulation of spin-up and spin-down electrons at different ages of the conducting channel, as takes place in the classical **Hall effect**, but without the influence of an external magnetic field. **Spin–orbit coupling** is considered to be the driving mechanism of the phenomenon. The process requiring electrons to scatter at impurities is referred to as the *extrinsic* spin Hall effect. In contrast, in the *intrinsic* spin Hall effect the effective magnetic field due to the spin–orbit interaction produces a torque on the electron spins that can also

spin–lattice relaxation

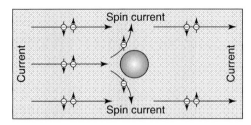

Figure S.17 Origin of a perpendicular spin current when spin-up and spin-down electrons are scattered at a defect in a conductor.

induce a spin current. The important distinction between these two effects is that the intrinsic one arises as a result of the crystal structure of the material, even in the absence of impurities, while the extrinsic Hall effect only occurs in the impurity scattering.

First described in: M. I. D'yakonov, V. I. Perel', *Possibility of orienting electron spins with current*, JETP Lett. **13**, 467–469 (1971) – theoretical prediction; J. E. Hirsch, *Spin Hall effect*, Phys. Rev. Lett. **83**(9), 1834–1837 (1999) – the term "Spin Hall effect" was introduced; Y. K. Kato, R. C. Myers, A. C. Gossard, D. D. Awschalom, *Observation of the Spin Hall effect in semiconductors*, Science **306**, 1910–1913 (2004) – first experimental observation.

More details in: V. Sih, Y. Kato, D. Awschalom, *A Hall of spin*, Physics World **18**(11), 33–36 (2005).

spin–lattice relaxation – a magnetic relaxation in which the excess potential energy associated with electron spin in a magnetic field is transferred to the lattice.

spinor – a vector with two complex components which undergoes a **unitary** unimodular transformation when the three-dimensional coordinate system is rotated. It can represent the spin state of a particle of spin 1/2. More generally, a spinor of order (or rank) n is an object with 2^n components, which transform as products of components of n spinors of rank one.

spin–orbit coupling (interaction) arises in an atom as a result of the interaction of the electron spin magnetic moment with the magnetic moment related to the orbital motion of the electron. It is accompanied by a shift in energy levels (due to the potential energy of the spin magnetic moment of the electron in the magnetic field) that it feels as it moves through the electric field of the nucleus. In solids, the spin–orbit interaction consists of two distinct contributions: the **Dresselhaus coupling**, which is due to bulk inversion asymmetry of the lattice, and the **Rashba coupling**, which is due to structure inversion asymmetry along a particular crystallographic direction. Both of these terms result in the splitting of electron energy levels and in the mixing of the electron spin states.

spin polarization in a particular material is defined in terms of the number of carriers with spin-up ($n_\uparrow$) and spin-down ($n_\downarrow$) as

$$P = \frac{n_\uparrow - n_\downarrow}{n_\uparrow + n_\downarrow}.$$

The most pronounced spin effects are reasonably expected for the materials with the highest fractions of spin-polarized electrons. These are materials that have only one occupied spin band at the Fermi level. In practice, partially polarized materials are mainly used in device applications. Such materials, which are metals and their alloys, oxides, magnetic semiconductors, and related bulk electron spin polarizations, are listed in Table S.1.

Note that spin polarization is sensitive to sample fabrication conditions and impurities in the material; therefore the highest polarizations experimentally observed are summarized in Table S.1 .

An electron current in solid-state structures composed of differently spin-polarized materials is indeed dependent on the particular spin orientations in the regions through which electrons pass. Electrons originating from one spin state at the Fermi level of the injection electrode would be accommodated only by unfilled states of the same spin at the Fermi level of the accepting material. An electron spin polarized in the injection electrode undergoes many collisions, which modify its momentum until eventually its spin flips. For practical applications we need to know how long electrons remember their spin orientation, characterized by **spin relaxation length**.

Table S.1 Maximum conduction electron spin polarization observed in different materials.

Material	Co	Fe	Ni	$Ni_{80}Fe_{20}$*	CoFe	NiMnSb	$La_{0.7}Sr_{0.3}MnO_3$	CrO_2
Polarization, %	42	46	46	45	47	58	78	90

* The alloy $Ni_{80}Fe_{20}$ is called permalloy (Py).

Material	$Zn_{0.9}Be_{0.07}Mn_{0.03}Se$	$Zn_{0.94}Mn_{0.06}Se$
Polarization, %	90 (at ~5 K, 1.5 T)	70 (at 4.2 K, 4–5 T)

spin-polarized low energy electron microscopy (SPLEEM) – an imaging technique based on the use of an **electron microscope** and on the fact that the reflectivity of a magnetic surface for a spin-polarized low-energy electron beam depends on the relative orientation of the magnetization and the **spin polarization** of the incoming electron beam. Is is linked to the investigation of **ferromagnetic** structures and cannot provide magnetic contrast images from **antiferromagnetic** samples. Because of the very short penetration depth in materials of low-energy electrons, the technique is sensitive to the surface properties, but lacks chemical selectivity since the spin-resolved low-energy electron reflectivity of a solid surface spin relaxation length

is determined by the empty electron states just above the **Fermi level** in the **conduction band**. The spin polarization of the electron beam can be oriented in all spatial directions. A quantitative determination of unknown magnetic domain microstructures, magnetic configurations and structure–property relationships is possible.

First described in: E. Bauer, T. Duden, R. Zdyb, *Spin-polarized low energy electron microscopy of ferromagnetic thin films,* J. Phys. D: Appl. Phys. **35**(19) 2327–2331 (2002).

spin relaxation length – the mean distance traveled by an electron in solids before its spin flips. It depends on the spin independent mean free path, which reasonably should be the inelastic mean free path controlled by inelastic scattering, according to $l_s = (l_{in} v_F \tau_{\uparrow\downarrow})^{1/2}$, where v_F is the Fermi velocity and $\tau_{\uparrow\downarrow}$ is the **spin relaxation time**. The spin relaxation length is mainly defined by spin–orbit and exchange scattering. In materials of one chemical composition the spin relaxation length is shorter in the amorphous phase than the crystalline phase. Also note that holes have much shorter spin relaxation length than electrons because of their stronger spin–orbit coupling. A typical spin relaxation length of electrons in solids is estimated to be above 100 nm.

spin relaxation time → **spin relaxation length**.

spin–spin coupling – a reciprocal magnetic interaction between nuclei in a molecular system facilitated by the binding electrons of the molecule.

spin transfer torque – transfer of angular momentum from charge carriers in a spin-polarized current to a ferromagnet through which they are passing resulting in an alteration of the magnetization of the ferromagnet. The phenomenon can take place even in the absence of an applied magnetic field.

First described in: L. Berger, *Emission of spin waves by a magnetic multilayer transversed by a current,* Phys. Rev. B **54**(13), 9353–9358 (1996); J. C. Slonczewski, *Current-driven excitation of magnetic multilayers,* J. Magn. Mater. **159**(1), L1–L7 (1996).

spintronics – the field of science and engineering, which studies and aims to exploit the subtle and mind-bendingly esoteric spin properties of the electron to develop electronic devices that make use of the electron spin as well as the electron charge. The term was coined by S. A. Wolf in 1996 as a name for an initiative launched by the Defense Advanced Research Projects Agency of USA for the development of novel magnetic materials and devices.

spin valve – a thin film structure consisting of two ferromagnetic layers. It is constructed so that the magnetic moment of one of the ferromagnetic layers is hard to reverse by an externally applied magnetic field, whereas the moment of the other layer is very easy to reverse. This magnetically soft layer then acts as the valve control, being sensitive to manipulation with an external magnetic field. A typical change of the resistance is about 1% per Oersted. Spin valves fabricated within conventional microelectronic technology are used for the monitoring of magnetic fields, magnetic recording of information and for other magnetic device applications. For more details → **giant magnetoresistance effect**.

spin-valve transistor – a three-terminal device analogous to a metal base transistor. It is schematically shown in Figure S.18 with the related energy band diagram. The base region of the transistor contains a metallic **spin valve** sandwiched between two *n*-type silicon regions acting as a current emitter and collector. It employs hot electron transport across the spin valve.

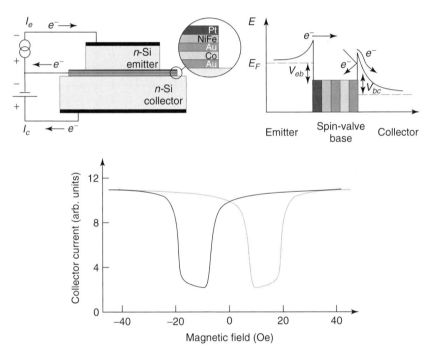

Figure S.18 Schematic cross-section and energy diagram of the spin-valve transistor with the Si/Pt emitter and Si/Au collector Schottky barriers and the NiFe/Au/Co spin-valve base. After P. S. Anil Kumar, J. C. Lodder, *The spin-valve transistor*, J. Phys. D: Appl. Phys. **33**(22), 2911–2920 (2000). The graph shows the collector current in the transistor as a function of the external magnetic field applied.

The base is designed as an exchange decoupled soft spin-valve system, where two **ferromagnetic** materials, namely NiFe and Co, which have different coercivities, are separated by a nonmagnetic spacer layer (Au). The NiFe and Co layers have well-separated coercivity, such that clear parallel and antiparallel alignment is obtained over a wide temperature range. The ferromagnetic layers can be individually switched by the application of a suitable magnetic field. At the interfaces between the metal base and the semiconductors, **Schottky barrier**s are formed. In order to obtain the desired high-quality barrier with good rectifying behavior, thin layers of Pt and Au are incorporated at the emitter and collector side, respectively. These also separate the magnetic layers from direct contact with the silicon. Since the Si/Pt contact forms a high Schottky barrier, it is used as an emitter. The collector Schottky diode is defined in such a way that it has a lower barrier height than that of the emitter diode. The Si/Au contact, whose barrier height is nearly 0.1 eV less than the Si/Pt one suits this condition very well. A specially developed technique that comprises metal deposition onto two silicon wafers and their subsequent *in situ* bonding under ultrahigh vacuum conditions is used for the fabrication of such a spin-valve transistor.

The transistor operates as follows. A current is established between the emitter and the base (the emitter current I_e), such that electrons are injected into the base, perpendicular to the layers of the spin valve. Since the injected electrons have to go over the Si/Pt Schottky barrier, they enter the base as nonequilibrium hot electrons. The hot-electron energy is determined by the emitter Schottky barrier height, which is typically between 0.5 and 1 eV depending on the metal/semiconductor combination. As the hot electrons traverse the base they are subjected to inelastic and elastic scattering, which changes their energy as well as their momentum distribution. Electrons are only able to enter the collector if they have retained sufficient energy to overcome the energy barrier at the collector side, which is chosen to be somewhat lower than the emitter barrier. Equally important, the hot-electron momentum has to match with the available states in the collector semiconductor. The fraction of electrons that is collected, and thus the collector current I_c, depends sensitively on the scattering in the base, which is spin dependent when the base contains magnetic materials. It is tuned by switching the base valve from its aligned low-resistance state to its antialigned high-resistance state. The total scattering rate is controlled with an external applied magnetic field, which for example changes the relative magnetic alignment of the two ferromagnetic layers of a spin valve.

The magnetic response of the spin-valve transistor, called magnetocurrent (MC), is defined as the change in the collector current normalized to its minimum value, that is

$$MC = \frac{I_c^p - I_c^{ap}}{I_c^{ap}},$$

where the superscripts p and ap refer to the parallel and antiparallel state of the spin valve, respectively.

The most important property of the spin-valve transistor is that its collector current depends sensitively on the magnetic state of the spin valve in the base. A typical dependence of the collector current on the magnetic field applied is shown in Figure S.18. At large magnetic fields, the two magnetic layers have their magnetization directions aligned parallel. This gives the largest collector current.

When the magnetic field is reversed, the different switching fields of Co (22 Oe) and NiFe (5 Oe) create a field region where the NiFe and Co magnetizations are antiparallel. In this state the collector current is drastically reduced. The relative magnetic response is indeed huge, providing the magneto current about 300% at room temperature and more than 500% at 77 K. Note that different scattering mechanisms of hot electrons may contribute to the decrease in the magnetic response. Scattering by thermal spin waves and thermally induced spin mixing look to be the most prominent.

The collector current and the magneto current are virtually independent of a reverse bias voltage applied across the collector Schottky barrier, when the leakage current of the collector is negligible. The reason for this behavior is that a voltage between base and collector does not change the maximum of the Schottky barrier when measured with respect to the Fermi energy in the metal. In other words, the energy barrier seen by hot electrons coming from the base is not changed. Similarly, a change in the emitter base voltage, or equivalently in the emitter current, does not affect the energy at which hot electrons are injected into the base. The result is that the collector current is simply linearly proportional to the emitter current, while the value of the collector current is a few orders of magnitude lower than the emitter current.

An important advantage of the spin-valve transistor is the huge relative magnetic effect, attained at room temperature, so that only small magnetic fields of a few Oersted are required. In spite of low current gain, it is a unique **spintronic** device with great prospects for magnetic memory and magnetic field sensor applications where current gain is not a critical issue.

First described in: D. J. Monsma, J. C. Lodder, T. J. A. Popma, B. Dieny, *Perpendicular hot electron spin-valve effect in a new magnetic field sensor: the spin-valve transistor*, Phys. Rev. Lett. **74**(26), 5260–5263 (1995).

SPLEEM – an acronym for a **spin-polarized low energy electron microscopy**.

split-gate structure – is used to "cut" electrostatically **quantum wires** and **quantum dots** from a **two-dimensional electron gas**. Its schematic design is presented in Figure S.19.

Metal gate electrodes are deposited on top of the semiconductor with a buried two-dimensional electron gas, which can be formed, for instance, in **modulation-doped** and **delta-doped** structures. When a reverse bias is applied to the gates the regions beneath are depleted of electrons by the **Schottky effect**, and current can only flow in the narrow ungated region. If the reverse bias is increased the fringing fields at the edges of the gates effectively squeeze the electron gas between them, and in this way the geometry can be varied within a single device. The particular configuration of the spin-gate electrodes defines the possibility to "cut" one-dimensional (**quantum wire**) or zero-dimensional (**quantum dot**) structures. Although the spin-gate approach is successful when applied to modulation-doped **heterojunction**s, it can hardly be used in the case of delta-doped structures. This

spontaneous emission

Figure S.19 Split gate arrangement for electrostatic cutting of one-dimensional and zero-dimensional elements from a two-dimensional electron gas.

is presumably because the gate isolation would begin to break down at the high electric fields required to deplete regions with an electron concentration close to or above 10^{13} cm^{-2}.

spontaneous emission – the radiation emitted when the internal energy of a quantum mechanical system drops from an excited upper state to a lower level without regard to the simultaneous presence of similar radiation. As a result, the population of the upper level is decreased and the population of the lower level increases. The excited system loses energy via the creation and emission of **photon**s.

sputtering – removal of surface atoms, atomic clusters or molecules by surface bombardment with energetic atomic particles, usually ions.

SQUID – an acronym for a **superconducting quantum interference device**.

SSIM – an acronym for a **saturated structured illumination microscopy**.

SSM – an acronym for a **scanning SQUID microscopy**.

stacking fault – a defect in a **face-centered cubic** or hexagonal crystal in which there is a change from a regular sequence of positions of atomic planes.

standing wave – a wave in which a ratio of an instantaneous value at one point to that at any other point does not vary with time.

Stark effect – modifications in emissions spectra which result from the application of a uniform external electric field to the source of light. In the atomic spectra from such a source many lines are appreciably displaced asymmetrically from their normal positions and commonly split into polarized components. Even more striking is the appearance of many new lines, which also have asymmetric displacements yet join with those of the normal spectrum to form groups with all-over patterns of marked symmetry. This effect is in many cases so great that in the green region the eye can readily detect the change in color. Modification of the electron orbits and related changes in their energy induced by the external electric field applied to atoms and molecules are responsible for the effect.

Suppression of the confined energy level by an electric field is also typical for low-dimensional structures, which has been named the **quantum confined Stark effect**. It is commonly observed in **heterostructures** and **quantum wells**. As the electric field F applied to the well has no impurity or defect dependence, then the change in the energy to second order $\triangle E^{(2)}$ is found to be proportional to F^2.

First described in: J. Stark, *Beobachtungen über den Effekt des elektrischen Feldes auf Spektrallinien*, Sitzungsber. Preuss. Akad. Wiss. (Berlin) 932–946 (1913) – the Stark effect; D. A. Miller, D. S. Chemla, T. C. Damen, A. C. Gossard, W. Wiegmann, T. H. Wood, C. A. Burrus, *Band edge electroabsorption in quantum well structures*, Phys. Rev. Lett. **53**(22), 2173–2176 (1984) – the quantum confined Stark effect.

Recognition: in 1919 J. Stark received the Nobel Prize in Physics for his discovery of the Doppler effect in canal rays and the splitting of spectral lines in electric fields.

See also www.nobel.se/physics/laureates/1919/index.html.

Stark–Einstein law states that one photon is absorbed by each molecule responsible for the primary photochemical process.

First described in: A. Einstein, *Thermodynamische Begründung des photochemischen Äquivalentgesetzes*, Ann. Phys. **37**, 832–838 (1912); J. Stark, *Über die Anwendung des Planckschen Elementargesetzes auf photochemische Prozesse. Bemerkung zu einer Mitteilung des Hrn. Einstein*, Ann. Phys. **38**, 467–469 (1912).

statics → **dynamics**.

STED – an acronym for **stimulated emission depletion microscopy**.

Stefan–Boltzmann constant → Stefan–Boltzmann law.

Stefan–Boltzmann law states that the total radiation emitted by a heated body increases rapidly with an increase in its temperature. For a perfect radiator, the radiating energy is proportional to the fourth power of the absolute temperature, so that for a body with surface area S it is $E = S\sigma T^4$, where σ is the **Stefan–Boltzmann constant**.

First described in: J. Stefan, *Über die Beziehung zwischen der Wärmestrahlung und der Temperatur*, Sitzungsber. Kaiser. Akad. Wiss. **79**, 391 (1879) – experiment, L. Boltzmann, *Ableitung des Stefanschen Gesetzes, betreffend die Abhängigkeit der Wärmestrahlung von der Temperatur aus der elektromagnetischen Lichttheorie*, Ann. Phys. **22**, 291–294 (1884) – theory.

step function – the function $\theta(a - b)$, which is equal to 1 for $a \geq b$ and zero for $a < b$.

Stern–Gerlach effect – splitting of an atomic beam passing through a strong in-homogeneous magnetic field into several beams. It is a demonstration of the quantization of angular momentum of electrons.

First described in: W. Gerlach, O. Stern, *Der experimentelle Nachweis der Rich-tungsquantelung im Magnetfeld*, Z. Phys. **9**(6), 349–352 (1922).

Recognition: in 1943 O. Stern received the Nobel Prize in Physics for his contribution to the development of the molecular ray method and his discovery of the magnetic moment of the proton.

See also www.nobel.se/physics/laureates/1943/index.html.

stereoisomers → isomers.

sterols → lipids.

stiffness constants → **elastic moduli of crystals.**

stiffness matrix → Hooke's law.

Stillinger–Weber potential – the interatomic many-body potential including empir-ical pair and triplet contributions:

$$V = \sum_i \sum_{j,i} V(r_{ij}) + \sum_i \sum_{j,i} \sum_{k,j} V(i,j,k),$$

with

$$V(r_{ij}) = A[B(r_{ij}/\sigma)^{-p} - r_{ij}/\sigma)^{-q}]exp(r_{ij}/\sigma - a)^{-1},$$

and

$$V(i,j,k) = h_{jik} + h_{ijk} + h_{ikj},$$

$$h_{jik} = \lambda exp[\gamma(r_{ij}/\sigma - a)^{-1} + \gamma(r_{ik}/\sigma - a)^{-1}][\cos\theta_{jik} + 1/3]^2.$$

Here r_{ij} is the distance between atoms i and j. It is implied that r_{ij} and r_{ik} are smaller than $a\sigma$. If not, the corresponding term $V(r_{ij})$ or h_{jik} vanishes. The angle θ_{jik} is the triplet angle subtended at atom i. The quantity σ sets the length scale and has a fixed value of 0.20951 nm. The values of the other parameters are fitted for a particular atomic system.

This potential gives a fairly realistic description of crystalline silicon. However, its built-in tetrahedral bias creates problems in other situations. Note that: it cannot predict the correct energies of the non-tetrahedral polytypes found under pressure; the coordination of the liquid is too low; surface structures are not correct.

First described in: F. H. Stillinger, T. A. Weber, *Computer simulation of local order in condensed phases of silicon*, Phys. Rev. B **31**(8), 5262–5271 (1985).

stimulated emission – the radiation emitted when the internal energy of a quantum mechanical system drops from an excited upper state to a lower level when the process is induced by the presence of radiant energy at the same frequency. It is illustrated in Figure S.20 for the case of stimulated one-electron transition from an upper to a lower energy level. The process requires that the energy of stimulating photon given by the **Planck's law** be equal to the energy separation of the participating pair of quantum energy states. The emitted photon has the same energy as the original photon. The total radiation intensity is increased. When a sizable population of electrons resides in the upper level, it sets the stage for stimulated emission of multiple photons. The process can be thought of as an optical amplification. It forms a principle basis for **maser** and **laser operation** → **Einstein's coefficients A and B**.

First described in: A. Einstein, *Strahlungs-Emission and Absorption nach der Quantentheorie*, Verh. Deutsch. Phys. Ges. **18**, 318–323 (1916); A. Einstein, *Zur Quantentheorie der Strahlung*, Phys. Z. **18**, 121–128 (1917).

$$E_{photon} = h\nu = E_2 - E_1$$

Figure S.20 Stimulated emission of photons.

stimulated emission depletion (STED) microscopy → **fluorescence nanoscopy**.
   First described in: S. W. Hell, J. Wichmann, *Breaking the diffraction resolution limit by stimulated emission: stimulated-emission-depletion fluorescence microscopy*, Opt. Lett. **19**(11), 780–782 (1994); S. W. Hell, M. Kroug, *Ground-state depletion fluorescence microscopy, a concept for breaking the diffraction resolution limit*, Appl. Phys. B **60**, 495–497 (1995).

STM – an acronym for a **scanning tunneling microscopy**.

stochastic process – a process in which one or more variables take on values according to some, perhaps unknown, probability distribution.

stochastic optical reconstruction microscopy (STORM) → **fluorescence nanoscopy**.
   First described in: M. J. Rust, M. Bates, X. Zhuang, *Sub-diffraction-limit imaging by stochastic reconstruction optical microscopy (STORM)*, Nature Methods **3**, 793–795 (2006).

Stokes equation → **Airy equation**.

Stokes law (of photoluminescence)

Stokes law (of photoluminescence) states that the wavelength of the light-emitted by light-excited solids is always longer than the wavelength of the exciting radiation. This is because of the energy conservation, which means that the energy of an absorbed photon is greater than the energy of the emitted photon and the difference in the energy is dissipated as heat in exciting lattice vibrations. This energy difference is called the **Stokes shift**.

First described in: G. G. Stokes, Proc. R. Soc. **6**, 195 (1852).

Stokes line → **Raman effect**.

Stokes parameters → **Poincaré sphere**.

First described in: G. G. Stokes, *On the composition and resolution of streams of polarized light from different sources*, Camb. Phil. Soc. Trans. **9**, 399–423 (1852).

Stokes process → **Raman effect**.

First described in: G. G. Stokes, *On the dynamical theory of diffraction*, Camb. Phil. Soc. Trans. **9**, 1–62 (1856).

Stokes shift – a shift between the peak absorption and peak emission wavelengths.

Stone–Wales defect – the defect on the surface of a **carbon nanotube**, which consists of a local $\pi/2$ rotation of a C–C bond creating two pentagons and heptagons as illustrated in Figure S.21.

Figure S.21 Surface of (5,5) carbon nanotube with a Stone–Wales defect (indicated in light colour). From V. Gayathri, R. Geetha, J. Phys. Conf. Ser. *34*, 824–828 (2006).

It is also called the "5-7-7-5 defect". Once formed, the pentagon/heptagons could move along the structure, creating either dislocation centers in regions of positive (pentagons) or negative (heptagons) Gaussian curvature, which ultimately lead to the closing of the nanostructure. The defect causes little change to the diameter and chirality of the nanotube because the deformation effect is rather local. Meanwhile, the defect-related transformation effectively elongates the tube in the strain direction, releasing the excess strain energy.

Stone–Wales defects can also appear on boron nitride (BN) nanotubes, in which B–B and N–N bonds are formed.

First described in: A. J. Stone, D. J. Wales, *Theoretical studies of icosahedral C$_{60}$ and some related species*, Chem. Phys. Lett. **128**(5–6), 501–503 (1986).

More details in: R. E. Smalley, M. S. Dresselhaus, G. Dresselhaus, P. Avouris, *Carbon Nanotubes: Synthesis, Structure, Properties and Applications* (Springer, Berlin, 2001).

Stone–Wales transformation – a $\pi/2$ rotation of two carbon atoms with respect to the midpoint of the C–C bond in a **graphene**-like structure. A sequence of such rearrangements supports coalescence processes of **fullerenes** or **carbon nanotubes**. Moreover, by this transformation, four hexagons are changed into two pentagons and two heptagons thus forming a **Stone–Wales defect**.

First described in: A. J. Stone, D. J. Wales, *Theoretical studies of icosahedral C$_{60}$ and some related species*, Chem. Phys. Lett. **128**(5–6), 501–503 (1986).

Stoner–Wohlfarth model deals with the equilibrium state of the total magnetic moment of a single-domain **ferromagnetic** particle with uniaxial anisotropy. This particle is called a **Stoner–Wohlfarth particle**. The model assumes that all the magnetic moments in the particle are always parallel. The magnetization processes of the particle are effected through coherent rotation of all these moments. At any moment the directions of the magnetic moments within the particle are parallel. Any motion and pinning of domain walls are ignored in the model.

First described in: E. C. Stoner, E. P. Wohlfarth, *A mechanism of magnetic hysteresis in heterogeneous alloys*, Philos. Trans. R. Soc. Lond. A 240, 599–642 (1948).

Stoner–Wohlfarth particle → Stoner-Wohlfarth model.

STORM – an acronym for **stochastic optical reconstruction microscopy**.

strained superlattice → superlattice.

Stranski–Krastanov growth mode (of thin films) – a combined initial layer-by-layer thin film growth with subsequent droplet formation from the deposited material, as illustrated in Figure S.22. A continuous layer forms first, but then for one reason or another the system gets tired of this and switches to islands. For more details → **self–organization (during epitaxial growth)**.

First described in: I. N. Stranski, L. von Krastanow, *Zur Theorie der orientierten Ausscheidung von Ionenkristallen aufeinander*, Sitzungsber. Akad. Wiss. Wien, Math.-Naturwiss. Kl. IIb **146**, 797–810 (1938).

Strěda formula

Figure S.22 Stranski–Krastanov growth mode during thin film deposition.

Strěda formula expresses the **Hall** conductivity as a sum of two contributions. Thus, components of the conductivity tensor σ are $\sigma_{ij} = \sigma_{ij}^{I} + \sigma_{ij}^{II}$. The first term depends on the structure of the solid and crystallographic orientation. Within the classical free-electron model with arbitrary energy-dependent self-energy $\sum(E)$ it is $\sigma_{xy}^{I} = -\omega\tau\sigma_{xx}$. Here ω is the **cyclotron frequency** and $\tau = \hbar/2(-\mathrm{Im}\sum(E_F))$ the lifetime, with E_F the **Fermy energy**.

The second term represents a quantum contribution. It is $\sigma_{xy}^{II} = -\sigma_{yx}^{II} = ec(\partial N(E)/\partial B)_{atE=E_F}$, where $N(E)$ is the number of states below the energy E and B is the induction of magnetic field. It has no classical analogy. In the classical limit the density of states is not a field-dependent quantity and σ_{xy}^{II} vanishes. Generally, this term depends on material constants only through the number of particles. It does not depend on the crystallographic orientation and the type of carrier scattering.

The formula shows that, if the **Fermi level** falls within the energy gap, the Hall conductivity can be given by the charge–density response to a magnetic field from orbital rather than Zeeman coupling. It can be used to calculate the conductivity of an electron gas in the presence of an additional periodic potential.

First described in: P. Středa, *Theory of quantised Hall conductivity in two dimensions,* J. Phys. C **15**, L717–L721 (1982).

Strehl ratio – the ratio of the illuminance at the peak of the diffraction pattern of an aberrated point image to that at the peak of an aberration-free image as formed by the same optical system.

First described in: K. Strehl, *Aplanatische und fehlerhafte Abbildung im Fernrohr,* Z. Instrumkde **15**, 362–370 (1895); K. Strehl, *Über Luftschlieren und Zonenfehler,* Z. Instrumkde, **22**, 213–217 (1902).

More details in: V. N. Mahajan, *Zernike annular polynomials for imaging systems with annular pupils,* J. Opt. Soc. Amer. **71**, 75–85 (1981).

structure-based engineering – the use of detailed knowledge of the structure and function of protein in order to produce desired changes and desired functions. Also called **rational design**.

More information in: J. A. Brannigan, A. J. Wilkinson, *Protein engineering 20 years on,* Nat. Rev. Molec. Cell Biol. **3**, 964–970 (2002).

structured illumination microscopy (SIM) → **fluorescence nanoscopy**.

First described in: M. G. L. Gustafsson, *Surpassing the lateral resolution limit by a factor of two using structured illumination microscopy*, J. Microsc. **198**, 82–87 (2000).

sub-band → **quantum confinement**.

superadditivity law states that the whole is more than the sum of its parts.
First described by Aristotle in his Metaphysics.

supercell method allows the use of the Bloch theorem also in the case of surfaces, clusters, molecules and defects calculations. An artificial periodicity is imposed on the calculation cell. The supercell is repeated over all space using periodic boundary conditions. Thus, in the case of a crystal surface the supercell will contain a crystal slab and a vacuum region repeated over all space. The vacuum region must be large enough so that the faces of the crystal slabs do not interact across the vacuum region and the crystal slab must be thick enough that the two surfaces of each slab do not interact through the crystal volume. For cluster or a molecule one assumes that the cluster or the molecule is enclosed in a sufficiently large box, in order to avoid interactions, and treated as a periodic system. For a defect, the supercell contains the defect surrounded by a sufficiently thick region of bulk crystal.
More details in: M. C. Payne, M. P. Teter, D. C. Allan, T. A. Arias, J. D. Joannopoulos, *Iterative minimization techniques for ab-initio total-energy calculations: molecular dynamics and conjugate gradients*, Rev. Mod. Phys. **64**(4) 1045–1097 (1992).

superconducting quantum interference device (SQUID) – the electronic device using the properties of electron-pair wave coherence and **Josephson effects** to detect very small magnetic fields. The central element of a SQUID is a ring of superconducting material with one or more weak links. An example is shown in Figure S.23 with weak-links at points W and X (Josephson junctions). Critical current at these points is much less than the critical current of the main ring. This produces a very low current density making the momentum of the electron-pairs small. The wavelength

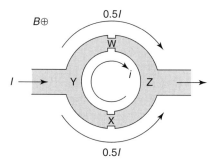

Figure S.23 Superconducting quantum interference device.

of the electron-pairs is thus very long leading to little difference in phase between any parts of the ring.

When a magnetic field B is applied perpendicular to the plane of the ring, a phase difference is produced in the electron-pair wave along the path XYW and WZX. A small current i is also induced to flow around the ring, producing a phase difference across the weak links. Normally the induced current would be of sufficient magnitude to cancel the flux in the hole of the ring but the critical current of the weak links prevents this.

The quantum condition that the phase change around the closed path must equal $2\pi n$ (n is an integer) can still be met by large phase differences across the weak links produced by even a small current. An applied magnetic field produces a phase change around the ring.

The phase difference due to the circulating current can either add to or subtract from that produced by the applied magnetic field but it is more energetically favorable to subtract. In this case a small anti-clockwise current i_- has the value $|i_-| = i_c \sin(\pi \Phi_a / \Phi_0)$, where i_c is the critical current, Φ_a is the flux produced in the ring by the applied magnetic field, and $\Phi_0 = h/2e$.

As the flux in the ring is increased from 0 to $0.5\Phi_0$ the magnitude of i_- increases to a maximum. As the flux is increased to $> 0.5\Phi_0$ it is now energetically favourable for a current, i_+, to flow in a clockwise direction, decreasing in magnitude to 0 as the flux reaches $0.5\Phi_0$. The circulating current has a periodic dependence on the magnitude of the applied field, with a period of variation of $0.5\Phi_0$, a very small amount of magnetic flux. Detecting this circulating current enables the use of a SQUID as a magnetometer. The device has a noise level as low as 3×10^{-15} $\mathrm{T \cdot Hz^{-1/2}}$ providing measurements of very small magnetic fields.

First described in: R. Jaklevic, J. Lambe, A. Silver, J. Mercereau, *Quantum interference from a static vector potential in a field-free region*, Phys. Rev. Lett. **12**(11), 274–275 (1964) – DC SQUID); J. E. Zimmerman in 1965 – RF SQUID.

More details in: J. Clarke, A. I. Braginski, *The SQUID Handbook: Fundamentals and Technology of SQUIDs and SQUID Systems* (Wiley-VCH, Weinheim, 2004); J. Clarke, A. I. Braginski, *The SQUID Handbook: Applications of SQUIDs and SQUID Systems* (Wiley-VCH, Weinheim, 2006).

superconductivity – disappearance of the electrical resistance of a material when it is cooled down. This phenomenon is characterized by a set of critical parameters, which includes critical temperature usually denoted T_c, critical magnetic field denoted H_c, and critical current denoted I_c. The critical temperature is the temperature at which electrical resistance of the material disappears. For most pure metals it is a few degrees above absolute zero.

The superconducting state in a material can be destroyed by an external applied magnetic field. It happens when the strength of the field reaches the critical value H_c, which is called the critical magnetic field. These fields are generally of the order of a few hundred Oersteds. It was also found experimentally that there exists for a current-carrying superconductor a critical current I_c above which the superconductivity is destroyed. The role of the current magnitude on the superconducting state of a material is recognized to be not a primary effect but rather due to the magnetic field produced by this current.

The number of known superconducting pure elements, compounds and alloys runs into thousands. Some 26 of the metallic elements are known to be super-conductors in their normal forms, and another 10 become superconducting under pressure or when prepared in the form of highly disordered thin films. Semiconductors with very high densities of charge carriers are superconducting; others such as Si and Ge have high-pressure metallic phases, which are superconducting. Many elements, which are not themselves superconducting, form compounds which are. Classical superconductors are represented by pure metals, their binary compounds and alloys, which have critical temperatures not exceeding 40 K (the highest critical temperature 39 K is observed for MgB_2, 23 K for a specially prepared alloy of Nb, Al, Ge and 18.1 K for Nb_3Sn). Complex copper oxides (cuprates), discovered as superconductors by J. G. Bednorz and K. A. Müller in 1986, have much higher critical temperatures, for instance $YBa_2Cu_3O_7$, 92 K; $Tl_2Ca_2Ba_2Cu_3O_{10}$, 125 K. These materials are usually called **high-T_c superconductors**.

The different character of penetration of an external magnetic field into super-conducting materials when they are heated up around critical the temperature is used to distinguish type I and type II superconductors. This is illustrated in Figure S.24. **Type I superconductors** in an external magnetic field below H_c at temperatures below T_c push out the magnetic field from their interior. In these conditions they behave like a perfect diamagnetic. If the applied field is increased above H_c, the entire superconductor reverts to the normal state and the flux of the magnetic field penetrates completely. The curve H_c versus temperature is thus a phase boundary in the magnetic field temperature plane separating a region where the superconducting state is thermodynamically stable from the region where the normal state is stable. This curve for type I superconductors is approximately parabolic: $H_c = H_0[1 - (T/T_c)^2]$, where H_0 is the value of H_c at absolute zero.

Type II superconductors are characterized by two critical magnetic fields: the lower critical field H_{c1} and the upper critical field H_{c2}. In applied fields less than H_{c1}, the superconductor completely excludes the fields, just as a type I superconductor does below H_c. At fields above H_{c1}, however, magnetic flux begins to penetrate the superconductor in microscopic filaments. Each wire consists of a normal core in which the magnetic field is large, surrounded by a superconducting region in which flows the vortex of the persistent supercurrent maintaining the field in the core. The diameter of the filament is typically of the order of 0.1 nm. In sufficiently pure and defect-free type II superconductors such wires tend to arrange themselves in a regular lattice. This vortex state of the superconductor is known as the mixed state. It exists for magnetic fields between H_{c1} and H_{c2}. At H_{c2} the superconductor becomes normal and the magnetic field penetrates completely.

Superconductivity was given a comprehensive explanation in the theory proposed by J. Bardeen, L. N. Cooper and J. R. Schrieffer, which bears their names (it is also abbreviated as the **BCS theory**). The basic idea behind the theory is that two electrons can attract each other and bind together to form a bound pair, called a **Cooper pair**, if they are in the presence of a high-density fluid of other electrons, superconductivity

Type I superconductor

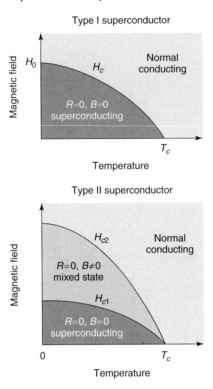

Figure S.24 Effect of temperature and external magnetic field on the superconducting state in type I and type II superconductors.

no matter how weak the interaction is. A Cooper pair is formed by electrons with opposite spins giving the total spin of the pair to be zero. Such pairs follow Bose–Einstein statistics so that all of them condense to the same ground state at low temperatures.

The pairing interaction is so weak that the two members of a pair are separated by a distance which turns out to be just the coherence length. The average distance between two paired electrons not involved in the same pair, however, is a hundred times smaller. When an electric field is applied, the electrons in a couple must move together. Consequently, electrons between the Cooper pairs have to follow them as far as they feel each other by the exciting interaction. A localized perturbation, which might deflect a single electron in a normal state and thus give rise to some resistance, cannot do so in the superconducting state without affecting all the electrons participating in the superconducting ground state at once. That is not impossible, but extremely unlikely, so that a collective drift of the coherent superconducting electrons, corresponding to an electric current, will be dissipationless.

The BCS theory gives three parameters essentially determining the critical temperature via $k_B T_c = 1.13 \hbar \omega_D \exp[-1/VN(E_F)]$. These are the **Debye frequency**

ω_D, the strength of the coupling between the electrons and the phonons in the solid V, and the electron density of states at the Fermi level $N(E_F)$.

A number of previous theories of superconductivity proposed by F. London and H. London, V. L. Ginzburg and L. D. Landau and others have found logical explanation within the BCS theory. Nevertheless, this theory meets some difficulties in the interpretation of high-T_c superconductivity in cuprates.

First described in: H. Kamerlingh-Onnes, *Disappearance of the electrical resistance of mercury at helium temperature*, Communication no. 122b from the Phys. Lab. Leiden, Electrician **67**, 657–658 (1911) – experimental observation; V. L. Ginzburg, L. D. Landau, *Theory of superconductivity*, Zh. Exp. Teor. Fiz. **20**(12), 1064–1082 (1950) – theory of superconductivity (in Russian); A. A. Abrikosov, *An influence of the size on the critical field for type II superconductors*, Doklady Akademii Nauk SSSR **86**(3), 489–492 (1952) – theoretical prediction of type II semiconductors (in Russian); J. Bardeen, L. N. Cooper, J. R. Schrieffer, *Theory of superconductivity*, Phys. Rev. **108**(5), 1175–1204 (1957) – a comprehensive theory; J. G. Bednorz, K. A. Müller, *Possible high-T_c superconductivity in the Ba-La-Cu-O system*, Z. Phys. B **64**(2), 189–193 (1986) – experimental observation above liquid nitrogen temperature (77 K).

More details in: M. Cyrot, D. Pavuna, *Introduction to Superconductivity and High-T_c Materials* (World Scientific, Singapore, 1992);

http://www.suptech.com/knowctr.html.

Recognition: in 1913 H. Kamerlingh-Onnes received the Nobel Prize in Physics for his investigations on the properties of matter at low temperatures which led, inter alia, to the production of liquid helium; in 1972 J. Bardeen, L. N. Cooper and J. R. Schrieffer received the Nobel Prize in Physics for their jointly developed theory of superconductivity, usually called the BCS-theory; in 1987 J. G. Bednorz and K. A. Müller received the Nobel Prize in Physics for their important breakthrough in the discovery of superconductivity in ceramic materials; in 2003 A. A. Abrikosov, V. L. Ginzburg and A. J. Leggett received the Nobel Prize in Physics for pioneering contributions to the theory of superconductors and superfluids.

See also www.nobel.se/physics/laureates/1913/index.html.

See also www.nobel.se/physics/laureates/1972/index.html.

See also www.nobel.se/physics/laureates/1987/index.html.

See also www.nobel.se/physics/laureates/2003/index.html.

superelastic scattering – a scattering process in which a particle scattered in a solid absorbs energy from an excited atomic system and therefore possesses a greater kinetic energy after collision than before, while the previously excited atom returns to its ground state without giving out any net radiation.

superfluidity – the frictionless flow of liquid helium at a temperature below the lambda point (2.178 K).

First described in: P. L. Kapitza, *Viscosity of liquid helium below λ-point*, C. R. (Doklady) Acad. Sci. USSR **18**(1), 21–23 (1938) – experiment; L. Landau, *Theory of the superfluidity of helium II*, Phys. Rev. **60**(4), 356–358 (1941) – theory.

superlattice

Recognition: in 1962 L. D. Landau received the Nobel Prize in Physics for his pioneering theories for condensed matter, especially liquid helium; in 1996 D. Lee, D. Osheroff and R. Richardson received the Nobel Prize in Physics for their discovery of superfluidity in helium-3; in 2003 A. A. Abrikosov, V. L. Ginzburg and A. J. Leggett received the Nobel Prize in Physics for pioneering contributions to the theory of superconductors and superfluids.

See also www.nobel.se/physics/laureates/1962/index.html.
See also www.nobel.se/physics/laureates/1996/index.html.
See also www.nobel.se/physics/laureates/2003/index.html.

superlattice – a monocrystalline film of one material reproducing the lattice constant of the monocrystalline substrate made of another material. When both materials have identical or very close lattice constants they produce so-called **pseudomorphic superlattices**. There are only a few such lattice-matched systems among semiconductors. Meanwhile, identical lattice constants are not a necessary condition for pseudomorphic growth of one semiconductor material on another.

Within a limited thickness of the film it is possible to force deposited atoms to occupy positions corresponding to the lattice constant of the substrate even though they may be different in the bulk. A **strained** but otherwise perfect superlattice is formed. It is schematically illustrated in Figure S.25.

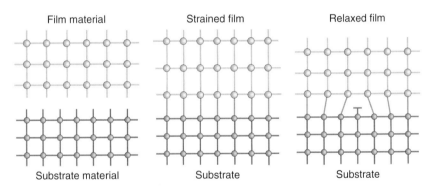

Figure S.25 Formation of strained and relaxed epitaxial films.

As the strain energy increases with the film thickness, beyond some critical thickness the film can lower its total energy by relieving the strain via the creation of misfit dislocations. The deposited material takes up its natural lattice constant, yielding a relaxed film. The critical thickness obviously depends on the lattice mismatch and elastic parameters of the contacting materials at the deposition temperature. In principle, by keeping the deposited film thinner than the critical thickness, it is possible to grow a strained superlattice from any two semiconductors with the same type of crystalline structure without regard to their particular lattice constants.

In practice, the smallest possible lattice mismatch and difference in electronic properties are desired for perfect superlattices creating heterojunctions with an appropriate potential barrier. There is a choice of semiconductors for this. Figure S.26 presents the energy band gaps of a number of semiconductors with **diamond** and **zinc blende structures** versus their lattice constants.

Figure S.26 Low-temperature band gaps and lattice constants of semiconductors with diamond and zinc blende crystalline structures. Hexagonal nitrides are plotted for consistency in terms of their α lattice constants.

The shaded vertical regions show the groups of semiconductors with similar lattice constants. Materials within the same shaded region but having different band gaps can be combined to form **heterojunctions** with the potential barriers corresponding to their band offsets. The choice of band offsets can be extended by the use of binary (such as SiGe), ternary (such as AlGaN, AlGaAs) and quaternary (such as GaInAsP) compounds. The solid lines in the figure joining some of the semiconductors indicate that these materials form stable intermediate compounds over the entire composition range. Among the semiconductors shown, the nitrides have lattice constants smaller than other materials because of the small size of the nitrogen atom.

superposition principle states that if a physical system is acted upon by a number of independent influences, the resultant effect is the sum (vector or algebraic, whichever is appropriate) of the individual influences. This principle is applied only to linear systems, that is systems the behavior of which can be expressed as linear differential equations.

supramolecular chemistry – the chemistry beyond the molecule, bearing on organized entities of higher complexity that result from the association of two or more chemical species held together by intermolecular forces (after J.-M. Lehn). In more generalized form the chemistry of systems made of two or more molecular components.

More details in: J.-M. Lehn, *Supramolecular Chemistry : Concept and Perspectives* (VCH, Weinheim, 1995); V. Balzani, M. Venturi, A. Credi, *Molecular Devices and Machines: A Journey into the Nanoworld* (Wiley-VCH, Weinheim, 2003).

supramolecule – a system of two or more molecular entities held together and organized by means of intermolecular (noncovalent) binding interactions.

More details in: K. Ariga, T. Kunitake, *Supramolecular Chemistry – Fundamentals and Applications: Advanced Textbook* (Springer, Berlin, 2006).

Recognition: in 1987 D. J. Cram, J. M. Lehn and C. J. Pedersen received the Nobel Prize in Chemistry for their development and use of molecules with structure-specific interactions of high selectivity.

See also www.nobel.se/chemistry/laureates/1987/index.html.

surface reconstruction – lowering of the symmetry of real surfaces of crystals as a result of the surface atomic relaxation processes. In fact, solid surfaces are usually of lower symmetry than they would be if the crystals were simply terminated at a plane and all remaining atoms remained at their original positions.

A highly unstable or metastable state occurs when interatomic bonds are broken at a surface by cleavage. The surface (and subsurface) atoms use considerable elastic distortive energy in order to reach a structure that facilitates new bond formation. It initiates the reconstruction of the surface yielding geometrical structures that are much more complex than the ideal surface termination.

Special notations are used to indicate reconstructed surfaces. When translation vectors of the surface **unit cell** are $n\mathbf{a}$ and $m\mathbf{b}$, with $\mathbf{a}$ and $\mathbf{b}$ being the translation vectors of corresponding plane of the underlying undisturbed solid and n and m positive integers, the surface structure is labeled p($n \times m$). The letter "p", meaning primitive, is often omitted. This simple notation is convenient when the surface unit cell is in registry with the substrate unit cell. The same scheme is used to describe square or rectangular lattices (with basis vectors $n\mathbf{a}$ and $m\mathbf{b}$) having two atoms per unit cell, the second atom occupying the center of the square or the rectangle. Such a structure is denoted c($n \times m$), where "c" means centered. More complex notation, like $(x \times y)\mathrm{R}\theta$ is applied when the surface unit cell is rotated with respect to the substrate unit cell or when the ratio of the dimensions of these cells is not an integer. It means that the surface structure is obtained from the substrate unit cell by multiplying the length of the basis vectors by x and y and rotating the cell through the angle θ.

More details in: W. Mönch, *Semiconductor Surfaces and Interfaces* (Springer, Berlin, 1995); F. Bechstedt, *Principle of Surface Physics* (Springer, Berlin, 2003).

surfactant – a surface active agent easily absorbing onto surfaces and reducing the surface tension. Detergents or cleaning agents are prominent examples for solid surfaces. They reduce the surface tension of the water resulting in a better cleaning ability or wetting property.

Sutton–Chen potential – the interatomic potential adding long-range interactions to the **Finnis–Sinclair potential**. It has the form

$$V(r_{ij}) = \epsilon[\sum_{ij}(\frac{a}{r_{ij}})^n - C\sum_i \sqrt{\sum_{j \neq i}(\frac{a}{r_{ij}})^m}],$$

where ϵ is a parameter with dimensions of energy, C is a dimensionless fitting parameter, a is the lattice constant, r_{ij} is the distance between atoms i and j, and m and n are positive integers with $n > m$.

This potential has been developed for computer simulations of atomic structures in which **van der Waals** type interactions between well-separated atomic clusters are as important as the description of metallic bonding at short range. The potential always favors **face-centered cubic (fcc)** and **hexagonal close-packed (hcp)** structures over the **body-centered cubic (bcc)** structure. It displays convenient scaling properties for both length and energy, and a number of properties of the perfect crystal may be derived analytically.

The potential provides a reasonable description of various bulk properties, with an approximate many-body representation of the delocalized metallic bonding. However, it does not include any directional terms, which are likely to be important for transition metals with partially occupied d shells.

First described in: A. P. Sutton, J. Chen, *Long-range Finnis-Sinclair potentials*, Phil. Mag. Lett. 61, 139–146 (1990).

More details in: G. A. Mansoori, *Principles of Nanotechnology: Molecular-based Study of Condensed Matter in Small Systems* (World Scientific, Singapore, 2005).

SWAP gate → quantum logic gate.

swarm intelligence – the emergent, collective intelligence of social insect colonies. Individually, one insect may not be capable of much, but collectively social insects are capable of achieving great things. As an example, a colony of ants can collectively find out where the nearest and richest food source is located, without any individual ant knowing it. By discovering the shortest path to a food source, the ants solve collectively an optimization problem using emergent computation. Learning from this has helped, for example, in developing algorithms for solving routing problems in telecommunications networks. Swarm intelligence looks to be a mindset rather than a technology, initially inspired by how social insects operate. It is a **bottom-up approach** to controlling and optimizing distributed systems, using resilient, decentralized, self-organized techniques.

More details in: E. Bonabeau, M. Dorigo, G. Theralauz, *Swarm Intelligence: From Natural to Artificial Systems* (Oxford University Press, 1999).

symmetry

symmetry that is exploited in solid state physics is defined in terms of symmetry operations. An operation is said to be a symmetry operation if the body looks exactly the same after the operation as it did before the operation was carried out. The operation could be, for example, rotations through certain angles or reflections in various mirror planes.

symmetry group – a collection of symmetry elements which satisfy the properties listed in the Table S.2.

More details in: S. J. Joshua, *Symmetry Principles and Magnetic Symmetry in Solid State Physics* (Adam Hilger, Bristol, 1991).

Table S.2 The basic type of symmetry elements in Schoenflies notation.

Symmetry element	Symbol	Symmetry operation(s)
Identity element	E	no operation; like multiplication by 1
Mirror plane	σ	reflection in a mirror
Center of inversion or inversion symmetry	I	inversion of all coordinates through the centre; it changes coordinates (x, y, z) to $(-x, -y, -z)$, which can be described as a reflection in a point
Proper rotation axis	C_n	n-fold rotation about an appropriate axis. An n-fold axis generates n operations, for example the symmetry element C_6 implies the operations $C_6^1, C_6^2, C_6^3, C_6^4, C_6^5, C_6^6 = E$
Improper rotation axis	S_n	n-fold rotation about an axis followed by a reflection in a plan perpendicular to the axis of rotation, that is the presence of S_n implies the independent existence of both C_n and σ elements

synergetics – an interdisciplinary science studying and explaining the formation and self-organization of patterns and structures in open systems far from thermodynamic equilibrium.

More details in: H. Haken, *Synergetics* (Springer, Berlin, 2004).

T

From Talbot's law to type II superconductors

Talbot's law states that the brightness of an object that is examined through a slotted disc, rotating over a critical frequency, is proportional to the angular aperture divided by the opaque sectors.

First described in: H. F. Talbot, *Experiments on light*, Phil. Mag. **5**, 321 (1834).

Tamm levels also called surface states, appear in solids at their surfaces as a result of the rupture of chemical bonds and distortion of the lattice there. In semiconductors, they usually locate in the forbidden energy gap.

First described in: I. E. Tamm, *Über eine mögliche Art der Elektronenbindung an Kristalloberflächen*, Z. Phys. **76**(11/12), 849–850 (1932).

tandem solar cell = **multijunction solar cell**.

Taylor criterion states that in interferometers in which the separation of the maxima is equal to the half-value width, a slight drop in intensity between fringes is distinguishable and the fringes can be considered resolved.

Taylor series corresponding to a function $f(x)$ at a point x_0 is the infinite series whose n-th term is $(1/n!)f^{(n)}(x_0)(x - x_0)^n$, where $f^{(n)}(x)$ denotes the n-th derivative of $f(x)$.

First described in: B. Taylor, *Methodus Incrementorum Directa et Inversa* (London, 1715).

TEEEM – an acronym for a **tip-enhanced electron emission microscopy**.

Teller–Redlich rule states that for two isotopic molecules the product of the frequency ratio values of all vibrations of a given symmetry type depends only on the geometrical structure of the molecule and the masses of the atoms and not on the potential constants.

First described in: O. Redlich, Z. Phys. Chem. B **28**, 371 (1935).

TEM – an acronym for a **transmission electron microscopy**.

What is What in the Nanoworld. Victor E. Borisenko and Stefano Ossicini
Copyright © 2008 WILEY-VCH Verlag GmbH & Co. KGaA, Weinheim
ISBN: 978-3-527-40783-5

template

template – a pattern or a structure which is used to guide or to frame a reaction or **adsorption** on solid surfaces. Its function principle has much in common with a lettering stencil.

tera- – a decimal prefix representing 10^{12}, abbreviated T.

Tersoff potential – the empirical interatomic many-body potential having the form of the **Morse potential**, but with the bond-strength parameters depending upon local environment including angle dependency upon bond-lengths: $V(r_{ij}) = f_c(r_{ij})[A \exp(-ar_{ij}) - B \exp(-br_{ij})]$, where r_{ij} is the distance between atoms i and j, A, a, B, b are positive constants, f_c is the cut-off function of the potential. The first term reproduces the repulse interaction, and the second term describes covalent bonding.

First described in: J. Tersoff, *New empirical model for the structural properties*, Phys. Rev. Lett. **56**(6), 632−635 (1986).

Tersoff–Hamann theory describes electronic interaction between the tip of a **scanning tunneling microscope (STM)** and a sample in terms of their **wave functions**. In this interacting system, schematically shown in Figure T.1, the tunneling current is supposed to depend on the overlap of the wave functions of the tip and the sample atoms, which decay exponentially. Thus, only the orbitals localized at the outermost tip atom are of importance for the tunneling process.

The main point of the theory is the choice of the wave function of the tip. It is modeled as a locally spherical potential with curvature R about the center r_0. For this case the tip wave function is chosen to have the form of an s wave:

$$\Psi_t(r) = kR \frac{\exp(-k|\gamma - \gamma_0|)}{k(r - r_0)} \exp(kR),$$

where $k = [2m\phi/\hbar^2]^{1/2}$ with m the effective mass of an electron and ϕ the **work function** of the surface. Finally, the current from the tip to the sample under the applied bias V has been found to be $I \sim VN_t(E_F)LDOS(r_0, E_F)$, where $N_t(E_F)$ is

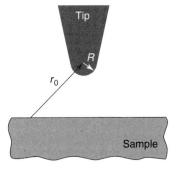

Figure T.1 Tip-sample configuration and related parameters in the Tersoff-Hamann model.

density of states at the **Fermi level** in the tip, and $LDOS(r_0, E_F)$ is the local density of states of the sample at the Fermi level evaluated at the tip center.

Theory predicts that the basic quantity imaging with STM is the local density of states of the sample. It is used in the interpretation of experimental STM images.

First described in: J. Tersoff, D. R. Hamann, *Theory and application for the scanning tunneling microscope*, Phys. Rev. Lett. **50**(25), 1998–2001 (1983).

tetragonal lattice – a **Bravais lattice** in which the axes of a unit cell are perpendicular and two of them are equal in length to each other but not to the third axis.

thermal noise in electrical circuits is a consequence of thermally induced charge fluctuations in it. It is mathematically described by the **Nyquist formula**.

thermionic emission – the emission of electrons from a hot solid. It is usually described by the **Richardson equation**.

thermistor – a solid-state semiconducting structure that changes electrical resistance with temperature. Materially, some kind of ceramic composition is used. It has much higher electrical resistance than identical in function metallic bolometers and hence requires much higher voltages to become useful.

thermocouple – a device composed of dissimilar metals that, when welded together, develop a small voltage dependent upon the relative temperature between the hotter and colder junctions. Banks of thermocouples connected together in parallel/series make up a multiple thermocouple, or a thermopile. Either may be thought of as a weak battery that converts radiant energy to electrical energy.

thermodynamic density of states –

$$\frac{\partial n}{\partial \mu} = \int n(E) \frac{\partial f(E, \mu)}{\partial \mu} dE = \int n(E) \left[\frac{\partial f(E, \mu)}{\partial E} \right] dE,$$

where $n(E)$ is the ordinary density of states and $f(E, \mu)$ is the distribution function.

thermogravimetry – a technique to measure automatically the weight change as a material is heated either isothermally or in a programmed manner. It is used to study thermal stability of materials and volatile contents, as well as to perform residue or purity analysis.

thermoluminescence – production of light by heating a sample.

thermopower = **Seebeck coefficient** → **Seebeck effect** describes the voltage induced by a temperature difference between two contacts $\alpha(T) = V/(T_2 - T_1)$.

thiols

thiols – organic compounds that contain the –SH group. The simplest example is methanethiol CH_3SH.

Thomas–Fermi equation – a differential equation of the form

$$x^{1/2}(d^2y/dx^2) = y^{3/2}$$

that is used in calculations of the potential and density of electrons within the Thomas–Fermi atom model (**Thomas–Fermi theory**) in which electrons are treated as independent particles and electron–electron interaction energy is considered to arise solely from the electrostatic energy. The dimensionless variables are defined as

$$x = \left[\left(\frac{9\pi^2}{128Z} \right)^{1/3} \frac{\hbar^2}{me^2} \right] r \quad \text{and} \quad y = \frac{r|\phi|}{Ze},$$

where Z is the number of electrons in the atom, m is the electron mass, r is the distance, ϕ is the electrostatic potential. A physically meaningful solution has to satisfy the boundary conditions $y(0) = 1$ and $y(\infty) = 0$.

Thomas–Fermi screening length – a penetration depth of an impurity charge screened by mobile electrons in metals $L = [(e^2/\epsilon_0)/D(E_F)]^{-1/2} = (3e^2n/2\epsilon_0 E_F)^{-1/2}$, where $D(E_F)$ is the density of states at the **Fermi level** E_F and n is the electron concentration. The quantity $(L)^{-1}$ is denoted as the **Thomas–Fermi screening wave number**.

Thomas–Fermi screening wave number $\rightarrow$ Thomas–Fermi screening length.

Thomas–Fermi theory – a simple many-body theory, sometimes called "statistical theory", proposed shortly after E. Schrödinger invented his quantum-mechanical wave equation. For a system with a large number of interacting electrons, the **Schrödinger equation**, which would give the exact electron density and ground state energy, cannot be easily handled. The basic idea of the Thomas–Fermi theory is to find the energy of electrons in a spatially uniform potential as a function of the density. Then one uses this function of the density locally, even when the electrons are in the presence of an external potential. The theory contains two basic ideas: the first electrostatic, the second quantum statistical. From the first we have that the electrostatic potential satisfies the **Poisson equation** that links potential and charge density. With quantum statistics we can connect the momentum of the electrons to the energy, the potential and the electron density. Thus, one obtains the Thomas–Fermi equations that provide a one-to-one implicit relation between the density and the external potential.

$$n(r) = \frac{1}{3\pi^2} \left(\frac{2m}{\hbar^2} \right)^{3/2} [\mu - v_{eff}(r)]^{3/2}, \quad v_{eff} \equiv v(r) + \int \left(\frac{n(r')}{|r - r'|} \right) dr',$$

where $n(r)$ is the electron density distribution, $v(r)$ is the applied external potential, μ is the r-independent chemical potential and the potential given by the integral is just the classically computed electrostatic potential times (-1), generated by the electron density distribution. The theory apply best for systems of slowly varying density and it was useful for describing qualitative trends like total energies of atoms.

The Thomas–Fermi theory was the first theory of electronic energy in terms of the electron density distribution. The idea of handling density functionals later became very successful in the treatment of many-body problems $\rightarrow$ **density functional theory.**

First described in: L. H. Thomas, *The calculation of atomic fields,* Proc. Camb. Phil. Soc. **23**, 542–548 (1927); E. Fermi, *Un metodo statistico per la determinazione di alcune proprietá dell'atomo,* Rend. Accad. Naz. Lincei **6**, 602–607 (1927).

More details in: E. H. Lieb, *Thomas–Fermi and related theories of atoms and molecules,* Rev. Mod. Phys. **53**(4), 603–641 (1981).

Thomas precession – the precession of a spinning object when it accelerates. It is a consequence of the special relativity consideration of the system. It gives a correction to the **spin–orbit interaction,** which takes into account the relativistic time dilatation between the electron and the nucleus in hydrogenic atoms.

First described in: L. H. Thomas, *The motion of the spinning electron,* Nature **117**, 514 (1926); L. H. Thomas, *The kinematics of an electron with an axis,* Phil. Mag. **3**(7), 1–22 (1927).

More details in: W. Rindler, *Relativity Special, General and Cosmological, Second edition* (Oxford University Press, Dallas, 2006).

See also www.mathpages.com/rr/s2-11/2-11.html.

Thomson effect – heat liberation or absorption when an electric current traverses a region where a temperature gradient $\triangle T$ exists along a single homogeneous conductor. The rate of the reversible heat generation or evolution is $Q = \tau I \triangle T$, where τ is the **Thomson coefficient** and I is the electric current.

Thomson relations $\rightarrow$ **Kelvin relations.**

Thomson scattering – scattering of electromagnetic radiation by free or very loosely bound charged particles. The energy is taken away from the primary radiation as the charged particles are accelerated by the transverse electric field of the radiation.

three-body-scattering (3BS) approach is based on the many-body **Hamiltonian,** containing on-site core–valence Coulomb interactions, being projected on a set of states obtained by adding a finite number of electron–hole pairs to the ground state of the single-particle band Hamiltonian. The expansion is truncated to include just one electron–hole pair, that is configurations with just three particles (two holes and one electron). The method can be seen as an extension to the solid state of the configuration–interaction scheme commonly used for finite systems. It fully includes the itinerant character of the valence states and treats the on-site many-body electron-electron interaction beyond a mean field. It is based on a non-perturbative solution of a multi-band **Hubbard Hamiltonian** and provides spin- and k-dependent self-energies, spectral functions, quasiparticle energies and lifetimes. It applies to all correlation regimes, from the low correlation regime up to the atomic limit, reproducing energy renormalization, band narrowing, quasiparticle quenching and, for sufficiently large Hubbard interaction, opening of a **Mott–Hubbard** gap. The method has been successfully applied in transition metals, transition-metal oxides, and cuprates.

First described in: J. Igarashi, *Three-body problem in transition metals: Application to nickel*, J. Phys. Soc. Jpn. **52**(8), 2827–2837 (1983); C. Calandra, F. Manghi, *Three-body scattering theory of correlated hole and electron states*, Phys. Rev. B **50**(4), 2061–2074 (1994); F. Manghi, C. Calandra, S. Ossicini, *Quasiparticle band structure of NiO: The Mott–Hubbard picture regained*, Phys. Rev. Lett. **73**(23), 3129–3132 (1994).

tight-binding approach assumes that valence electrons are completely delocalized when the solid is formed. In contrast, core electrons remain very localized and, thus, discrete core levels of the atoms are only very slightly broadened in the solid. The corresponding wave functions are not very different, in the vicinity of each atom, from the atomic wave functions.

time-dependent Schrödinger equation → Schrödinger equation.

time-independent Schrödinger equation → Schrödinger equation.

tip-enhanced electron emission microscopy (TEEEM) – like **scanning tunneling microscopy** this technique measures the effect of a sample surface on the electrical current flowing through a probe tip; here, however, the current is stimulated by a pulsed laser beam focused on the tip. Due to the fact that the **laser** field is affected by any kind of sample that is introduced near to the probe, the device acts in three dimensions. Moreover, the current flow scales highly nonlinearly with changes in the laser field, making the imaging extremely sensitive to whatever is out near the probe tip. Ultrafast time resolution is another advantage of this technique.

First described in: C. Ropers, D. R. Solli, C. P. Schulz, C. Lienau, T. Elsaesser, *Localized multiphoton emission of femtosecond electron pulses from metal nanotips*, Phys. Rev. Lett. **98**, 043907 (2007).

More details in: H. Batelaan, K. Uiterwaal, *Tip-top imaging*, Nature **46**, 500–501 (2007).

***T*-matrix** = transfer matrix. It describes a transfer process from one region of a nonhomogenous system to another usually accounting for incident and reflected fluxes. For instance, if a system contains a barrier separating it into right and left side regions, wave propagation through this barrier from left to right can be expressed by the matrix **T** satisfying

$$\left(\begin{array}{c} C \\ D \end{array} \right) = \mathbf{T} \left(\begin{array}{c} A \\ B \end{array} \right),$$

where A and B are the amplitudes of the incident and reflected waves, respectively, at the left side of the barrier and C and D are the amplitudes of the transmitted and scattered back to the barrier waves, respectively, at the right side.

Toffoli gate $\rightarrow$ quantum logic gate.

Tomonaga–Luttinger liquid = Luttinger liquid.

Tomonaga–Luttinger model = Luttinger model.

Tomonaga–Schwinger equation – an equation of motion of the state vector of a quantized field expressed in a covariant form. This is achieved by laying down a method for finding the field on a space-like surface provided this field is known on a similar surface initially close to that space-like surface. For an arbitrary space-like surface σ it is: $i\hbar c \partial \Psi(\sigma)/\partial\sigma(x) = \mathbf{H}(x)\Psi(x)$, where the interaction Hamiltonian is $\mathbf{H}(x) = -c^{-1}j(x)\mathbf{A}(x)$. Here $j(x)$ is the electric current density of the electrons, and $\mathbf{A}(x)$ is the electromagnetic vector potential.
   First described in: S. Tomonaga, *On a relativistically invariant formulation of the quantum theory of wave fields*, Prog. Theor. Phys. **1**(2), 27–42 (1946); J. Schwinger, *Quantum electrodynamics. I. A covariant formulation*, Phys. Rev. **74**(10), 1439–1461 (1948).
   Recognition: in 1968 S. Tomonaga and J. Schwinger shared with R. P. Feynman the Nobel Prize in Physics for their fundamental work in quantum electrodynamics, with ground-breaking consequences for the physics of elementary particles.
   See also www.nobel.se/physics/laureates/1968/index.html.

top-down approach – one of two ways to fabricate micrometer and nanometer size elements of integrated electronic circuits. It pre-supposes that the manufacturing starts at the wafer level and employs a variety of sophisticated lithographic and etching techniques in order to pattern the substrate. The alternative is a **bottom-up approach**.

torsion – a twisting deformation in a solid about an axis in which lines that were initially parallel to the axis become helices.

Touschek effect – the effect whereby two electrons lose synchronism with the accelerating field and are lost during synchronous radiation. The effect is produced by the scattering of the electrons that are oscillating in the equilibrium orbit, which results in the transferral of transverse momentum into longitudinal momentum.

This effect is one of the limiting mechanism for present-day synchrotron radiation sources.

First described in: C. Bernardini, G. F. Corazza, G. Di Giugno, G. Ghigo, J. Haissinski, P. Marin, R. Querzoli, B. Touschek, *Lifetime and beam size in a storage ring*, Phys. Rev. Lett. **10**, 407–409 (1963).

trace operation denoted Tr. For a vector **A** it means $\mathrm{Tr}\mathbf{A} = \sum_k \langle k|\mathbf{A}|k\rangle$, where $\{|k\rangle\}$ is any orthonormal basis for the **Hilbert space**. The numerical result is independent of the basis choice.

transcription – the central dogma of molecular biology stating that the information required to build a living cell or organism is stored in **DNA**. This information must be transferred from the DNA to the **proteins**. These processes are called transcription and translation and are all executed by biomolecular components, mostly proteins and nucleic acids.

transient spectral hole-burning → spectral hole-burning.

transistor – a three-terminal semiconductor device for switching and amplification of electric signals.

First described in: J. Bardeen, W. H. Brattain, *Three-electrode circuit element utilizing semiconductive materials*, US Patent no. 2524035 (1948); J. Bardeen, W. H. Brattain, *The transistor, semiconductor triode*, Phys. Rev. **74** (1), 230–231 (1948); W. Shockley, *The theory of p-n junctions in semiconductors and p-n junction transistors*, Bell Syst. Tech. J. **28**(3), 435–489 (1949).

Recognition: in 1956 W. Shockley, J. Bardeen and W. H. Brattain received the Nobel Prize in Physics for their researches on semiconductors and their discovery of the transistor effect.

See also www.nobel.se/physics/laureates/1956/index.html.

transition metal – an element which has only partially filled d orbitals in its electronic structure. Transition metals and the number of electrons in the d orbitals are listed in the Table T.1.

Table T.1 Number of d electrons in a free atom of transition metals.

Groups ↓ Number of electrons →	1	2	3	4	5	6	7	8	9	10
3d elements	Sc	Ti	V		Cr,Mn	Fe	Co	Ni		
4d elements	Y	Zr		Nb	Mo,Tc		Ru	Rh		Pd
5d elements	La–Lu	Hf	Ta	W	Re	Os	Ir		Pt	
6d elements	Ac–Lr	Th, Ku								

transmission electron microscopy (TEM) → electron microscopy.

transverse acoustic mode → phonon.

transverse optical mode → phonon.

trefoil knot – the simple closed curve in three-dimensional space shown in Figure T.2. It can be a molecular architecture consisting of a single macrocycle.

Figure T.2 A trefoil knot.

tribo – a prefix meaning pertaining to or resulting from friction.

tribology – the science of interacting material surfaces in relative motion. It is specially focused on the friction and wear properties of the materials and the control of these processes through lubrication. The word tribology derives from the Greek "tribo" meaning "rubbing".

triboluminescence – luminescence that arises from friction and that usually occurs in crystalline materials.

triclinic lattice – a **Bravais lattice** with a unit cell in which axes are not at right angles and are not equal.

trigonal lattice – a **Bravais lattice** in which three axes of a unit cell are of equal length and the three angles between axes are the same and are not right angles. It is also know as **rhombohedral lattice**.

trimer – a molecule composed of three identical molecular subunits (**monomers**) linked together.

triplet state – a system of atomic terms (energy levels) with maximum multiplicity 3. Triplet terms occur if the spin quantum number of electrons is unity. It can be formed by two electrons with parallel spins. Two spins add together giving a nonzero total spin.

tRNA

tRNA – transfer RNA → **RNA**.

truth table → **logic operator**.

Tsuneyuki–Tsukada–Aoki–Matsui potential – the interatomic many-force-field-pair potential in the form $V(r_{ij}) = V_{ij}^{\text{Coulomb}} + f_0(b_i + b_j)\exp[(a_i + a_j - r_{ij})/(b_i + b_j)] - c_i c_j/r_{ij}^6$ in which Coulomb interaction via V_{ij}^{Coulomb}, as well as **Born–Mayer** type repulsion and dispersive interaction are included subsequently. Here r_{ij} is the interatomic distance, and a_i, a_j (effective radius), b_i, b_j (softness parameter), and c_i, c_j are the fitting parameters. The potential has been shown to be effective in **molecular dynamics** simulations for silica polymorphs (α-quartz, α-cristobalite, coesite, stishovite).

First described in: S. Tsuneyuki, M. Tsukada, H. Aoki, Y. Matsui, *First-principles interatomic potential of silica applied to molecular dynamics*, Phys. Rev. Lett. **61**(7), 869–872 (1988).

tunneling – the term referring to a particle transport into and through a classically forbidden region defined by a potential barrier with any energy larger than the total energy of the incident particle. This is illustrated in Figure T.3, where a particle of energy E approaches a rectangular barrier of height $U_0 > E$.

In classical mechanics, such particle has to be completely reflected from the barrier. This is not the case for quantum particles. Quantum mechanically, the motion of a particle around a step-like potential barrier is described by the **Schrödinger equation**, which for the one-dimension case is

$$-\frac{\hbar^2}{2m}\frac{d^2\psi(x)}{dx^2} + U(x) = E\psi(x).$$

Expecting the barrier to be rectangular with a finite height U_0 one has

$$U(x) = \begin{cases} 0 & \text{for} \quad x < x_0 \\ U_0 & \text{for} \quad x \geq x_0 \end{cases}$$

In case of $0 < E \leq U_0$ the solution of the Schrödinger equation is

$$\psi(x) = \begin{cases} A\exp(ik_1 x) + B\exp(-ik_1 x) & \text{for} \quad x < x_0 \\ U_0 & \text{for} \quad x \geq x_0 \end{cases}$$

with $k_1 = (1/\hbar)\sqrt{2mE}$ and $k_2 = (1/\hbar)\sqrt{2m(U_0 - E)}$. As the value and the slope of the wave function $\psi(x)$ must be continuous everywhere in the space, matching at the step $x = x_0$ gives the relations between amplitudes $A, B, C,$ and D

$$A + B = C + D \quad \text{and} \quad k_1(A - B) = k_2(C - D).$$

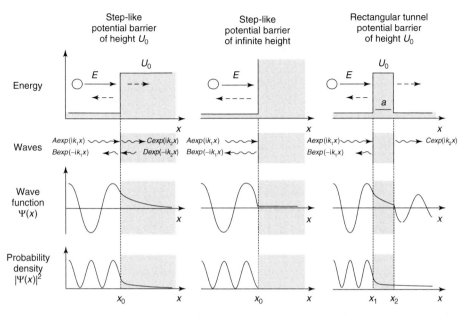

Figure T.3 Interaction of quantum particle having total energy E with potential barriers of finite (U_0) and infinite heights.

Thus, if the amplitudes of the waves on the left, A and B, are known, the amplitudes on the right are

$$C = \frac{1}{2}\left(1 + \frac{k_1}{k_2}\right) A + \frac{1}{2}\left(1 - \frac{k_1}{k_2}\right) B,$$

$$D = \frac{1}{2}\left(1 - \frac{k_1}{k_2}\right) A + \frac{1}{2}\left(1 + \frac{k_1}{k_2}\right) B,$$

which give transmission T and reflection R coefficients in the form

$$T = \frac{4k_1 k_2}{(k_1 + k_2)^2} \quad \text{and} \quad R = \left(\frac{k_1 - k_2}{k_1 + k_2}\right)^2 \quad \text{where} \quad R + T = 1.$$

As a result, one can see that the wave $A \exp(ik_1 x)$, representing a quantum particle of mass m with energy E, coming to the step-like potential barrier of height U_0, is reflected as the wave $B \exp(-ik_1 x)$. It also penetrates the barrier, decaying in the barrier region. The penetration, characterized by the transmission coefficient T, increases as E approaches U_0. The function $|\psi(x)|^2$ shown in Figure T.3 characterizes the probability to find the incident quantum particle. It is oscillating at the low energy side of the barrier and exponentially decreases inside the barrier. On the other hand, if the potential barrier is infinite in height, or at least $U_0/E \gg 1$, there is no penetration into the barrier region. The transmission coefficient becomes zero and the reflection coefficient reaches unity. Thus, perfect reflection takes place with the interference of incident and reflected waves at the left side of the barrier. The interference gives rise to the oscillations of the probability density at the barrier. Both, penetration of quantum particles into a classically forbidden region and oscillating behavior of the probability to find a quantum particle at a potential barrier are particular manifestations of quantum mechanics, which have no analogy in classical mechanics.

More mystery arises when a quantum particle passes over the step-like potential barrier, so $E > U_0$. From the point of view of classical mechanics, no particle reflection can be expected. But quantum mechanically, the reflection coefficient in this case is not absolute zero. As a result, the wavelength $\lambda_1 = h/(2mE)^{1/2}$ representing the quantum particle approaching the barrier is transformed into $\lambda_2 = h/[2m(E - U_0)]^{1/2}$ when the particle crosses the border at $x = x_0$ and moves over the barrier.

Although potential barriers in the form of steps are important to confine electrons within a particular region, barriers of a certain thickness allowing electrons to tunnel through them into another neighboring region are more often used in nanoelectronic devices. Consider electron transport through the rectangular tunnel potential barrier shown in Figure T.3, which has a height U_0 and thickness $a = x_2 - x_1$. A classical particle with total energy $E < U_0$ approaching such a barrier has no chance to pass through it. It will be reflected at the so-called classical turning point. The **turning point** is a point with a coordinate x at the boundary of a potential barrier where the total energy of the particle E equals the potential energy at the barrier $U(x)$. For the classical particle it means that its velocity becomes zero at this point and the particle starts moving in the opposite direction. For the rectangular tunnel barrier the turning points have coordinates (x_1 and x_2) coinciding with the boundary of the barrier.

For a quantum particle, there is a certain nonzero probability to find it, or the wave representing that particle, at the opposite side of the barrier. The probability function remains constant at the right side of the barrier, while at the left, incident, side it oscillates reaching values even lower than in the region after the barrier. The transparency of a symmetric rectangular tunnel barrier is characterized by the transmission coefficient

$$T = \left[1 + \frac{U_0^2}{4E(U_0 - E)} \sin h^2(ak_2)\right].$$

The reflection coefficient $R = 1 - T$. For many practical cases related to electron tunneling the product ak_2 is large enough to make the term with $\sinh^2(ak_2)$ dominating, thus providing a simplified representation of the transmission through the barrier as

$$T \approx \frac{U_0^2}{4E(U_0 - E)} \exp\left(-\frac{a}{\hbar}\sqrt{2m(U_0 - E)}\right).$$

There is also a useful representation of the rectangular barrier in the form of δ-function barrier. It appears when the height U_0 goes to infinity while the thickness a is reduced to zero, such that the product $S = aU_0$ remains constant. The transmission coefficient of this barrier is

$$T = \left(1 + \frac{2mS^2}{4\hbar^2 E}\right)^{-1}.$$

The tunnel transparency of potential barriers of an arbitrary shape $U(x)$ can be estimated with,

$$T \approx \exp\left(-\frac{2}{\hbar} \int_{x_1}^{x_2} \sqrt{2m(U(x) - E)}dx\right)$$

where x_1 and x_2 are turning points defined by the condition $U(x_1) = U(x_2) = E$. Figure T.4 shows qualitatively variation of the transmission coefficient for tunnel barriers of different shapes as a function of the energy E of an incident electron weighted to the barrier height U_0.

Figure T.4 Transmission coefficient versus electron energy E reduced by the barrier height U_0 for differently shaped tunnel potential barriers.

Surprisingly, electron waves passing over a symmetric rectangular barrier, so that $E > U_0$, experience non-monotonous, in fact resonance, behavior. The maximum over-the-barrier transmission, corresponding to $T = 1$, occurs only for electrons with particular energies

$$E = U_0 + \frac{\pi^2\hbar^2}{8ma^2}n^2, \quad \text{with} \quad n = 1, 2, 3, \ldots.$$

Thus, a rectangular barrier has no influence on over-the-barrier passing electron waves if their wavelength is $\lambda = a/2, a, 2a, 4a, \ldots$. In other cases, incident electron waves are partially reflected. The over-the-barrier resonance is also seen in other systems, such as microwaves.

Electron tunneling is a rather general phenomenon for solid-state structures. However, there are two specific issues that distinguish tunneling in nanostructures from that in bulk systems. One is the granular nature of the electron charge, which evidences itself in the **single-electron tunneling** phenomenon. The second involves discrete energy levels induced in nanostructures by quantum confinement. Tunneling via energetically equivalent levels occurs with conservation of energy and momentum of electrons representing the **resonant tunneling** phenomenon. Moreover, additional effects arise when spin-polarized electrons participate in tunneling → **tunneling magnetoresistance**. All these phenomena are widely employed in **nanoelectronic** devices.

First described in: M. A. Leontovich and L. I. Mandelshtam, *On the theory of the Schrödinger equation*, Z. Phys. **47**, 131 (1928).

tunneling magnetoresistance effect – external magnetic field dependence of the tunnel current between two differently magnetized **ferromagnetic** layers separated by a thin insulator. Particular magnetization of the ferromagnetic layers is provided by their deposition in a magnetic field. The tunneling process is strongly mediated by spin orientations of the carriers in the electrodes controlled by the ferromagnetic material magnetization. As a general feature, the structure has high resistance when two ferromagnetic layers are magnetically antialigned and the resistance is significantly reduced when they become aligned in an external magnetic field. As in case of the **giant magnetoresistance effect**, the phenomenon is often described in terms of the tunneling junction magnetoresistance.

Typical junctions for spin-polarized tunneling are composed of ferromagnetic layers of Co, CoCr, CoFe, NiFe or other ferromagnetic alloys separated by Al_2O_3, MgO, Ta_2O_5 up to a few nanometers thick. Magnetoresistance of the junctions is a function of the bias, magnetic field and temperature. In the absence of the magnetic field, a tunnel junction, if it is perfect, has nearly constant conductance at low bias, that is in the mV range. At higher voltages, a close to parabolic dependence of the conductance on the applied voltage is observed.

The variation of the magnetoresistance with direction and strength of the magnetic field is illustrated in Figure T.5, where representative experimental data for $CoFe/Al_2O_3/Co$ junction, Co and CoFe films are depicted. Two curves shown for each structure correspond to two opposite starting directions of the magnetic field applied, namely black for starting from "+" and gray for starting from "−". At a starting high field, the junction magnetoresistance is low, because directions of magnetization of both ferromagnetic electrodes and consequently spin polarization there are aligned. It begins to increase as the strength of the magnetic field decreases toward zero. Upon reversing the field direction, the magnetoresistance rapidly rises, showing a peak. In the reversed magnetic field the magnetization of the electrode with a lower coercive force aligns itself in the new direction, while the second electrode with a higher coercive force remains magnetized in the starting field direction. The magnetization of the two electrodes appears to be antiparallel to each other. Note that the coercive force of a ferromagnetic film is easily controlled by the deposition conditions: the presence of a magnetic field, substrate temperature, a nucleating layer, thickness of the film, and so on.

Figure T.5 Magnetoresistance of two ferromagnetic thin films and their tunnel junction as a function of applied magnetic field at room temperature. The arrows indicate directions of magnetization in the films. After J. S. Moodera, L. R. Kinder, *Ferromagnetic-insulator-ferromagnetic tunneling: spin-dependent tunneling and large magnetoresistance in trilayer junctions*, J. Appl. Phys. **79**(8), 4724–4729 (1996).

With further increase in the field it becomes strong enough to align the magnetization of the second ferromagnetic electrode in the new field direction, resulting in parallel orientation. The magnetoresistance drops to its starting value. At high fields in either direction, the magnetizations of both electrodes are saturated and parallel to each other. When the magnetizations are parallel, the tunneling probability is the highest and tunneling current reaches its maximum, thereby yielding low junction resistance. In the antiparallel configuration, the tunneling probability and the current are lowest, resulting in higher resistance.

The model for spin-mediated tunneling between two differently magnetized ferromagnetic electrodes separated by a thin insulator assumes that spin is conserved in tunneling and tunnel current is dependent on the density of states of the two electrodes. Because of the uneven spin distribution of conduction electrons at the Fermi level in ferromagnetics, one can expect the tunneling probability to be dependent on the relative magnetization of the ferromagnetic films. The change in the tunnel resistance, which can be considered to be the highest tunneling junction magnetoresistance, is given by

$$\frac{\triangle R}{R} = \frac{R_a - R_P}{R_a} = \frac{2 P_1 P_2}{1 + P_1 P_2}.$$

Here R_p and R_a are the resistances with magnetizations of the electrodes parallel and antiparallel, respectively, and P_1 and P_2 are the conduction electron spin polarizations in the ferromagnetic electrodes. The highest room temperature tunneling junction magnetoresistance at low bias, that is 20–23%, has been experimentally achieved for Al_2O_3 insulating layers.

First described in: P. M. Tedrow, R. Meservey, *Spin-dependent tunneling into ferromagnetic nickel*, Phys. Rev. Lett. **26**(4), 192–195 (1971).

tunneling magnetoresistance nonvolatile memory – a thin-film memory device employed the **tunneling magnetoresistance effect**. It is schematically shown in Figure T.6. The random access memory is formed by two arrays of parallel ferromagnetic lines separated in space by a thin insulating layer and orthogonal in the plane. Each intersection of the lines acts as a magnetic tunnel junction. When the two opposite ferromagnetic regions are magnetically aligned, the tunnel resistance is lower than when they are antialigned. A resistance change of at least about 30% is needed for the memory application.

The high resistance of tunnel junctions precludes the sense-line scheme used in a giant magnetoresistance-based random access memory. Instead, essentially a four-point probe arrangement (two that provide current and two that permit an independent voltage measurement) is attached to every device. Furthermore, the leads can provide a dual service, because pulse currents, which are directed to run above and below rather than through the tunnel junction, can provide the necessary magnetic fields to manipulate the magnetization directions in the ferromagnetic regions. This configuration is similar to the addressing scheme of the **giant magnetoresistance memory**. However, one problem is that such an array is multiply shorted through the elements; that is, the electrical path from an input lead to an output lead can proceed through many elements, not just the one at the intersection. The solution is to place a diode at every intersection so that the current can pass in only one direction, which eliminates alternative paths. It is a technological challenge to fabricate these diodes in an integrated manner with tunnel junction storage elements. Its solution permits the fabrication of an extremely high-density memory.

Figure T.6 Schematic of random access memory constructed of magnetic tunnel junctions.

turbulence – a chaotic motion of a fluid. It usually occurs when the flow rate exceeds a particular threshold limit value that is mainly determined by the viscosity of the fluid.

Turing machine – a mathematical idealization of a computing machine similar in some ways to real computing machines. It is used to define the concept of computability. The machine has a finite set of states S, a finite alphabet of symbols A, and a finite set of instructions I. In addition, it has an external infinitely long memory tape. This is called an S-state, A-symbol Turing machine.

The states s_i correspond to the functioning modes of the machine. The machine is exactly in one of these states at any given time. The symbols in the alphabet serve to encode the information processed by the machine: they are used to code input/output data and to store the intermediate operations. The instructions are associated with the states in S, and they tell the machine what action to perform if it is currently scanning a certain symbol, and what state to go into after performing this action. There is a single halt state s_{halt} (halt, for short) from which no instructions emerge, and this halt state is not counted in the total number of states.

The elements S, A, I are physically arranged in three main components:

1. The tape, which is a doubly infinite tape divided into distinct sections or cells. Each cell can hold only one symbol.

2. A read/write head or cursor, which can read or write a symbol in each tape cell. The read/write head is capable of only three actions: writing on the tape or erasing from the tape only the cell being scanned, changing the internal state, and moving the head one cell to the left or right.

3. A control unit, which is a device that controls the movements of the head according to the current state of the machine and the content of the cell currently scanned by the head.

The operation of a Turing machine is governed by the set of instructions I. These are rules that describe the transition from an initial pair of s_i and a_i to a final pair of s_f and a_f plus the movement of the read/write head.

First described in: A. Turing, *On computable numbers, with an application to the Entscheidungsproblem*, Proc. Lond. Math. Soc. **42**, 230–265 (1936); correction, ibid. **43**, 544–546 (1937).

turning point → **tunneling**.

twin (crystallographic) consists of two crystals of the same composition and structure, in intimate atomic contact with each other and possessing orientations that are related to each other in a special way. The orientations of the two crystals may be related by a reflection in a lattice plane (the twinning plane) common for both crystals, or they may be related by rotation, usually of $180°$, about an axis (the twinning axis) common for both crystals. These types are termed reflection twins and rotation twins, respectively. When the crystal is of high symmetry, a twin may be of both types simultaneously. A given twinning plane or axis defines a twin law. There may be several twin laws for one substance.

two-dimensional electron gas (2DEG) – an ensemble of electrons in **quantum films**.

two-dimensional electron gas field effect transistor (TEGFET)

two-dimensional electron gas field effect transistor (TEGFET) → high electron mobility transistor (HEMT).

type I quantum wells → quantum well.

type II quantum wells → quantum well.

type I superconductors → superconductivity.

type II superconductors → superconductivity.

U

From ultraviolet photoelectron spectroscopy (UPS) to Urbach rule

ultraviolet photoelectron spectroscopy (UPS) – the technique for studying electronic states in valence and conduction bands in solids using energy analysis of photoelectrons emitted under illumination of the sample with ultraviolet radiation. The energy of the photons ranges from 10 eV to 50 eV.

Umklapp process – an interaction of three or more waves in a solid, such as lattice waves or electron waves, in which the sum of the wave vectors is not equal to zero but, rather, is equal to a vector in the reciprocal lattice.

unbound state (of an electron in an atom) – a state to which an electron is raised when it is ejected from the atom by a high-energy collision or phonon. The energies of an electron in an unbound state are always positive. They are not quantized and form continuum states of the atom. The alternative is a **bound state**.

uncertainty principle → Heisenberg's uncertainty principle.

unitary group – a group of **unitary transformations** on k-dimensional complex vector space.

unitary matrix – a matrix whose inverse is identical with its conjugate transpose.

unitary transform – a linear transform on a vector space, which preserves inner products and norms. Alternatively, it is a linear operator whose adjoint is equal to its inverse.

unit cell – the smallest three-dimensional cell which being translated into three directions reproduces the crystalline lattice of the solid (see Figure U.1).

It is described by six lattice parameters: length of lattice translation vectors $\mathbf{a} = a$, $\mathbf{b} = b$, $\mathbf{c} = c$ and interaxial lattice angles $\mathbf{b} \wedge \mathbf{c} = \alpha$, $\mathbf{a} \wedge \mathbf{c} = \beta$, $\mathbf{a} \wedge \mathbf{b} = \gamma$. The unit cell completely defines the entire lattice.

What is What in the Nanoworld. Victor E. Borisenko and Stefano Ossicini
Copyright © 2008 WILEY-VCH Verlag GmbH & Co. KGaA, Weinheim
ISBN: 978-3-527-40783-5

universal ballistic conductance

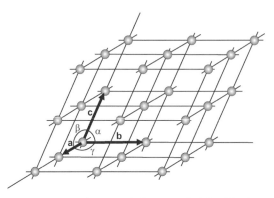

Figure U.1 Space lattice with the unit cell defined by the vectors **a**, **b** and **c**.

universal ballistic conductance → **ballistic conductance**.

universal conductance fluctuations – the root mean square fluctuations of the conductance in low-dimensional conductors are of the order e^2/h, independently of the size of the average conductance.

Variation of conductance is a result of the phase interference of electron waves in conventional low-dimensional conductors, which are normally long and thin. Impurities or other defects inside such conductors create potential hills around which electron waves must pass. This is schematically illustrated in Figure U.2 for one impurity atom (or other type point defect) disturbing the coherent propagation of electrons.

At low temperatures the material of a conductor is usually degenerate, and only the carriers with **Fermi energy** take part in transport processes. The Fermi energy can be varied either by some potential applied to the gate covering the conductor or by a magnetic field, both of which deplete the conduction band. As a result, the detailed Fermi surface upon which the carriers travel appears to be slightly shifted. It is thus possible for a carrier that moves on one side of the defect (path A) to modify the trajectory so that it travels on the other side of the defect (path B) after the shift of the Fermi energy. The change in the trajectory is equivalent to coupling an **Aharonov–Bohm** loop composed of the A and B arms. As a result, fluctuations of conductance occur in this process. It is quite static in time and depends on the particular configuration of scattering centers in the sample.

In order to observe quantum changes in conductance related to the phase interference of electron waves, the sample must be comparable in size to the

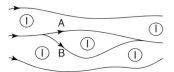

Figure U.2 Electron wave trajectories in the sample around impurity atoms.

phase coherence length, which is controlled by the density of scattering centers in the sample material. As the sample grows larger, the quantum fluctuation of conductance is smoothed out by ensemble averaging.

First described in: B. L. Altshuler, *Fluctuations of residual conductivity in non ordered semiconductors*, Pisma Zh. Exp. Teor. Fiz. **41**(12), 530–533 (1985); P. A. Lee, A. D. Stone, *Universal conductance fluctuations in metals*, Phys. Rev. Lett. **55**(15), 1622–1625 (1985).

unsaturated hydrocarbons → **hydrocarbons**.

up- and down-conversion solar cell – a **solar cell** in which the sun spectrum is converted into a smaller energy range by converting two or more low-energy **photon**s into a high-energy photon or converting one high-energy photon into two or more low energy-photons. Photon shifting can also be used where, by a **luminescence** process, a high-energy photon is converted into a single lower energy photon. Energetically, the latter process is not favorable, but there might still be some gain at solar cell level for industrial cells, which feature a rather poor response for short-wavelength light.

More details in: A. Shalav, B. S. Richards, T. Trupke, K. W. Krämer, H. U. Güdel, *Application of $NaYF_4:Er_{3+}$ up-converting phosphors for enhanced near-infrared silicon solar cell response*, Appl. Phys. Lett. **86**, 013505 (2005); V. Svrcek, A. Slaoui, J.-C. Muller, *Silicon nanocrystals as light converter for solar cells*, Thin Solid Films **451–452**, 384–388 (2004).

UPS – an acronym for **ultraviolet photoelectron spectroscopy**.

Urbach rule – the exciton absorption $\alpha(\omega)$ for frequencies below the exciton resonance at ω_0 decreases exponentially: $\alpha(\omega) = \alpha_0 \exp[-(\omega_0 - \omega)/\sigma]$.

First described in: F. Urbach, *The long-wavelength edge of photographic sensitivity and of the electronic absorption of solids*, Phys. Rev. **92**(5), 1324–1324 (1953).

V

From vacancy to von Neumann machine

vacancy – a point defect in the form of an unoccupied lattice position in a crystal.

valence band – an uppermost, filled electron energy band in solids. It is most commonly available for conduction of electric current by holes in solids.

van der Pauw configuration (sample) – the four-contact geometry shown in Figure V.1.

It is usually used for measurements of a conductivity of semiconductors.

First described in: L. van der Pauw, *A method of measuring specific resistivity and Hall effect of discs of arbitrary shape*, Philips Res. Rep. **13**(1), 1–9 (1958).

Figure V.1 van der Pauw sample.

van der Waals force – an attractive force between two atoms or nonpolar molecules arising as a result of a fluctuating dipole moment in one molecule that induces a dipole moment in the other whereupon the two dipole moments then interact.

First described in: J. D. van der Waals, *Over de Continuiteit van den Gas-en Vloeistoftoestand (on the continuity of the gas and liquid state)*, (Ph.D. thesis, Leiden, The Netherlands, 1873).

Recognition: in 1910 J. D. van der Waals received the Nobel Prize in Physics for his work on the equation of state for gases and liquids.

See also www.nobel.se/physics/laureates/1910/index.html.

Van Hove singularities – singularities in density of states of electron and phonon bands in solids at points where $|\nabla_k(E)|$ vanishes. These points are known as critical points.

What is What in the Nanoworld. Victor E. Borisenko and Stefano Ossicini
Copyright © 2008 WILEY-VCH Verlag GmbH & Co. KGaA, Weinheim
ISBN: 978-3-527-40783-5

Van Vleck equation

Assuming that $k = 0$ is a critical point in the three-dimensional space, $E(k)$ can be expanded as a function of k about the critical point:

$$E(k) = E(0) + a_1 k_1^2 + a_2 k_2^2 + a_3 k_3^2 + \dots .$$

The singularities are classified according to the number of negative coefficients a. In three dimensions there are four kinds of Van Hove singularities, labeled as M_0, M_1, M_2, and M_3 critical points. M_0 critical point has no negative as and therefore represents a minimum in the band separation. M_1 and M_2 are known as saddle points, since the plots of their energies versus wavevector resemble a saddle. M_3 critical point represents a maximum in the band separation.

First described in: L. Van Hove, *The occurrence of singularities in the elastic frequency distribution of a crystal,* Phys. Rev. **89**(6), 1189–1193 (1953).

Van Vleck equation shows the temperature dependence of susceptibility. It was obtained using the fact that atoms, ions or molecules are distributed in a magnetic field among various allowed energy levels according to the **Boltzmann distribution** in the form:

$$\chi = N \frac{\sum_i [(E_i')^2 / k_B T - 2 E_i''] \exp(-E_i / k_B T)}{\sum_i \exp(-E_i / k_B T)},$$

where N is the number of elementary dipoles, and E_i is the energy of a spin level developed into a **Taylor series** as a function of magnetic field as $E = E_i + E_i' H + E_i'' H$.

By the perturbation theory one has that $E_i' = < i | \mu_H | i >$, $E_i'' = \sum_i [| < i | \mu_H | j > |^2 / h \nu_{ij}]$ $(j \neq i)$, where $h \nu_{ij}$ is the energy interval $(E_i - E_j)$ spanned by the matrix element of the magnetic moment in the direction of the field H. Thus, the susceptibility behaves thus:

1) χ is proportional to $1/T$, if all $|h\nu_{ij}| \ll k_B T$

2) χ is independent of T, if all $|h\nu_{ij}| \gg k_B T$

3) $\chi = A + BT$, if all $|h\nu_{ij}|$ are either $\gg k_B T$ or $\ll k_B T$

4) no simple dependence of χ on T, if $|h\nu_{ij}|$ is comparable with $k_B T$.

The temperature-independent paramagnetic contribution is termed **Van Vleck paramagnetism**. Note that if the susceptibility follows the classical **Curie law** → also **Curie–Weiss law**) then it will display a linear dependence on the reciprocal of the temperature.

First described in: J. H. Van Vleck, *On dielectric constants and magnetic susceptibilities in the new quantum mechanics. Part I. A general proof of the Langevin–Debye formula,* Phys. Rev. **29**(5), 727–744 (1927); J. H. Van Vleck, *On dielectric constants and magnetic susceptibilities in the new quantum mechanics. Part III. Application to dia- and paramagnetism,* Phys. Rev. **31**(4), 587–613 (1928).

More details in: J. H. Van Vleck, *The Theory of Electric and Magnetic Susceptibilities* (Clarendon, Oxford, 1932).

Recognition: in 1977 P. W. Anderson, N. F. Mott and J. H. Van Vleck received the Nobel Prize in Physics for their fundamental theoretical investigations of the electronic structure of magnetic and disordered systems.

See also www.nobel.se/physics/laureates/1977/index.html.

Van Vleck paramagnetism $\rightarrow$ **Van Vleck equation.**

varactor – a semiconductor diode that exhibits change in capacitance with a change in applied voltage; used as a voltage-variable capacitor.

variance – the square of the standard deviation.

variational method – a method of calculating an upper bound on the lowest energy level of a quantum mechanical system and an approximation for the corresponding wave function. In the integral representing the expectation value of the **Hamiltonian operator**, one substitutes a trial function for the true wave function and varies parameters in the trial function to minimize the integral.

variational principle states that the lowest energy obtained is the closest approximation to the true state of the system. If an arbitrary wave function is used to calculate the energy, then the value calculated is never less than the true energy. The procedure based on this principle is also known in the literature as **Ritz procedure.**

Varshni law – the semiempirical expression describing the temperature dependence of the energy gap in semiconductors in the form $E(T) = E(0) - aT^2/(b + T)$, where a and b are the fitting parameters.

First described in: Y. P. Varshni, *Temperature dependence of the energy gap in semiconductors*, Physica (Amsterdam) **34**, 149–154 (1967).

V-based molecular ring $\rightarrow$ **molecular nanomagnet.**

VCSEL – an acronym for a **vertical cavity surface emitting laser.**

Vegard's law states that the lattice constant of an intermediate compound is given by linear interpolation between lattice constants of its constituents. It is valid for substitutional solid solutions of atoms with a similar electronic structure and close size. Thus, the lattice constant a of new compound can be calculated as $a(x) = xa_1 + (1 - x)a_2$, where x is the atomic fraction of the substituted atoms, and a_1 and a_2 are the lattice parameters of the virgin compounds between which the new one is formed.

First described in: L. Vegard, *Die Konstitution der Mischkristalle und die Raumfüllung der Atome*, Z. Phys. **5**(1), 17–26 (1921).

Verdet constant

Verdet constant → **Faraday effect**.

*First described in:*M. E. Verdet, *Recherches sur les propriétés optiques développées dans les corps transparents par l'action du magnétisme*, Ann. Chim. **41**, 370–412 (1854).

vertical-cavity surface-emitting laser (VCSEL) – a semiconductor laser that has an optical resonator formed by two Bragg reflectors parallel to the surface of the wafer in which it has been fabricated. It is schematically shown in Figure V.2.

The reflectors sandwiching the active light-emitting region are composed of multilayers of semiconductors or dielectrics. Carrier injection is performed through a pair of electrodes in the upper and lower surfaces to excite cylindrical active region between them, which is typically about 10 μm in diameter and 1–3 μm thick. The generated light is emitted normal to the surface.

The simple difference in the cavity orientation results in many advantages of the VCSEL over the edge-emitting lasers in beam characteristic, scalability, optoelectronic design, fabrication and array configurability. In an edge-emitting laser the output beam is highly astigmatic whereas with the laser emission perpendicular to the wafer surface one has a symmetrical beam cross section, with small beam divergence. VCSEL can be fabricated into high-density two-dimensional arrays using low-cost batch processing techniques. Very short cavity lengths dictated by the vertical orientation lead to lasing in a single longitudinal mode. VCSEL mirrors are fabricated during epitaxial growth eliminating cleaving or dry etching processes. However, the reduction in the gain length must be compensated by a high-Q cavity. Materials used to manufacture VCSEL include GaAs, AlGaAs, InGaAsP, and InGaAsN.

VCSELs are indeed useful for monolithic integration to form and two-dimensional densely packed laser arrays. Other advantages over conventional edge-emitting lasers are the feasibility of stable single longitudinal mode operation because of the large mode spacing due to the short optical cavity length, and the attainability of a narrow circular output beam.

Figure V.2 Schematic sketch of vertical-cavity surface-emitting laser.

First described in: H. Soda, K. Iga, C. Kitahara, Y. Suematsu, *GaInAsP/ InP surface emitting injection lasers,* Jpn. J. Appl. Phys. **18**(12), 2329–2330 (1979).

More details in: W. W. Chow, K. D. Choquette, M. H. Crawford, K. L. Lear, G. R. Hadley, *Design, fabrication, and performance of infrared and visible vertical-cavity surface-emitting lasers,* IEEE J. Quantum Electron., **33**(10), 1810–1824 (1997).

Verwey transition – a metal–insulator transition accompanied by a structural phase transition. It takes place in mixed-valent systems. The driving force for the phenomenon is strong electron–electron and electron–lattice interactions in the system resulting in ordering of formal valence states.

The transition is also reflected by anomalies in many other physical properties, such as a jump in the magnetic susceptibility, a sharp peak in the specific heat and a change of the crystal symmetry above a certain temperature.

First described in: E. J. W. Verwey, *Electronic conduction of magnetite (Fe₃O₄) and its transition point at low temperatures,* Nature **144**, 327–328 (1939).

vibronic coupling – an interaction between electronic and nuclear vibrational motions in molecules. This type of coupling contradicts the **Born–Oppenheimer approximation** by considering the motion of electrons and nucleus in a molecule to be separable. Many effects, in particular at an intense excitation of molecules, are due to the coupling of electronic and nuclear degrees of freedom.

More details in: G. Fischer, *Vibronic Coupling – The Interaction between the Electronic and Nuclear Motions* (Academic Press, New York, 1984).

vicinal surface – a surface with periodic succession of terraces and steps of monoatomic height obtained by cutting a crystal along a plane making a small angle ($\leq 10^{\circ}$) with a low index plane. This surface is not an equilibrium shape of the crystal (**Wulff's theorem**).

Villari effect – a change in the magnetic induction within a ferromagnetic substance placed in a magnetic field when the substance is subjected to mechanical stress.

First described in: E. Villari, *Über die Änderungen des magnetischen Moments, welche der Zug und das Hindurchleiten eines galvanischen Stroms in einem Stabe von Stahl oder Eisen hervorbringen,* Ann. Phys. **126**, 87–122 (1865).

virus – a small biological particle which can infect other biological organisms. They are obligate intracellular parasites, meaning that they can only reproduce by invading and taking over other cells as they lack the cellular machinery for self-reproduction. The name virus usually refers to those particles which infect **eukaryote**s, these being multi-celled organisms and many single-celled organisms, whilst the term **bacteriophage** or **phage** is used to describe those infecting **prokary-ote**s (bacteria and bacteria-like organisms). Typically these particles carry a small amount of nucleic acid (either **DNA** or **RNA**) surrounded by some form of protective coat consisting of **protein** or protein and **lipid**.

First described by: The term virus was introduced by M. J. Beijerinck in 1898. M. J. Beijerinck, *Über ein Contagium vivum fluidum als Ursache der Fleckenkrankheit der Tabakblätter*, Verh. Kon. Akad. Wetensch. **65**, 3–21 (1898).

Voigt configuration – the term used to indicate a particular orientation of a **quantum film** structure with respect to the magnetic field applied, that is when the field direction is parallel to the film plane. The alternative orientation is labeled as a **Faraday configuration**.

Voigt effect – double refraction of light passing through a substance placed in a magnetic field perpendicular to the light propagation.

First described in: W. Voigt, *Weiteres zur Theorie des Zeemaneffektes*, Gött. Nachr. (1898) and Ann. Phys. **67**, 352–364 (1899).

Volmer–Weber growth mode (of thin films) – droplet (island) formation during thin film deposition as illustrated in Figure V.3.

It appears when the deposited atoms are more strongly bound to each other than they are to the substrate.

First described in: M. Volmer, E. Weber, *Keimbildung in übersättigten Gebilden*, Z. Phys. Chem. **119**, 277–301 (1926).

Figure V.3 Volmer–Weber growth mode during thin film deposition.

von Neumann machine – a principle of computing automation based on the sequential execution of the programs registered in the memory of the computer. A computer realizing this principle has to be composed of two main components, the processor and the memory.

The processor is the active part of the computer in which the information contained in the programs is processed step by step. It is in turn divided into three main parts:

1. Control unit, which is the unit that controls all the parts of the computer in order to carry out all the operations requested by other parts, such as extracting data from the memory, executing and interpreting instructions, and so on.

2. Register, a very fast memory unit inside the processor, which contains that part of the data currently being processed.

3. Arithmetic and logic unit, which is the unit devoted to the actual computations such as sums, multiplications, logic operations, and so on, executed on the data supplied by the registers or memory upon demand by the control unit.

The memory is the part of the computer devoted to the storage of the data and instructions to be processed. It is divided into individual cells, which are accessible by means of a number called the address.

The functioning of the machine is cyclic. One of these cycles contains the following operations: the control unit reads one program instruction from the memory, which is executed after being decoded. Depending on the type of instruction, a piece of data can either be read from or written into the memory, or an instruction can be executed. In the next cycle to be performed, the control unit reads another program instruction, which is precisely next in the memory to the one processed in the previous cycle. The simplicity of this sequentially operating model is advantageous for many purposes because it facilitates the design of machines and programs.

First described in: J. von Neumann, *The principles of large scale computing machines* (1946); reprinted in: Ann. Hist. Comput. **3**(3), 263–273 (1981).

From Waidner–Burgess standard to Wyckoff notation

Waidner–Burgess standard – a standard of luminous intensity evaluated as the luminous intensity of 1 cm^2 of a black body at the melting point of platinum.

First described in: C. W. Waidner, G. K. Burgess, *Note on the primary standard of light*, Elec. World **52**, 625 (1908).

wall-in-plug efficiency → **quantum efficiency.**

Walsh functions – the complete system of orthogonal functions defined as: $f_0(x) = 1, f_n(x) = f_{n_1}(x)f_{n_2}(x)\ldots f_{n_q}(x)$ for $n = 2^{n^1} + 2^{n^2} + \ldots + 2^{n^q}$, where the non-negative integers n_i are uniquely defined by the condition $n_{i+1} < n_i$.

First described in: J. L. Walsh, *A closed set of normal orthogonal functions*, Am. J. Math. **55**(1), 5–24 (1923).

Wang–Chan–Ho potential – the interatomic pair potential for a given configuration of N atoms $\{r_1, r_2, \ldots, r_N\}$, expressing the total potential energy (per atom) of the system as $E_{tot}\{r_1, \ldots, r_N\} = E_{bs}\{r_1, \ldots, r_N\} + U\{r_1, \ldots, r_N\}$, where $E_{bs} = N^{-1} \sum_{j=1}^{N} E_j$ is the "band-structure" energy (per atom) consisting of the sum of the eigenvalues E_j for the occupied part of the electronic band structure; $U\{r_1, \ldots, r_N\} = (1/2N) \sum_{i,j=1, j\neq i}^{N} V(r_{ij})$ is a short-ranged two-body potential representing the sum of ion–ion repulsion and the correction to the double counting of the electron–electron interaction in the band-structure energy E_{bs}.

First described in: C. Z. Wang, C. T. Chan, K. M. Ho, *Empirical tight-binding force model for molecular-dynamics simulation of Si*, Phys. Rev. B **39**(12), 8586–8592 (1988).

Wannier equation describes the relative motion of an electron and a hole interacting via the attractive **Coulomb potential** $V(r)$:

$$-\left[\frac{\hbar^2 \nabla^2}{2m_r} + V(r)\right] \psi_\nu(\mathbf{r}) = E_\nu \psi_\nu(\mathbf{r}),$$

What is What in the Nanoworld. Victor E. Borisenko and Stefano Ossicini
Copyright © 2008 WILEY-VCH Verlag GmbH & Co. KGaA, Weinheim
ISBN: 978-3-527-40783-5

Wannier excitation

where one uses the inverse reduced mass $1/m_r = 1/m_e + 1/m_h$ with m_e and m_h to be masses of the electron in the conduction band and hole in the valence band, respectively. It has the form of a two-particle **Schrödinger equation**.

Wannier excitation – the excitation of electron–hole pairs with a distance between electron and hole, which is the exciton Bohr radius, larger than the length of the lattice **unit cell**. Such excited pairs are called **Wannier excitons**. These are typical for most $A^{II}B^{VI}$, $A^{III}B^V$ and group IV semiconductors. An alternative case is represented by **Frenkel excitation** and **Frenkel excitons**.

Wannier excitons → **Wannier excitation.**

Wannier functions – Fourier transforms of the **Bloch functions** $\psi_{nk}(\mathbf{r})$ in the form $W_n(\mathbf{r}, \mathbf{R}_i) = N^{-1/2} \sum_{\mathbf{k}} \exp(-i\mathbf{k} \cdot \mathbf{R}) \psi_{nk}(\mathbf{r})$. For a crystal, N is the number of unit cells in the crystal, $\mathbf{R}_i$ is the lattice vector, and n is the band index. The wave vector $\mathbf{k}$ covers the first **Brillouin zone**. While Bloch functions are indexed by the wave vectors in reciprocal lattice space, Wannier functions are indexed by lattice vectors in real space. Bloch functions are convenient for representing extended states in solids and Wannier functions are more appropriate for localized states.

More details in: G. Wannier, *Elements of Solid State Theory* (Cambridge University Press, Cambridge, 1959).

Wannier–Mott exciton – an **exciton** with the excitation energy spread over a large number of lattice cells. It can be visualized as a correction to the transition energy between valence and conduction band due to the interaction potential between the extra electron in the conduction band and the hole in the valence band. The energy of the bound state will be lower that the transition energy by $E_{ex} = (e^2/\epsilon)/2a_0^*$, where ϵ is the dielectric constant of the crystal material and $a_0^* = \epsilon(m/m^*)a_B$ is the Bohr radius of the exciton, m and m^* being the electron and the reduced effective mass respectively and a_B the electron Bohr radius. This description is expected to be particularly convenient for crystals with a high dielectric constant and consequently weak electron–hole interaction.

In nanostructures a characteristic size effect is the confinement of the Mott–Wannier exciton, where there is an interesting interplay between the attractive electron–hole interaction and the boundary repulsion.

First described in: G. H. Wannier, *The structure of electronic excitation levels in insulating crystals*, Phys. Rev. **52**(3), 191–197 (1937); N. F. Mott, *Conduction in polar crystals: II. The conduction band and ultra-violet absorption of alkali-halide crystals*, Trans. Faraday Soc. **34**, 500–506 (1938).

More details in: D. L. Dexter, R. S. Knox, *Excitons* (Wiley, New York, 1965).

watched-pot effect → **Zeno paradox.**

Watson–Crick base pairs → **DNA.**

Watson–Crick double-helix model of DNA → DNA.

wave equation describes the wave ψ traveling with velocity v:

$$\nabla^2 \psi = \frac{1}{v^2} \frac{\partial^2 \psi}{\partial t^2}.$$

wave function has been introduced in order to describe the position of a quantum particle, which is distributed through space like a wave, in place of the trajectory.

waveguide – a device or structured material designed to confine and direct electromagnetic waves in a direction determined by its physical boundaries.

wavelet – a waveform-like function with a fast tail decay, oscillatory and localized. Such functions comprise the family of translations and dilatations of a single function. Wavelets are used to transform a signal into another representation that carries the information in a more useful form.
   More details in: Y. Mayer, *Wavelets and Operators* (Cambridge University Press, Cambridge, 1992).

wavelet signal processing – the transform of signals using a single function $w(t)$ localized both in time and frequency: $F(m, n) = (1/\sqrt{a}) \int w^*[(t - n)/m] f(t) dt$, in contrast to traditional signal processing based on **Fourier transform** analysis, where the signal is decomposed into complex sinusoidal basis functions localized in time.
   The function $w(t)$ can be thought of as a band-pass filter. From one side, the fine temporal analysis is performed using high-frequency contracted versions of the **wavelet**. On the other side, the fine frequency analysis is made using dilated versions of the wavelet. The localization in both time and frequency domains is one of the main advantages of wavelet applications, together with the existence of fast discrete wavelet transform algorithms that render the wavelet method particularly suitable for multiresolution signal decomposition. This approach is different from the traditional signal processing based on Fourier transform analysis, where the signal is decomposed into complex sinusoidal basis functions localized in time.
   First described in: I. Daubechies, *Orthonormal bases of compactly supported wavelets*, Commun. Pure Appl. Math. **41**(7), 909–996 (1988); S. Mallat, *A theory for multiresolution signal decomposition: the wavelet representation*, IEEE Trans. On Pattern analysis and Machine Intelligence **11**, 674–693 (1989).
   More details in: I. Daubechies, S. Mallat, A. S. Willshy (editors), *Special Issue on Wavelet Transforms*, IEEE Trans. Inform. Theory, **38**(2), part II (1992).

wavelet transform = wavelet signal processing.

wave number

wave number – the number of waves per unit length. It is the reciprocal of the wavelength: $k = 1/\lambda = v/c$, v is the frequency and c is the velocity of the wave propagation.

wave packet – a wave restricted to a finite space by an envelope. It is illustrated in Figure W.1.

The **wavelets** inside the packet move along at the phase velocity $v_{ph} = \omega/k = \hbar k/2m = p/2m$, while the envelope moves at the **group velocity** $v_{gr} = d\omega/dk = \hbar k/m = p/m$, where ω is the circular frequency of the wave representing the particle, k is the wave number, m is the mass of the particle described as a wave, and p is the particle momentum.

If the packet represents a particle, for instance an electron, one is usually interested in the behavior of the **wave function** as a whole rather than its internal motion. The group velocity is often the appropriate one, and it is a relief that this agrees with the classical result.

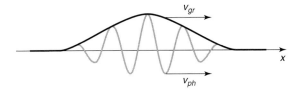

Figure W.1 A wave packet.

wave plate – an optical element having two principal axes, slow and fast, that resolve an incident polarized beam into two mutually perpendicular polarized beams. The emerging beam recombines to form a particular single polarized beam. Wave plates produce full-, half- and quarter-wave retardations. Also known as a retardation plate.

wave train – a continuous group of waves that persists for a short time only.

weak localization – a set of electronic phenomena in solids characterized by weak (logarithmic) decrease of the conductivity to zero at low temperatures.

Weiss oscillations – the magnetoresistance oscillations in a two-dimensional electron gas with a lateral periodic modulation in one direction. The oscillations are periodic in B^{-1}. The effect results from an oscillatory dependence of the bandwidth of the modulation-broadened **Landau levels** of the two-dimensional electron system on the band index leading in high-mobility samples to a strongly anisotropic oscillating contribution to the conductivity tensor. The oscillations, periodic in B^{-1}, reflect the commensurability of cyclotron diameter at the **Fermi energy** and modulation period.

First described in: D. Weiss, K. v. Klitzing, K. Ploog, G. Weimann, *Magnetoresistance oscillations in a two dimensional electron gas induced by a submicrometer periodic potential*, Europhys. Lett. **8**(2), 179–184 (1989); R. R. Gerhards, D. Weiss, K. v. Klitzing, *Novel magnetoresistance oscillations in a periodically modulated two-dimensional electron gas*, Phys. Rev. Lett. **62**(10), 1173–1176 (1989).

Weiss zone law → **zone law of Weiss.**

Wentzel–Kramers–Brillouin (WKB) approximation supposes slow variation of coefficients α and β when one solves the differential equation

$$\frac{\mathrm{d}}{\mathrm{d}x}\left(\alpha\frac{\mathrm{d}\psi}{\mathrm{d}x}\right) + \alpha\beta^2\psi = 0.$$

If α and β^2 are real functions of x, the two independent approximate solutions are

$$\psi = A(\alpha\beta)^{-1/2}\exp\left(\pm\mathrm{i}\int_0^x\beta(x')\mathrm{d}x'\right),$$

where A is a constant. These solutions are good when α and β vary by only a small fraction of themselves in the interval $|\beta^{-1}|$. They break down therefore at values of x for which α or β becomes zero. However, solutions valid at some distance on either side of such points can be correctly matched.

   With $\alpha = \hbar^2/2m$ and $\beta = [(E-V)2m/\hbar^2]^{1/2}$ the above equation becomes the one-dimensional **Schrödinger equation** for the motion of a particle of mass m with energy E at the potential barrier of height V. If $(E-V) < 0$ between the points x_1 and x_2 (**turning points**), the particle cannot pass through this region from the classical point of view. Meanwhile, wave mechanics allows it to tunnel through such potential barriers. The transition probability obtained within the WKB approximation is $(1 + \exp 2K)^{-1}$ with $K = \int_{x_1}^{x_2}[(v-e)2m/\hbar^2]^{1/2}\mathrm{d}x$, which is the most frequently referred result of the approximation.

   First described in: G. Wentzel, *Eine Schwierigkeit für die Theorie des Kreiselelektrons*, Z. Phys. **37**(12), 911–915 (1926); H. A. Kramers, *Wellenmechanik und halbzahlige Quantisierung*, Z. Phys. **39**(10/11), 828–840 (1926); L. N. Brillouin, *La mécanique ondulatoire de Schrödinger; une méthode générale de résolution par approximations successives*, C. R. **183**, 24–27; L. N. Brillouin, *Sur un type général de problèmes, permettant la sèparation des variables dans la mécanique ondulatoire de Schrödinger*, C. R. **183**, 270–271 (1926).

Whittaker–Shannon theorem states that, when the sampling period in a recorded sample hologram is matched to the object spectrum, the resulting image does not suffer any loss of information content.

Wiedemann additivity law states that the mass (or specific) magnetic susceptibility of a mixture or solution components is the sum of the proportionate (by weight fraction) susceptibilities of each component in the mixture.

Wiedemann–Franz law → **Wiedemann–Franz–Lorenz law.**

Wiedemann–Franz–Lorenz law

Wiedemann–Franz–Lorenz law states that the ratio of thermal (χ) to electrical (σ) conductivity at room temperature is the same for all metals, which is the **Wiedemann–Franz law**. Lorenz extended the law showing that this ratio is proportional to the absolute temperature, so that $\chi/(\sigma T)$ is a constant independent of temperature. This constant is called the **Lorenz number**. Within the **Drude theory** of a gas of free electrons it has been shown that $\chi/(\sigma T) = 3(k_B/e)^2$. Meanwhile, assuming that only the electrons at the top of the **Fermi distribution** (over the energy range where the distribution function falls to zero) contribute to the electrical and thermal properties this relation is modified to $\chi/(\sigma T) = \pi^2/3(k_B/e)^2$.

First described in: G. Wiedemann, R. Franz, *Über die Wärme–Leitungsfähigkeit der Metalle*, Ann. Phys. **89**, 497–531 (1853).

Wien's displacement law states that the spectral emittance of a heated body has its maximum at the wavelength $\lambda_{max} = A/T$, where A is a constant, and T is the absolute temperature.

First described in: W. Wien, *Die obere Grenzen der Wellenlängen, welche in der Wärmestrahlung fester Körper vorkommen können; Folgerungen aus dem zweiten Hauptsatz der Wärmetheorie*, Ann. Phys. **49**, 633–637 (1893).

Wien's formula $\rightarrow$ **Wien's law states.**

Wien's law states that energy density per unit frequency interval for the light emitted by a heated body is defined as $w_\nu = A\nu^3 \exp(-B\nu/T)$, where A and B are constants, ν is the light frequency, and T is the absolute temperature. It is called **Wien's formula** and in fact appears to be the limiting case of Planck's radiation formula for high frequencies.

First described in: W. Wien, *Über die Energievertheilung im Emissionsspektrum eines schwarzen Körpers*, Ann. Phys. **58**, 662–669 (1896).

Recognition: in 1911 W. Wien received the Nobel Prize in Physics for his discoveries regarding the laws governing the radiation of heat.

See also www.nobel.se/physics/laureates/1911/index.html.

Wigner coefficients $\rightarrow$ **Clebsch-Gordan coefficients.**

Wigner crystal – a periodic spatial arrangement of electrons in matter appearing when a low-density electron liquid solidifies at low temperatures. The periodic structure is formed so that the energy due to mutual Coulomb repulsion is reduced. A two-dimensional Wigner crystal structure is experimentally observed above the surface of liquid helium, in a microchannel capillary filled with liquid helium, and at the interface of semiconductor heterostructures. When a small electric field is applied to a Wigner crystal, all electrons act in concert, driving the crystal with the electric field. Conductivity resonance and wave emission phenomena are typical for the system. The melting point of Wigner solids in liquid helium is around 0.5 K.

The Wigner crystals observed in semiconductors and on liquid helium have distinctive properties due to the difference in electron concentration in each system.

At the interface of a semiconductor heterostructure, there is typically a high concentration of about 10^{12} electrons cm^{-2}. To create a crystal, one needs to squeeze the in-plane wavefunction of each electron, for example by applying a strong magnetic field perpendicular to the interface. As an interface of a semiconductor heterostructure is not free of defects, at least one electron localized by a defect can bind the whole crystal. A small electric field applied to the heterostructure will only deform this pinned Wigner solid, rather than produce an electric current. Thus, Wigner crystallization forces the interface to behave like an insulator.

Peculiarities of Wigner crystallization and properties of the crystals formed are under investigation, keeping in mind their prospects for quantum computations and other applications for information processing.

First described in: E. Wigner, *On the interaction of electrons in metals*, Phys. Rev. **46**(11), 1002–1011 (1934).

Wigner–Eckart theorem states that the matrix element of a tensor operator can be factored into two quantities. The first one is a vector coupling coefficient. The second contains the information about the physical properties of the particular states and operator. It is completely independent of the magnetic quantum numbers.

First described in: E. P. Wigner, *Einige Folgerungen aus der Schrödingerschen Theorie für die Termstrukturen*, Z. Phys. **43**(5–6), 624–652 (1927); C. Eckart, *The application of group theory to the quantum dynamics of Monatomic Systems*, Rev. Mod. Phys. **2**, 305–380 (1930).

Wigner–Dyson distribution originated from the study of matrices with random entries used in nuclear physics. Later, the **random matrix theory** was found to have many other successful applications in different physical systems like molecules and solids, in particular in the field of quantum chaos.

In nuclear physics, the problem was to understand the energy level spectra of complex nuclei when model calculation failed to explain experimental data. E. P. Wigner made the assumption that the **Hamiltonian** of a heavy nucleus had, in the matrix representation, elements that could be assumed to be mutually independent random numbers. In fact, an infinite Hermitian matrix that possessed random matrix elements related to the level widths was involved in the problem. Finally, it was found that if level spacing distribution of a classical integrable system is Poissonian (the probability density of the nearest-neighbor spacing s is given by $P_P(s) = \exp(-s)$), the level statistics in a random system is given by the distribution $P_{WD}(s) = (\pi s/2) \exp(-\pi s^2/4)$.

The transition of the level spacing statistics from the **Poisson distribution** to the Wigner–Dyson distribution is one of the most direct indications of the emergence of quantum chaos.

First described in: E. P. Wigner, *On the statistical distribution of the width and spacings of nuclear resonance levels*, Proc. Camb. Phil. Soc. **47**(4), 790–798 (1951); F. J. Dyson, *Statistical theory of energy levels of complex systems I, II, and III*, J. Math. Phys. **3**(1), 140–175 (1962).

Wigner rule

More details in: M. L. Metha, *Random Matrices* (Academic Press, Boston, 1991); A. Crisanti, G. Paladin, A. Vulpiani, *Products of Random Matrices in Statistical Physics* (Springer-Verlag, Berlin, 1993).

Recognition: in 1963 E. P. Wigner received the Nobel Prize in Physics for his contributions to the theory of the atomic nucleus and elementary particles, particularly through the discovery and application of fundamental symmetry principles.

See also www.nobel.se/physics/laureates/1963/index.html.

Wigner rule states that for all possible transfers of energy between two atoms, the one most likely to occur is that in which the total resultant spin remains constant.

Wigner–Seitz method approximates a band structure of a solid. **Wigner–Seitz primitive cells** surrounding atoms in the solid are approximated by spheres. Band solution of the **Schrödinger equation** for one electron are estimated by using the assumption that an electronic wave function is the product of a plane wave function and a function whose gradient has a vanishing radial component at the sphere's surface.

First described in: E. Wigner, F. Seitz, *On the constitution of metallic sodium*, Phys. Rev. **43**(10), 804–810 (1933); E. Wigner, F. Seitz, *On the constitution of metallic sodium. II*, Phys. Rev. **46**(6), 509–524 (1934).

Wigner–Seitz primitive cell displays the entire symmetry of the lattice. For a two-dimensional lattice constructed as illustrated in Figure W.2.

Draw lines to connect a given lattice point to all nearby lattice points and then at the midpoint and normal to these lines, draw new lines or planes. In three dimensions the cell is constructed by drawing perpendicular bisector planes in the reciprocal lattice from the chosen center to the nearest equivalent reciprocal lattice sites. In a crystal, all space may be filled by these cells.

First described in: E. Wigner, F. Seitz, *On the constitution of metallic sodium*, Phys. Rev. **43**(10), 804–810 (1933); E. Wigner, F. Seitz, *On the constitution of metallic sodium. II*, Phys. Rev. **46**(6), 509–524 (1934).

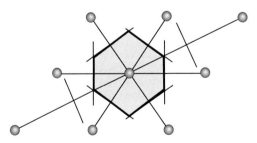

Figure W.2 Construction of the Wigner–Seitz cell for a two-dimensional oblique lattice.

Wigner–Seitz radius is equal to the ratio of the **Coulomb** energy per electron to the **Fermi** energy. It characterizes the strength of the Coulomb interaction between the charge carriers in solids.

First described in: E. Wigner, F. Seitz, *On the constitution of metallic sodium,* Phys. Rev. **43**(10), 804–810 (1933); E. Wigner, F. Seitz, *On the constitution of metallic sodium. II,* Phys. Rev. **46**(6), 509–524 (1934).

Wigner semicircle distribution – a probability function of the form $f(x) = (2/\pi R^2) \sqrt{R^2 - x^2}$, where R is the parameter and x is varied in the interval $[-R, R]$. If $x > R$ or $x < -R, f(x) = 0$. The graph of the function is a semicircle of radius R centered at $(0, 0)$. This distribution arises as the limiting distribution of eigenvalues of many random symmetric matrices as the size of the matrix approaches infinity.

Wilson–Sommerfeld quantization rule provides a method of quantizing action integrals of classical mechanics. A necessary condition is that each generalized coordinate q_k and its conjugate momentum p_k must be a periodic function of times. Then the action integral taken over one cycle of the motion is quantized: $\oint p_k dq_k = n_k h$.

First described in: W. Wilson, *On the quantum-theory of radiation and line spectra,* Phil. Mag. **29**, 795–802 (1915); A. Sommerfeld, *Zur Quantentheorie der Spektrallinien,* Ann. Phys. (Leipzig) **51**, 1–94, 125–167 (1916).

Wootters–Zurek theorem → **noncloning theorem**.

work function – an energy distance between the **Fermi level** in a solid and the electron energy level in a vacuum.

Wulff's construction → **Wulff's theorem**.

Wulff's theorem formulates the criterion determining the shape of a small crystal: the equilibrium shape of a crystal is that providing a minimum for its surface free energy for a given volume. Total free surface energy of a crystal with i faces is $U = \sum_i \gamma_i S_i$, where γ_i is the surface tension and S_i is the area of the i-th face. Thus, the crystal gets an equilibrium shape if the condition $dU = 0$ when $dV = 0$, where V is the volume of the crystal, is satisfied. It is called the **Curie–Wulff condition**.

The theorem links linear dimensions, surface area and the volume of crystals having the equilibrium shape. A geometrical construction based on this theorem is referred to as **Wulff's construction**. It uses the so-called **Herring's γ plot**, which is a polar diagram representing surface tension versus angular space orientation. The two-dimensional plot and construction procedure for an anisotropic solid are illustrated in Figure W.3.

At each point of this polar plot (gray curve) construct a plane (represented by the line in two dimensions) perpendicular to the radius vector at that point. The volume, which can be reached from the origin without crossing any of the planes,

Wyckoff notation

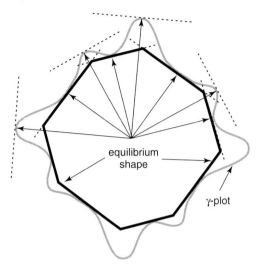

Figure W.3 Herring's γ plot and Wulff construction of the equilibrium shape of a crystal.

is geometrically similar to the ultimate equilibrium shape of the crystal, as it minimizes the total free surface energy of the crystal for a fixed volume.

While the proof of the theorem has been given only for the case in which the equilibrium shape is a polyhedron, it is easily generalized to apply to cases in which part or all of the equilibrium shape is bounded by smoothly curved surfaces. Also note that an equilibrium shape is of practical interest only for small crystals. For large crystals a significant alternation in shape can be achieved only by transporting a large number of atoms through a large distance, and the effort needed for this becomes very large in comparison with the minimization of the surface free energy.

First described in: G. Wulff, *Zur Frage der Geschwindigkeit des Wachstums und der Auflösung der Kristallflächen*, Z. Krist. Mineral. **34**, 449–530 (1901) and then extended in: C. Herring, *Some theorems on the free energies of crystal surfaces*, Phys. Rev. **82**(1), 87–93 (1951).

More details in: B. Mutaftschiev, *The Atomic Nature of Crystal Growth* (Springer, Berlin, 2001).

Wyckoff notation – a letter representing a set of symmetry-equivalent positions in a space group. A general point in a **unit cell** of a crystal belonging to any space group has coordinates x, y, z. However, if one or more of the coordinates has a specified value, the number of symmetry-equivalent positions is reduced. Most space groups have several sets of such points, and they are replaced by letters a, b, c, and so on, which are Wyckoff notation.

First described in: R. W. G. Wyckoff, *The Structure of Crystals* (New York, 1924) and *International Tables for X-ray Crystallography* (New York 1935).

From XMCD (X-ray magnetic circular dichroism) to XRD (X-ray diffraction)

XMCD – an acronym for **X-ray magnetic circular dichroism**.

XMLD – an acronym for **X-ray magnetic linear dichroism**.

XOR operator → logic operator.

XPS – an acronym for **X-ray photoelectron spectroscopy**.

X-ray – electromagnetic radiation with typical photon energies in the range 100 eV–100 keV.
   First described in: W. C. Röntgen, *Eine neue Art von Strahlen*, read before the Würzburg Physical and Medical Society in 1895.
   Recognition: in 1901 W. C. Röntgen received the first Nobel Prize in Physics in recognition of the extraordinary services he had rendered by the discovery of the remarkable rays subsequently named after him.
   See also www.nobel.se/physics/laureates/1901/index.html.

X-ray absorption near-edge structure (XANES) → extended X-ray absorption fine structure (EXAFS) spectroscopy

X-ray diffraction (XRD) – a technique to characterize the crystalline nature of a condensed matter. When a beam of monochromatic X-rays is directed at a solid material, diffraction of the X-rays can occur. The relationship between the wavelength of the X-rays (λ), the angle of diffraction (θ), and the distance between each set of atomic planes of the crystal lattice (d) is given by the Bragg equation: $n\lambda = 2d \sin \theta$, where $n = 1, 2, \ldots$ represents the order of diffraction. The interplanar distance is easily derived from this relationship. It is also possible to reconstruct the total molecular structure of a solid by an analysis of the intensity of the X-ray scattering from various crystalline planes.

What is What in the Nanoworld. Victor E. Borisenko and Stefano Ossini
Copyright © 2008 WILEY-VCH Verlag GmbH & Co. KGaA, Weinheim
ISBN: 978-3-527-40783-5

X-ray fluorescence spectroscopy

In the case of thin film analysis, XRD provides:

- estimation of lattice mismatch between the film and the substrate and related strain or stress in the thin film structure by precise lattice constant measurements derived from $2\theta - \theta$ scans

- study of the dislocation quality of the film by rocking curve measurements made by doing a θ scan at a fixed 2θ angle: the width of the rocking curve is inversely proportional to the dislocation density and is therefore used as a gauge of the quality of the film

- study of multilayered heteroepitaxial structures, which manifest themselves as satellite peaks surrounding the main diffraction peak: the heteroepitaxial film thickness and its quality can be deduced from the data

- determination of the thickness, roughness, and density of the film by glancing-incidence X-ray reflectivity measurements (the technique does not require a crystalline film and works even with amorphous materials).

First described in: W. Friedrich, P. Knipping, M. Laue, *Interferenz-Erscheinungen bei Röntgenstrahlen*, Sitzungsber. Bayer. Akad. Wiss. (Math. Phys. Klasse), 303–322 (1912).

Recognition: in 1914 M. Laue received the Nobel Prize in Physics for his discovery of the diffraction of Röntgen rays by crystals.

See also www.nobel.se/physics/laureates/1914/index.html.

X-ray fluorescence spectroscopy – a technique of nondestructive elemental analysis of condensed matter. When a material is bombarded by high-energy photons or electrons, some of the electrons in the atoms of the material are ejected. As the other electrons occupy the vacated levels, quanta of radiation, characteristic of the atom, are emitted. Such quanta in the form of X-rays can be detected by either a wavelength-dispersive or an energy-dispersive X-ray fluorescence spectrometer. Thus, particular atoms can be recognized. The technique makes it possible to detect and analyze all atoms down to boron with a lower limit of detection down to a few parts per million.

X-ray magnetic circular dichroism (XMCD) and X-ray magnetic linear dichroism (XMLD) consist in measuring the absorption cross section near a core level threshold as a function of the circular or linear **polarization** of the photons, respectively. The dependence of the absorption spectra on the direction of polarization is determined by the expectation values of odd-order (for magnetic circular dichroism) or even-order (for magnetic linear dichroism) local multipole magnetic moments of the ground state of the material under investigation. Even-order multipoles can have a non-zero value even in the absence of long-range magnetic order; thus magnetic circular dichroism can yield information about the magnetic structure even in **antiferromagnetic** systems, where no net magnetization is present. Both techniques can be employed with large externally applied magnetic fields and are characterized by the existence of sum rules allowing one to extract ground-state magnetic properties directly from the experimental spectra. The expectation value of the site and orbital projected orbital and spin magnetic moment can be obtained, in a large variety of cases, from the magnetic circular dichroism spectra, while the spin-orbit anisotropy can be estimated from the magnetic linear dichroism spectra. A drawback of magnetic linear dichroism is that dichroism effects in linearly polarized X-ray absorption spectroscopy may be determined not only by local exchange fields and magnetic moments, but also by the local crystal field symmetry, and it could be difficult to separate the magnetic from the non-magnetic contribution to the spectra. Moreover, in magnetic metallic systems the linear signal is much lower (about one order of magnitude) than the circular one.

First described in: B. T. Thole, G. van der Laan, M. Fabrizio, *Magnetic ground-state properties and spectral distributions. I. X-ray-absorption spectra*, Phys. Rev. B **50**(16), 11466–11473 (1994); B. T. Thole, G. van der Laan, *Magnetic ground-state properties and spectral distributions. II. Polarized photoemission*, Phys. Rev. B **50**(16), 11474–11483 (1994).

More details in: M. Finazzi, L. Duó, F. Ciccacci, *Magnetic properties of interfaces and multilayers based on thin antiferromagnetic oxide films*, Surface Sci. Rep. **62**(9), 337–371 (2007).

X-ray photoelectron spectroscopy (XPS) – the technique for studying composition and electronic states in solids using photo-ionization and energy dispersive analysis of photoelectrons emitted under X-ray illumination of the sample. It is based upon single photon in/electron out processes. The energy of the X-ray photons ranges from 100 eV to 10 eV.

In XPS the photon is absorbed by an atom in a molecule or solid, leading to ionization and the emission of a core (inner shell) electron. In contrast to **ultraviolet photoelectron spectroscopy** (UPS), the photon interacts with valence electrons leading to ionization by their removal from atoms or molecules. The kinetic energy distribution of the emitted photoelectrons can be measured using any appropriate electron-energy analyzer, and a photoelectron spectrum can thus be recorded.

Photoionization of an atom A can be considered as $A + h\nu \rightarrow A^+ + e^-$. Energy conservation requires $E(A) + h\nu = E(A^+) + E(e^-)$. Since the energy of the photoelectron (e^-) is present solely as kinetic energy (E_{kin}) this can be rearranged to give the following expression for it: $E_{kin} = h\nu - (E(A^+) - E(A))$. The final term in brackets, representing the difference in energy between the ionized and neutral atoms, is generally called the binding energy (E_{bin}) of the electron. This then leads to the following commonly quoted equation: $E_{kin} = h\nu - E_{bin}$. Note that the binding energies in solids are conventionally measured with respect to the **Fermi level** rather than the vacuum level. This involves a small correction to the equation given above in order to account for the **work function** of the solid.

Each chemical element has a characteristic binding energy associated with each core atomic orbital, that is, each element will give rise to a characteristic set of peaks in the photoelectron spectrum at kinetic energies determined by the photon energy and the respective binding energies. The presence of peaks at particular energies therefore indicates the presence of a specific element in the sample. Furthermore, the intensity of the peaks is related to the concentration of the element within the sampled region. Thus, the technique is capable of yielding a quantitative analysis and is sometimes known by the alternative acronym, **ESCA** that is **electron spectroscopy for chemical analysis**. Its sensitivity is about 1 %, and the depth resolution ranges from 1 to 10 nm.

Recognition: in 1981 K. M. Siegbahn received the Nobel Prize in Physics for his contribution to the development of high-resolution electron spectroscopy.

See also www.nobel.se/physics/laureates/1981/index.html.

XRD – an acronym for a **X-ray diffraction** analysis.

Y

From Yasukawa potential to Yukawa potential

Yasukawa potential – a development of the **Tersoff potential** taking into account the effects of charge transfer in the materials containing ions of different species. For the interatomic distance r_{ij} between atoms i and j charged to q_i and q_j respectively, it is $V(r_{ij}) = \sum_i V_i = \sum_i [E_i^S + 0.5 \sum_{i \neq j} V_{ij}(r_{ij}, q_i, q_j)]$. Here V_i is the potential for the i-th atom. $E_i^S(q_i) = \chi_i q_i + 0.5 J_i q_i^2$ is the self-energy term for the i-th atom with χ_i being its electronegativity (a purely atomic, environment independent quantity), and J_i being the self-Coulomb repulsion (atomic hardness). The interaction potential energy V_{ij} consists of four components: the repulsive energy, the short-range attractive energy, the ionic bond energy, and the van der Waals energy.

The dynamic charge transfer intrinsic to the potential function allows the interface properties to be captured automatically,

First described in: A. Yasukawa, *Using an extended Tersoff interatomic potential to analyze the static-fatigue strength of SiO$_2$ under atmospheric influence*, JSME Int. J., Ser. A **39**(3), 313–320 (1996).

More details in: J. Yu, S. B. Sinnott, S. R. Phillpot, *Charge optimized many-body potential for the Si/SiO$_2$ system*, Phys. Rev. B **75**(8), 085311 (2007).

Young's modulus – the ratio of a simple tension stress applied to a material to the resulting strain parallel to the tension.

Yukawa potential describes the strong short-range interaction between **nucleons**: $V(r) = (A/r) \exp(-r/b)$, where r is the distance, and A and b are constants giving measures of the strength and range of the force, respectively.

First described in: H. Yukawa, *Interaction of elementary particles*, Proc. Phys. Math. Soc. Japan **17**, 48–57 (1935).

Recognition: in 1949 H. Yukawa received the Nobel Prize in Physics for his prediction of the existence of mesons on the basis of theoretical work on nuclear forces.

See also www.nobel.se/physics/laureates/1949/index.html.

What is What in the Nanoworld. Victor E. Borisenko and Stefano Ossicini
Copyright © 2008 WILEY-VCH Verlag GmbH & Co. KGaA, Weinheim
ISBN: 978-3-527-40783-5

Z

From Zeeman effect to Zundel ion

Zeeman effect – splitting of energy levels of electrons in an external magnetic field related to different spins of electrons. It is characterized by the so-called Zeeman term $g\mu_B B$ with the free-electron spin **g-factor** of about 2, where B is the induction of the external magnetic field applied.

The normal Zeeman effect is the observation of three lines in the emission spectrum where, in the absence of the field, there was only one. In the **anomalous Zeeman effect** the original line splits into more than three components. It relates to the anomalous magnetic moment of electron spin, and when spin is present the spin and orbital magnetic moments interact with the field in a more complicated way than in its absence.

First described in: P. Zeeman, Verslag. Koninkl. Akad. Wet. Amsterdam **5**, 181–242 (1897); P. Zeeman, *On the influence of magnetism on the nature of the light emitted by a substance*, Phil. Mag. **43**, 226 (1897); P. Zeeman, *The effect of magnetisation on the nature of light emitted by a substance*, Nature **55**, 347 (1897).

Recognition: in 1902 P. Zeeman received the Nobel Prize in Physics shared with H. A. Lorentz in recognition of the extraordinary service they rendered by their researches into the influence of magnetism upon radiation phenomena.

See also www.nobel.se/physics/laureates/1902/index.html.

Zener double exchange occurs when the two isolated magnetic atoms have a different number of electrons in the magnetic shell and hopping through the intermediate nonmagnetic atom involves magnetic-shell electrons. The double exchange couples magnetic moments ferromagnetically. Parallel spin alignment is favored because it increases the hopping probability and therefore decreases the kinetic energy of spin-polarized electrons. An example of double exchange is $A^{III}B^V$ semiconductors doped with Mn, in which Mn acceptor states form an impurity band with a mixed spd character. In these systems electrical conduction and Mn–Mn exchange coupling are both realized through hopping within an impurity band.

First described in: C. Zener, *Interaction between the d-shells in the transition metals. II. Ferromagnetic compounds of manganese with perovskite structure*, Phys. Rev. **82**(3), 403–405 (1951).

What is What in the Nanoworld. Victor E. Borisenko and Stefano Ossini
Copyright © 2008 WILEY-VCH Verlag GmbH & Co. KGaA, Weinheim
ISBN: 978-3-527-40783-5

Zener exchange

Zener effect – interband tunneling, in which electrons tunnel from one band to another through the forbidden energy gap of the solid. The presence of an external electric field in a periodic lattice potential causes transitions between the allowed energy bands in the crystal.

First described in: C. Zener, *A theory of the electrical breakdown of solid dielectrics,* Proc. R. Soc. A **145**, 523–529 (1934).

Zener exchange or indirect exchange interaction assumes that the interaction between the incomplete d shells of transition elements is insufficient to disrupt the coupling between the d electrons in the same shells, and that the exchange interaction between adjacent d shells always has the same sign irrespective of distance of separation. The direct interaction between adjacent d shells then invariably leads to a tendency for an **antiferromagnetic** alignment of d spins.

The spin coupling between the incomplete d shells and the conduction electrons leads to a tendency for a ferromagnetic alignment of d spins. The occurrence of ferromagnetism and antiferromagnetism in the body-centered cubic structure of the transition metals, as well as more complex lattices of certain alloys, is readily interpreted in a straightforward manner.

First described in: C. Zener, *Interaction between the d shells in the transition metals,* Phys. Rev. **81**(3), 440–444 (1951).

Zeno paradox – motion is impossible because before reaching a goal one has to get halfway there. Before that, one had to get halfway to the midpoint, and so on

zeolite – a crystalline Al–O–Si material with regularly arranged cages of about 1 nm size. It contains some very closely held water, which can be expelled by heat and reabsoarbed from a moist atmosphere without affecting the crystalline structure of the mineral.

zepto- – a decimal prefix representing 10^{-21}, abbreviated z.

zinc blende structure can be considered as the **diamond structure** in which two interpenetrating **face-centered cubic** lattices contain different atoms. The compound ZnS gives its name to this type of lattice. Other materials crystallizing with the same lattice are $A^{III}B^{V}$ compounds.

zitterbewegung – a theoretical helical or circular motion of elementary particles, in particular **electrons**, which is responsible for producing their **spin** and magnetic moment. It comes as a result of the analysis of the **wave packet** solutions of the **Dirac equation** for **relativistic** electrons in free space. An **interference** between positive and negative energy states produces what appears to be a fluctuation (at the speed of light) of the position of an electron around the median, with a circular frequency of $2mc^2/\hbar$, or approximately 1.6×10^{21} Hz. This peculiar oscillation happens because the velocity is not a constant of motion even in the absence of any potential and has a fluctuating component, while the momentum commutes with the Hamiltonian.

First described in: E. Schrödinger, *Über die kräftefreie Bewegung in der relativistischen Quantenmechanik*, Berliner Ber., pp. 418–428 (1930); *Zur Quantendynamik des Elektrons*, Berliner Ber., 63–72 (1931).

More details in: J. J. Sakurai, *Advanced Quantum Mechanics* (Addison Wesley, Reading, 1967).

zone law of Weiss states that the condition for a crystal face (*hkl*) to lie in the zone [*UVW*] is $Uh + Vk + Wl = 0$.

First described in: C. S. Weiss, *De Indagando Formarum Crystallinarum Charactere Geometrico Principali* (Tauchnitz, Leipzig 1809).

Zundel ion – the smallest **protonated water cluster** $H_5O_2^+$.

First described in: G. Zundel, H. Metzger, *Energiebänder der tunnelnden Überschuss Protonen in flüssigen Säuren - eine IR-spektroskopische Untersuchung der Natur der Gruppierungen* $H_5O_2^+$, Z. Phys. Chem. **58**, 225–245 (1968).

A list and a presentation of Scientific Journals which contain the stem *Nano* in their title

One measure of the emergence of new fields of science and technology is the growth of the literature dedicated to the new disciplines. Nanoscience and nanotechnology have, in the last years, witnessed not only an impressive growth in the number of relevant classical scientific journals which have devoted, more and more, a consistent part of their published papers to "nano"-related researches, but also in the number of entirely new journals, which contain the stem "nano" in their title. Here a list and a short presentation of these journals is presented. Inevitably, we will have omitted and forgotten some title, and new titles will appear in the coming years, confirming the broad and interdisciplinary nature of nanoscience and nanotechnology.

More details in: for the role of Journals in defining a new research field see C. G. Mainley, T. S. Welsh, *Publishing patterns and core journals of the "Nano" research front*, Proceedings of the SBA Conference 2005; L. Leydesdorff, P. Zhou, *Nanotechnology as a field of science: Its delineation in terms of journals and patents*, Scientometrics **70**, 693–713 (2007).

ACS Nano is an international forum for comprehensive articles on synthesis, assembly, characterization, theory, and simulation of nanostructures, nanobiotechnology, nanofabrication, methods and tools for nanoscience and nanotechnology, and self- and directed-assembly. The journal serves researchers at the interfaces of chemistry, biology, materials science, physics, and engineering.
Started in 2007, and published in 5 issues per year.
More details in: //pubs.acs.org.

Current Nanoscience publishes authoritative reviews and original research reports, written by experts in the field on all the most recent advances in nanoscience and nanotechnology. All aspects of the field are represented including nano- structures, synthesis, properties, assembly and devices. Applications of nanoscience in biotechnology, medicine, pharmaceuticals, physics, material science and electronics are also covered.
Started in 2005, and published in 4 issues per year.
More details in: www.bentham.org.

What is What in the Nanoworld. Victor E. Borisenko and Stefano Ossini
Copyright © 2008 WILEY-VCH Verlag GmbH & Co. KGaA, Weinheim
ISBN: 978-3-527-40783-5

Fullerenes, Nanotubes, and Carbon Nanostructures – an international and interdisciplinary forum aimed at publishing peer-reviewed research of original work in all areas of fullerene research. Publishing high quality papers from all fields of scientific inquiry related to fullerenes, the journal intends to provide a means of communication between researchers who are interested in fundamental and applied fullerene science issues.

Started in 1993, and published in 6 issues per year.

More details in: www.tandf.co.uk.

IEEE Proceedings Nanobiotechnology covers all aspects of research and emerging technologies related to the convergence of nanotechnology with biotechnology, especially those aspects relating to the interactions between biomolecules and biomolecular assemblies with electronic structures or materials.

Started in 2003, and published in 6 issues per year.

More details in: //ieeexplore.ieee.org.

IEEE Transactions on NanoBioscience – this transaction reports on original, innovative and interdisciplinary work on all aspects of molecular systems, cellular systems, and tissues - including molecular electronics. The journal accepts basic and applied papers dealing both with engineering, physics, chemistry and computer science and with biology and medicine with respect to bio-molecules and cells.

Started in 2002, and published in 4 issues per year.

More details in: //ieeexplore.ieee.org.

IEEE Transactions on Nanotechnology – this transaction publishes novel and important results in engineering at the nanoscale. It focuses on nanoscale devices, systems, materials and applications, and on their underlying science. It is an interdisciplinary journal that covers all areas of nanotechnology.

Started in 2002, and published in 6 issues per year.

More details in: //ieeexplore.ieee.org.

IET Nanobiotechnology covers all aspects of research and emerging technologies related to the convergence of nanotechnology with biotechnology, especially those aspects relating to the interactions between biomolecules and biomolecular assemblies with electronic structures or materials.

Started in 2007, and published in 6 issues per year.

More details in: //ieeexplore.ieee.org.

International Journal of Nano and Biomaterials aims at promoting and coordinating developments in the field of nano and biomaterials. The international dimension is emphasized in order to meet the latest needs of accelerating theoretical knowledge and practical verification of new technologies and products.

Started in 2008, and published in 4 issues per year.

More details in: www.inderscience.com.

International Journal of Nanomanufacturing – the objectives of this journal are to establish an effective channel of communication on the latest developments in micro- and nanomanufacturing science, processes, technology and applications between engineers, scientists, industry, policy makers, academic and research institutions, and persons concerned with any aspect of nanomanufacturing. The interdisciplinary nature of nanomanufacturing will serve to break down cultural, international and academic barriers to meet the needs of accelerating technological developments in the international community for a global change.

Started in 2006, and published in 6 issues per year.

More details in: www.inderscience.com.

International Journal of Nanomedicine – an international, peer-reviewed journal focusing on the application of nanotechnology in diagnostics, therapeutics, and drug delivery systems throughout the biomedical field. Reflecting the growing activity in this emerging specialty, the aim of this journal is to highlight research and development leading to potential clinical applications in the prevention and treatment of disease.

Started in 2006, and published in 4 issues per year.

More details in: //dovepress.com.

International Journal of Nanoparticles focuses on the chemical, physical and biological phenomena and processes associated with nanoparticles and nanostructures ranging from molecular size to 100 nm which exhibit improved properties or have novel applications due to their size. It will include aspects of submicron nanoparticles in special cases.

Started in 2008, and published in 4 issues per year.

More details in: www.inderscience.com.

International Journal of Nanoscience – this inter-disciplinary, internationally reviewed research journal covers all aspects of nanometer scale science and technology. Articles in any contemporary topical areas are sought, from basic science of nanoscale physics and chemistry to applications in nanodevices, quantum engineering and quantum computing.

Started in 2002, and published in 6 issues per year.

More details in: www.worldscinet.com.

International Journal of Nanoscience and Nanotechnology is published by the Iranian Nanotechnology Society and is devoted to original scientific papers in the broad area of nanotechnology.

Started in 2007, and published in 4 issues per year.

More details in: www.nanosociety.ir.

International Journal of Nanotechnology offers a multidisciplinary source of information in all subjects and topics related to Nanotechnology, with fundamental, technological, as well as societal and educational perspectives.

Started in 2004, and published in 12 issues per year.

More details in: www.inderscience.com.

Internet Journal of Nanotechnology is devoted to the electronic publication of original high-quality medical articles on the Internet to distribute knowledge and research results.

Started in 2007, and published in 2 issues per year.

More details in: www.ispub.com.

Journal of Biomedical Nanotechnology – nanotechnology, dealing with functional structures and materials smaller than 100 nm, is emerging as a truly interdisciplinary research area spanning several traditional scientific disciplines. In keeping with the growing trend, there is a strong need for a platform to share original research related to applications of nanotechnology in biomedical fields. In order to fulfill this demand, this journal is being created as an international peer-reviewed periodical that covers applications of nanotechnology in all fields of life sciences.

Started in 2005, and published in 14 issues per year.

More details in: www.asppbs.com.

Journal of Bionanoscience – bionanoscience attempts to harness various functions of biological macromolecules and integrate them with engineering for technological applications. It is based on a bottom-up approach and encompasses structural biology, biomacromolecular engineering, material science, and engineering, extending the horizon of material science. The journal aims at publication of Letters, Reviews, Concepts, Rapid communications, Research papers, Book reviews, and Conference announcements in the interface between chemistry, physics, biology, material science, and technology.

Started in 2007.

More details in: www.asppbs.com.

Journal of Computational and Theoretical Nanoscience is an international peer-reviewed journal with a wide-ranging coverage, and consolidates research activities in all aspects of computational and theoretical nanoscience into a single reference source. This journal offers scientists and engineers peer-reviewed research papers in all aspects of computational and theoretical nanoscience and nanotechnology in chemistry, physics, materials science, engineering and biology.

Started in 2005, and published in 8 issues per year.

More details in: www.asppbs.com.

Journal of Experimental Nanoscience – international and multidisciplinary, this journal provides a showcase for advances in the experimental sciences underlying nanotechnology and nanomaterials. The journal exists to bring together the most significant papers making original contributions to nanoscience in a range of fields including biology and biochemistry, physics, chemistry, chemical, electrical and mechanical engineering, materials, pharmaceuticals and medicine. The aim is to provide a forum in which cross fertilization between application areas, methodologies, disciplines, as well as academic and industrial researchers can take place and new developments can be encouraged.

Started in 2006, and published in 4 issues per year.

More details in: www.tandf.co.uk.

Journal of Geoethical Nanotechnology explores economic, political, social and technological aspects of the developing technology of molecular nanotechnology. The journal invites submissions on controversial topics such as self-replication and governance.

Started in 2008, and published in 2 issues per year.

More details in: www.terasemjournal.org.

Journal of Laser Micro/Nanoengineering is an international online journal for the rapid publication of experimental and theoretical investigations in laser-based technology for micro- and nano-engineering. Access to the full article is provided free of charge.

Started in 2006, and published in 3 issues per year.

More details in: www.jlps.gr.jp.

Journal of Metastable and Nanocrystalline Materials is devoted to the rapid dissemination of recent high-quality research results concerning nanocrystalline, nanocomposite and metastable materials.

More details in: www.worldscinet.com.

Journal of Micro and Nano Mechatronics is devoted to the theory and applications of micro and nanotechnologies. The Journal promotes both theoretical and practical engineering research based on the analysis and synthesis from the micro level to the nano-level. The journal includes survey papers, technical papers, short papers and news items in the field of Micro and Nano Mechatronics.

Started in 2008, and published in 4 issues per year.

More details in: www.springerlink.com.

Journal of Micro/Nano Lithography, MEMS, and MOEMS – the Journal of Microlithography, Microfabrication, and Microsystems changed its title to the Journal of Micro/Nanolithography, MEMS, and MOEMS in January 2007.

Started in 2007, and published in 4 issues per year.

More details in: //spiedigitallibrary.aip.org.

Journal of Nano and Microsystem Technique – the task of the journal is to illuminate the modern status, perspectives and development trends of microsystem engineering, to consider the problems of development and embedding microsystems in

different areas of science, technology and production.
Started in 1999.
More details in: www.microsystems.ru.

Journal of Nanobiotechnology is devoted to the publication of research on all aspects of scientific and technological advances in the fields of medical, biological and nanoscale sciences.
Started in 2003, and published in 10 issues per year.
More details in: www.jnanobiotechnology.com.

Journal of Nano Education is a peer-reviewed international journal that aims to provide the most complete and reliable source of information on current developments in nanoscale science, technology, engineering, and medical education.
Started in 2008.
More details in: www.asppbs.com.

Journal of Nanoengineering and Nanosystems is dedicated to the particular aspects of nanoscale engineering, science, and technology that involve descriptions of nanoscale systems. This journal reflects the work on real devices and processes that arises from transitioning the scientific and engineering database.
Started in 2006, and published in 4 issues per year.
More details in: www.pepublishing.com.

Journal of Nanoelectronics and Optoelectronics is an international and cross-disciplinary peer-reviewed journal to consolidate emerging experimental and theoretical research activities in the areas of nanoscale electronic and optoelectronic materials and devices into a single and unique reference source. The aims are to facilitate the dissemination of interdisciplinary research results in the inter-related and converging fields of nanoelectronics and optoelectronics.
Started in 2006, and published in 3 issues per year.
More details in: www.asppbs.com.

Journal of Nanomaterials – the overall aim of this journal is to bring science and applications together on nanoscale and nanostructured materials with emphasis on synthesis, processing, characterization, and applications of materials containing true nanosize dimensions or nanostructures that enable novel/enhanced properties or functions. It is directed at both academic researchers and practicing engineers. Journal of Nanomaterials will highlight the continued growth and new challenges in nanomaterials science, engineering, and nanotechnology, both for application development and for basic research. All papers should emphasize original results relating to experimental, theoretical, computational, and/or applications of nanomaterials ranging from hard (inorganic) materials, through soft (polymeric and biological) materials, to hybrid materials or nanocomposites.
Started in 2006, and published in 1 issue per year.
More details in: www.hindawi.com.

Journal of Nanoneuroscience – advances in nanomaterials indicate that the central nervous system is susceptible to nanoparticle-induced alterations leading to functional or structural alterations. This knowledge is currently disseminated in vast array of journals dealing with broad subject areas related to pharmacology, toxicology, neuroscience or nanosciences.

Started 2008.

More details in: www.asppbs.com.

Journal of Nanoparticle Research – the objective of this journal is to disseminate knowledge of the physical, chemical and biological phenomena and processes in structures that have at least one lengthscale ranging from molecular to approximately 100 nm (or submicron in some situations), and exhibit improved and novel properties that are a direct result of their small size.

Started in 1999, and published in 6 issues per year.

More details in: www.springerlink.com.

Journal of Nanophotonics is an online-only journal and focuses on the fabrication and application of nanostructures that facilitate the generation, propagation, manipulation, and detection of light from the infrared to the ultraviolet regimes.

Started in 2007.

Journal of Nano Research is a multidisciplinary peer-reviewed journal, which publishes high quality work on all aspects of nanoscience and nanotechnology. This journal serves the global "nano" community in providing up-to-date information on all developments and progress being made in nanoscience and nanotechnology and the future predictions for this extraordinary technology.

Started in 2007.

More details in: www.scientific.net.

Journal of Nanoscience and Nanotechnology is an international and multidisciplinary peer-reviewed journal with a wide-ranging coverage, consolidating research activities in all areas of nanoscience and nanotechnology into a single and unique reference source. It is devoted to the publication of original full research articles, rapid communications of important new scientific and technological findings, timely state-of-the-art reviews with author's photo and short biography, and current research news encompassing the fundamental and applied research in all disciplines of science, engineering and medicine.

Started in 2001, and published in 10 issues per year.

More details in: www.asppbs.com.

Micro & Nano Letters is devoted to fundamental and applied research conducted at the forefront of micro- and nanoscale science and technology. The journal specializes in the express publication of short research papers concentrating on the latest advances in the science, engineering, technology and applications of miniature and ultraminiature structures and systems that have at least one dimension ranging from a few tens of micrometres to a few nanometres. This journal offers a rapid route for international dissemination of high quality research findings generated by researchers from both the micro and nano communities. Contributions addressing issues ranging from fundamental understanding to experimental demonstrations in the disciplines of engineering, physical, chemical, biological and material science should be submitted, with particular interest placed on manuscripts addressing cross-disciplinary activities and applications.

Started in 2006, and published in 4 issues per year.

More details in: www.ietdl.org.

Microfluidics and Nanofluidics is an international peer-reviewed journal that aims to publish papers in all aspects of microfluidics, nanofluidics and lab-on-a-chip science and technology. The journal broadly interprets microfluidics and nanofluidics as the studies of mass (including molecular and colloidal) and momentum transfer, heat transfer, and reactive processes coupled with transport in microscale and nanoscale systems.

Started in 2005, and published in 6 issues per year.

More details in: www.springerlink.com.

Microsystem Technologies. Micro- and Nanosystems Information Storage and Processing Systems was born on 1 January 2002, from the merging of Journal of Information Storage and Processing Systems and Microsystem Technologies. It is intended to provide rapid publication of important and timely results on electromechanical, materials science, design, and manufacturing issues of these systems and their components.

Started in 2002, and published in 6 issues per year.

More details in: www.springerlink.com.

Nano is an international peer-reviewed journal for nanoscience and nanotechnology that presents papers at the forefront of fundamental research and new emerging topics. It features timely scientific reports of new results and technical breakthroughs and also contains interesting review articles about recent hot issues.

Started in 2006, and published in 6 issues per year.

More details in: www.worldscinet.com.

NanoBiotechnology is an international peer-reviewed journal that publishes research papers covering all aspects of the emerging and rapidly growing research area at the intersection of nanotechnology, molecular biology and biomedical sciences. This journal serves as an authoritative forum presenting state-of-the-art, multidisciplinary research and technological advances in theory, instrumentation and methods as well as applications in a variety of areas of nanotechnology related to biology and medicine. The convergence of nanotechnology and biomedical sciences opens up a wide variety of biological research topics and medical uses at the molecular and cellular level.

Started in 2005, and published in 4 issues per year.
More details in: www.springerlink.com.

NanoBiotechnology Nanotechnology is a peer-reviewed scientific journal that provides a multidisciplinary forum for state-of-the-art nanobiotechnological methods, instrumentation, and research.

Started in 2005, and published in 4 issues per year.
More details in: www.humanapress.com.

Nanoethics – nanoscale technologies are currently surrounded by both hype and fear. Optimists suggest that these technologies are desperately needed to solve problems of terrorism, global warming, the shortage of clean water, land degradation and health. Pessimists fear the loss of privacy and loss of autonomy, "grey goo" and weapons of mass destruction, and serious environmental and health risks. There is also concern about the fair distribution of the costs and benefits of the technology. The field needs a forum for informed discussion of ethical and social issues related to nanotechnology to counterbalance popular discussions and fragmented representation of opinions and findings.

Started in 2007, and published in 3 issues per year.
More details in: www.springerlink.com.

Nano Letters – the leading forum for rapid communication of nanoscale research in a range of disciplines, from physical and materials chemistry to biotechnology and applied physics.

Started in 2001, and published in 12 issues per year.
More details in: www.acs.com.

Nanomedicine addresses the important challenges and advances in medical nanoscale-structured materials and devices, biotechnology devices and molecular machine systems and nanorobotics, delivering this essential information in concise, clear and attractive article formats, a valuable information source for all players in the field and across the scientific community.

Started in 2006, and published in 6 issues per year.
More details in: www.futuremedicine.com.

Nanomedicine: Nanotechnology, Biology, and Medicine is an international, peer-reviewed journal that delivers the latest advances in the innovative field of nano-scale medicine. The journal publishes articles on basic, diagnostic, experimental, clinical, engineering, pharmacologic and toxicologic nanomedicine. In addition, regular features address the commercialization of new technologies, ethics in the field, research funding, and other topics of interest to researchers and clinicians.

Started in 2005, and published in 4 issues per year.
More details in: www.sciencedirect.com.

Nanoscale and Microscale Thermophysical Engineering

Nanoscale and Microscale Thermophysical Engineering is a peer-reviewed journal covering the basic science and engineering of nanoscale and microscale energy conversion, transport, storage, mass transport, and reactions. In addition, the journal addresses their integration and uses in devices and systems in the fields of medicine, transportation, energy, environment, and information.

Started in 1997, and published in 4 issues per year.

More details in: www.tandf.co.uk.

Nanoscale Research Letters – as an open access journal, it is the first nanotechnology journal from a major commercial publisher to make all research results available to anyone with an internet connection. This Journal provides an interdisciplinary forum for open communication of scientific and technological advances in the creation and use of objects at the nanometer scale. The journal crosses all disciplines, and emphasizes research that seeks to uncover the underlying science and behavior of nanostructures. Nano Express are original research papers which publish approximately eight weeks following submission. Nano Reviews, Nano Ideas, and Nano Commentaries are also welcomed.

Started in 2006, and published in 12 issues per year.

More details in: www.springerlink.com.

Nano Science & Nano Technology is an Indian journal devoted to the rapid publication of fundamental research papers on all branches of the theory and practice of nanoscience and nanotechnology. It encompasses the understanding of the fundamental physics, chemistry, biology and technology of nanometer-scale objects.

Started in 2007, and published in 2 issues per year.

More details in: www.tsijournals.com.

Nanotechnology aims to publish papers at the forefront of nanoscale science and technology and especially those of an interdisciplinary nature. Here, nanotechnology is taken to include the ability to individually address, control, and modify structures, materials and devices with nanometer precision, and the synthesis of such structures into systems of micro- and macroscopic dimensions such as MEMS-based devices. It encompasses the understanding of the fundamental physics, chemistry, biology and technology of nanometer-scale objects and how such objects can be used in the areas of computation, sensors, nanostructured materials and nano-biotechnology.

Started in 1990, and published weekly.

More details in: www.iop.org.

Nanotechnology Law & Business is the source of information for professionals involved in micro- and nanotechnologies. This journal is devoted to the institutional investor seeking to understand how nanotechnology will impact on certain markets, to lawyers mapping the patent landscape surrounding a particular technology, or to companies navigating regulatory issues.

Started in 2004, and published in 4 issues per year.
More details in: www.nanolabweb.com.

Nanotechnology Perceptions is published to provide a stimulating forum for the exchange of ideas and information concerning nanotechnology (including bio-nanotechnology and nanomedicine) and ultraprecision engineering. Each issue contains essays on critical issues of the technologies themselves, as well as incisive appraisals of their impact on human life, society and its institutions. Articles are intended for a worldwide readership, and should be of especial interest to leaders in industry, commerce (including bankers and investors), the professions, politics and public administration, and all others who must make decisions based on a sober, impartial assessment of the potential and risks of nanotechnology in their areas of work.

Started in 2005, and published in 3 issues per year.
More details in: //pages.unibas.ch.

Nano Today is an international magazine for researchers with interests across the whole of nanoscience and technology. Through its unique mixture of peer-reviewed articles, the latest research news, and information on key developments, this journal provides comprehensive coverage of this exciting and dynamic new field.

Started in 2006, and published in 6 issues per year.
More details in: www.sciencedirect.com.

Nanotoxicology invites contributions addressing research relating to the potential for human and environmental exposure, hazard and risk associated with the use and development of nano-structured materials. In this context, the term nano-structured materials has a broad definition, including materials with at least one dimension in the nanometer size range. These nanomaterials range from nanoparticles and nanomedicines, to nano-surfaces of larger materials and composite materials. The range of nanomaterials in use and under development is extremely diverse, so this journal includes a range of materials generated for purposeful delivery into the body (food, medicines, diagnostics and prosthetics), to consumer products (for example, paints, cosmetics, electronics and clothing), and particles designed for environmental applications (such as remediation). It is the nano-size range of these materials which unifies them and defines the scope of of the journal. While the term "toxicology" indicates risk, the journal also aims to encompass studies that enhance safety during the production, use and disposal of nanomaterials. Well-controlled studies demonstrating a lack of exposure, hazard or risk associated with nanomaterials, or studies aiming to improve biocompatibility are welcomed and encouraged, as such studies will lead to an advancement of nanotechnology. Furthermore, many nanoparticles are developed with the intention to improve human health (for example, antimicrobial agents), and again, such articles are encouraged.

Started in 2007, and published in 4 issues per year.
More details in: www.tandf.com.

Nanotrends Journal

Nanotrends Journal is devoted to the publication of basic and applied papers dealing with latest trends in Nanotechnology and other scientific applications in the nanometer range.
Started in 2006, and published in 6 issues per year.
More details in: www.nstc.in.

Nature Nanotechnology is a research journal dedicated to all aspects of nanoscience and nanotechnology.
Started in 2006, and published in 12 issues per year.
More details in: www.nature.com.

Recent Patents on Nanotechnology publishes review articles by experts on recent patents on nanotechnology. A selection of important and recent patents on nanotechnology is also included in the journal. The journal is essential reading for all researchers involved in nanotechnology.
Started in 2007, and published in 3 issues per year.
More details in: www.bentham.org.

Open Nanoscience Journal is an Open Access online journal, which publishes research articles, reviews, and letters in the field of nanoscience and nanotechnology, encompassing biotechnology, medicine, pharmaceuticals, chemistry, physics, biology, materials science and electronics. This journal, a peer-reviewed journal, aims to provide the most complete and reliable source of information on current developments in the field. The emphasis will be on publishing quality papers rapidly and freely available to researchers worldwide.
Started in 2007.
More details in: www.bentham.org.

Photonics and Nanostructures: Fundamentals and Applications establishes a dedicated channel for physicists, material scientists, chemists, engineers and computer scientists who are interested in photonic crystals and photonic band gaps. The journal will help shed light on the latest developments in this growing field of science that will see the emergence of faster telecommunications and ultimately computers that use light instead of electrons to connect components. This journal features mainly original research work in experiment, theory and applications.
Started in 2003, and published in 6 issues per year.
More details in: www.sciencedirect.com.

Physica E (Low-dimensional systems and nanostructures) contains papers and review articles on the fundamental and applied aspects of physics in low-dimensional systems, including semiconductor heterostructures, mesoscopic systems, quantum wells and superlattices, two-dimensional electron systems, and quantum wires and dots. Both theoretical and experimental contributions are invited. Topics suitable for publication in this journal include optical and transport properties, many-body effects, integer and fractional quantum Hall effects, single-electron effects and devices, and novel phenomena.

Started in 1998, and published in 3 issues per year.

More details in: www.sciencedirect.com.

Precision Engineering–Journal of the International Societies for Precision Engineering and Nanotechnology is devoted to the multidisciplinary study and practice of high accuracy engineering, metrology, and manufacturing. The journal takes an integrated approach to all subjects related to research, design, manufacture, performance validation, and application of high precision machines, instruments, and components, including fundamental and applied research and development in manufacturing processes, fabrication technology, and advanced measurement science. The scope includes precision-engineered systems and supporting metrology over the full range of length scales, from atom-based nanotechnology and advanced lithographic technology to large-scale systems, including optical and radio telescopes and macrometrology. Precision Engineering was first published in January 1979; from January 2000, it assumes a new look, as the Journal of the International Societies for Precision Engineering and Nanotechnology.

Started in 2000, and published in 4 issues per year.

More details in: www.sciencedirect.com.

Synthesis and Reactivity in Inorganic, Metal-Organic, and Nano-Metal Chemistry is devoted to the rapid dissemination of original research papers of relevance to inorganic and metal-organic chemists and nano scientists engaged in synthesis. This journal publishes syntheses and reactivity of new compounds, new methods for preparation of known compounds, and new experimental techniques and procedures. The journal is a publication aimed at the rapid dissemination of original research papers of particular importance to inorganic and metal-organic chemists and nanotechnology professionals. This journal replaced in 2005 the journal Synthesis and Reactivity in Inorganic and Metal-Organic Chemistry. This new title reflects the rapidly growing fields of nanoscience and nanotechnology.

Started in 2005, and published in 10 issues per year.

More details in: www.tandf.com.

Virtual Journal of Nanoscale Science & Technology – this weekly virtual journal contains articles that have appeared in one of the participating source journals and that fall within a number of contemporary topical areas in the science and technology of nanometer-scale structures. The articles are primarily those that have been published in the previous week.

Started in 2000, and published weekly.

More details in: http://publish.aps.org.

Abbreviations for the scientific journals which appear as sources in the text

Abhand. Köngl. Böhm. Gesellsch. : Abhandlungen der Königlichen Böhmischen Gesellschaft der Wissenschaften.
Arch. Electrotech. : Archiv für Elektrotechnik.
Adv. Coll. Int. Sci. : Advances in Colloid and Interface Science.
Adv. Mater. : Advanced Materials.
Adv. Phys. : Advances in Physics.
Am. J. Math. : American Journal of Mathematics.
Am. J. Phys. : American Journal of Physics.
Am. J. Sci. : American Journal of Science.
Angew. Chem. : Angewandte Chemie.
Ann. Chim. Phys. : Annales de Chimie et de Physique.
Ann. Hist. Comput. : Annals of the History of Computing.
Ann. Phys. : Annalen der Physik.
Ann. Rev. Mater. Sci. : Annual Review on Materials Science.
Appl. Phys. : Applied Physics.
Appl. Phys. Lett. : Applied Physics Letters.
Appl. Surf. Sci. : Applied Surface Science.
Arch. Mikr. Anat. : Archiv für Mikroskopische Anatomie.
Astrophys. J. : Astrophysical Journal.
Bell Syst. Tech. J. : Bell Systems Technical Journal.
Ber. Dtsch. Bot. Ges. : Berichte der Deutschen Botanischen Gesellschaft.
Biochim. Biophys. Acta : Biochimica et Biophysica Acta.
Biophys. J. : Biophysical Journal.
Bull. Soc. Chim. Paris : Bulletin de la Société Chimique de Paris.
Bull. Natl. Bur. Std. : Bulletin of the National Bureau of Standards.
Can. J. Phys. : Canadian Journal of Physics.
Chem. Phys. Lett. : Chemical Physics Letters.
Chem. Rev. : Chemical Review.
Commun. Assoc. Comput. Mach. : Communications of the Association for Computing Machinery.
Commun. Pure Appl. Math. : Communications on Pure and Applied Mathematics.

What is What in the Nanoworld. Victor E. Borisenko and Stefano Ossini
Copyright © 2008 WILEY-VCH Verlag GmbH & Co. KGaA, Weinheim
ISBN: 978-3-527-40783-5

Abbreviations for the scientific journals which appear as sources in the text

C. R. : Comptes Rendus.
Contemp. Phys. : Contemporary Physics.
Cur. Opin. Colloid Interf. Sci. : Current Opinion in Colloid & Interface Science.
Curr. Opin. Struct. Biol. : Current Opinion in Structural Biology.
Disc. Faraday Soc. : Discussion of the Faraday Society.
Dokl. Akad. Nauk. SSSR : Doklady Akademii Nauk SSSR.
Electron. Lett. : Electronic Letters.
Electr. World : Electrical World.
Europhys. Lett. : Europhysics Letters.
Fund. Math. : Fundamentals of Mathematics.
Fiz. Tekh. Poluprovodn. : Fizika i Tekhnika Poluprovodnikov.
Fiz. Teverd. Tela : Fizika Tverdogo Tela.
Gött. Nachr. Math. Phys. Klass. : Nachrichten von der Akademie der Wissenschaften zu Göttingen-Mathematisch-Physikalische Klasse.
Helv. Phys. Acta : Helvetica Physica Acta.
IBM J. Res. Dev. : IBM Journal of Research and Development.
IEEE: Institute of Electrical and Electronics Engineers.
IEEE Acoustic, Speech, Sign. Process. Mag. : IEEE Acoustic, Speech, and Signal Processing Magazine.
IEEE Int. Conf. Comp. Syst. Signal. Proc. : IEEE International Conference Composition System Signal Proceedings.
IEEE J. Quantum Electron. : IEEE Journal on Quantum Electronics.
IEEE Trans. Electron. Dev. : IEEE Transactions on Electronic Devices.
IEEE Trans. Inform. Theor. : IEEE Transactions on Information Theory.
IEEE Trans. Appl. Superconductivity : IEEE Transactions on Applied Superconductivity.
Ind. J. Phys. : Indian Journal of Physics.
Int. J. Theor. Phys. : International Journal of Theoretical Physics.
J. Am. Chem. Soc. : Journal of the American Chemical Society.
J. Am. Stat. Ass. : Journal of the American Statistical Association.
J. Appl. Phys. : Journal of Applied Physics.
J. Chem. Phys. : Journal of Chemical Physics.
J. Colloid Interface Sci. : Journal of Colloid & Interface Science.
J. Compt. : Journal of Computer.
J. Cryst. Growth : Journal of Crystal Growth.
J. Ecole Polytechn. : Journal de l'Ecole Polytechnique.
J. Electrochem. Soc. : Journal of the Electrochemical Society.
J. Frank. Inst. : Journal of the Franklin Institute.
J. Low Temp. Phys. : Journal of Low Temperature Physics.
J. Magn. Magn. Mat. : Journal of Magnetism and Magnetic Materials.
J. Mat. Res. : Journal of Material Research.
J. Math. Phys. : Journal of Mathematical Physics.
J. Microsc. : Journal of Microscopy.
J. Opt. Soc. Am. : Journal of the Optical Society of America.

Abbreviations for the scientific journals which appear as sources in the text

J. Photochem. Photobiol. : Journal of Photochemistry and Photobiology.

J. Phys. : Journal of Physics.

J. Phys. Chem. Sol. : Journal of Physics and Chemistry of Solids.

J. Phys. Condens. Matter : Journal of Physics; Condensed Matter.

J. Phys. Soc. Jpn. : Journal of the Physical Society of Japan.

J. Sci. Inst. Elect. Engrs. : Journal of Scientific Institute of Electrical Engineers.

JSME Int. J. : Japan Society Mechanical Engineering International Journal.

J. Stat. Phys. : Journal of Statistical Physics.

J. Vac. Sci. Tech. : Journal of Vacuum Science and Technology.

Jpn. J. Appl. Phys. : Japanese Journal of Applied Physics.

Kgl. Danske Videnskab. Selskab – Mat.-fys. Medd. : Kongelige Danske Videnskabernes Selskab – Matematisk-Fysiske Meddelelser.

Kolloid Z. : Kolloid-Zeitschrift.

Math. Syst. Theory : Mathematical Systems Theory.

Mat. Sci. Techn. : Materials Science and Technology.

Math. Annal. : Mathematische Annalen.

Methods Enzymol. : Methods in Enzymology.

Mol. Phys. : Molecular Physics.

Monatsh. Chem. : Monatshefte für Chemie.

Nano Lett. : Nano Letters.

Nanomedicine: Nanotech. Biol. Med. : Nanomedicine: Nanotechnology, Biology, and Medicine.

Naturwiss. : Naturwissenschaften.

Opt. Lett. : Optics Letters.

Phil. Mag. : Philosophical Magazine.

Phil. Trans. R. Soc. London : Philosophical Transactions of the Royal Society of London.

Philips Res. Rep. : Philips Research Reports.

Photogr. J. : Photographic Journal.

Phys. A : Physica A.

Phys. Lett. : Physics Letters.

Phys. Rev. : Physical Review.

Phys. Rev. A : Physical Review A.

Phys. Rev. B : Physical Review B.

Phys. Rev. Lett. : Physical Review Letters.

Phys. Today : Physics Today.

Phys. Z. : Physikalische Zeitschrift.

PNAS : Proceedings of the National Academy of Sciences.

Proc. Am. Acad. Arts Sci. : Proceedings of the American Academy of Arts and Sciences.

Proc. Inst. Radio Eng. : Proceedings of the Institute of Radio Engineers.

Proc. K. Ned. Akad. Wet. : Verhandelingen der Koninklijke Nederlandse Akademie van Vetenschappen.

Proc. Phys. Soc. : Proceedings of the Physical Society.

Abbreviations for the scientific journals which appear as sources in the text

Proc. R. Acad. Sci : Proceedings of the Royal Academy of Science.

Proc. R. Soc. Lond. : Proceedings of the Royal Society of London.

Proc. Camb. Phil. Soc. : Proceedings of the Cambridge Philosophical Society.

Prog. Phot. Res. Appl. : Progress in Photovoltaics: Research and Applications.

Proc. Phys. Math. Soc. Jpn. : Proceedings of the Physico-Mathematical Society of Japan.

Prog. Theor. Phys. : Progress on Theoretical Physics.

Pure Appl. Math. : Pure and Applied Mathematics.

Rend. Accad. Naz. Lincei : Rendiconti dell'Accademia Nazionale dei Lincei.

Rep. Prog. Phys. : Reports on Progress in Physics.

Rev. Mod. Phys. : Review of Modern Physics.

Rev. Sci. Instrum. : Review of Scientific Instruments.

Schultzes Arch. Mikr. Anat. : Schultzes Archiv für Mikroskopische Anatomie.

Semicond. Sci. Technol. : Semiconductor Science and Technology.

SIAM J. Comp. : Society for Industrial and Applied Mathematics Journal on Computing.

SIGACT News : Association for Computing Machinery – Special Interest Group on Algorithm and Computational Theory News.

Sitzungsber. Akad. Wiss. Wien, Math. Naturw. Kl. : Sitzungsberichte der Akademie der Wissenschaften in Wien, Mathematisch-Naturwissenschaftliche Klasse.

Sitzungsber. Bay. Akad. Wiss. : Sitzungsberichte der Bayerischen Akademie der Wissenschaften.

Sitzungsber. Preuss. Akad. Wiss. : Sitzungsberichte der Preussischen Akademie der Wissenschaften.

Skr. Norsk. Vid. Akademi, Oslo, Mat. Nat. Kl. : Tidskrifte Norske Videnskaps-Akademi i Olso, Matematisk-Naturvitenskapelige Klasse.

Solid State Commun. : Solid State Communications.

Sov. J. Nucl. Phys. : Soviet Journal of Nuclear Physics.

Sov. Phys. JETP : Soviet Physics – Journal of Experimental and Theoretical Physics.

Sov. Phys. Semicond. : Soviet Physics – Semiconductors.

Surf. Sci. : Surface Science.

Surf. Sci. Rep. : Surface Science Reports.

Symp. Quant. Biol. : Quantum Biology Symposium.

Theory Prob. Appl. : Theory of Probability and its Applications.

Trans. AIEE : Transactions of the American Institute of Electrical Engineers.

Trans. Conn. Acad. : Transactions of the Connecticut Academy of Arts and Sciences.

Trans. Faraday Soc. : Transactions of the Faraday Society.

Trans. Opt. Soc. : Transactions of the Optical Society.

Trans. R. Irish Acad. : Transactions of the Royal Irish Academy.

Usp. Fiz. Nauk. : Uspekhi Fiziceskih Nauk.

Verh. Deutsch. Phys. Ges. : Verhandlungen der Deutschen Physikalischen Gesellschaft.

Z. Instrumkde : Zeitschrift für Instrumentenkunde.

Z. Kristallogr. : Zeitschrift für Kristallographie.

Z. Krist. Miner. : Zeitschrift für Kristallographie und Mineralogie.

Abbreviations for the scientific journals which appear as sources in the text

Z. Naturforsch. : Zeitschrift für Naturforschung.

Z. Phys. : Zeitschrift für Physik.

Z. Phys. Chem. : Zeitschrift für Physikalische Chemie.

Zh. Exp. Theor. Fiz. : Zhurnal Experimentalnoy i Theoreticheskoy Fiziki.

Zh. Russkogo Fiz.-Khim. Obsch. : Zhurnal Russkogo Fiziko-Khimicheskogo Obschestva.

Appendix – main properties of intrinsic (or lightly doped) semiconductors

Table 1 Group IV semiconductors.

Material	C	Si	Ge	α Sn
Crystalline structure	cubic – diamond			
Lattice parameter at 300 K, nm	0.35668	0.54311	0.56579	0.64892
Nature of the E_{gap} (d – direct, i – indirect)	i	i	i	i
Fundamental E_{gap} at 0 K, eV	5.48	1.166	0.74	0.09
Fundamental E_{gap} at 300 K, eV	5.47	1.11	0.67	
Refractive index		3.44	4.00	
Rel. permittivity (static dielectric constant)	5.7	11.7	16.3	24
Mass of electrons, in m_0 units	0.2	0.980.19	1.580.08	0.023
Mass of holes, in m_0 units	0.25	0.52	0.3	0.20
Mobility of electrons at 300 K, cm^2/(V·s)	1800	1500	3900	1400
Mobility of holes at 300 K, cm^2/(V·s)	1200	450	1900	1200

What is What in the Nanoworld. Victor E. Borisenko and Stefano Ossicini
Copyright © 2008 WILEY-VCH Verlag GmbH & Co. KGaA, Weinheim
ISBN: 978-3-527-40783-5

Appendix – main properties of intrinsic (or lightly doped) semiconductors

Table 2 Semiconducting SiC.

Compound	α–SiC	β–SiC
Crystalline structure	6H (hexagonal)	3C (cubic)
Lattice parameters at 300 K (a or a,c), nm	0.30806	0.43596
	1.51173	
Nature of the E_{gap} (d – direct, i – indirect)	i	i
Fundamental E_{gap} at 0 K, eV		
Fundamental E_{gap} at 300 K, eV	2.86	2.2
Refractive index	2.65	2.66
Rel. permittivity (static dielectric constant)	$9.66_{\perp c}, 10.03_{\parallel c}$	9.72
Mass of electrons, in m_0 units	$0.25_{\perp c}, 1.5_{\parallel c}$	0.647
		0.24
Mass of holes, in m_0 units	1.0	
Mobility of electrons at 300 K, cm^2/(V·s)	260	900
Mobility of holes at 300 K, cm^2/(V·s)	50	

Table 3 Semiconducting $A^{III}B^{V}$ nitrides.

Compound	AlN	GaN	InN
Crystalline structure	hexagonal – wurtzite		
Lattice parameters at 300 K (a,c), nm	0.3111	0.3189	0.3544
	0.4978	0.5185	0.5718
Nature of the E_{gap} (d – direct, i – indirect)	d	d	d
Fundamental E_{gap} at 0 K, eV		3.50	
Fundamental E_{gap} at 300 K, eV	6.2	3.44	1.89
Refractive index	2.15	2.4	2.56
Rel. permittivity (static dielectric constant)	8.5	8.9	15.3
Mass of electrons, in m_0 units	0.48	0.2	0.11
Mass of holes, in m_0 units	0.471	0.259	
Mobility of electrons at 300 K, cm^2/(V·s)	135	1000	3200
Mobility of holes at 300 K, cm^2/(V·s)	14	30	

Appendix – main properties of intrinsic (or lightly doped) semiconductors

Table 4 Semiconducting $A^{III}B^V$ phosphides.

Compound	AlP	GaP	InP
Crystalline structure		cubic – zinc blende	
Lattice parameter at 300 K, nm	0.54635	0.5450	0.58687
Nature of the E_{gap} (d – direct, i – indirect)	i	i	d
Fundamental E_{gap} at 0 K, eV	2.51	2.4	1.42
Fundamental E_{gap} at 300 K, eV	2.45	2.25	1.35
Refractive index		3.37	3.37
Rel. permittivity (static dielectric constant)	9.8	11.1	12.1
Mass of electrons, in m_0 units	0.166	0.82	0.077
Mass of holes, in m_0 units	0.20	0.60	0.64
Mobility of electrons at 300 K, cm^2/(V·s)	80	110	4600
Mobility of holes at 300 K, cm^2/(V·s)		75	150

Table 5 Semiconducting $A^{III}B^V$ arsenides.

Compound	AlAs	GaAs	InAs
Crystalline structure		cubic – zinc blende	
Lattice parameter at 300 K, nm	0.5660	0.5653	0.6058
Nature of the E_{gap} (d – direct, i – indirect)	i	d	d
Fundamental E_{gap} at 0 K, eV	2.229	1.519	0.43
Fundamental E_{gap} at 300 K, eV	2.153	1.424	0.36
Refractive index		3.4	3.42
Rel. permittivity (static dielectric constant)	10.1	12.5	12.5
Mass of electrons, in m_0 units	0.1	0.07	0.028
Mass of holes, in m_0 units	0.15	0.5	0.33
Mobility of electrons at 300 K, cm^2/(V·s)	294	8500	33000
Mobility of holes at 300 K, cm^2/(V·s)		400	460

Appendix – main properties of intrinsic (or lightly doped) semiconductors

Table 6 Semiconducting $A^{III}B^{V}$ antimonides.

Compound	AlSb	GaSb	InSb
Crystalline structure	cubic – zinc blende		
Lattice parameter at 300 K, nm	0.61355	0.6095	0.64787
Nature of the E_{gap} (d – direct, i – indirect)	i	d	d
Fundamental E_{gap} at 0 K, eV	1.686	0.81	0.235
Fundamental E_{gap} at 300 K, eV	1.61	0.69	0.17
Refractive index		3.9	3.75
Rel. permittivity (static dielectric constant)	14.4	15	18
Mass of electrons, in m_0 units	0.12	0.045	0.0133
Mass of holes, in m_0 units	0.98	0.39	0.18
Mobility of electrons at 300 K, cm^2/(V·s)	200	5000	80 000
Mobility of holes at 300 K, cm^2/(V·s)	420	850	1250

Table 7 Semiconducting $A^{I}B^{VII}$ chlorides.

Compound	γCuCl	AgCl
Crystalline structure	cubic – zinc blende	cubic – NaCl
Lattice parameter at 300 K, nm	0.54057	0.55023
Nature of the E_{gap} (d – direct, i – indirect)	d	i
Fundamental E_{gap} at 0 K, eV	3.95	3.25
Fundamental E_{gap} at 300 K, eV		
Refractive index	2.1	2.1
Rel. permittivity (static dielectric constant)	7.9	11.1
Mass of electrons, in m_0 units	0.43	
Mass of holes, in m_0 units	4.2	
Mobility of electrons at 300 K, cm^2/(V·s)		
Mobility of holes at 300 K, cm^2/(V·s)		

Appendix – main properties of intrinsic (or lightly doped) semiconductors

Table 8 Semiconducting $A^I B^{VII}$ bromides.

Compound	γ CuBr	AgBr
Crystalline structure	cubic – zinc blende	cubic – NaCl
Lattice parameter at 300 K, nm	0.56905	0.57748
Nature of the E_{gap} (d – direct, i – indirect)	d	i
Fundamental E_{gap} at 0 K, eV	3.07	2.68
Fundamental E_{gap} at 300 K, eV		
Refractive index		
Rel. permittivity (static dielectric constant)	8.0	11.8
Mass of electrons, in m_0 units	0.21	0.22
Mass of holes, in m_0 units	23.2	0.52
Mobility of electrons at 300 K, cm^2/(V·s)		60
Mobility of holes at 300 K, cm^2/(V·s)		2

Table 9 Semiconducting $A^I B^{VII}$ iodides.

Compound	γCuI	β–AgI
Crystalline structure	cubic – zinc blende	hexagonal-wurtzite
Lattice parameters at 300 K (a or a,c), nm	0.60427	0.4592 0.7512
Nature of the E_{gap} (d – direct, i – indirect)	d	d
Fundamental E_{gap} at 0 K, eV	3.12	3.02
Fundamental E_{gap} at 300 K, eV		
Refractive index		
Rel. permittivity (static dielectric constant)	6.5	7.0
Mass of electrons, in m_0 units	0.3	
Mass of holes, in m_0 units	1.4	
Mobility of electrons at 300 K, cm^2/(V·s)		
Mobility of holes at 300 K, cm^2/(V·s)		

Appendix – main properties of intrinsic (or lightly doped) semiconductors

Table 10 Semiconducting $A^{II}B^{VI}$ sulfides.

Compound	ZnS	CdS	HgS
Crystalline structure	cubic – zinc blende		
Lattice parameter at 300 K, nm	0.541	0.582	0.58517
Nature of the E_{gap} (d – direct, i – indirect)	d	d	
Fundamental E_{gap} at 0 K, eV	3.84	2.58	−0.2
Fundamental E_{gap} at 300 K, eV	3.68	2.50	
Refractive index	2.37	2.5	
Rel. permittivity (static dielectric constant)	8.9		
Mass of electrons, in m_0 units	0.40	0.2	
Mass of holes, in m_0 units			
Mobility of electrons at 300 K, cm^2/(V·s)	165	340	
Mobility of holes at 300 K, cm^2/(V·s)	5	50	

Table 11 Semiconducting $A^{II}B^{VI}$ selenides.

Compound	ZnSe	CdSe	HgSe
Crystalline structure	cubic – zinc blende		
Lattice parameter at 300 K, nm	0.5667	0.608	0.6085
Nature of the E_{gap} (d – direct, i – indirect)	d	d	
Fundamental E_{gap} at 0 K, eV	2.80	1.85	−0.22
Fundamental E_{gap} at 300 K, eV	2.58	1.75	
Refractive index	2.89		
Rel. permittivity (static dielectric constant)	8.1	10.6	23
Mass of electrons, in m_0 units	0.21	0.13	0.05
Mass of holes, in m_0 units	0.6		0.02
Mobility of electrons at 300 K, cm^2/(V·s)	500	800	
Mobility of holes at 300 K, cm^2/(V·s)	30		

Appendix – main properties of intrinsic (or lightly doped) semiconductors

Table 12 Semiconducting $A^{II}B^{VI}$ tellurides.

Compound	ZnTe	CdTe	HgTe
Crystalline structure	cubic – zinc blende		
Lattice parameter at 300 K, nm	0.6101	0.6477	0.6453
Nature of the E_{gap} (d – direct, i – indirect)	d	d	
Fundamental E_{gap} at 0 K, eV	2.39	1.60	−0.28
Fundamental E_{gap} at 300 K, eV	2.2	1.50	−0.015, 0.14
Refractive index	3.56	2.75	3.7
Rel. permittivity (static dielectric constant)	9.7	10.9	20
Mass of electrons, in m_0 units	0.15	0.11	0.029
Mass of holes, in m_0 units	0.2	0.35	−0.3
Mobility of electrons at 300 K, cm^2/(V·s)	340	1050	
Mobility of holes at 300 K, cm^2/(V·s)	100	100	

Table 13 Semiconducting $A^{IV}B^{VI}$ tin compounds.

Compound	SnS	SnSe	SnTe
Crystalline structure	orthorhombic distorted – – distorted NaCl		
Lattice parameters at 300 K (a, b, c), nm	1.157	1.120	
	0.419	0.399	
	0.446	0.434	
Nature of the E_{gap} (d – direct, i – indirect)	d	i	
Fundamental E_{gap} at 0 K, eV			
Fundamental E_{gap} at 300 K, eV	1.09	0.9	0.36
Refractive index			
Rel. permittivity (static dielectric constant)	32	45	
Mass of electrons, in m_0 units			
Mass of holes, in m_0 units	0.2	0.15	0.07
Mobility of electrons at 300 K, cm^2/(V·s)			
Mobility of holes at 300 K, cm^2/(V·s)	90		840

Appendix – main properties of intrinsic (or lightly doped) semiconductors

Table 14 Semiconducting $A^{IV}B^{VI}$ lead compounds.

Compound	PbS	PbSe	PbTe
Crystalline structure		cubic – NaCl	
Lattice parameter at 300 K, nm	0.5936	0.6117	0.6462
Nature of the E_{gap} (d – direct, i – indirect)	d	d	d
Fundamental E_{gap} at 0 K, eV	0.29	0.15	0.19
Fundamental E_{gap} at 300 K, eV	0.41	0.28	0.31
Refractive index			
Rel. permittivity (static dielectric constant)	170	210	~1000
Mass of electrons, in m_0 units	0.25	0.04	0.17
Mass of holes, in m_0 units	0.25		0.20
Mobility of electrons at 300 K, cm^2/(V·s)	600	1	6000
Mobility of holes at 300 K, cm^2/(V·s)	700		4000

Table 15 Semiconducting Group II silicides.

Compound	Mg$_2$Si	Ca$_2$Si	Ca$_3$Si$_4$	BaSi$_2$
Crystalline structure	cubic – CaF$_2$	orthorhombic	hexagonal	orthorhombic – BaSi$_2$
Lattice parameters at 300 K (a, b, c), nm	0.63512	0.7691 0.4816 0.9035	0.8541 1.4906	0.8942 0.6733 1.1555
Nature of the E_{gap} (d – direct, i – indirect)	i	d	i	i
Fundamental E_{gap} at 0 K, eV	0.65	0.35, 1.0	0.35	0.83
Fundamental E_{gap} at 300 K, eV	0.78			1.10 1.30
Refractive index				
Rel. permittivity (static dielectric constant)				
Mass of electrons (m_x, m_y, m_z), in m_0 units		0.64 0.10 0.71	>1 0.7 1.0	0.60 0.37 0.30
Mass of holes (m_x, m_y, m_z), in m_0 units		0.16 2.21 2.30	0.46 >1 0.83	0.31 0.73 0.67
Mobility of electrons at 300 K, cm^2/(V·s)	406			
Mobility of holes at 300 K, cm^2/(V·s)	56		~50	

Appendix – main properties of intrinsic (or lightly doped) semiconductors

Table 16 Semiconducting Group VI silicides.

Compound	CrSi$_2$	MoSi$_2$	WSi$_2$
Crystalline structure		hexagonal – CrSi$_2$	
Lattice parameters at 300 K (a, c), nm	0.44281	0.4596	0.4614
	0.63691	0.6550	0.6414
Nature of the E$_{gap}$ (d – direct, i – indirect)	i	i	i
Fundamental E$_{gap}$ at 0 K, eV	0.35	0.07	0.07
Fundamental E$_{gap}$ at 300 K, eV			
Refractive index			
Rel. permittivity (static dielectric constant)			
Mass of electrons (m_x, m_y, m_z), in m_0 units	0.69	0.38	0.33
	0.66	0.33	0.30
	1.49	0.73	0.65
Mass of holes (m_x, m_y, m_z), in m_0 units	1.10	0.43	0.39
	1.20	0.55	0.57
	0.82	0.50	0.45
Mobility of electrons at 300 K, cm^2/(V·s)			
Mobility of holes at 300 K, cm^2/(V·s)	10	200	220

Table 17 Semiconducting Group VII silicides.

Compound	MnSi$_{2-x}$	ReSi$_{1.75}$
Crystalline structure	tetragonal	triclinic
		$\alpha = 89.90°$
Lattice parameters at 300 K (a, b, c), nm	$a = 0.5530$	0.3138
	$c = 1.7517$	0.3120
	4.7763	0.7670
	6.5311	
	11.79	
Nature of the E$_{gap}$ (d – direct, i – indirect)	i	i
Fundamental E$_{gap}$ at 0 K, eV	0.7–0.8	
Fundamental E$_{gap}$ at 300 K, eV	0.7–0.9	0.15
Refractive index		
Rel. permittivity (static dielectric constant)		
Mass of electrons (m_x, m_y, m_z), in m_0 units	8.17	0.35
	8.17	0.32
	3.40	0.37
Mass of holes (m_x, m_y, m_z), in m_0 units	1.64	0.27
	1.64	0.27
	5.72	11.82
Mobility of electrons at 300 K, cm^2/(V·s)		
Mobility of holes at 300 K, cm^2/(V·s)	230	370

Appendix – main properties of intrinsic (or lightly doped) semiconductors

Table 18 Semiconducting Group VIII disilicides.

Compound	β-FeSi$_2$	RuSi$_2$	OsSi$_2$
Crystalline structure	orthorhombic – β-FeSi$_2$		
Lattice parameters at 300 K (a, b, c), nm	0.98792	1.0053	1.0150
	0.77991	0.8028	0.8117
	0.78388	0.8124	0.8223
Nature of the E_{gap} (d – direct, i – indirect)	d/i	?	i
Fundamental E_{gap} at 0 K, eV			
Fundamental E_{gap} at 300 K, eV	0.78–0.87	0.35–0.52	1.4
Refractive index			
Rel. permittivity (static dielectric constant)			
Mass of electrons (m_x, m_y, m_z), in m_0 units	$>>1$		
Mass of holes (m_x, m_y, m_z), in m_0 units	0.21		
	0.27		
	0.27		
Mobility of electrons at 300 K, cm^2/(V·s)	900		
Mobility of holes at 300 K, cm^2/(V·s)	200		

Table 19 Semiconducting Group VIII silicides (except disilicides).

Compound	Ru$_2$Si$_3$	Os$_2$Si$_3$	OsSi	Ir$_3$Si$_5$
Crystalline structure	orthorhombic-Ru$_2$Si$_3$		cubic-FeSi	monoclinic $\alpha = 89.90°$
Lattice parameters at 300 K (a, b, c), nm	1.1057	1.1124	0.4729	0.6406
	0.8934	0.8932		1.4162
	0.5533	0.5570		1.1553
Nature of the E_{gap} (d – direct, i – indirect)	d	d		i
Fundamental E_{gap} at 0 K, eV				
Fundamental E_{gap} at 300 K, eV	0.7–1.0	2.3	0.34	1.2
Refractive index				
Rel. permittivity (static dielectric constant)				
Mass of electrons (m_x, m_y, m_z), in m_0 units	3.28			
	2.85			
	0.61			
Mass of holes (m_x, m_y, m_z), in m_0 units	0.47			
	0.15			
	0.45			
Mobility of electrons at 300 K, cm^2/(V·s)				
Mobility of holes at 300 K, cm^2/(V·s)	2–3			2–3